AUDITORY DISPLAY

AUDITORY DISPLAY

SONIFICATION, AUDIFICATION, AND AUDITORY INTERFACES

Editor

Gregory Kramer
Clarity
Nelson Lane
Garrison, NY 10524
and
Santa Fe Institute
1399 Hyde Park Road
Santa Fe, NM 87501

Proceedings Volume XVIII

Santa Fe Institute
Studies in the Sciences of Complexity

Addison-Wesley Publishing Company
The Advanced Book Program

Reading, Massachusetts   Menlo Park, California   New York
Don Mills, Ontario   Wokingham, England   Amsterdam   Bonn
Sydney   Singapore   Tokyo   Madrid   San Juan
Paris   Seoul   Milan   Mexico City   Taipei

Publisher: *David Goehring*
Editor-in-Chief: *Jack Repcheck*
Production Manager: *Michael Cirone*
Production Supervisor: *Lynne Reed*

Director of Publications, Santa Fe Institute: *Ronda K. Butler-Villa*
Publications Assistant, Santa Fe Institute: *Della L. Ulibarri*

This volume was typeset using T_EXtures on a Macintosh II computer. Camera-ready output from a Hewlett Packard Laser Jet 4M Printer.

Copyright © 1994 by Addison-Wesley Publishing Company, The Advanced Book Program, Jacob Way, Reading, MA 01867

All rights reserved. No part of this publication may be reproduced, stored in a retrieval system, or transmitted in any form or by any means, electronic, mechanical, photocopying, recording, or otherwise, without the prior written permission of the publisher. Printed in the United States of America. Published simultaneously in Canada.

ISBN 0-201-62603-9 (Hardcover)
ISBN 0-201-62604-7 (Paperback)

1 2 3 4 5 6 7 8 9 10–MA–97969594
First printing, April 1994

About the Santa Fe Institute

The *Santa Fe Institute* (SFI) is a multidisciplinary graduate research and teaching institution formed to nurture research on complex systems and their simpler elements. A private, independent institution, SFI was founded in 1984. Its primary concern is to focus the tools of traditional scientific disciplines and emerging new computer resources on the problems and opportunities that are involved in the multidisciplinary study of complex systems—those fundamental processes that shape almost every aspect of human life. Understanding complex systems is critical to realizing the full potential of science, and may be expected to yield enormous intellectual and practical benefits.

All titles from the *Santa Fe Institute Studies in the Sciences of Complexity* series will carry this imprint which is based on a Mimbres pottery design (circa A.D. 950–1150), drawn by Betsy Jones. The design was selected because the radiating feathers are evocative of the outreach of the Santa Fe Institute Program to many disciplines and institutions.

Santa Fe Institute Editorial Board
June 1993

Dr. L. M. Simmons, Jr., *Chair*
Vice President for Academic Affairs, Santa Fe Institute

Prof. Kenneth J. Arrow
Department of Economics, Stanford University

Prof. W. Brian Arthur
Dean & Virginia Morrison Professor of Population Studies and Economics,
Food Research Institute, Stanford University

Prof. Michele Boldrin
MEDS, Northwestern University

Dr. David K. Campbell
Head, Department of Physics, University of Illinois and
Director, Center for Nonlinear Studies, Los Alamos National Laboratory

Dr. George A. Cowan
Visiting Scientist, Santa Fe Institute and Senior Fellow Emeritus, Los Alamos National Laboratory

Prof. Marcus W. Feldman
Director, Institute for Population & Resource Studies, Stanford University

Prof. Murray Gell-Mann
Division of Physics & Astronomy, California Institute of Technology

Prof. John H. Holland
Division of Computer Science & Engineering, University of Michigan

Prof. Stuart A. Kauffman
School of Medicine, University of Pennsylvania

Dr. Edward A. Knapp
President, Santa Fe Institute

Prof. Harold Morowitz
University Professor, George Mason University

Dr. Alan S. Perelson
Theoretical Division, Los Alamos National Laboratory

Prof. David Pines
Department of Physics, University of Illinois

Prof. Harry L. Swinney
Department of Physics, University of Texas

Santa Fe Institute
Studies in the Sciences of Complexity

Lectures Volumes

Vol.	Editor	Title
I	D. L. Stein	Lectures in the Sciences of Complexity, 1989
II	E. Jen	1989 Lectures in Complex Systems, 1990
III	L. Nadel & D. L. Stein	1990 Lectures in Complex Systems, 1991
IV	L. Nadel & D. L. Stein	1991 Lectures in Complex Systems, 1992
V	L. Nadel & D. L. Stein	1992 Lectures in Complex Systems, 1993

Lecture Notes Volumes

Vol.	Author	Title
I	J. Hertz, A. Krogh, & R. Palmer	Introduction to the Theory of Neural Computation, 1990
II	G. Weisbuch	Complex Systems Dynamics, 1990
III	W. D. Stein & F. J. Varela	Thinking About Biology, 1993

Reference Volumes

Vol.	Author	Title
I	A. Wuensche & M. Lesser	The Global Dynamics of Cellular Automata: Attraction Fields of One-Dimensional Cellular Automata, 1992

Proceedings Volumes

Vol.	Editor	Title
I	D. Pines	Emerging Syntheses in Science, 1987
II	A. S. Perelson	Theoretical Immunology, Part One, 1988
III	A. S. Perelson	Theoretical Immunology, Part Two, 1988
IV	G. D. Doolen et al.	Lattice Gas Methods for Partial Differential Equations, 1989
V	P. W. Anderson, K. Arrow, D. Pines	The Economy as an Evolving Complex System, 1988
VI	C. G. Langton	Artificial Life: Proceedings of an Interdisciplinary Workshop on the Synthesis and Simulation of Living Systems, 1988
VII	G. I. Bell & T. G. Marr	Computers and DNA, 1989
VIII	W. H. Zurek	Complexity, Entropy, and the Physics of Information, 1990
IX	A. S. Perelson & S. A. Kauffman	Molecular Evolution on Rugged Landscapes: Proteins, RNA and the Immune System, 1990
X	C. G. Langton et al.	Artificial Life II, 1991
XI	J. A. Hawkins & M. Gell-Mann	The Evolution of Human Languages, 1992
XII	M. Casdagli & S. Eubank	Nonlinear Modeling and Forecasting, 1992
XIII	J. E. Mittenthal & A. B. Baskin	Principles of Organization in Organisms, 1992
XIV	D. Friedman & J. Rust	The Double Auction Market: Institutions, Theories, and Evidence, 1993
XV	A. S. Weigend & N. A. Gershenfeld	Time Series Prediction: Forecasting the Future and Understanding the Past
XVI	G. Gumerman & M. Gell-Mann	Understanding Complexity in the Prehistoric Southwest
XVII	C. G. Langton	Artificial Life III
XVIII	G. Kramer	Auditory Display

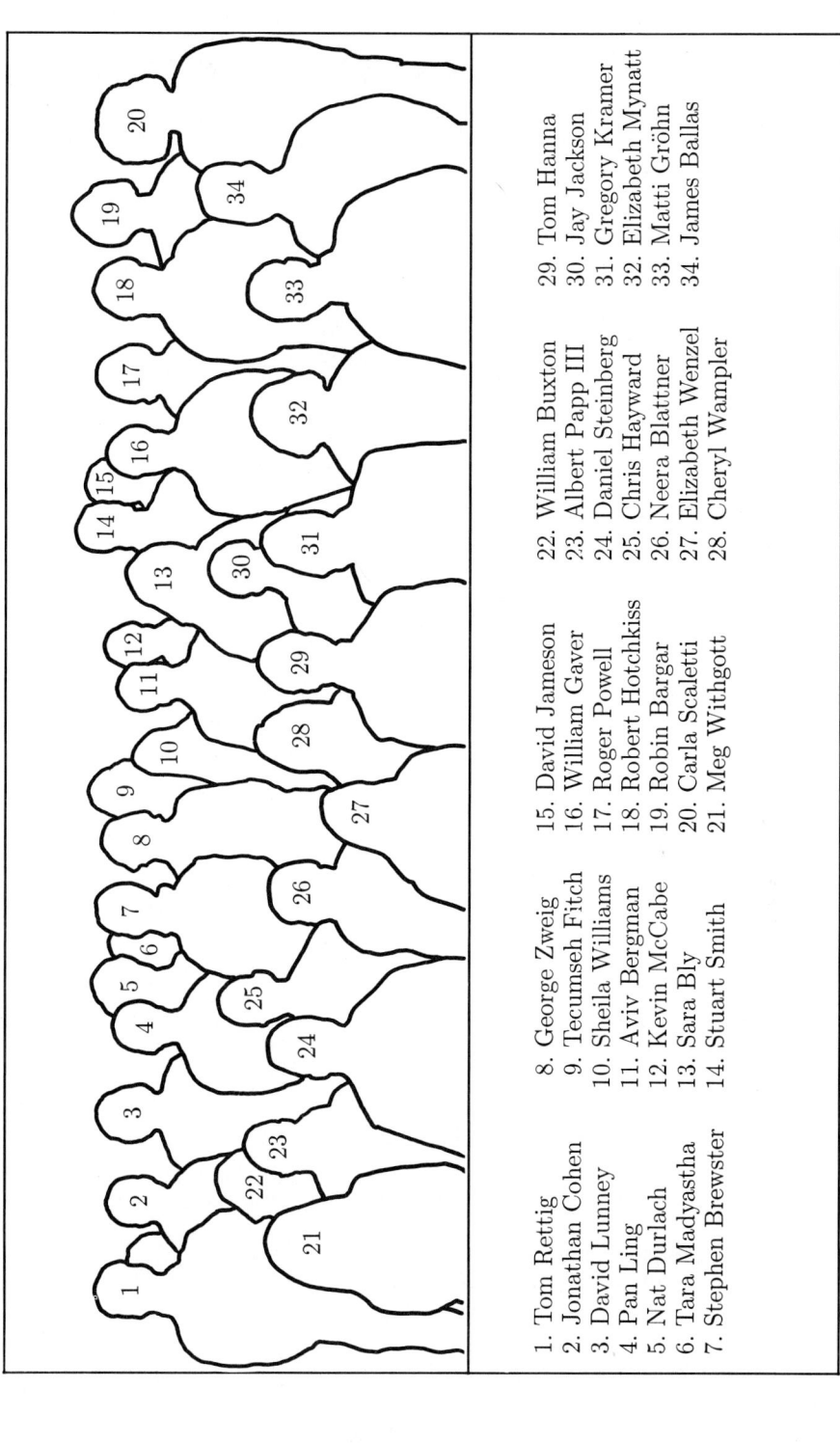

Contents

Foreword
    Dr. Albert Bregman   xv

Preface
    Gregory Kramer   xxiii

An Introduction to Auditory Display
    Gregory Kramer   1

Delivery of Information Through Sound
    James A. Ballas   79

Perceptual Principles in Sound Grouping
    Sheila M. Williams   95

Spatial Sound and Sonification
    Elizabeth M. Wenzel   127

Pattern and Reference in Auditory Display
    Robin Bargar   151

Environments for Exploring Auditory Representations of Multidimensional Data
    Stuart Smith, Ronald M. Pickett, and Marian G. Williams   167

Some Organizing Principles for Representing Data with Sound
    Gregory Kramer   185

Sound Synthesis Algorithms for Auditory Data
Representations
    Carla Scaletti   223

Sonnet: Audio-Enhanced Monitoring and Debugging
    David H. Jameson   253

A Framework for Sonification Design
    Tara M. Madhyastha and Daniel A. Reed   267

Synchronization of Visual and Aural Parallel Program
Performance Data
    Jay Alan Jackson and Joan M. Francioni   291

Sonifying the Body Electric: Superiority of an Auditory
over a Visual Display in a Complex, Multivariate System
    W. Tecumseh Fitch and Greogry Kramer   307

Auditory Display of Computational Fluid Dynamics Data
    Kevin McCabe and Akil Rangwalla   327

Musical Structures in Data from Chaotic Attractors
    Gottfried Mayer-Kress, Robin Bargar, and Insook Choi   341

Listening to the Earth Sing
    Chris Hayward   369

Multivariate Data Mappings
    Sara Bly   405

Using and Creating Auditory Icons
    William W. Gaver   417

Sonic Enhancement of Two-Dimensional Graphic Displays
    Merra M. Blattner, Albert L. Papp III, and
    Ephraim P. Glinert   447

A Detailed Investigation into the Effectiveness of Earcons
    Stephane A. Brewster, Peter C. Wright, and
    Alistair D. N. Edwards   471

Monitoring Background Activities
    Jonathan Cohen   499

Auditory Presentation of Graphical User Interfaces
   Elizabeth D. Mynatt   533

Appendix I: Comments on ICAD 92   557

Appendix II: Resources   577

Appendix III: Annotated Bibliography   583

Appendix IV: Annotations to Audio Examples   623

Index   661

Foreword

In October 1992, a group of researchers met for the first time to discuss their work on an intriguing question: how can nonverbal sound be used to convey meanings to people working with information?

The chapters in the present volume provide a fascinating glimpse of how they imagined that the information workers of the future would make use of sound. Reading the chapters reminds me of the science fiction novel, *Neuromancer* by William Gibson, which describes an information thief operating inside a virtual reality in which data sets can be seen as objects and information-processing events heard as sounds. When I first read this story, I thought the hero's exploits in the virtual-reality world of information were great fun but described a process that I would never be part of.

The present volume has convinced me otherwise. Putting together the ideas from many papers in the present volume, I see in my mind's eye (and ear) an executive of a shoe company in a modern office only a few years from now. She faces a computer screen on which there is a graphic interface. By using a pointer to select an icon, she chooses to copy a file. A clinking sound verifies that the file has been selected, and then a pouring sound continues until the copying is finished. A bit later, the computer generates the sound of a creaking door to tell her that a remote user has signed onto her machine to examine data files and the sound of a slamming door when that person signs off again.

She decides to examine some data about the sales of shoes over the past year, using her "acoustic viewer." To get a general idea of the trends, she opens up a visual map containing a dot for every regional sales office. Each time she selects one of them, she hears an auditory summary of the sales for the past twelve months, but not in the form of spoken words. Instead, she hears a rapid sequence of tones in which the pitch of each tone represents the number of sales in a week and the complexity of the note's timbre represents the total revenue for that week. The sequence of tones reflects any trends or changes over the twelve months.

It takes her about two minutes to notice that, in many offices, the sales have been roughly steady in numbers, but revenues are going down (that is, she hears a relatively narrow range of pitches, gradually becoming purer in timbre). She concludes that people are beginning to buy cheaper models of shoes. On a hunch, she brings up, in an adjacent window on the screen, a map of unemployment data for the country and, by drawing a few lines, sets up her acoustic viewer for listening to month-by-month figures, arranging it so that the higher the employment, the more complex the timbre of tones (the pitches are used to display some other feature). She alternates between selections of regional office sales and unemployment figures and hears the same sort of timbral patterns, confirming her hypothesis. Since her boss is visually impaired, and normally wants original sound patterns as a supplement to voice summaries, she includes the patterns in her otherwise spoken report to him, and sends it off by electronic mail.

Toward the end of this task, a muted trumpet call from the computer announces the arrival of electronic mail. She recognizes the fanfare of one of her coworkers. She touches the mail icon and sees a list of old and new mail. Selecting the new message, she reads its summary and classifies it as requiring 30 minutes to answer. Then she slides a pointer down the list of names and, as the pointer crosses each one, it causes it to emit a tone whose height represents how urgent the message is, its loudness, how long the message has been there unanswered, and its duration, her previously estimated time for answering it. Seeing that it is about lunch time, and that no high pitches "pop out" of the tone stream evoked by her moving pointer, she selects a short, loud one. Her eyes are tired, so she chooses to hear it as spoken words. It is a question about sales force size from her boss.

She's still tired of reading, so putting on a special pair of goggles, and handling a joy-stick, she sees and hears herself entering the virtual "stacks" of the company's data files. She turns right at accounting (where she can hear the coins jingling), and passes through the door marked "personnel" into a bigger room (her footsteps start to sound more resonant). She leafs through some map racks, hearing a creaking noise as each one is turned over, until she finds one for sizes of sales force in different regional offices, double clicks it, and takes off her goggles. The map is now there on her screen in miniature.

Her coworkers do not complain about the squeaks and whistles issuing from her machine, because these exist only in a miniature speaker seated in her left ear. She always wears it there so she can tell immediately which sounds are environmental and which from her information world. The sounds from her computer are always heard as emanating from a spatial position very close to her left ear, whereas environmental sounds come from "out there." She also wears the speaker in her left ear because the world still has telephones and she needs her right ear for them. Unlike the information thief in *Neuromancer*, who totally disappears for long periods into his virtual world, she chooses to keep true and virtual realities concurrent, but distinct.

In this scenario, the executive works with sound as well as numbers, pictures, and words (both written and spoken). The only part of this account that present-day information workers will find novel is the heavy involvement of sounds.

This volume is the report of a historical first—the First International Conference on Auditory Display. The researchers who have contributed to this volume are all pioneers in the design of acoustic signals for conveying the types of information that appeared in my fanciful scenario. Nothing in my example was made up, but was derived from the chapters. The technology already exists to accomplish everything I mentioned, and the chapters in this book represent important steps toward making nonverbal sound easy to use and to interpret.

Many questions come up when one begins to think about using sound in an information-handling system. For example, is it really a good idea for the sounds from the computer to be presented only to a single ear? If the executive had worn a pair of headphones, current technology could have made different sounds appear to be coming from different locations, and the location of the sound could be added to the list of qualities that could be used to represent information. Another question is whether to use sounds that are familiar, such as a creaking door, as signals whose meaning can be learned with little effort, or to invent a language built from a vocabulary of short melodic sequences which can carry much more detailed information, but are harder to learn. Does it depend on vocabulary size or on the average length of message? At what point do spoken sentences become more efficient? If ears and eyes are to be used together, what tasks should be assigned to each?

These are not trivial problems, and one might think that given the countless existing studies of auditory perception, researchers in the field of auditory displays should be able to build on a strong scientific foundation. Unfortunately this is not so. Until very recently, research in auditory perception has been focused either on the perception of basic, meaningless sounds at one extreme, or on music and speech at the other. There is almost no research on how humans come to understand the sorts of complex sounds that we encounter in everyday life. And, apart from a small amount of work in speech perception, there is little research on how the senses collaborate to understand the environment.

The acoustic scenery of our everyday world has a very intricate structure because it is a direct consequence of the complex happenings around us. Listeners are able to "read" the sound directly and to hear, in it, the events that gave rise to it. The Gestalt psychologists argued that when we perceive, we do not simply experience a mixture of sensations with different qualities. We experience the things and events around us. In the case of audition, they would argue that, instead of hearing mere sounds, one hears a car pull up and stop, a person get out, a door slam, and the person walk away. The sounds give us an immediate awareness of patterns of events that have transpired. Modern cognitive science believes that to be able to "read" sound in this way, we must have some inner understanding of how the properties of physical events are reflected in the sounds that they make.

Yet how do we know that it was a car door slamming and not a house door, or even that the event involved slamming at all? Do we have to learn the sound of each type of slamming door as a separate association or can we learn how the impact of slamming causes different kinds of resonances in different materials? Despite the number of fascinating questions that are evoked when we see how people actually use the sounds around them, there has been, until very recently, almost no research on the perception and understanding of environmental sounds.

The stimulus to explore these issues has come from the arrival of new technologies, which offer the possibility of combining visually presented information with sound. The need to design better computer systems to extend the range of human thought has encouraged people who understand this new acoustic technology to provide more and better information to the user. Many of the contributors to this volume are employed by companies specializing in computers or information, and are in the business of turning science fiction into reality. Others are researchers for government and industry and yet others carry out their work in the universities.

One of the first questions that these auditory enthusiasts have to face is why we should use sound at all. After all, we do very well, thank you, with the visual information that a computer screen typically offers. Why make matters more complicated? Are there any jobs for which sound has been nominated that can't be done better by a visual display? These sorts of questions lead us to delve into the differences between the uses of eyes and ears.

One reason for wanting to use sound is the desire that different sources of information in a complex task should collaborate rather than interfere with one another. In my example, the executive used her eyes to keep track of where she was on a map of regional offices while she used her ears to analyze the data for each region. Had there been a pattern of differences from east to west across the map, she might have been able to detect it, because her eyes never lost sight of where she was when she was hearing each summary. If, instead of sounds, a chart of sales had popped up when she selected each location, the act of looking at it might have engaged her spatial awareness in a new task, the reading of charts, thereby distracting it from keeping track of where she was on the map. It appears from research on cognition that there is less interference between tasks carried out using different senses than between those that use the same sense.

A second reason for using sounds came out only briefly in my scenario. Individuals who are deprived of vision can still inspect bodies of data both in detail and globally if the data can be listened to. They can be aware of ongoing processes in the computer, their own options, and the results of their own actions, if these are signaled by sounds, both verbal and nonverbal.

In searching for the particular uses of sound, we soon come to realize that it has an insistent quality that makes it especially suitable for alarms and for signals such as the arrival of mail. Indeed, the word "alarm" virtually implies that we are speaking about sound. Graphic displays seem, in contrast, to have a take-it-or-leave-it quality.

Sound has another important talent. Since we are used to hearing things we cannot see, it is suited for signaling the ongoing status of background activities (such as a remote user checking our data files) while we focus our eyes on a foreground task on the screen.

Nonverbal sound can also provide a sense of place. The resonant footsteps as she entered the virtual reality of the personnel database, told our executive that she was in the right place and evoked her memories of how to interact with personnel files. The technique of providing different sounds in different "places" not only serve to confirm that the worker is in the right place, but also protect the memories and skills appropriate to that place from interference from ones that have been learned in other acoustic contexts. Experimental psychologists have long understood how important it is to ensure that the stimuli in different contexts are as different as possible if one wants to ensure that the memories and skills of one context do not spark over inappropriately into a different one.

You will encounter much talk about "events" in this volume. Environmental sound is inherently concerned with events (the physical interactions between things) such as impacts, sliding, one thing brushing against another, bouncing, or scraping. These sorts of events are hard to assess accurately with the eyes, from which the events are often hidden, or too transitory to be registered. But they are directly accessible to the ears, which pick up the acoustic energy that they create. Because sounds are specialized for telling us about events in everyday life, it is natural to use them to symbolize events in the world of information: the copying of files, the arrival of a "guest" among your data files, and so on.

Even if readers accepted my argument about the value of sounds, they might still reply that if you want to use sound, why not just use words? After all, we are able to record speech and play it back, and are becoming increasingly adept at synthesizing it as needed. To see why words are not always the best choice, recall the speed at which our executive was able to alternate among map locations getting a rapid summary of the trends. If the reports had been verbal, it would have taken much longer. Furthermore, the trends would have been predigested into words that spoke of aspects of the data that someone else had thought was important. Words are inherently conceptual. In our example, the executive was able to listen to the raw data stream itself so that she might examine it with her own ears and impose her own concepts on it.

The ears are very good for certain types of data analysis. They are expert at picking up correlations and repetitions of various sorts. If our ears receive data in which earlier events tend to be repeated at a certain delay, even if they are mixed with other events, we will hear a pulsing rhythm, or at higher rates of playback, a pitch. The pitch will become clearer as the delayed repetition becomes more regular. This sensitivity to repetition can be helpful, not only for data analysis, but for keeping track of programs running in the background. If you tune your acoustic viewer into a program that has got itself into an infinitely repeating loop, this fact will be easily audible in its sound.

Finally, nonverbal sound may be preferred to words in cases where there is already a heavy demand being placed on the verbal capacities of information workers. A computer screen may be full of text. Providing some simple information in the form of a nonverbal sound instead of words may allow users to cope with the new information without losing their place in the text. This is another example of trying to provide more information without increasing interference.

Given that modern technology allows us to present any or all combinations of words, text, visuals, or sounds, it is essential to discover the best uses of each. One picture may be worth a thousand words, but it is equally true that one word is worth a thousand pictures. Think, for example, of transmitting the idea of "unless" by means of any number of pictures. All ways of presenting information have their own strengths, and it is up to the research community to sort them out. We must bear a simple principle in mind as we design information systems: use each sensory modality for what it does best.

This volume describes many creative ideas for using patterns of sound to convey information, and for studying the human capacities that make it

Foreword

possible. It also intrigues us with a new vocabulary. Words such as sonification, audification, auralization, audiolization, earcon, and auditory icon all press forward, demanding to be heard. When outsiders no longer raise an eyebrow when they hear these words, the authors will know that their field is mature.

—Dr. Albert Bregman
McGill University

December 1993

Preface

In October 1992, the first International Conference on Auditory Display (ICAD) convened in Santa Fe, New Mexico under the sponsorship of the Santa Fe Institute. ICAD brought together 36 researchers, nearly all working with issues of how nonspeech audio can be used to convey information. Their backgrounds included, amongst other disciplines: computer science, music, experimental psychology (in general), auditory perception (in particular), chemistry, mathematics, speech recognition, theoretical biology, geology, and physics. Their work included diverse applications, including such topics as: monitoring computer system states, analyzing physical structures, numerical representations of physical, social, financial phenomena, simulation and immersive interfaces, to name a few.

Following my own interest in auditory data display, I spent years searching for papers in the field. The articles were to be found in journals ranging from those with obvious direct relevancy (e.g., *Journal of the Acoustical Society of America*) to those of peripheral relevance vis-à-vis auditory display (e.g., Supercomputing proceedings). While it is true that occasionally several papers would appear together in a specialized publication or as a subtopic in a workshop on comprehending data, I continued to be haunted by the feeling that "There must be a mother lode of information somewhere, and I'm just not finding it."

Not wanting to reinvent the wheel, I continued my search for more research on auditory data display. Gradually I began to realize that there was no such untapped vein of research findings and that this field was more in its infancy than I had expected.

So I began to seek out the people doing the research and inquiring as to where their work had led them to date. It now became clear that, as frequently happens in an infant science, a small group of people in the field knew each other but were not well known outside of their bailiwick, while a number of "outliers" were doing related work but had not found a research community with which to share ideas.

ICAD grew out of a desire to pull together researchers in auditory data display and to fill this mutual need for sharing results, stimulating new ideas, and identifying the field as a whole.

Over the next year and a half I began contacting people. The threads I followed were, at times, thin indeed. It appeared for quite some time that I would have trouble getting the minimum number of people necessary to constitute a meaningful workshop. It seemed even less likely that I would find enough people with new research suitable for publication. Moreover, my own conception of our subject matter was changing.

Since my own interest in audition, and my function as a member of the Santa Fe Institute, were centered around auditory display of data, I had been focusing my search in areas that explicitly had to do with *quantitative* data display. What I had not appreciated is the fact that *there is no distinct line between auditory data display and auditory interfaces.* In other words, the distinction between displaying quantitative information and qualitative information in an auditory environment is actually a continuum.

Sonification, or data-controlled sound, and audification, or the direct playback of data samples, are the general means used to display quantitative information. Auditory interfaces for general computing environments, teleoperation, virtual reality, and other "environmental" displays generally convey qualitative information.

It became clear that general user interfaces for computers not only drew upon the same issues of technology and auditory perception as the auditory data analysis but, indeed, they were both often trying to accomplish identical tasks. For example, an auditory display peripheral to my computer system might tell me that my cursor has encountered a text file (a qualitative, general interface issue). The auditory display might also tell me how large that file is (a quantitative issue). Furthermore, it is possible that both these pieces of information would be delivered simultaneously, as part of the same display. If the file were a database and I used the data to control the variables of a sound by routing the data to a sonification system, I will have shifted from a general interface with an auditory component to a data sonification.

Similarly, if a virtual-reality-type interface is being used to display spatially indexed medical data, the line between objects (e.g.. the liver) and data (the densities at different points in the liver) will be, and should be, easily crossed.

So ICAD became a forum for research in a spectrum of areas, all tied together under the rubric of what seemed to be the most general title I could come up with: Auditory Display. It was a forum to present current research, but perhaps it was, above all, a place where fellow researchers could find people with common interests and understandings.

As the topic area broadened, so did the population of investigators interested in our forum. The conference room at the Santa Fe Institute only held 36 people. This fact made it necessary to expand to a larger venue or to turn some people away. Major funding that would allow us to significantly expand the workshop turned out to not be available. While a collaboration with a larger conference may have been possible, I felt it was important to maintain a sharp focus for this first gathering of the field. So I was obliged to limit the attendance in order to preserve the focused quality of ICAD and thus provide a solid foundation upon which we can all pursue our work.

Certain research topics that are clearly related to auditory display were intentionally not well represented at ICAD. Perhaps the most obvious of these is speech, the preeminent research and development area involving audio and computers. My rationale was that there are a great many forums for speech-related research and a vast body of literature available. While there are common issues between speech and nonspeech auditory display design, and unique issues that arise when both are employed simultaneously, the need to focus this meeting was seen as paramount to the need to include the broadest possible audience.

The reasoning was virtually identical in the consideration of data visualization research. Visualization has developed farther than sonification has, and the volume of research into visual perception has far outweighed auditory perception research. We were fortunate enough to have many present at ICAD who used visual and auditory displays in tandem. All present readily acknowledged that the design considerations aroused in visualization research will inform and help guide research into sonification. Nevertheless, in the interests of focus, limits were necessary.

Finally, experts in particular data types and in mathematics and statistics were underrepresented. This was one shortcoming that was unintentional and unfortunate. Some sonification researchers were familiar with their data sets, such as David Lunney, (chemistry) and Jay Jackson (parallel processor diagnosis). Others were trying to solve problems with auditory display, such as Tom Rettig (information and engagement in consumer software) and Kevin McCabe (audification display of an aircraft rotor). Regrettably the type of familiarity that comes with daily, intimate work with a data type or certain statistical analysis was underrepresented, or at least underdiscussed.

The feeling attending we three dozen researchers as we moved into our round-the-table introductions was one of excitement and relief. Excitement that we would soon be learning novel concepts and approaches to auditory display, and relief because we would not have to justify our very existence, as we had so often been obligated to do at other conferences or lectures. This was a group of people who, while professional and rigorous, were sufficiently open to the possibilities of auditory display, and ready to challenge flimsy assumptions or sloppy work.

As the representatives from many major computer manufacturers, software manufacturers, universities, national laboratories, and other institutes explained their work and their background, I was struck by how many of them had been involved in music. Some were hobbyists, others had successful careers as musicians. All obviously expressed a respect for the communicative properties of sound. Many present expressed their frustration at getting their colleagues or managers to share the vision of where this work could lead. Their coworkers often tended to express the point of view that vision is a better modality for data display, and that there are already a number of good tools available for doing most data comprehension or system monitoring tasks.

I think it is safe to say that conference attendees concurred in the belief that auditory displays can provide effective tools for the meaningful exploration of data, monitoring of processes and design of comprehensible user interfaces in a wide variety of environments. If this vision was not unanimous, at least we were all open to the possibility. We are all, in the words of Bill Buxton, addressing the "impedance mismatch" between the world of machinery and the human sensorimotor system.

TERMINOLOGY

As in any new field, there are many words being used to describe the same idea. In the course of editing I did not attempt to impose my suggested terminology on the authors, although the title of this volume mentions the three primary areas of investigation. I will therefore offer a brief summary of the terms used and remind the reader that differences in wording do not necessarily reflect differences in the research itself.

"Sonification" is perhaps the most common designation for the audible display of data. Scaletti's paper gives a clear definition of this word which is more or less echoed in my paper. Sonification has been used informally for several years.

Where Scaletti refers to any auditory data representation as sonification, I make a distinction between data-controlled sound (sonification) and data samples played back directly as sound. The direct playback of data samples I refer to as "audification." This research area is even less explored than sonification and is represented in the volume primarily by the works of Hayward, McCabe et al., and Mayer-Kress et al.

"Auralization" is a word sometimes used for sonification, probably because of its kinship with "visualization," the word that has taken hold as referring to the visible representation of data. So "auralization" refers to the auditory representation or "imaging" of data. This use is found in Jackson and Francioni's paper, and in the review of the working session by Sara Bly (see Appendix I).

Other words and phrases that have been, and in some cases continue to be, used are *audiolization, acoustic display, virtual audio worlds,* and *virtual acoustic display*. All of these terms have arguments in their favor and may, in fact, become dominant "terms of the art" in one or many areas of auditory display. As is usually the case, popular convention will determine, for better or worse, which terms are finally used. Put another way, the marketplace will decide.

THE STRUCTURE OF THIS BOOK

It has been my attempt with this volume to provide more than a collection of papers, however important those papers may be. It is hoped that the conference and documentation of its proceedings will act as a focal resource in the broad and emerging area of auditory display (AD) research and development.

We are fortunate to have a foreword written by Dr. Albert Bregman, one of the foremost researchers in auditory streaming. Dr. Bregman begins his foreword with a fanciful but not so unrealistic set of scenarios involving the use of auditory display techniques in the future. He goes on to point up some of the salient perception issues that will work both to the advantage and disadvantage of auditory interfaces and data display schemes. Our request to Dr. Bregman to write this foreword (and the early appearance in the volume of auditory perception papers) belies our respect for the perception issues that will underlie any meaningful progress in auditory display design.

The Introduction to this volume is designed to provide an overview of the field, including its history and current dilemmas. Beginning with a review of the role of audition in everyday life and progressing through the ways in which we harness our hearing for alarms and auditory pattern-recognition capabilities in service of comprehending complex systems, the introduction also raises some virtually unexplored questions, such as the role of aesthetics in auditory display design and the relationship of different AD techniques in terms of knowledge representation.

The papers, described below, range from the theoretical to the applied, from underlying issues to solutions for specific problems.

In the interest of providing the most useful reference material, I decided to include the following appendixes:

"Appendix I: Comments on ICAD '92" is our attempt to provide a forum for more informal reflections on ICAD and the issues raised there. It grew out of a submission by Nat Durlach, Dan Ling, and George Zweig that was not really a formal paper but had too many valuable comments to omit from the volume. Nat said that a more informal approach might be refreshing, after a whole book of more rigorous and cautious contributions. This led me to solicit other ICAD participants for their comments, particularly those who did not submit papers but contributed a great deal to the conference.

"Appendix II: Resource List": This brief list is intended to steer the researcher new to this field towards printed and on-line resources that may provide essential background material. It enumerates some books that will provide an overview of auditory perception, music, sound computation, and human interface. It also lists several forums, journals, and conferences in these areas.

"Appendix III: Annotated Bibliography" is the result of contributions from many ICAD participants and other researchers in this field. In an interdisciplinary study such as auditory display, covering perception, sound synthesis, computer science, mathematics, and the many data-creating fields it serves, a bibliography could be so huge as to be of little use. One would be better served by simply reading the listings in the Library of Congress. To make this a more meaningful resource, all available sonification, audification, and auditory interface research is listed and, in related disciplines, contributors were asked to select and annotate what they considered to be representative publications. Fear not: if you read all of the books listed in Appendix B, you would be very well informed, indeed.

"Appendix IV: Annotations to Audio Examples": This appendix will provide the reader with a guide through the audio CD that accompanies this volume. This CD presents a more direct experience of the research discussed in the chapters. Ranging from Sheila Williams' simple tones arranged to provide first-hand experiences of numerous auditory perception phenomena to more complex sonification examples associated with the papers, the CD is an instantiation of current research. It is as much a marker telling us what needs to be done as it is a record of what has been done already.

Finally, an index is provided to assist the reader in navigating this volume.

SELECTION OF PAPERS

The process of forming ICAD had an immediate link to the formation of this volume. All of the papers here were invited and presented, in an earlier form, at ICAD. However, because of space considerations, because some presenters were not able to formalize their presentations, and in the interests of providing current and representative work, not all of the work presented at ICAD is included in this volume.

I attempted to strike a balance between providing a broad view of the field, which necessitated the inclusion of some previously published research, and providing the type of immediacy that is expected of a scientific publication. Wherever already published research is presented, it is augmented with the current research by the author and/or additional findings. The objective was to provide the reader with the most recent seminal research in the area of auditory display.

I must apologize to those whose papers were not included and thank the authors for their patience with my editing.

THE PAPERS

The papers in this volume are broadly grouped by research focus. Following my introduction, the individual contributions run the gamut from a survey of how we hear in the world to a consideration of the underlying mechanisms for extracting meaning from sound. General theoretical concerns are followed by descriptions of various data display systems, some addressing specific problems in auditory display and others describing generalized sonification tools. Next are several papers that emerge directly from particular disciplines. The final five papers cover research into auditory interface design.

Auditory display is a young field. As in any field, it takes time to establish a solid foundation of rigorous research. Certain subsets of this auditory domain, such as sonification, have seen nearly all of their research published within the last 12 years. Few of the papers at ICAD included the actual running of subjects along with the experimental design necessary for a more rigorous proof of display effectiveness. As in any new field, theories are presented and systems are developed, which then prime the pump for future movement towards maturity and verification. The papers in this volume will no doubt seed much productive research by these authors and others about to enter this emerging field.

When nearly identical sounds are played in different contexts, the effects of these contexts on identification of the sounds are considerable. In our first paper of the volume, Jim Ballas discusses the effects of context and expectancy in identifying sounds. When the cause of a sound can't be identified, Ballas refers to this condition as "causal uncertainty." Using a framework of linguistic analogies, including exclamation, deixis, simile, metaphor, and onomatopoeia, Dr. Ballas provides an overview of how sound can deliver information. By introducing the reader to essential aspects of how we identify and extract information from sound in our natural environments, his provocative contribution provides us with a much-needed foundation in creating effective auditory displays.

Sheila Williams offers a broad overview of classical perception research regarding grouping and gestalt formation, relating these important principles to the area of auditory display. In providing an overview of how perceptual groupings occur, Williams covers analytic and synthetic listening, proceeding thence to an overview of the fundamental gestalt principles in audition. After describing the effects of such variables as proximity, familiarity, common fate, and closure on Pragnanz (a symmetrical and stable condition), Sheila goes on to explain the important implications these effects will have for designing auditory displays.

After providing this overview, Williams discusses her recent research on mapping the auditory scene and describes *Streamer*, a tool for exploring low-level auditory grouping phenomena. Her paper alerts the reader to appreciate the richness of auditory perception phenomena. Her observations suggest, I think, a prudent humility as we approach the complexity of the auditory world.

Between telepresence/virtual reality and auditory data representation are techniques developed to exploit our auditory spatialization capacities. Beth Wenzel's discussion of three-dimensional auditory perception provides a bridge between these two. After discussing the integration of spatialization and sound synthesis technologies, Dr. Wenzel analyzes the advantages of what she refers to as *virtual acoustic displays*. She explains how the advantages of audition, such as spatial orientation and temporal resolution, make these displays well suited to some tasks while indicating how weaknesses can be minimized by coupling auditory with visual displays. Telerobotic control and aeronautical displays are treated in some depth in this chapter. Shuttle launch communications are suggested as another emerging applications area.

The frames of reference that a composer constructs establish guidelines for listeners. At the same time meaning may be attributed to sound via the use of signs or symbolic constructs. In "Pattern and Reference in Auditory

Display" Robin Bargar helps provide a vocabulary for different ways to manipulate sound in auditory displays. Generalizing principles from musical composition to sonification, Bargar explains how combining an understanding of the structures of composition with semiotics (attribution of meaning to signs) can yield a vast array of techniques for conveying information coherently. Examples of the "symbolic constructs" that Bargar considers are sequence, inflection, meter, rhythm, register, tonal center, polyphony, spatial location and environmental ambiance. In relating elements of music to messaging concepts, Bargar points to theories that are fundamental to organizing both sound and information.

Stuart Smith and his colleagues Ron Pickett and Marian Williams, point out that using even as few as ten data variables to control ten sound variables will result in over 3 million possible pairings. Considering the exponential expansion of possible interaction effects, one of the key challenges facing the field of auditory display pertains to how we weigh and quantify the effectiveness of our displays. Stuart et al. address this question, arguing that formal psychometric testing is necessary in order to learn how to restrict the universe of possible sound attributes to those most effective for data representation.

The authors go on to consider display evaluation procedures and the need for tools that will help in the management of a sonification system's potential configurations. In doing so they describe their own tightly coupled visual/auditory display system, Exvis, and discuss some of the underlying perceptual processes that may be at work in this display. Providing an eight-step experimental design procedure, Stuart and his colleagues lay out a sound scientific basis for studying this vast representation space that has been approached heretofore only on a less systematic basis.

My own paper offers a conceptual structure for understanding a range of sonification techniques. The paper categorizes different approaches in terms of "levels of directness" by which I mean the degree of mediation between the data source and sound generation. Distinguishing between paradigms along this continuum of "directness," the paper goes on to compare and contrast the various approaches, discussing important differences between them. I then describe a number of techniques I have developed for extending the dimensionality and comprehensibility of sonification displays. *Parameter nesting*, a technique for stretching the information carrying capacity of sound parameters, uses differences in time scale to maximize display dimensionality.

Whenever several variables are manipulated simultaneously, by this or any other technique, problems of orthogonality (overlap and independence) may occur. My paper discusses balancing the compellingness, or force, of different auditory variables used in a display and suggests some heuristics for minimizing or exploiting interaction effects. I then present the concept of a "beacon" which one may use to orient oneself within a changing auditory environment. In an attempt to provide some overall guidelines for auditory display design, my paper suggests matching data hierarchies with display hierarchies, and making maximal use of both metaphor and affect for reaching an intuitively satisfying, meaningful display.

Beginning with an introduction to sound computation, Carla Scaletti provides an overview of sound computation techniques for data sonification in her paper, "Sound Synthesis Algorithms for Auditory Data Representation." Scaletti suggests that given the enormous possibilities in the selection of sound synthesis algorithms and the huge variety of data types, we should look for structural correlations between the dynamics of the data and the dynamics of the synthesis algorithms.

Scaletti describes how a visual sound-specification language that she designed, and which runs on personal computers, is used in conjunction with special-purpose hardware to provide an environment for trying a wide variety of synthesis techniques. In doing so, she provides numerous data-to-sound mappings and sonification schemes. Techniques described include the direct mapping of data to an A-to-D converter, multiplication, addition, frequency modulation, transfer functions, filtering, physical models, and sampled sound. Perhaps more important than the description of these techniques is the clever ways Scaletti suggests employing the unique qualities of a synthesis technique to represent similarly unique data relationships. This tight coupling, I think, can speed up the design process and contribute to the efficiency of the display itself. Scaletti's sound examples provide precise and compelling examples of relationships between certain data inspection problems and auditory representation techniques.

After first providing an introduction to some recent work using sound to represent software processes, David Jameson describes an audio-enhanced debugger he has designed to help programmers attain an overview of their code. A salient guideline Jameson employs in his work is that use of sound be simple to implement and easy to understand. Stating as one of his goals that users of Sonnet should be able to predict what sound they would expect to hear, he goes on to describe a number of tools in the software, providing examples of how they might be used for debugging code. The author closes by suggesting that auditory representations of data structures may provide a valuable global view to the programmer, noting that there is some cause for optimism as tools become more fully developed.

Tara Madyastha and Daniel Reed describe a system that allows the user to configure sound devices and generate sonifications independent of the sound synthesis hardware. In creating *Porsonify*, a generalized toolkit providing a uniform network interface to sound devices, the authors hope to encourage broad experimentation with different sonification techniques and thus provide a means of investigating perceptual and applications issues. An important aspect of this system is that it has been coupled with the *Pablo Performance Analysis Environment*. Not only did their software allow Madyastha and Reed to use sonification techniques on the analysis of parallel systems and other data sets, but it also provided a means for integrating sonification and visualization environments.

Performance analysis of parallel computers, mentioned in Madyastha and Reed's paper, is dealt with in more detail in Jay Jackson and Joan Francioni's *Synchronization of Visual and Aural Parallel Program Performance Data*. Jackson and Francioni discuss the properties of parallel programs that may profitably be evaluated by auditory inspection. They then describe other research directed towards sonification, or auralization, of parallel processes. The essence of the authors' auditory display techniques is five event-to-sound mappings. Two of these, *Send-Receive* and *Idle-Busy*, are selected for an in-depth presentation. Like the other display methods described in this paper, these two techniques attempt to employ the acute temporal abilities of the auditory system to look for patterns or problems in time-sensitive data. Mapping note-on messages to sends and receives and playing tones corresponding with processor usage provide the analyst with tools to observe the macrostructure of highly parallel processes. Combined graphical and auditory displays were informally found to offer performance superior to either display alone.

Sonification, like many new areas of inquiry, has suffered from a paucity of demonstrations as to the applicability and performance capability of the new techniques under working conditions. In "Sonifying the Body Electric: Superiority of an Auditory over a Visual Display in a Complex, Multivariate System," Tecumseh Fitch and Gregory Kramer describe the design and testing of a system for application in the world of the surgical operating theater. In addressing this issue they report some surprising results.

The study describes an eight-variable auditory display which was designed to monitor the status of a "digital patient" under the care of an anesthesiologist. Subjects were trained to recognize a number of operating room emergencies using both visual and audio feedback. They were then tested with visual-only, audio-only, and audio-visual feedback. The authors found, contrary to their expectations, that the auditory display alone surpassed not only the visual display but the combined auditory and visual displays as well. Fitch and Kramer attribute their results to, amongst other things, the ways the visual and auditory systems evolved to deal with the visible world serially and the audible world in parallel.

In the next paper, Kevin McCabe and Akil Rangwalla describe their work with two fundamentally different auditory display techniques. The first has to do with modeling the pressure samples of a rotor over time and audifying the results. The authors were faced with the problem that visualization of computational fluid dynamics may yield insights in three static dimensions, but adding the temporal dimension presents formidable problems. Current technology and our visual systems both have difficulties with the volumes of data and rapidity of change presented by CFD. The authors address these problems with what they refer to as "direct auditory simulation." By sending the data samples through an A-to-D converter, McCabe and Rangwalla were able to listen to their model. If the sound is not what they expected, then they looked for problems in the model or in the design.

In the second auditory display technique described in this chapter, these same authors took a different approach. Their task involved the addition of an auditory component to the visualization of an artificial heart. By mapping time critical events to percussive sounds and continuous data to pitch, they provided a means of tracking certain events, such as the opening and closing of a valve, while focusing attention on selected aspects of a concurrent visualization.

In "Musical Structures in Data from Chaotic Attractors," Mayer-Kress, Bargar, and Choi investigate auditory representations of chaotic dynamics. In doing so they employ techniques that range in their "level of directness" from direct audification of chaotic data (the use of chaotic data to synthesize sound), through low-level parameter mappings, such as data to pitch, to mapping statistically derived quantities into auditory variables. At each level they discuss the ways in which these display techniques reveal key aspects of the chaotic systems. The authors then extend their report of these fundamental approaches, considering the presence of analogous structures in music composition. Thus the reader is provided with a range of time scales from minutes (musical compositions) to milliseconds (sound synthesis). It is suggested that each display technique may reveal or obscure different levels of structure in the data. Thus the authors underline the importance of selecting the most appropriate time scale for one's representation.

Most of the research in this volume involves the playback and control of a sound-generating means under the control of data or in response to a user's action. In a significant departure from this, Chris Hayward's "Listening to the Earth Sing" describes how the direct manipulation and playback of data samples has enabled him to listen to earthquakes, nuclear explosions, and other seismic events. After providing a background in the demands and procedures of seismology, Hayward goes on to review the literature of audified seismic data, or audio seismograms. Given the waveform nature of the data (it is, after all, a vibrating body), and the fact that these data obey the elastic wave equation, we are surprised to find only two papers published on the technique of audifying, or making audible, these very low frequency waveforms.

Hayward's paper is structured around the sound examples themselves. He provides a survey of the techniques he employed to work with the data, including frequency shifting, time compression, automatic gain control, audible markers, and looping. He then relates his concerns as a seismologist to the sound examples, describing features of the data and how those features are reflected in the audification. His extensive sound examples, accompanied by thorough annotations, provide us with a broad survey of techniques and applications. This paper lays the groundwork for extensive follow-on research.

By way of a challenging exercise in sonification, prior to ICAD, Sara Bly designed a matrix of data, and then invited several conference members to produce auditory representations of the data set and detect embedded patterns within it. "Multivariate Data Mappings" describes the process and results of this exercise.

The data consisted of a static six-dimensional data set such as might be found in a discriminant analysis problem. Attendees were presented with the sonification efforts of three researchers and asked to use their displays to decide whether each entry in a test data set belonged to a target data set, as determined, presumably, by discriminating and recognizing some unspecified pattern embedded in the data. The techniques of each researcher are described and the results of the group's discriminations are presented. The results ranged from no better than chance to noticeably better than chance, suggesting that, if carefully designed and implemented, sonification may contribute to data classification.

This session brought up an important issue related to preprocessing data for sonification. One of the researchers used extensive preprocessing of the data prior to the sonification. This resulted in a display that made the actual means of representing the data, i.e., the particular sonification scheme, almost trivial. The categorization of the data became essentially a binary choice. This researcher's efforts showed that when mathematical methods can contribute to the analysis, they should be used. However, many data sets will not lend themselves as well as Bly's to understanding by the use of mathematical techniques. Clearly one must judge when to apply what analysis techniques, and know when to leave one's innate perceptual system alone to do the processing.

Bill Gaver has undertaken ground-breaking work in auditory interfaces, including the development of auditory icons and the SonicFinder, SoundShark, and the manufacturing model ARKola. In "Creating and Using Auditory Icons," Gaver provides a review of his past work and discusses the creation of auditory icons. He also describes some of the possibilities and limitations of sampled and synthesized sounds.

Gaver points towards a means of flexibly using everyday sounds to represent a wide variety of computer events. He applies sound synthesis techniques to the fabrication of such sounds as: impact sounds; breaking, bouncing, and spilling sounds, scraping sounds, and machine sounds. With his sharp focus on the importance of everyday listening, Gaver provides us with a systematic approach to navigating the huge space of possible auditory representations.

From Gaver's auditory icons we go to Blattner's *earcons*. These are tones or sequences of tones for building messages. The construction and manipulation of earcons form the body of the paper by Blattner, Papp, and Glinert. Like Gaver, Blattner et al. seek to represent a finite universe of events and system states in general user interfaces. While Gaver focuses on the mnemonic and perceptual advantages of familiar sounds, Blattner et al. focus on the large array of messages that it is possible to build using hierarchically structured abstract sounds.

After providing a foundation of the prior work on earcons and auditory icons and looking at some of the perceptual issues associated with manipulating sound, the authors provide a sample implementation of how earcons may be used in the computer-based navigation of two-dimensional representations of buildings.

Brewster, Wright, and Edwards build on the work of Blattner et al. and describe psychometric testing they performed on some of Blattner's earlier work with earcons. Brewster et al. demonstrate that earcons can be an effective method for communicating information with sound. They also investigate whether there are significant differences between musicians and nonmusicians in using such a display. In addition to testing, however, the authors made changes to Blattner's fundamental techniques. They changed such elements as timbres and timings, then conducted another series of tests on subjects using earcons. The net result is some guidelines for designers to use when creating earcons that will help in the comprehensibility and learnability of the displays.

Jonathan Cohen describes a software system employing visual and auditory cues designed to inform a computer user of background activities taking place on his or her system. Cohen's work explores the strength of auditory displays to provide information via the ears while the eyes are busy with a foreground task. Cohen's displays proved, in user testing, somewhat disruptive and not harmonious with the user's environment. Cohen notes that effective selection and molding of the informing sound is deceptively difficult, but to do so but is crucial to the success of a display.

Elizabeth Mynatt discusses the design of a general interface she developed for blind computer users which, using X Windows, provides the underlying structures for an auditory interface. Beth has produced a prototype implementation which uses auditory cues designed to mimic certain graphical cues (e.g. low-pass filtering a sound cue is analogous to "greying out" a written textual display)

Using *Mercator*, Mynatt seeks to efficiently convey information which users of graphical interfaces can gain from other means, such as identifying file types by auditory rather than visual icons. The author reported difficulties due to poor sound design, echoing Cohen's findings as to the import and difficulty of this task. Nevertheless, both sighted and nonsighted users found the Mercator navigation scheme easy to use. Mynatt's future directions are suggestive of a more comprehensive immersive environment, including spatialized sound and tactile displays.

Rounding out the formal papers is Appendix A, which contains comments and informal observations, ICAD attendees, concerning both the state of auditory display in general and of ICAD '92 in particular. These range from the brief (Bly) to the extended (Durlach et al.) and from the enthusiastic (Powell) to the cautionary (Buxton). In all cases they provide a fresh perspective on where we are and suggestions as to where we need to go.

ACKNOWLEDGMENTS

ICAD '92 and this volume have been supported by funding from a small group of forward-thinking institutions. In hosting ICAD, the Santa Fe Institute expanded their scientific mandate, the study of complex systems, to include the tools that may help scientists comprehend complexity. Their endorsement was backed up by their assumption of the overhead costs associated with the workshop. I also want to express my appreciation to the Center for Nonlinear Studies and Advanced Computation Laboratory (both at Los Alamos National Laboratory) and to Microsoft Corporation. Their financial support made ICAD '92 possible.

I also want to acknowledge the invaluable assistance of the Santa Fe Institute staff in pulling the conference and proceedings together. Andi Sutherland provided administrative and housing assistance. Ginger Richardson gave fund-raising and administrative support; while Scott Yelich moved computers around and made them all work, and then followed up with the establishment of the ICAD e-mail list. The copyediting and patient advice on the proceedings from Ronda Butler-Villa were crucial to the production of a quality product, and her guiding hand helped me navigate the sometimes touchy process of editorial decision making.

I would especially like to thank Mike Simmons, SFI's Vice President of Academic Affairs, for his willingness to validate the vision that eventually brought ICAD to fruition. By following his intuition in support of the workshop, he extended the Institute's definition of *interdisciplinary*. I hope the results of our efforts bear out the wisdom of his risk.

I want to thank my colleagues on the ICAD Steering Committee, Bill Buxton and Stu Smith, for providing input on the direction of ICAD '92 and assisting me in locating individuals whose research and thoughts helped to make ICAD such a stimulating forum. The Papers Review Committee, by submitting informal comments on the papers, provided me with a useful perspective during my editing process. Beth Mynatt, Carla Scaletti, and

Stuart Smith took the time to read all of the papers and offer suggestions for groupings and, in some cases, comments on particular passages in the texts. I also must thank all of the ICAD attendees for their generosity with their scientific efforts and patience with the demanding editorial process.

Finally, I am indebted to my support system here in Garrison, New York, for providing me with perspective, grounding, and pragmatic advice over the year of planning and additional year of editing that ICAD demanded. My brother, R. Jonathan Kramer, was an indispensable foil and check throughout the editorial process. Elizabeth Gillespie has been my intrepid administrative assistant. The rest of my family, including my wife and three sons, all made certain that I kept my feet on the ground one day and out of the mud on the next. ICAD '94 already looms large, but the groundwork has been laid.

—Gregory Kramer
Clarity and the Santa Fe Institute

December 1993

AUDITORY DISPLAY

Gregory Kramer
Clarity, Nelson Lane, Garrison, NY 10524; e-mail: kramer@santafe.edu and Santa
Fe Institute, 1399 Hyde Park Road, Santa Fe, NM 87501

An Introduction to Auditory Display

In (sound) the wise can interpret the secret and the nature of the working of the whole universe.
—Hazrat Inayat Kahn

1. INTRODUCTION

Auditory display research applies the ways we use sound in everyday life to the human/machine interface and extends these uses via technology. The function of an auditory display (AD) is to help a user monitor and comprehend whatever it is that the sound output represents. If the interface medium between user and system is speech, then the display is exploiting a learned repertoire of language and cognitive meaning. If the display medium is nonspeech sound, the AD will exploit evolutionarily acquired environmental adaptations, including cognitive and preattentive cues. While the International Conference on Auditory Display (ICAD), and this volume, focus on nonspeech audio, auditory display rightly includes all uses of sound at the interface.

In order to develop an effective auditory display, we must understand certain key aspects of the system we are working with. This system can be defined schematically as the information generator, the communicative medium, and the information receiver (see Figure 1).

Information generators include databases of nearly every conceivable type, from atomic to astrophysical, geologic to sociologic. Also included are real-time data generators, such as computer models and factory machinery, Wall Street's stock markets and Main Street's market stock.

The communicative medium includes all aspects of an auditory display system. This includes (i) the data-receiving means, (ii) intermediary structures such as the means of mapping the data to sound, and (iii) the sound-generating means. Receiving the data is a housekeeping function, a translation from the vast world of data to the specific realm of the display system. The intermediary structure may be as simple as a look-up table linking text characters to a speech generator or a routing system for mapping selected signals from a computer interface to sampled sound files. On the other hand, the intermediary structure may be as complex as a system for preprocessing the incoming data, scaling and routing that data to an intermediary model and, thence, to the sound generator. The sound-generating means can, again, be as simple as a sound file that is played back without modification each time it receives a message, or as complex as a sound synthesis system with many input variables, each of which effects the auditory result in a number of ways.

The final *information receiver* is, of course, the listener. The human auditory system is the ultimate destination of the acoustic signal; that signal is, to some extent, preprocessed by the auditory system. Finally, the mind derives meaning from the experience.

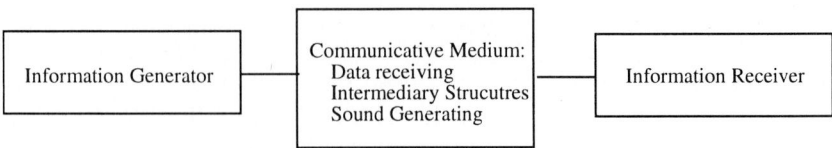

FIGURE 1 Schematic of an Auditory Display System.

We must appreciate the interdisciplinary nature of auditory display research. Clearly no individual can realistically hope to master the entire encyclopedic scope of the undertaking. AD can, after all, encompass the disciplines of cognitive and experimental psychology, psychoacoustics, communication, education, computer science, mathematics, statistics, linguistics, psychomusicology, sound analysis and synthesis, music composition, and all of the data generating fields, such as seismology, physics, economics, complex systems, machinery and instrumentation, and sociology. Information-generating sources, sound-generating and manipulating techniques, human perception and comprehension, all must be thoroughly comprehended in order to assure effectiveness in the field.

Obviously this line of endeavor will require alliances that allow each participant to contribute their most refined insights while broadening their understanding of the field as a whole. To begin this process we would do well to summarize the role that sound plays in our lives now. We can then take this as a jumping-off point to consider the advantages of developing auditory displays. This will lead us to a brief look at the history of AD and related fields. Finally, we will consider some of the common research themes and current issues in the field, such as theories, applications, and platforms.

2. SOUND IN EVERYDAY LIFE

Sound is a crucial source of information about our physical environment. The air vibrations, which upon striking our eardrums are interpreted as sound, contain information about the nature of the mechanical source of those vibrations, the space we occupy, and the direction and possibly the distance of the sound source.

Prior to the development of sound recording and broadcasting technology, sound always originated from physical interactions of elements in our immediate environment. This information enabled us to develop an idea as to what generated the sound and, more importantly, to determine whether the sound was important enough to warrant further attention and/or to base our immediate actions upon it. There is no reason to think that these mechanisms do not continue to operate now much as they have since their inception. If the sound is the accelerating rattle of a snake or that of an approaching truck, we may act to get out of the way. If we scan the environment and hear the sigh of a mate or the cry of a child, we may act to get closer.

These simple scenarios point to aspects of our auditory systems that have a direct bearing on auditory display design. Firstly, we may hear what we cannot see (e.g., the snake). We know the direction of the snake without seeing it. We separate the sound of the approaching truck from all of the other traffic sounds around us, having relegated those sounds to background status. We are able to judge that the individual clicks of the rattle are getting faster. We sweep our attention through the auditory scene, searching for a human sigh. We recognize the cry of our child as distinct from other children and we understand its emotional message.

In hearing what we could not see, we are taking advantage of the omnidirectional nature of auditory perception. Vision focuses in a single direction and provides enormous detail about what is in that direction. Using the auditory sense we can monitor all directions simultaneously.

How can we know the direction of the snake if we cannot see it? Our acute perception of differences in loudness, phase (time of arrival), and spectrum of the sound form the basis of this localization ability. These same abilities are also used to obtain a level of detail about sound that we usually take for granted. Just as we perceive textures of objects visually, we perceive detailed textures aurally. Rather than these textures providing information about color and surface quality, they provide us with very subtle information about the nature of the physical process we are observing.

In separating the sound of the approaching truck from the rest of the auditory scene, we are using skills that are highly complex and as yet not fully understood. We are using cues such as frequency region, the direction of frequency changes, spatial location, intensity fluctuations, and spectrum, or timbre.[14] We are also employing our memory, identifying a coherent auditory stream quickly because we have heard it before.

Inherent in auditory stream segregation is the more general phenomenon of grouping. Grouping includes isolating the sound of a single approaching truck, as well as creating groupings from many simultaneous traffic sounds. We form a cohesive whole, or perceptual gestalt, out of the simultaneously occurring traffic sounds. This whole can then be dealt with as a few grouped parts or even a single entity. In relegating the sounds of traffic to the background, we are employing the elemental capacity of our nervous system to attend to some changes and to ignore others. While sound only exists as fluctuations of a medium over time and, therefore, always implies change, some sounds are relegated to the background and other sounds are not. Whether it is the hum of a refrigerator or of an automobile engine, or even a forest full of bird songs, we acclimate to consistent elements of our acoustic environment and only attend to them via conscious effort or when stimulated by change.

In hearing the acceleration of the snake's rattle, we are employing the remarkably fine temporal resolution of our auditory systems. This resolution has many scales, as is evidenced by how we perceive two audible events occurring sequentially. As the time between onsets increases, we do not necessarily perceive this increase as a continuum. We hear the phase difference of a sound arriving at each ear at a different time, a time difference of a couple of milliseconds. The result of this small difference has an impact on where in space we believe a sound has originated.[1] A small step up to 5 or 10 milliseconds and we are using the different onset times of two sounds to discern that there are two sources, rather than one, for that sound. When we get to 20 milliseconds or so, we begin to hear independent onsets and, in the case of repeated sounds, rhythmic irregularities. This perception of rhythm takes us up to the range of 100 to 2000 ms, where rhythmic irregularities are easily perceived. By the time we get up to more than a few seconds, it seems we maintain rhythmic accuracy only by mentally subdividing the time lapses. (It is interesting to speculate whether the stepped nature of color perception, wherein a gradual increase in the frequency of light results in the perception of distinct color bands, might bear any relation to the distinct percepts from the a gradual increase in time between acoustic events. Both involve incremental differences in the frequency of the perceived event.)

In sweeping the auditory scene for a mate's sigh, we are using the attentional capacities of the auditory system. There is the capacity to attend to different sounds much as we look at different objects, one after another. The hearing mechanism itself adjusts its frequency response as we search for more specific sound cues.[105,106] These facts portend an even more profound aspect of our auditory systems. We can selectively attend to a single source of auditory information but, even while doing so, there is some analysis of other sources in parallel.[21]

Our ability to recognize the particular cry of our child, as distinct from nearly identical cries of other children, bears special mention. Obviously, we use our memory of certain sounds to identify those sounds when they recur in our environment. An astonishingly subtle difference in the harmonic structure of an individual's voice is sufficient to distinguish it from that of another person. In this regard, our ability to detect subtle timbral differences between one voice and another is in a class by itself.

Likewise, a human sigh, through subtle changes of the pitch or loudness shapes, conveys considerable emotional information. We respond to this emotional message consciously or unconsciously, perhaps feeling fear or love. We also have affective responses to the snake and the traffic sounds (alertness, revulsion, attraction). Using sound to actively convey emotions and arouse affective response are fundamental activities of music. Affective responses to environmental sounds often exceed our expectations.

---

[1] If the sequential events are repeated continuously, they will create a perception of pitch over the broad range of .05 ms (20 kHz) to 50 ms (20 Hz). A 1-ms delay between pulses will yield a 1000-kHz tone.

3. A SUMMARY OF ASSETS AND PROBLEMS

The above examples point to some advantageous aspects of auditory displays. We can also consider some problems that may arise from the limitations of our auditory capacities.

TABLE 1 Benefits of Auditory Displays

Quality	Application/Advantage
Eyes free	Monitoring where other variables or tasks must be observed visually; complex or quickly changing visualizations; interface is vision impaired.
Rapid detection	Monitoring; high-stress environments.
Alerting	Monitoring; high-stress environments; general interfaces.
Orienting	Data exploration, indicates areas of interest. In monitoring tells the eyes where to look.
Backgrounding	Monitoring or exploring very large data sets.
Parallel listening	Exploring high-dimensional systems; monitoring multiple processes; comparing multiple data sets.
Acute temporal resolution	Time-sequenced data; broad dynamic range (milliseconds to several thousand milliseconds).
Affective response	Ease of learning; engagement; convey subtle qualitative information.
Auditory gestalt formation	Discerning overall relationships or trends in data; picking out meaningful events or states in a stream of data.

3.1 GENERAL BENEFITS OF AUDITORY DISPLAYS

The benefits of auditory displays may be summarized as (i) eyes-free use, (ii) rapid detection of acoustic signals, (iii) alerting, (iv) orienting, (v) backgrounding, (vi) parallel listening, (vii) acute temporal resolution, (viii) affective response, and (ix) auditory gestalt formation. These are summarized in Table 1.

Eyes-free displays are essential in circumstances where a system operator has to maintain visual contact with other elements of his environment and where vision might be dangerously occluded or impaired. High-information/high-stress environments such as operating theaters, aerospace control and cockpit operation, network-monitoring environments, and factory floors come to mind in this regard, as do complex or quickly changing data visualizations.

Research has shown that response time to an acoustic signal can be shorter than to a visual signal.[66] The capacity for *rapid detection* enables acoustic signals to play a key role in interface design.

The *alerting* role of auditory displays, long used for alarms in countless applications, is intimately connected with the *eyes free, rapid* detection, and *orienting* qualities mentioned elsewhere. Because we do not need to be looking in any particular direction to hear an alarm, because detection of the source is rapid, and because the alarm orients us, acoustic signals are singularly well suited to alerting system users of important events. The relevance to monitoring and high-stress environments is obvious. In fact, the very word "alarm" is used to refer to both a sudden fear and an auditory warning signal.

In monitoring tasks, sound can be used to indicate the importance of a variable, even though the details of that variable may be delivered visually. This *orienting* capacity is particularly apparent when monitoring spatially indexed data with spatially indexed sound. One hears the cue "over there" and looks in that direction. As Wenzel puts it, the ears tell the eyes where to look (see Perrott et al.[70]). Orienting also occurs, in a somewhat different manner, in data exploration tasks. A sonification may indicate an area of interest that the analyst may explore further by other means (e.g., statistical analysis).

The *backgrounding* capacity of the auditory system is its ability to relegate some sounds to a low attentional priority while maintaining sufficient awareness of these sounds so that any significant change will forcefully draw one's attention. This capability allows the system user to attend to monitoring tasks or the exploration of large data sets while still maintaining contact with the data.

Parallel listening,[2] or the ability to keep track of several voices simultaneously, allows the display user to monitor multiple processes (for example, see Gaver[42]). Likewise, an analyst could explore and compare multiple data sets simultaneously, possibly finding correlations that would otherwise be missed. Because several data variables can be imbedded in each voice, parallel listening also opens up the possibility of designing displays capable of representing high-dimensional data sets.

As mentioned above, the temporal acuity of the auditory channel ranges from a few milliseconds to several thousand milliseconds. This provides the capacity to experience time-sequenced data encompassing a very wide dynamic range while maintaining excellent resolution. *Acute temporal resolution* is one of the greatest strengths of the auditory system.[66]

Our *affective response* to sound, if properly used, may make displays easier to learn, more engaging, and capable of conveying subtle, qualitative information. By associating our affective response to changes in a sound with our affective response to changes in the data (see Kramer, in this volume), we can create a more intuitively meaningful display. Also, the use of soundtracks in computer software, games, and so on offers these media advantages identical to film soundtracks. Fear, anticipation, calm, humor, and other emotions can all be suggested.

Auditory gestalt formation allows one to discern overall relationships or trends in data (see Williams, in this volume). As our perceptual systems strive to make sense of a complex sound field, we may perceive the sound as a whole, without necessarily directing attention to its component parts. These gestalts allow us to pick out meaningful events or states in a stream of data.

3.2 BENEFITS OF AUDITORY DISPLAY IN CONJUNCTION WITH OTHER DISPLAYS

Many of the above-mentioned benefits of auditory displays translate into a related set of benefits when combined with visual displays. These advantages may be summarized as (i) nonintrusive enhancement, (ii) increase in perceived quality, (iii) superior temporal resolution, (iv) high dimensionality, (v) engagement, (vi) complementary gestalts, (vii) intermodal correlations, (viii) realism, (ix) synesthesia, (x) enhanced creativity, and (xi) lower computational requirements. These are summarized in Table 2.

---

[2] It is unresolved whether there is strictly parallel listening or rapidly multiplexed serial listening.

TABLE 2 Benefits of AD in conjunction with other displays

Quality	Advantage
Nonintrusive enhancement	Augments visual displays without interfering with existing tools and skills.
Increase in perceived quality	Affordable and easily implemented enhancement.
Superior temporal resolution	Time series data. Shorter duration events can be detected with auditory displays.
High dimensionality	Adds to (and exceeds) dimensionality of visual displays.
Engagement	Decrease learning times, reduce fatigue, and increase enthusiasm.
Complimentary pattern recognition capabilities	AD provides the opportunity to bring new and different capacities to the detection of relationships in data.
Intermodal correlations	Reinforcement of sensed experiences; veridical representations.
Enhanced realism	Immersive, interactive interfaces, (e.g., VR) become more realistic.
Synesthesia	Replacement of insufficient or inappropriate cues from other sensory channels.
Enhanced learning and creativity	Provide a presentation modality suited to the student's learning style; encourage fresh interpretations not suggested by traditional representation techniques; imagined extensions of data when "audiating."
Lower computational requirements	Efficient use of CPU and memory resources for display tasks (temporarily meaningful).

Auditory displays may be integrated with visual displays without interfering with existing tools and skills. The large installed base of graphics-oriented software and visualization packages may thus benefit from the *nonintrusive enhancement* offered by an acoustic element, without serious negative side effects.

Adding high-quality audio to the interface can *increase the perceived quality* of the interface.[68] This not only provides a cost effective, if incremental, increase in satisfaction with visual data displays, but it offers general computer interfaces, games, and other applications the potential for greater refinement and subtlety.

The *superior temporal resolution* described above enables auditory displays to present events that visual displays cannot because our visual perception systems are unable to respond quickly enough. The 30-frames-per-second refresh rate of a CRT display points to temporal resolution limits of visual displays in representing events of less than 33-ms duration. Indeed, films are usually projected at 24 fps and produce an illusion of uninterrupted motion. A synchronized auditory/visual display may also offer temporal detail and enhanced temporal pattern recognition.

High dimensionality, or the ability to display multiple variables simultaneously, is very difficult to attain in visual displays. By adding auditory dimensions to visual dimensions, display dimensionality can be increased. It is also possible that auditory displays alone can represent higher dimensional systems than visual displays alone.

Engagement stemming from a multisensory interface may decrease learning times, reduce fatigue, and increase enthusiasm. Seeing a film without a sound track points this out. The addition of music and sound effects to scientific visualizations and computer games is a testament to the value of the relatively unstudied quality of engagement. The scientific community stands to benefit greatly from the accomplishments of the arts and entertainment communities, where audition is used explicitly as an integral part of the creative task.

Because the raw material and sense organs of vision are so different from those of audition, it can be expected that auditory processing will provide a rich complement of pattern recognition, gestalt formation, and preattentive capacities to those of vision. The superior temporal acuity of audition mentioned above provides one provocative suggestion of audition's unexplored potential to afford *complimentary pattern recognition capabilities*.

Intermodal correlations provide the user of an interface with the kinds of convergence between the senses found in the course of everyday life. Correlations of vision and hearing can verify, for example, that two objects have come in contact with one another.

Intermodal correlations result in covarying operations and, thus, redundancy. This provides the foundation for *enhanced realism* in virtual environments and teleoperation. (Closing a visualized door or pushing one virtual object into another can hardly achieve a high degree of realism without an audible consequent.)

In virtual environments and teleoperation applications, some sensory experiences are difficult to display or unacceptably intrusive. *Synesthesia*, the substitution of one sensory modality for another, is one technique for addressing this problem. A common example of synesthesia in current VR systems is the use of a sound cue to provide feedback that, in the absence of tactile cues, the system user has grasped an object.

For individuals who have an inclination towards auditory learning (or a corresponding weakness for visual learning), employing auditory displays may result in *enhanced learning and creativity*. Also, presenting information in novel ways may engender the potential to produce novel results. Where a nonintuitive sonification may be difficult to learn, for example, it may also provide unexpected insights into one's data by jogging the mind into a fresh approach. While *audiation*, the manipulation of auditory imagery, is not well understood, it is conceivable that we manipulate auditory images differently than visual images. If so, perhaps audiation would help us to generate new insights into otherwise perplexing problems.

CPU demands and the cost of graphical displays reflect the high-frequency/high-data-rate nature of graphical displays and suggest that cost-effective solutions to certain problems may be found in auditory displays. One minute of uncompressed video might take 14 gigabytes to store; one minute of uncompressed, high-quality stereo audio takes 10 megabytes. Extending the interface via sound currently has an advantage in relationship to graphical displays because of its *lower computational requirements*.[3] On the other hand, this gain may not be as important in the future, once the rapid progress of computer hardware and software are taken into account.

3.3 DIFFICULTIES WITH AUDITORY DISPLAYS

AD also has its limitations and difficulties. These complications, in some cases, are the direct result of how our nervous system processes sound, while other problems can be associated with the task presented or the environment in which the AD is used. Difficulties with AD may be summarized as (i) low resolution of some auditory variables, (ii) limited spatial precision, (iii) lack of absolute values, (iv) lack of orthogonality, (v) annoyance, (vi) interference with speech communication, (vii) not bound by line of sight, (viii) absence of persistence, (ix) no printout, and (x) user limitations. These are summarized in Table 3.

---

[3] Considering the expense associated with graphics, one can add some AD capabilities to a system at relatively low cost. However, if complex spatialization, sound diffraction, and diffusion for virtual auditory environments are taken into account, the computation and storage requirements can be substantial.

In the display of quantitative information, the *low resolution* with which we perceive many sound variables falls far short of the resolution required for certain tasks. Most auditory parameters are not suitable for high-resolution display.[4] For example, a sonification using brightness (spectral centroid) to represent a data variable can unambiguously represent a limited number of values. Compare this representation to the precision afforded by numbers in a simple printout. Psychometric tests have yet to be conducted that will quantify the resolution of many acoustic variables (see Smith, in this volume).

A case of low resolution that bears particular mention is the *limited spatial precision* of our auditory systems. Within the arc covered by foveal vision (about 2 degrees) we have the capacity to resolve very fine detail, typically one minute of arc (as few as 2 seconds for selected tasks).[12,101] The capacity of the auditory system to pinpoint a location in space is much coarser, with 1 degree of resolution in front of the listener and 5–10 degrees to the side (see Wenzel, in this volume). When one is representing spatially indexed data, this limitation is of particular importance.

Spatial precision is only part of the problem, however, when attempting to represent volumetric data. Another AD problem is that sound can only define spatial extent by extension in time or by crudely defining edges. For example, the edges of an "audible object"[54] are defined when a sound appears to move from one location to another, or a subject moves through the sound. Either way, the boundary of the defined space must be determined by some movement.

One may also use transient sounds to indicate the top, bottom, and sides of an object.[5] The limitation here is that edges are not persistent, so the need for movement remains if the boundaries are to be refreshed in memory. We cannot readily judge the size of an audible object, for example, or compare two such objects, if they can only exist in a temporal domain. Scanning a single object or looking back and forth between objects is a simple task visually; judging spatial extent by listening to multiple sounds moving from edge to edge is not.

It is worth noting here that the problems associated with representing spatial information in sound refer to direct, isomorphic mappings. If one intends to represent volumes, a discrete set of surface textures, numbers of corners or edges, and other such variables, it is possible to represent these by mapping numbers representing them to auditory variables other than spatial location.

Data sonifications are relative. While we may detect minute variations in the pitch of a sound, for example, we know only if it went up or down,

---

[4] It is worth noting that pitch resolution offers a very wide dynamic range and that the temporal resolution of audition is unparalleled by other senses (see above).

[5] Distance cues are an as-yet unsolved problem.[59]

not the absolute value of what is represented. Compare this to a simple XY graph, where the precise values of X and Y at any given point are easily discerned. Those gifted with absolute pitch may, in this instance, have a strong framework for determining absolute values when pitch is used to represent data. However, we do not know of any similar gift which might be called "absolute spectral centroid" or "absolute buzziness." The *lack of absolute values* in auditory representations will limit their application.

Most auditory parameters are not perceptually independent; a change in one may interfere with or contaminate the perceived values of other variables. This *lack of orthogonality* contributes to the precision problem mentioned above and may make auditory displays prone to display-induced anomalies.

Because we have no "ear-lids," ADs are difficult to block from our awareness. As Steinberg points out (in the Comments appendix of this volume), this may contribute to the fact that our auditory space is particularly personal. Intrusions into this space may be *annoying*, causing users to turn off or turn down the auditory icons or eliminate the sonification despite its potential contribution to the user's tasks.

It is a major concern of auditory interface designers that their work may *interfere with speech communication* in the work environment. As discussed by Patterson (see Section 9), this can create resistance to the displays and limit their use in certain applications.

Because sound is *not bound by line-of-sight*, the use of ADs can be a problem in open work areas. One can hardly imagine tolerating a busy insurance office, for example, where each cubicle has someone sonifying a spreadsheet or navigating a database using earcons. While followers of John Cage may have enjoyed the sonic experience, it can hardly be thought of as a good environment for concentrated work. The use of headphones can address this problem, but headphone use will not always be desirable and may be uncomfortable or interfere with other tasks. Note that the virtual environment community has similar problems with head-mounted displays.

Perceiving at time $t+1$ what was displayed at time t is only possible via some form of memory. Information present auditorily at any given moment will only yield a partial reconstruction later on. Simultaneous comparisons and reminiscences are problematical. Put another way, ADs are necessarily time contingent, while visual displays persist until they are removed from the field of view (or shortly thereafter). We refer to this problem as *absence of persistence*. By way of example, consider a simple XY visual graph, where the X axis represents time. By scanning the eyes left and right, data from different periods can be reviewed and compared. While one could "shuttle" through a sonification, or replay a sound sequence generated by a series of GUI (Graphical User Interface) actions, any given moment will yield only the representation of a selected time step. Simultaneous comparisons may produce cacophonous and incomprehensible results.

TABLE 3 Difficulties with auditory displays

Quality	Explanation
Low resolution of many auditory variables	Precise representation of quantitative data is difficult.
Limited spatial precision	Special case of low resolution; volumetric displays are poorly represented.
Lack of absolute values	Sonifications are relative; auditory icons indicate general system states, not precise values.
Lack of orthogonality	Most auditory parameters are not perceptually independent.
Annoying	No "ear-lids" so we can't casually tune out the display.
Interferes with speech communication	Can make group processes difficult.
Not bound by line-of-sight	Bleeds to nearby listeners; loss of privacy to user and others.
Absence of persistence	Reviewing or comparing two data regions can be far more difficult.
No printout	Distribution of results is problematic.
User limitations: The aural equivalent of color blindness	As with visual displays, some people will have less acuity for the display variables, such as timing, pitch, or timbre.

The fact that ADs produce *no printout* may seem to be a fairly trivial problem, particularly given the dawning of the all-digital multimedia age. However, until this transformation is complete, the inability to include ADs in magazines, printed reports, newspapers, and books will pose problems. Imagine trying to take an interactive sonification to the beach along with your newspaper. Given the lack of standardized sound playback hardware and software, distribution difficulties, even over computer networks, are a real problem. Absence of hard copy represents a major impediment to AD use and acceptance. In fact, many major government institutions will not accept research proposals consisting of anything other than text, drawings, and photographs. It could be argued that the omnipresence of the telephone delivery system, which enables audio to exceed video in its distribution reach, mitigates this problem. We are not aware, however, of any significant publishing activities that use telephone lines to facilitate sonification home deliveries. At any rate, such publishing would eliminate the interactivity of a good AD system.

As with visual displays, *user limitations* will always present problems. While color blindness is a relatively common but usually recognized deficit, similar shortcomings in individual's auditory perceptual systems are poorly understood. AD researchers are often asked if a tone-deaf person could use their displays. People are generally timid and defensive about their auditory capabilities. Many of the problems anticipated with individual user differences may prove no more problematic than similar differences have proven with vision. However, until we understand the variables in this area, our uncertainty will limit the reliability and acceptance of auditory displays.

3.4 WHERE THE PROS AND CONS ARE NOW

This review of the benefits and difficulties associated with auditory displays cannot cover the topic from every viewpoint. Given the paucity of implementations, it is difficult to project where audio will shine and where it will fall short. The nascent state of the field and our limited understanding of how we process complex sounds contribute to the difficulty of assessing AD. The huge diversity of applications, from telepresence and robotics to seismology and meteorology, with their particular demands on an AD system, will undoubtedly bring to light advantages and difficulties we have not considered. This review may serve, however, as a starting point for considering the use of AD in a variety of applications.

4. TASK DEPENDENCE

In considering this review of the advantages and difficulties of auditory displays, it becomes apparent that the best use of sound will be dependent upon the task to be accomplished. In particular, there are two types of tasks that seem to require somewhat different approaches: *template matching* and *data exploration*. Although there is a great deal of overlap, these tasks tend to be quite distinct and correlate well with the two general task types: template matching is generally a *monitoring* task while data exploration is pretty much an *analysis* task.

In monitoring real-time events, whether seismic audifications or medical sonifications, we are generally looking for known patterns in the sound. Designing displays for template matching ("listening in search"), therefore, presents a far more tractable problem than that of data exploration since the universe of meaningful sounds in template matching is far narrower. Even if the display user must recognize many system states, she has the assurance that if the sound points to an important feature of the system, she will know it. Certainly in the case of auditory icons and earcons, the sound and the system being represented are well understood.

When using sonification and audification for data exploration, the system user does not know precisely what to listen for. If users are familiar with their display systems, they may create data-to-sound mappings that highlight what they believe to be important elements of the data. They may then attend to those variables more carefully. However, the possibility that previously unknown features will "pop out" of the display is a key motivation for using auditory (and visual) data representations for analysis. By definition, the display user cannot precisely anticipate what will be heard.

It is likely that even in analysis tasks the analyst is using template-matching skills, albeit for less precise templates. For example, determining if the data is monotonic may involve relating the current display to a previously experienced monotonic data set. Or perhaps one may recognize a region of a data set as somehow similar to another region with very wide swings in the data values. Precisely what accounts for this experience of recognition is difficult to define. The key point here is that when the task is one of analysis, perception of deviations from (possibly subcognitively detected) patterns is the norm rather than the exception.

Different tasks call for different optimizations of display design and training. Monitoring tasks involving template matching necessitate a limited but unambiguous display of the data. In order to emphasize meaningful features, the display can be designed with important data variables mapped to multiple auditory parameters, creating a more "forceful" perceptual experience of the relevant data and, thus, "loading" the display. Naturally, one must be able to identify the important data variables before one can "load" a display to detect them. Certainly, novel *and* significant states may arise in the course of a monitoring task, but this is anomalous. In a system where the important interactions are unknown, the neutrality of the display should be emphasized.

Another task-contingent issue to be considered by the AD designer is the way absence of psychophysical linearity is employed for purposes of loading a specialized display. By their very nature, some parameters do not bear a linear relationship to their psychophysical intensity. Referring to Stevens' psychophysical power laws,[93] Rogowitz et al. point out that linear representations of variations in objective intensity may result in nonlinear (or even quantized) variations in perceived intensity. Due to their often precognitive character, Rogowitz refers to these distortions of linearity as "preattentive." Some tasks, says Rogowitz, benefit from these weighted representations while, in other tasks, visualization techniques that produce preattentive perceptual experiences will lead to serious misinterpretations.[78]

Training paradigms for template-matching tasks will generally be quite different from those employed in the learning of data exploration tasks. In the former, a user must be trained to recognize a known universe of possible auditory results and to know what those results indicate in the data. In an analysis task involving data exploration, preknowledge of recognized states may also be possible, and even necessary. On the other hand, preknowledge could, conceivably, act as an obstacle to fresh insights. Data exploration is likely to necessitate hearing the data in multiple ways. This is akin to remapping colors or rotating an object in a data visualization. Thus, a deeper comprehension of the display becomes more important as the need to change mappings or synthesis techniques becomes part of the analytical process. Generally speaking, the monitoring task requires the design of a single optimal display. The analysis task favors a "soft wired" and interactive user interface which allows user access to data scaling and routing, data preprocessing, selecting regions of interest, statistical analysis, and other operations.

Template matching and data exploration are essentially different but interdependent tasks. In order to discover the cues or signatures of significant features in a given dataset, we are obliged to use data exploration techniques. We then can design a display to optimize those cues for use in seeking similar patterns in similar data sets. One can describe template matching as the application of the knowledge gained through exploration. Conversely, experience with monitoring one's data in real time and becoming familiar with specific system states may improve the data exploration process by providing verifiable system states as references. Regardless what the application may be, we are apt to alternately find ourselves performing matching and seeking operations.

5. A BRIEF LOOK AT SPECIFIC APPLICATIONS

While the two designations of monitoring and analysis described above may not necessarily encompass all possible applications, these categories can serve as a basis from which we make sense of the widely varying existing and potential uses of auditory displays.

The applications of AD to date have been largely experimental. Exceptions consist essentially of monitoring tasks such as games, alarms, speech interfaces, and devices for blind people (e.g., thermometers). These applications are real-time and convey the status of a system in rather elementary terms. The monitoring tasks selected are sufficiently simple that little training is necessary, the tools are readily available or can be inexpensively built into the delivery device or environment, and how we hear and make sense of the display is relatively straightforward. Indeed, some of the most fruitful research in AD has involved dedicated, real-time applications yielding immediate benefits and/or using sound in familiar ways. Using a Geiger counter, for example, is not unlike listening to beads landing in a box, with more beads representing more radiation.

Auditory analysis of a data set, on the other hand, presents formidable problems akin to those encountered monitoring a highly complex and multivariate system. Using data exploration tools often requires extensive training and, it is widely believed, more sophisticated tools than are yet generally available. Also, the thicket of difficult perceptual issues, such as auditory streaming, parameter orthogonality, and gestalt formation, immediately bring the AD researcher to the thorny problems at the frontiers of current auditory perception knowledge. There is a good reason why simple monitoring tasks comprised the first AD applications. The tasks selected were easily accomplished and provided obvious utility.

5.1 MONITORING APPLICATIONS

Simple ADs, including the beeps of the heart monitor and the alarms of ICU's, have been used in medical monitoring for years. The high-stress/high-information and eyes-busy nature of medical environments make AD eminently well suited to enhanced monitoring applications. More sophisticated displays, such as the anesthesiologist's workstation, have been proposed by Fitch and Kramer (in this volume). Tasks demanding heightened attention to complex environments have driven the development of AD in aircraft flight decks,[69] air traffic control (see Begault[5] and Wenzel, in this volume), financial trading desks (See Kramer, in this volume), and, of course, sonar. Other environments with similar demands would also be prime candidates for effective AD implementations. Factories with human process control monitors,[42] such as paper and steel mills and assembly lines, come to mind. Applications in this area could range from nuclear power plants to commercial kitchens. Readers are invited to reflect on the considerable possibilities.

Instrumentation designed for vision-impaired users, such as Lunney et al.'s chromatograph,[61] Smith Kettlewell's Auditory Oscilloscope,[82] and the navigation system of Loomis et al.[60] point to another rich set of possible applications for AD. While text-to-speech converters have been developed to serve the blind, one can envision auditory indicators (speech and otherwise) in public transportation, more sophisticated instrumentation, and data-searching tools for blind spreadsheet users.

Auditory interfaces on instrumentation for sighted users, such as the Geiger counter and metal detector, mostly center around eyes-busy tasks. The beeps in electronic circuit testers, notification tones in automobiles, planes and boats, and the variety of medical monitor interfaces mentioned above are a few instances that come to mind. In most cases these interfaces are simple alarms; in a few, they provide continuous information on the status of a system.

Several papers delivered at ICAD '92 described work in debugging software (see Jameson, in this volume) and analysis of processor performance in parallel computers (see Jackson and Madyastha, both in this volume). While the parallel processor work falls into the analysis category, it is not difficult to imagine the tools and principles developed in this work being applied to real-time monitoring in the near future. Extensions of this work might include monitoring and/or analyzing functions in such domains as large computer-controlled transportation or broadcast networks, power or fluid distribution systems, or even human resource allocation systems.

General computer interfaces also received a lot of attention at ICAD and at computer-human interfaces conferences prior to ICAD (see Cohen and Mynatt, both in this volume, and others[7,17,40]). User notification strategies developed by researchers in this area will no doubt have extensive applications in future computers imbedded in home entertainment centers, electronic books, special-purpose office equipment, automated teller machines, and shopping guides in malls or at home. Extensions of these notification strategies may well apply to human communication with software agents, easing interaction where graphical interfaces are insufficient or impractical.

Graphical user interface [GUI] strategies using audio are already being implemented in computer-shared cooperative work (CSCW) environments.[41] Also, researchers can employ GUI-inspired techniques to robotics and telepresence interfaces, along with other means to provide situational awareness and user feedback, such as microphones and synthesized and spatialized sound cues.[26] In addition to CSCW, the telepresence applications being researched at the University of Toronto by Baecker et al.,[3] EuroPARC,[18,43] and elsewhere are using audio to provide speech communication and improved veridicality.

Virtual environments share many of the problems and solutions of teleoperation. The VR system user needs to be informed of the status of a remote or fabricated environment. To this end, synthesized, sampled, and spatialized cues serve to make the VR world more realistic, informative, and engaging (see Astheimer[2] and Wenzel, in this volume).

So extensively has auditory display been developed and utilized in modern games and entertainment, its utility today has come to be taken for granted in these areas. Game companies have teams of sound designers adding speech, music, and sound effects to their products. Films, music videos, CD-ROM, and other multimedia products rely so heavily on sound that no product is considered complete without it. While the development path in multimedia cannot be foretold, without a doubt novel approaches to electronic information presentation will suggest new uses of audio, as technological developments suggest approaches not yet imagined.

5.2 DATA EXPLORATION APPLICATIONS

Analytical data exploration using sound is a new and challenging task for AD research. Several of the monitoring tasks mentioned above have parallels in data exploration, including financial[53,64] and medical[86] data analysis, and parallel computer and network analysis. Other data types targeted for auditory inspection and discussed at ICAD and/or described elsewhere in papers are seismic (see Hayward, in this volume), census (see Smith,[86] and Madyastha, in this volume), environmental,[79] physical,[76] mathematical (see Kramer and Ellison[52] and Mayer-Kress, in this volume), astrophysical,[94] geographical and map (see Blattner, in this volume, Weber[99] and Krygier[55]) neurological,[102] sorting algorithms,[16] and simulations including artificial hearts and rotors (see McCabe, in this volume) and predator/prey.[50,51]

Extensions of the auditory data analysis endeavors may well reach into areas of application under investigation worldwide by data visualization researchers. To date great effort has been invested in developing tools for visually exploring meteorological, immunological, physical, chemical, and other large and sometime intractable data sets. Auditory data analysis might well provide natural extensions of such visualization research. Complex simulations and computational processes, such as artificial life, genetic algorithms, and neural networks, may all benefit from auditory analysis. While the prospects in auditory data analysis are exciting and provocative, a widely recognized and truly convincing practical demonstration has yet to be realized.

5.3 THE FUTURE

AD is an embryonic field and the pace of development has accelerated rapidly. Tracking each and every possible use of AD is not practical. Rather than producing an exhaustive list of AD applications, it is the intention here to suggest possibilities. Hopefully the reader will find the above survey of existing and potential applications suggestive.

Some hold the theory that a *Grand Challenge* type of application, one that will tackle a difficult and heretofore unsolvable problem with stunning success, is needed and can be expected from AD. Others suggest that it will be a gradual increase in applications of AD to more straightforward problems that will lead to ubiquitous and increasingly sophisticated solutions for everyday problems. Whatever the case, increased interest in, and funding for, AD research can be expected to provide the knowledge and practical tools for addressing increasingly complex and demanding challenges in the future.

6. ANALOGIC AND SYMBOLIC REPRESENTATION

In order to better understand how auditory displays can be designed and used, we can discuss AD research in terms of two broad categories, which we will refer to as *analogic* and *symbolic* information representation.[85][6] A symbolic representation categorically denotes the thing being represented while the analogic representation directly displays relationships. Representation techniques such as those presented at ICAD can be placed along a continuum defined in terms of direct and connotative qualities, where "connotative" implies a less direct, more associative or suggestive correspondence between the thing itself and its sonic representation.

An *analogic* representation is one in which there is an immediate and intrinsic correspondence between the sort of structure being represented and the representation medium. The relations in the representation medium are a structural homomorph of the relations in the thing being represented.[46] They are the same sort of structure and admit the same sorts of operations (e.g., movement in a given dimension), but the representation is a simplification of the reality. In this sense, there is a directness about the representation, possibly even a one-to-one relationship between changes in the medium and changes in the thing being represented.

By way of example, a Geiger counter produces an analogic representation. A change in the representation medium, the speed of clicks, has a direct correspondence with the thing being represented, radiation. Increases and decreases in the acoustic signal are a simple analog of increases and decreases in radiation. The spatial aspect of two-dimensional maps are one of the more common examples of an analogic visual representation (see below).

By *symbolic* representation we refer to those display schemes in which the representation involves an amalgamation of the information represented

---

[6] In his paper, "Afterthoughts on Analogical Representations,"[80] Sloman uses the terms *Fregean* and *analogical*. He specifically decries the use of the less precise *symbolic*. However, for purposes of our discussion, I feel that *symbolic* will more easily convey the concept to the reader. An in-depth analysis of how the work by Gottleb Frege relates to AD research cannot be provided here.

into discrete elements and the establishment of a relationship between information conveying elements that does not reflect intrinsic relationships between elements of what is being represented. As Sloman says, "In a complex Fregean symbol the structure of the symbol corresponds not to the structure of the thing denoted, but to the structure of the procedure by which that thing is identified or computed."[84]

An alarm is one example of a symbolic auditory representation. The alarm represents a (possibly complex) event or family of events. There need be no direct relationship between the structure of the alarm and the structure of the event. Other symbols include words, denotative circles and triangles on a road map, and the sound signatures and graphic logos employed by businesses.

It will become apparent that symbolic and analogic representations are points on a continuum, and shifts from one to the other are not necessarily discrete. The end points of that continuum may be understood to be spoken language on the symbolic end and sound produced directly as a result of mechanical events on the analogic end.

6.1 SYMBOLIC AND ANALOGIC AUDITORY DISPLAY TECHNIQUES

Speech interfaces are explicitly symbolic. A word or set of words denote a thing, a concept, or a relationship via a set of learned sounds which we call language. Blattner's earcons are a tone-based symbol set, wherein the verbal language is replaced with combinations of pitch and rhythmic structures, each of which denotes something in the data or computer environment. Earcons have the languagelike characteristics of rules and internal references. When these references are learned, they are understood to refer to very specific items or events. However, the construction of earcons does not mimic speech, nor can earcons be used for the type of complex information conveyed by speech. The content, construction, and complexity of earcons parallels that of icons. Basic building blocks (sound fragments) are used to create more complex structures.

Gaver's auditory icons, in their early implementations, employed a form of language. A caricature of the thing or event being represented was symbolically and metaphorically presented in sound. For example, the idealized sound of a trash can represent the successful deletion of a file. Because of the naturalness of this language, we have to examine it closely to ascertain the symbolic nature of the final sound product. Generally speaking, an auditory icon represents a discrete item or event.

Gaver's more recent work, presented in this volume, extends this discrete, symbolic representation by suggesting that the symbol (auditory icon) be controlled by relevant data to represent continuous changes in that data. For example, a caricature of a machine sound may indicate that a computer process is currently running. The speed at which the process runs may be metaphorically represented by the changes in the speed of repetition of a rhythmic element of the sound. In this example, the auditory symbol manifests continuous changes and these continuous changes analogically represent speed. If the changing speed of the machine were used to represent, by analogy, changes in a country's population, this might be considered a more explicitly analogic representation. As this example points out, Gaver's work reaches towards the middle of the continuum, revealing that the boundaries between analogic and symbolic representations are not discrete.

Cohen's use of sound for notification of system states is akin to Gaver's auditory icon work. Discrete events are represented symbolically by familiar sounds. For example, a user login to a computer causes a door-knocking sound. Different levels of a system event may then be mapped to discrete sounds that have a symbolic relationship. For example, lower and higher percentages of CPU cycles being employed by a guest on the computer may be represented by walking, jogging, and running sounds. In these examples we have the beginnings of a data continuum being represented by a sound continuum fabricated from discrete, symbolic events.

Like Gaver and Cohen, Mynatt uses realistic sounds as symbols of events within the computer. She suggests conveying information as to the state of the event being represented by changing the "state" of the representative sound. For example, she used the sound of a push button to represent software buttons that are displayed graphically on the screen. Then, the sound may be low-pass filtered to represent that the button is "greyed-out," or not available for operation. In this way, she is nesting symbols within symbols and the user learns to navigate the system by learning the language of symbols she has constructed.

The sonification work by Jackson and Francioni, along with the parallel processor analysis sonifications by Madyastha and Reed, indicate the on-off status of elements of a multinode system by the presence and absence of tones. This may be seen as hinting at a progression from symbolic to analogic representation. The binary simplicity of the thing being represented and the representation technique are analogs of one another. Given the lack of continuous representation, however, the representation can also be understood as a very complex alarm, where portions of the system under analysis present their overall state as a group of tones and the tones themselves have no movement in their component variables. As such, these schemes may tend to induce the user to associate a small set of possible system states (all processors off, many on, etc.) with a corresponding set of auditory states (silence, dense sound, etc.) in a symbolic manner.

The sonification work by Kramer, Smith, Mayer-Kress et al., McCabe et al., Scaletti, Jameson (all in this volume) and others, is more explicitly

24 Gregory Kramer

ANALOGIC ←———————————————————————→ SYMBOLIC

AUDIFICATION

SONIFICATION: DATA TO SOUND PARAMETER MAPPINGS

PARAMETERIZED AUDITORY ICONS   AUDITORY ICONS   EARCONS

GEIGER COUNTER   AUDITORY OSCILLOSCOPE, ETC.

SPATIALIZATION

AS ANALOG TO REAL SPACE
AND VOLUMETRIC DATA

AS PARAMETER REPRESENTING
NON-SPATIAL DATA

An Introduction to Auditory Display

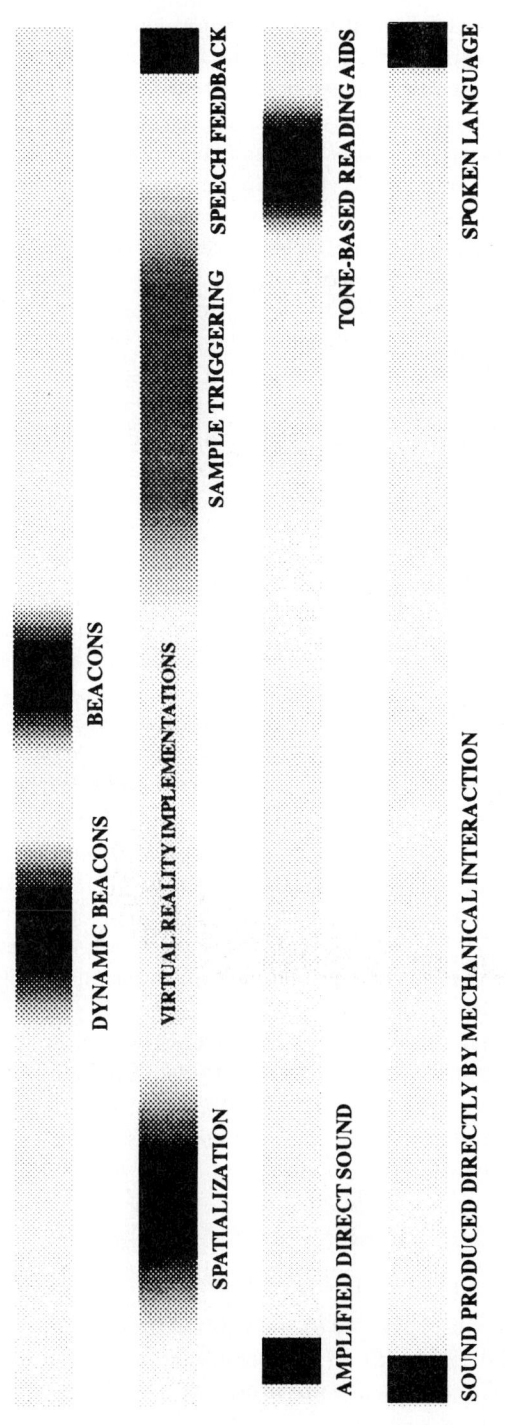

FIGURE 2 The Analogic/Symbolic Continuum.

analogic. When Madyastha and Reed use pitch to represent changes in arts expenditures, or when Kramer represents a change in bond market data with a change in the brightness of a sound, they are creating an analogy between the data and the sound that the analyst can use to directly track a variable. There are few mediating translations necessary for comprehension, as changes in the data may be understood to map homomorphically to changes in the sonic representation.

A hybrid between the metaphorical symbols by Gaver and the analogic mapping by Madyastha and others is the work by Fitch and Kramer. In using a caricature of a breath sound to identify the sound stream symbolizing breath variables, these investigators bypass the cognitive translations necessary in the work of others (Smith, etc.) In using the speed of the breath sound to represent the speed of the patient's breathing, metaphor and analog are combined. Embedding within the breath sound a mapping of body temperature to the pitch area of that sound is solidly analogic and nonmetaphorical. A one-to-one mapping is created between the representation and the thing represented. Increase and decrease in pitch now become an analog of decrease and increase in body temperature.

Sound spatialization provides a unique opportunity to integrate symbol and analogy. In a two-dimensional graphical map, cities and roads are represented by symbols. The relationships between one place and another are represented analogically. That is, the two-dimensional space of the map is an analog of the two-dimensional space of the surface of the geographical area being represented. In a similar way, spatialized sound can, with limitations, be used to analogically represent three-dimensional volumetric data. The exploitation of this analogy is perhaps the greatest power of spatialized sound, in that it presents spatially indexed data in a way that is highly intuitive and, therefore, instantly understood. A less intuitive but nevertheless analogic use of spatial location would be its use to represent two or three nonspatial data variables, for example, average temperature and biomass in an ecosystem. In this way, the localization functions much as changing the pitch or brightness of a sound does.

Combining spatialization with symbolic representations occurs in AD much as it does with the geographic map mentioned above. The symbols for cities on a graphic map, in one example, are replaced by sounds representing, say, types of computer installations with earcons (see Blattner et al., in this volume), while the spatial locations of those sounds are indexed to the relative locations in two-dimensional space of the installations. Wenzel's use of auditory objects, or icons, in telerobotic control involves using sounds fabricated in their Audio Cue Editor to symbolize selected hand gestures. By coordinating the perceived spatial location of the sound with the spatial location of the target real-world object (e.g., "I have grasped *that* object on the right side of the remote room"), the audio cue becomes instantly correlated with the environment. Begault's work with spatialized sound in a Traffic Collision Avoidance System (TCAS) also illustrates this symbolic/analogic nature. The word "traffic" is used to (symbolically/linguistically) represent air traffic. The perceived spatial location of the spoken word is analogous to the real-world location of the traffic. That is to say, it came from the same direction of the traffic, establishing a scale model of the air traffic within the TCAS control room.

Perhaps the most direct mapping is to be found in the audification work by Hayward. In this case the sonic representation is highly analogic, yet the sound we hear is sped up and so, strictly speaking, is not the earth as we actually hear it. In using the direct mapping of data samples to sound, however, Hayward is assured that the physical laws that governed the formation of those waves (as represented by the data samples) still govern the waves once they are made audible. Audification is unique in that the display technique directly determines its position on the symbolic/analogic continuum. Although the increased rate of playback required by this technique is the thread that ties this technique to the domain of analogy, we are very close indeed to a nonanalogic, truly direct representation.

6.2 THE ANALOGIC/SYMBOLIC CONTINUUM

Having placed a variety of AD work in the perspective provided by this analysis, we can examine the way that our precognitive and cognitive processes effect the symbolic or nonsymbolic nature of a representation. Firstly, let us look at the ways that the mechanics of the display can effect a crossover from analogic to symbolic or vice versa. We will see how processing by our auditory systems and cognitive faculties make these categories that we have so carefully constructed melt away.

Smith and his colleagues speak of using sonification (and visualization) to find what they refer to as *signatures* in the data. By a signature they mean a feature or set of features that "pop out" to the analyst largely as a result of precognitive processing. In a similar vein, Kramer refers to auditory *gestalts*, wherein integrated structures are perceived in a complex sound. Once features are discovered, the analyst presumably takes some action, either making a decision based upon the perception or perhaps turning the relevant section of data over to mathematical analysis.

The reason for relying on human processing is simple: we are very good at it. For thousands of years our sensory systems have developed pattern recognition skills that, for many tasks, still dwarf even the most powerful technological means. These pattern recognition skills have a key role in the shift from analogic to symbolic use of data representations. Simply put, as we learn to use a representation technique and become familiar with its gestalts, we recognize gestalts as signatures of specific events. If the universe of possible events is sufficiently limited, we in effect learn the "language" of the display, wherein each class of gestalts symbolizes a general category of data event or system state.

By the learning sequence just described, precognitive and cognitive processing facilitate an evolution from analogic to symbolic display use, *without any change in representation technique*. If the universe being represented subsequently expands (or if our skills for some reason degrade), we will be once again thrust into perceiving the representation analogically until we again learn to categorize the system's gestalts. If the system is sufficiently dynamic, the gestalts we perceive in any representation of the system may still indicate important features but, since they lack the consistency that would make them valuable tools for categorization, our use of them remains primarily analogic.

In sonar, a signature is a specific auditory object representing a specific physical object or process. In electronic circuit testing, a signature is the combination of variables (e.g., voltage waveforms) possessing some discernible "quality" indicating whether or not the circuit is working. This differs from Smith's use of the term signature in one important way. Smith's signature is like the sonar signature, an event-created object, identifiable and categorizable. However, it also has a temporal aspect, or "gradient." We should understand that symbolic (the signature) and analogic (the temporal gradient) coexist. For example, we may perceive meaningful features in census data indicating the onset of a recession. The perception of the recession's signature involves a gestalt formation. Particular changes in population, spending habits, and so on can be heard as gradients within the signature. These latter representations are heard analogically within the learned symbol of the former.

Kramer's *beacons* are an explicit use of the acquired capacity to effect an analog-to-symbol conversion. Static beacons employ gestalt formation when they "take a snapshot" of an auditory data representation (i.e., capture an analog). When these snapshots are learned by the system user, such as the operator of a machine, they become symbolic of an event. Their original analogic quality can be nearly ignored. *Dynamic beacons*, on the other hand, employ a hybrid symbolic/analogic functionality. Rather than a snapshot, a short sequence of the representation is extracted. If the dynamic beacon user is familiar with his or her data, the dynamic beacon functions as a symbol of a type of event, while the movement within the extracted sequence acts as an analogy of movements within the data being represented.

So we can see that while some auditory display systems build signatures into the system, such as earcons and auditory icons, others provide the system user with the discretion to form his or her own gestalts. The use of these gestalts as symbols, then, depends upon training. If the training has brought the user to a place where little or no cognitive effort is required for the link between the data and the sonic representation, so that the representation leads to a "second nature" categorization of the underlying events, we can say that the analogic display is now being used symbolically. Musical training neatly reflects this, as when a student learns to identify certain chords or intervals from what was previously a sea of notes rising and falling.

In a final twist on this near-circle, sequences of symbols may lead to the reintroduction of analogic display use. For example, if we employ a series of beacons, each of which represents a known system state, to create a sort of "meta-animation," we no longer just interpret the display as a collection of beacons. We perceive *relationships* between the beacons. It may be conceived as analogous to stroboscopic photos of an object moving through an n-dimensional space. The sound of each beacon represents its position in n-space, where each position has become, via learning, a symbol of a system state. The collection of beacons, then, is an analog of the movement through that space.

7. OTHER USES OF NONSPEECH AUDIO TO CONVEY INFORMATION

Before providing a brief history of auditory display, we should take a look at a few ways that sound has been used over the years to convey the status of natural and human made entities. While there are countless examples, I have selected a few that may provide us with stimulating curiosities and perhaps indications as to other ways in which sound may be used.

7.1 SONAR

Perhaps the most thoroughly investigated practical use of nonspeech audio is sonar. (A broad definition of the term "practical" might encompass music, in which case music would win the title for most prodigiously investigated.) While much of what has been learned is classified and, therefore, unavailable to the general public, something of value can be gained from considering how complex sound has been used to understand events unfolding in a visually impenetrable environment.

Broadly speaking, sonar is the use of sound for determining the nature of underwater entities. Sonar systems may be divided into two basic types: listening systems and echo-ranging systems. Applications of the systems include determining distances, locating targets (such as boats or fish), signaling, and so forth.

7.1.1 HISTORY OF SONAR In 1490 Leonardo da Vinci reported hearing ships at a long distance by submerging a long tube under the water and placing the other end of the tube near his ear. This "passive sonar" technique remained in use as late as World War I, when a second tube was added providing a modicum of directionality.[97] Four hundred years after da Vinci, the carbon button microphone, developed in the late nineteenth century, became the basis for the hydrophone, or underwater microphone. Its first practical application was for determining distance. By noting the time interval between hearing a submarine bell and a simultaneously sounded fog horn, the distance from ship to the shore could be calculated.

In 1912, five days after the Titanic collided with an iceberg, a patent was filed for an echo-ranging device. This lead to a number of echo-ranging and signaling devices that were widely used during W.W.I. These systems used frequencies of around 500 and 1000 Hz.

Spurred by the war, echo ranging and location quickly lead to the development of an electrostatic projector and, later, a piezoelectric device consisting of a quartz-steel sandwich which was used to project the sonar beam. In 1917 vacuum tubes were used as amplifiers for this system, probably constituting the first application of electronics to underwater sound equipment. Passive sonar, meanwhile, had moved on to arrays of listening tubes incorporating interactive control of array directionality. Operating in groups of 2 or 3, ships with these systems could achieve a surprising degree of accuracy regarding the bearing of submarine contacts.

Advances in electronics during the interwar period, including amplification, signal processing, and visual display of waveforms, had a significant impact on sonar. Fainter signals could be monitored, electronic methods for filtering out noise were experimented with, ultrasonic frequencies for active sonar enabled an increased directionality, multimodal displays were designed, and nonmilitary applications were developed. Depth sounding, acoustic speedometers, fish finding systems, diver's aids, and position markers all grew out of underwater sound technology.

7.1.2 SONAR SYSTEM TYPES. As mentioned above, sonar systems may be divided into two basic types; passive listening systems and echo-ranging systems. Listening systems make use of the sounds emitted by objects in the water, such as propeller noises, which are in the audible frequency range, from 20 Hz to 20 kHz. Supersonic listening systems pick up sounds above 20 kHz and shift the sounds down into the audible range. Echo-ranging systems transmit pulses of sound into the water and make use of echoes received from objects in the path of the transmitted pulses. For enhanced directionality, echo-ranging system pulses are above 20,000 Hz.[67]

In both listening and echo-ranging systems, the sound energy is picked up by an array of hydrophones and transduced to electronic signals. These

An Introduction to Auditory Display

are then amplified and transduced back to audible signals for direct playback or they are transduced to light energy and displayed visually. Additionally, the electronic signals may be digitized and subjected to a variety of signal-processing techniques to enhance the auditory or visual displays.

In a direct listening system the sounds may often be intuitively related to their causes. For example, a propeller may sound like churning water. The supersonic listening systems have a more artificial character, being more like bursts of noise. The echo-ranging systems are primarily tonal in nature, with a noise background. Pulses with a reverberant quality, caused by the scattering effect as the sound reflects off the surface of the water or small particles in the water, are followed by pulses of slightly greater intensity. If the object causing the reflection is moving, the pitch of the pulses will vary.

In a listening system, the sonar operator will attempt to identify the source of a sound by listening to the rate and quality of the rhythmic patterns. The size of the ship being monitored will effect the pitch area, speed and quality of the signal. Marine animals, such as croakers and crackling shrimp, will also produce audible results. Changes in rhythmic patterns, rates, and so on may indicate the types of maneuvers being undertaken by the target.

The operator of an echo-ranging system must first identify a "pip," which is a short tonal pulse appearing against a background of noise and an irregularly modulated tone. Movement of the target will be indicated by a Doppler shift in the "pip." A target at right angles to the sound beam striking it will return an echo that is louder, briefer, and "sharper" than one that is headed towards or away from the transmitted pulses. Echoes from stern-on targets will usually be longer in duration and "mushier" in quality.

7.1.3 SONAR AND AUDITORY DISPLAY: COMMON RESEARCH THEMES.

AD and sonar clearly share many auditory perception and human factors issues. What are the relative benefits and drawbacks of auditory, visual, and bimodal displays?[58] How can a display be designed to minimize the problems associated with masking? What is the impact of listener fatigue on the efficacy of the display? Which elements of a display should be made interactive and which should be optimized and left alone?

The broader research issues that face sonar researchers also confront AD designers. The limitations of reductionalistic auditory perception studies based upon the use of pure tones have taught us a great deal about the mechanics of the auditory system, but have proved of limited use in efforts to understand how we deal with complex acoustic phenomena. The lack of a language to classify and describe sounds is also a common problem. Sonar operators, like electronic music composers, often use real-world references to describe sounds. For example, one might say, "It sounds like an electric razor."[7] Another common question is where to use preprocessing of the raw data, whether it is the acoustic signals of sonar or the numerical data of sonification, and where to let our perceptual systems do the processing and detection.

The results of well-funded sonar research, including a relatively long history of broad applications, has resulted in a field with much to contribute to the comprehensive study of auditory display. As AD research and development matures, it will no doubt enable reciprocal contributions towards advances in sonar.

7.2 OF TALKING DRUMS, SOUNDING BRASS, AND MORSE CODE

Of course, sound as a communicative medium has a rich history. A brief look at the way nonspeech sound has been used by different cultures and in different situations serves to provide a broader context for a consideration of current AD research. Languagelike encoding, as exemplified by the talking drums cited by Blattner,[8] may predate Morse Code by hundreds of years, but the means and the end were nearly identical. Combinations of sounds or durations were strung together to communicate over a long distance. They were, of course, the precursor to the telephone, which itself will probably be the precursor to a broader band audio/visual communications medium.

Higher level encoding, a form of shorthand, may be found in the use of brass instruments in the military. Going back to the Romans, in a mapping of timbre to event, three instruments were used to signal distinct military actions. By the 1300s, in the attack on the European city of Basle, fife and drums were used as an extension of the human voice.[8] Drums, in particular, were used to convey nearly every maneuver used in battle as well as many maneuvers or other activities back at the garrison.

Because of their function, the most important feature of military signals or calls is lack of ambiguity. Drums contained nearly all of the information, while fifes were mostly for color. The calls included orders such as advance, retreat, halt, lie down, load, fire, and so on. Some calls could, like earcons, be combined with other calls to create a more complex call. For example, a call for a particular group, such as reserves, could be combined with a call for a particular rally. This "Rally on the Reserve" would then indicate a very specific and complex maneuver, such as "form a hollow square, one soldier in depth, fix bayonets and face them outward."

[7] Private correspondance with Thomas Hanna.
[8] Private correspondence with George Carroll, Fellow of Company of Military Historians.

7.3 COMPUTERS AND NEURONS

Some of the following examples cross the line from the generation of sound to the structuring of situations in which sound will be produced as an intended byproduct of another event. The end point in this continuum, which we will not investigate in detail, is the use of listening as a means of obtaining insights into the observed system. We will not, for example, examine how automobile engines are diagnosed via audition, or machine tools and other factory equipment are attended by skilled operators who glean remarkable insights into the status of their tools by listening to them. Physicians who listen to the heart or a cough, neonatologists who listen to in utero sounds, and other skilled professionals who listen diagnostically to a wide variety of acoustic signals will only be mentioned here. We, no doubt, have a lot to learn about attention, directed listening and auditory pattern recognition from studying these highly developed skills.

7.3.1 COMPUTER MONITORING. A variety of schemes have been used as expedient means for monitoring computer activities. While these may be seen as a precursor to the work by Jameson, Jackson and Francioni, and Madyastha and Reed (in this volume), as we will see, the technology of the implementations are far more crude. One can still appreciate, however, the elegance of the connection between the problem and the solution.

In the early days of computing, intrepid researchers were often obligated to attend to their computer for hours as it crunched code. Amongst other things, they needed to know if the software contained infinite loops. It was also helpful for them to monitor the progress of their programs and be alert to early warnings that things were not going as expected. They discovered that if a cheap A.M. radio was placed near the processing unit (it *had* to be a cheap one), the interference generated by the computer would cause predictable patterns in the radio's sound. For example, infinite loops would make a continuous pitched sound that was distinct from the noiselike quality of normal functionality.[9]

In a similar vein, when LISP was first installed on the PDP-6, register 1 was reserved for garbage collection computations. Thus, if one monitored this register as sound via a DAC, one would hear bursts of noise whenever garbage collection took place and silence otherwise. This would give the user a crude idea of how much of his clock time was going into garbage collection, rather than the computing he intended. Steve Smoliar reports, "Contemporary systems now tend to do things like modify the cursor when garbage collection is taking place, but I still tend to prefer the old auditory cue."[10]

---

[9] Private conversations with Nick Metropolis, John Holland, and Creve Maples.
[10] Private correspondence with Steve Smoliar.

7.3.2 NEURONS. Neurophysiologists have listened to neurons firing for at least thirty years. The practice was well engrained in the culture of the Neural Encoding Laboratory at the Johns Hopkins University under Dr. Eric Young. At the Eaton Peabody Laboratory in Massachusetts, under Nelson Kiang, listening to neurons to determine the responsiveness of auditory nerve fibers was in place in the early 1960s.[11] Neurophysiologists still use amplifiers and speakers to listen to the sound of spiking neurons. Since neurons spike at a rate from < 1 Hz to 1000 Hz, their spike trains are pitched, easy to listen to, and useful for the discernment of different neuron types.

In a jump from listening directly to neurons to a simple sonification of intracellular neuron activity, at least one researcher has used a voltage-controlled oscillator to monitor the resting potential of a neuron. A neurophysiologist knows if she has penetrated a cell when the voltage drops from 0 to the neuron's resting potential. Also, she can tell if she is coming out of the cell if the resting potential begins to rise towards 0. Using an oscillator to monitor these changes leaves one's eyes free to look through the microscope or change stimulus conditions.[12]

7.4 AEROSPACE ANECDOTES

Perhaps it is the high-information, high-stress nature of aerospace control environments that make them fertile soil for creative uses of audio. The eyes-free nature of sound interfaces makes AD attractive in such circumstances. Whatever the reason, precursors to the work by Wenzel, Begault, Furness and others, as well as antecedents to that work, are worthy of mention.

7.4.1 AUDITORY-GUIDED INSTRUMENT FLYING. In 1945, Forbes conducted a series of experiments on the use of auditory signals for instrument flying.[31] After finding that combinations of tones created a confusing and awkward display, Forbes turned to one signal in which a number of data variables were represented by multiple auditory variables. Citing the 1936 work by de Florez as a precedent, Forbes describes an AD in which as many as four auditory indications can be followed without interfering with pilot understanding of radio and interphone communication.

Turning of the craft was indicated, in this early sonification, by a "sweeping type of motion of the signal from left to right" (pan). An auditory cue for banking, or "apparent 'tilt,' was produced by pitch variations."

---

[11] Private correspondence with Herb Voigt.
[12] Private corresponence with Susan Volman. Volman also notes that "the constant hum of a VCO can be somewhat annoying—the trick is to use a comfortable frequency range. The tones definitely have to be below about 800 Hz."

The speed of the aircraft was represented by the rate of occurrence of "a 'putt' sound." Forbes also used a speech interface for the altimeter, which simply indicated the altitude by speaking messages such as "four thousand four hundred feet."

A Link Trainer, equipped for testing the AD, was used by Forbes to discern how well subjects could learn and use the display. Using the same subjects, the results from the AD corresponded well with performance using the visual display after the same period of practice. As expected, some subjects had more difficulty learning to use the display than others. Pilots who used the display were, after only an hour of training, confident that they could use these auditory indicators to fly.

The general principles that Forbes suggested should guide the display designer were:

1. Pilots have certain habitual methods of thinking about the airplane, and the signals must be designed to fit these habits of thought.
2. Because most fliers are accustomed to using visual indicators, the auditory indicators must be as simple and self-explanatory as possible.
3. When multiple signals are used, there is a tendency for one signal to "capture" the attention of the pilot, to the exclusion of the other signals. This phenomenon should be avoided.
4. The display should be designed to fit the capabilities of the average pilot, not a talented few, and it should be subjected to unbiased psychological testing.

7.4.2 SYNESTHESIA FOR SPACESUITS. A system employing auditory feedback was developed at United Technologies Research Center for use in spacesuits. Since the gloves on the suits do not give much tactile feedback to the astronauts, astronauts were having trouble doing some tasks with power tools while on spacewalks. The system provides an audio cue which is tied to RPM of the tool.[13]

7.3.3 PLASMA WAVE AND THE RINGS OF SATURN. In 1979 Fred Scarf, the Principal Investigator of the Voyager Plasma Wave Experiment, used a primitive synthesizer attached to an Apple II computer to translate the 8-channel plasma wave data into audible sounds.[103] Since the frequencies of the AC electric fields were in the audio range, it was straightforward to map each spectral channel to a particular synthesized frequency range. Anecdotally, this was a fruitful experiment. Certain types of wave structures, such as ion-cyclotron waves and electron plasma waves, had audio signatures that could easily be detected despite the background noise. Well-defined

---

[13] Private corrsepondence with Dan Quinlan.

events, such as a bow-shock crossing, could also be picked out because of their auditory signature.

Auditory inspection of plasma waves remained primarily a curiosity, with the possible exception of the 1981 analysis of the Voyager 2 crossing of the Saturn ring lane. Then a concern arose about the possibility that the dense micrometeoroid environment of the rings could damage the spacecraft, since the meteoroids were moving past the spacecraft at several tens of kilometers per second. During the ring crossing, there were some strange events observed on board the spacecraft, but nothing that specifically pinpointed micrometeoroids as the cause of the disturbances. When Scarf and his colleagues listened to the plasma waves, a connection became evident.

Theory indicated that the meteoroids inside the amin ring structure of Saturn were negatively charged due to the plasma embedded in Saturn's magnetosphere. The number of electrons per particle varied, but could range upwards toward several thousand electrons per particle. The researchers reasoned that if a micrometeoroid with these properties hit the spacecraft, they would splatter, emitting copious electromagnetic waves in the process. The visual data contained so much noise that they were useless in assessing the situation. However, when Scarf played the audio data from the AC electric fields, a very clear "machine-gunning" sound could be heard which correlated precisely with the few seconds that the spacecraft spent in the region of Saturn's rings where the largest dust concentration was.

8. HISTORY

The history of auditory representations of data could be said to include the research by Pythagoras, Ptolemy, Kepler, Mozart, and Dufay, as described in *Sonification and Music* (below). It could also be said to include the 1914 development of the tone-based interfaces for blind users described in *Auditory Interfaces for Blind Users* (below). Also included would be the history of sonar, which is briefly described elsewhere in this chapter. The history of alarms and other auditory interfaces are also a part of the broad heritage of auditory display.

For the sake of brevity, the account presented here will limit itself to an overview of the uses of sound to represent data in the second half of the twentieth century (for another account, see Frysinger[37]). It was at this time that the technologies developed during World War II began both to demand richer interfaces and to supply the means for providing these interfaces. Information theory, powerful and affordable computers, and sound synthesis technology all contributed to the acceleration of interest in auditory data display.

8.1 PIONEERING EFFORTS

In 1954 Pollack and Ficks published a paper detailing research into the use of abstract auditory variables to convey quantitative information.[73] Using alternating tone and noise bursts, they designed a display that presented eight binary variables encoded as the pitch area of the noise, the loudness of the noise, the pitch of the tone, the loudness of the tone, the pitch/noise alternation rate, the temporal ratio of tone to noise, the total duration of the display, and the stereo location of the display. They also created a display without the noise bursts which yielded six binary variables.

Pollack and Ficks found that displays using multiple parameters of sound generally outperformed selected unidimensional displays measured elsewhere. Their research also indicated that the subdivision of display dimensions into finer levels does not improve information transmission as much as increasing the number of display dimensions does. Much current sonification research employs continuous changes in selected variables. This presents different problems from those of the binary displays designed by Pollack and Ficks. The possibility of methodically maximizing the balance of display resolution and dimensionality has not been thoroughly investigated.

In 1961, Speeth reported the results of experiments that used audification of seismic data to determine if subjects could differentiate earthquakes from underground bomb blasts.[92] Because of their complexity, seismograms that resulted from these events were difficult to understand and categorize. By speeding up the recordings of the seismic data, the complex wave was shifted into the audible range. For over 90% of the trials, subjects were able to correctly classify seismic records as either bomb blasts or earthquakes. Additionally, by speeding up the playback of the data, analysts could review 24 hours of data in about five minutes.

Inspired by Speeth's research, Frantii[33] had 21 observers classify 200 time-compressed audified seismic events as either earthquakes or explosions. They correctly categorized the events 66% of the time (50% corresponds to a chance performance). Fantii's work included experiments designed to determine the receiver operating characteristics of listeners, the effect of training on performance, the effect of epicentral distance, and the effect of dual (horizontal and vertical) component playback. Among the significant conclusions was that the observers reached plateau performance after about 1500 decisions, and that the performance could be improved by using multiple component (stereo) playbacks. While Fantii's and Speeth's results would seem to indicate that audification of seismic data offers significant benefits, very little research has since been published in this or related areas. (For an example of related research, see Smith Kettlewell's 1977 research on control of a tone generator with seismic data in *Interfaces for Blind People*, below.)

Chambers, Mathews, and Moore[19] designed a three-dimensional auditory display at AT&T Bell Laboratories. In an auditory enhancement of a scatterplot, they encoded three data variables in pitch (quantized chromatically), timbre (by adding formants), and amplitude modulation. While no formal testing was conducted, they found that the auditory representation did assist in the classification of the data.

Meanwhile, in industry, two German inventors working for M.A.N.-Roland Druckmaschinen Aktiengesellschaft had developed a scheme for auditory monitoring of printing presses.[49] In a striking but unnoticed departure from traditional alarms, Kaiser and Greiner designed a system by which an arbitrary number of parameters of the printing press could each be routed to signal generators to control some acoustic characteristic. This is, perhaps, the first patent filed for multidimensional auditory data representation.

In its simplest implementation, the invention was basically a sophisticated alarm: switches turned on the signal generators (which had different timbres) when specified thresholds were reached. More sophisticated implementations included (i) prioritization of the different parameters, (ii) switching off other signal generators when one parameter goes too far out of bounds, and (iii) continuous pitch control of the signal generators based upon the urgency of the situation, presumably a function of how critical a given factor was and how far out of range it was becoming.

In 1979 Fred Scarf of TRW used sound to explore data from the Voyager-2 Plasma Wave Experiment. This research is described above in *Other Uses of Nonspeech Audio to Convey Information*. Correlating plasma waves, generated when micrometeoroids collided with the spacecraft, with fluctuations in sensor measurements, posed a problem that Scarf addressed utilizing AD techniques. The work is noteworthy in that it produced scientifically valuable results that were elusive via conventional visual analysis techniques. That these successes did not appear to generate follow-on research is a measure of the pervasive resistance to using sound for data analysis.

In 1982, chemist Edward Yeung developed a sonification technique for displaying experimental data from analytical chemistry. Like current researchers, he looked for auditory variables that had some degree of independence. He selected two pitch ranges, loudness, decay time, stereo location, duration, and silences between events. Using these variables and no more than two training sessions per subject, Yeung asked the subject to classify detected levels of metals in a given sample. A given data point could belong to one of four categories and Yeung's subjects attained a 98% correct classification rate.

Preceding, following, and concurrent with this research were some notable efforts at designing displays of various sorts for blind users. These efforts included the research on instrumentation with acoustic feedback at Smith Kettlewell in the late 1970s, the work by Mansur et al. on representing graphs using sound, and the chemistry displays by Lunney et al. While these are described in more detail in the section on interfaces for blind users, the reader should keep in mind that these events paralleled and interacted with display developments for sighted users.

In what was to become the most frequently cited sonification paper of the 1980s, Sara Bly's doctoral thesis[9] explored the classification of nonordered multivariate data sets. Each n-dimensional data point was represented by a discrete auditory event in which n sound parameters were controlled by the data. The parameters Bly used were pitch, loudness, duration, timbre (a continuum from sinusoid to buzzy), attack time, and fifth and ninth harmonic waveshape. The waveshapes of the fifth and ninth harmonics were the first to be abandoned when fewer data dimensions were used.

Bly was interested in presenting three kinds of data: multivariate, logarithmic, and time-varying. She selected a multivariate data problem that involved classifying different species of flowers using four measurements per plant. Bly found that, using sound, most observers could correctly classify most of the samples.

The logarithmic relationship between frequency and pitch provided the basis for Bly's representation of logarithmic data. Encoding the exponential variable of quake magnitude in pure frequency (without quantizing it to a chromatic scale), she also mapped quake magnitude to the loudness and duration of the sound. Bly reported that the trials were "positive indications" that sound could be used to highlight relevant features in seismic displays.

In the same paper, Bly described how she mathematically generated a battle simulation between two opponents and aurally represented the resultant time-varying multivariate data using the pitch and loudness of multiple tones. Different waveforms were used for each tone, though waveform was not used to represent data. One tone (sinusoidal or noisy) was used to represent each side, with pitch representing the number of units that side had at the battlefront, and the loudness representing the number of units in transit to the front. Listeners were able to use the resultant "battle songs" to generally distinguish different battles which had the same outcome but which evolved differently. With practice the listeners became more adept at using the display, although hearing the two sides independently was consistently a problem.

Bly conducted a series of formal experiments on multivariate data displays using sound only, graphics only, and bimodal displays. She also experimented with changes in the data-to-sound mappings and changes in training methods. Subjects were asked to classify an unknown test sample as belonging to one of two sets, which differed in a well-defined way and which the subjects were trained beforehand to discern. Bly's results indicate that the auditory display was as effective as the visual display, and that the combined display outperformed them both.

Integrating animated graphics with data-controlled sounds, Mezrich, Frysinger, and Slivjanovski attempted to produce dynamic representations of oil well log data.[64] The authors developed a dynamic representation of the log data employing both auditory and visual components for multivariate time-series displays. Later, due to the proprietary nature of such data, the authors ended up using the statistically similar data of economic indicators. In the techniques used, each n-dimensional data point was presented as a visual "frame" and n tones. The visual display represented data values by location and size, while simultaneously sounding musical notes changed in pitch. All other sound parameters were held constant, although the user could elect to employ one of two waveforms for each voice. In this scheme, the analyst is confronted at any moment with one sample from the time series, rather than the whole data set. These multivariate samples are displayed in succession, forming "frames" of data analogous to frames in a movie. The authors found that using identical waveforms seemed to promote global pattern recognition. When a subject was hearing globally, the dynamic display allowed the subject to detect correlation with fewer points than were required when the subject was using the overlaid display (and fewer points than a correlation detection algorithm).[14]

Frysinger later tested the display to determine the sensitivity of the previous results of Mezrich et al. to time-series, dimensionality, and signal detection task, as well as to determine the degree to which an auditory display was enhanced by a simultaneous visual display.[37] He found that the subjects' data interpretation performance depended upon the detection task. For correlation detection, time-series dimensionality was a significant variable in display performance.

Like Bly and other researchers, Frysinger found that the combined auditory/visual display proved superior to the auditory-only display. For trained-pattern detection, dimensionality was not a factor, and the performance of the auditory/visual display was essentially the same as the auditory-only display, indicating that the visual display did not contribute significantly to trained pattern detection. This result may be important, since trained pattern detection is a common data analysis task. If this

---

[14] Private discussion with Frysinger. The precise correlation detection algorithm is unknown.

result holds more generally, it suggests that auditory displays offer new capabilities rather than a mere enhancement of visual displays.

In 1985 the small cadre of researchers involved in auditory data representation got together to present their work at the CHI '85 (Computer Human Interface) conference. This "sonification junta" consisted of Bly, Frysinger, Lunney, Mansur, Mezrich, and Morrison.[10,11] The panel was moderated by Bill Buxton. This is the first time that a national conference session was dedicated exclusively to the general use of nonspeech audio for data representation. (In 1984 Dave Lunney organized a smaller regional gathering of some of these researchers to investigate auditory interfaces for vision-impaired people.)

There was a feeling amongst those gathered at CHI '85 of being onto something important. While the panel session itself did not cover any new research issues, the group spent a lot of time during the in-between times that can make conferences truly interesting trying to figure out where to go next. Perhaps more than research issues, participants discussed whether they should hold a special workshop on the topic of AD, how they might get support within their institutions for this research, and how they might make a business out of this nascent field.

Steve Frysinger referred to this period as the "prefrustration" time, during which one experiences the first flush of excitement that comes with seeing the possibilities of the field and before the really difficult research questions draw the work to a slower and sometimes less rewarding pace. This is also a time when very little research funding is available, making progress even more difficult. It is a pattern that continues to this day: researchers new to the field of auditory data representation are impressed with how easy it is to get started and see some results. However, in looking beyond the easy answers (such as the use of data to control the pitch of a tone generator), an imposing set of problems emerge, including: the perception issues, the sound computation demands, and the applications. Overcoming these obstacles requires time and funding that are not always available. From the 1985 meeting through the end of the decade, progress was sporadic.

In 1986 a team of engineers, working at Fiat Auto, S.p.A., developed and patented a sonification system for continuous monitoring of various automobile parameters.[38] Like their German predecessors, a plurality of sensor devices were used as control signals for a group of tone generators. Once again the task was real-time monitoring, not analysis, and psychometric tests were not conducted to determine the efficacy of the display system. It is not clear whether this system was ever implemented.

From the mid-1980s until present there was a parallel effort to develop nonspeech auditory elements for general computer interfaces. Initiated by Gaver's SonicFinder[40] for the Macintosh and pursued since that time at

Apple Computer's Advanced Technology group, these researchers have conducted a series of experiments using realistic sounds to inform the user about events in the user's computer environment. Gaver, working with Smith and O'Shea,[42] developed a simulated manufacturing plant called ARKola to test the uses of audition in a process control application. They found that subjects could easily monitor off-screen processes and improve their assessment of processes displayed on-screen. Similarly, research at Xerox into Computer Supported Cooperative Work by Bly, Buxton, Gaver, and others explored the role of audio in telepresence applications.[43]

8.2 THE PACE PICKS UP

The relatively slow pace of development in sonification began to accelerate around 1989. Stuart Smith, at the University of Lowell (now University of Massachusetts/Lowell), began work with a team on Exvis.[86] Exvis is a tightly coupled auditory/visual display tool for representing multidimensional (up to seven-dimensional) data. The data variables were encoded simultaneously as the geometric attributes of graphic elements called "icons" and as the attributes of a synthesized sound. The icons produced data-driven visual textures, and the auditory display was triggered by moving the mouse cursor through the graphical representation.

At about the same time, Gregory Kramer began work at the Santa Fe Institute on sonification of complex systems[50,51] and Clarity's Sonification Toolkit. In searching for ways to enable our perceptual systems to more fully contribute to comprehending complexity, Kramer's work, meanwhile, was pushing the limits of dimensionality. Using data supplied by the mathematician Mayer-Kress, Kramer attempted to represent nine-dimensional chaotic systems (ten-dimensional including time) in an auditory display.[52] He also worked with Apple Computer's ACOT group to produce sonifications of predator/prey models for education purposes,[50] using both realistic and abstract sounds to represent the dynamic system.

The burgeoning interest in and capacities of computer music composition were developing along with data sonification. While a history of this music is beyond the scope of this paper, some work bears particular mention. Brian Evans had been working through the late '80s on combined visualizations/sonifications, with the intention to produce an integrated art form. The equations that generated the visuals also controlled the sound tracks. However, aesthetic impact generally took precedence over a comprehensible representation of the data. Various composers have been working with chaotic equations to generate waveforms or provide sources of control[6,20,74] and many pieces have been composed using data from natural phenomena as sources for control. For a summary review of the relationship between sonification and music, see Section 11.

In 1990, Scaletti and Craig, working at the National Center for Supercomputing Applications, produced a series of sonifications to accompany scientific visualizations developed there.[79] Their work added sonification to create a sophisticated sonified data visualization. The data represented both aurally and visually included ozone levels, swinging pendula, and forestry data. By displaying these video tapes to the robust computer graphics community, a new and broader audience became aware of sonification.

At the same time, Rabenhorst was working with some colleagues at IBM's Watson Labs on an auditory and visual representation of three scalar fields associated with electron density, hole density, and potential throughout the volume of a semiconductor.[76] Like Exvis, the user could use a mouse to select the region to be displayed. In the IBM work, two volumetric variables were visualized in high resolution while one was sonified.

The next two years saw the acceleration of the research pace continue. Work has been done or is in progress at computer companies, electronic game companies, supercomputing centers and national laboratories, medical instrumentation manufacturers, and, of course, the parent of them all, aerospace and defense companies. While the field of sonification is still quite young, it is emerging as an interesting and viable research area. Applications cannot be too far behind.

9. ALARMS

Alarms are symbolic auditory representations of discrete events. They are generally associated with an urgent situation and as such are designed to stand out in the prevailing acoustic ecology.

Alarms, as generally defined, are ubiquitous. They include doorbells, ringing telephones, buzzers on washing machines, and car horns. Sounds we explicitly identify with alarms include alarm clocks, burglar alarms, smoke detectors, and, of course, the numerous implementations of alarms in professional environments such as factories and flight decks. There are literally hundreds of patents on various types of alarms.

Of course, the study of alarms can be understood as a subset of AD in general. It is also clear that the body of knowledge, largely empirical, that has developed around alarms, can inform designers of general auditory interfaces, multimedia environments, and sonifications. A thorough examination of the history, perceptual issues, and design considerations of alarms is beyond the scope of this paper. However, an examination of one key research study may suggest a link between current AD research and the corpus of experience associated with the broadly installed base of alarms.

9.1 PATTERSON'S 1982 STUDY

Flight decks are an environment where more than a dozen (sometimes several dozen) alarms might be used. In the early 1980s Roy Patterson set out to develop a set of guidelines for the design and evaluation of the auditory warning systems used in commercial aircraft. Earlier studies had found that pilots felt an aversion for many of the alarms, leading them to turn them off as quickly as possible or, in response to the intensity of the alerting, to respond inappropriately. For example, a common reaction to a very loud and intrusive alarm is to focus attention not on the cause of the alarm, but on searching for the cancellation button for the alarm.

In 1982 the Civil Aviation Authority in London published a paper by Patterson entitled "Guidelines for Auditory Warning Systems on Civil Aircraft."[69] Patterson addressed (i) the overall sound level, (ii) the temporal characteristics, (iii) the spectral characteristics, and, briefly, (iv) the ergonomics of auditory warnings. He also considered the use of speech displays integrated with the warning sounds.

Patterson established three priority levels, including (i) *emergency*, which might number 4–6, (ii) *abnormal*, as many as 10 and frequently including a voice warning, and (iii) *advisory*. After introducing this classification system (to what apparently was a potentially messy acoustic environment), Patterson then looked at some of the problems extant in the then-current warning systems. These included abrupt onsets that induce startle reactions, confusing similarities of alarms, and excessive on time, to name a few.

As regards overall level, Patterson found that continued loud sounds tend to incapacitate. This led him to a series of tests and the subsequent conclusion that warnings should be at least 15 dB above the masked threshold to ensure they will be noticed, and that they should not be more than 30 dB above threshold or they may disrupt verbal communication. Patterson's findings that "when a signal is 10 dB above threshold, it is easy to hear and, when it is 15 dB above threshold, it is difficult to miss" has obvious implications for all types of auditory interfaces.

As regards temporal characteristics of warning signals, Patterson noted that most onsets and offsets were too abrupt, temporal patterns were too similar, and the ratio of on-time to silent interval was far too high. Onset times of 10 ms or less tend to startle and 20 to 30-ms rise times are preferable. The duration of the basic pulse of an alarm should be long enough to ensure detection, but not longer than necessary, to avoid reducing the intelligibility of the speech it interrupts and limiting the diversity of temporal patterns that can be used. As a suitable on-time 100 ms was suggested. Warnings with the same temporal patterns were likely to be confused even when there were gross spectral differences between the warning sounds (earcon designers take note!) and a vocabulary of temporal patterns, each using at least 5 pulses, is suggested. Finally, pulse rate effected the sensed urgency of the warning, with faster pulses inducing a higher sense of urgency.

Regarding spectral characteristics, Patterson found that if the first five harmonics have a significant proportion of the energy, the note will sound smooth, sonorous, and full. As the energy shifts to higher harmonics, the note sounds sharper and has more "edge." Sounds with more edge imply greater urgency. The incorporation of nonharmonic partials tends to give the sound a harsher or more shrill quality, and should be reserved for even higher priority warnings. The upper bound for the fundamental pitch of warning sounds, Patterson suggested, is 1000 Hz, and the lower bound is 143 Hz. This allows the use of lower and upper harmonics described to fall within the suggested spectral range of .5–5.0 kHz.

As regards learning a vocabulary of warning sounds, Patterson found that naive listeners were able to learn 4–6 different warnings very quickly but, thereafter, the rate of learning slows considerably. Up to seven warning sounds were reasonably quickly learned and, beyond that, learning was slow but steady (up to ten distinct signals).

Speech warnings, Patterson notes, are versatile, highly redundant, and reliable. However, they may interfere with existing conversations while not standing out as warnings due to a lack of perceptual contrast with speech on the flight deck. Also, speech occupies the entire auditory communication channel and it does so for a relatively long period of time. So immediate action voice warnings should be brief and use a key-word format, awareness warnings should use a full-phrase format, and, due to the wide dynamic range of speech, the appropriate level for voice warnings should be near the maximum level outlined above.

A remarkable aspect of this treatise is that it points, indirectly, to the earlier neglect of many important human factors issues in designing auditory interfaces. On a more positive note, it is a landmark study of the problems and solutions associated with a complex auditory interface. As such it serves as a good example of what current AD research can learn from prior efforts. Fore more recent studies of auditory alarms, readers are referred to the work of Robert Sorkin et al.[88,89,90,91]

9.2 RELATING THIS RESEARCH TO OTHER AUDITORY DISPLAYS

In the above discussion of Patterson's research, some explicit findings relevant to AD design are mentioned, such as maximum suggested loudness levels, and how confusion resulting from similar temporal patterns has implications for earcon design. Depending upon the type of display, various elements of Patterson's comprehensive study, along with alarm studies by other researchers, are bound to prove beneficial to future AD designers.

Establishing different priority levels for notifications is important to the design of auditory implementations for GUIs (graphical user interfaces). Avoiding unwanted startle reactions is clearly in the best interests of all AD designers. Careful use of variety, for example, variety of temporal patterns, will reduce confusion in notifications. The sense of urgency associated with faster pulses or harsher sounds described above can be used to good effect (or ignored for potentially ill effect) in the design of affectively coherent displays. The learning curve that Patterson describes addresses a key issue in any interface: how difficult is it to learn and how well will that learning be retained. Finally, the effects of using speech on the acoustic ecology of the display and on the interference of the work environment of the display user, translate directly from alarms to general AD design.

10. INTERFACES FOR BLIND PEOPLE

The first reading machine with an audible output was built by Mr. Fornier D'Albe in 1914.[56] The *Optophone* had a six-tone code and was significantly improved in 1922. In the 1950s the Veteran's Administration began more concentrated work on reading aids. The VA-Battelle Optophone had nine tones which were generated as the user moved a column of photo sensors across the page. Letter shapes translated into tone patterns. The Optophone was replaced in the early 1960s with the Visotoner, a device similar to its predecessor. The penultimate in this line of reading aids was the Stereotoner, made by Mauch Laboratories. It was a ten-channel device with stereo output.

The most current implementation of a tone-encoding reading device at this writing is the experimental *Optaudicon*, from Lauer and Mowinski at the VA Hospital in Hines, Illinois. The Optaudicon was a child of the Optacon, developed at SRI in the 1960s.[57] The Optaudicon could produce 12, 20, and 24 tones and had much finer resolution than its predecessors. Users preferred the 20-tone units and found that pure tones worked better than tones with harmonics.

Initial research and development on reading machines provided some guidelines for contemporary auditory display designers. For one thing, the use of half-step intervals was found to be discordant and, therefore, undesirable. Furthermore, a tactile interface joined with the auditory interface was found to yield higher accuracy and greater ease of use.

It is only relatively recently that speech-output reading aids have become technologically practical. Devices such as the Kurzweil Reading Machine opened up scanned text to vision-impaired users. Meanwhile, software was being developed that made computer screens accessible; a number of products to accomplish this are now on the market, for example, *Outspoken* from Berkeley Systems. Speech, however, has a low information transmission rate for continuously changing variables relative to the bandwidth of the human auditory system. It may be desirable in some cases to use nonspeech encoding, perhaps a variation on the tonal interfaces described above, to rapidly scan a page for overall qualities of the text. One may, for example, gain an overview by assigning numbers to one kind of tone and letters to another kind. (Note that this assumes the display device knows what the characters are, for example, when reading a computer file. This approach may not be practical with an optical reader.) More generally, one could get an idea as to how much blank space was on a page from the silences.

A shift in display use, from decoding particular cues to hearing larger features of the system, can be made by minor changes in display technique. For example, at the Smith Kettlewell Eye Research Institute research has been conducted on encoding numbers read after decimal points with a higher pitch, thus eliminating the need for enunciation of the word "point."[13] By encoding additional information in "piggybacked" auditory variables, such as pitch of the voice, or by using nonspeech tones to represent selected elements, the speed of information presentation can be increased.

To place this spoken, pitch-encoded display in the context of other techniques discussed in this volume, let us consider it in terms of the analogic/symbolic continuum. Sonic representations of broad categories of characters indicate a shift from pure language towards nonlinguistic symbol. In mapping numbers or letters to different pitch areas, a slight shift in display use towards the analogic end of the continuum was made, entailing only minor changes to the display technique but significant changes in how the person *used* the display. In these cases, both linguistic and novel forms of symbolic information are represented simultaneously.

A primary problem in using speech to represent rapidly changing data is that by the time the speech output is complete, the value has changed. Nonspeech encoding can eliminate this problem and provide the additional benefit of a more intuitive display. Continuously changing acoustic variables are always self-referenced. That is to say, they are rising, falling, or staying the same. Thus trends in the data can be quickly perceived and values from one moment to the next can be easily compared. Using sound to directly display quantitative information for blind users takes advantage of these qualities.

One example of this type of analogic data display may be seen in the nonspeech interfaces for instrumentation developed at Smith Kettlewell. In their auditory oscilloscope, a combination of tactile and auditory presentation is used to display the X and Y axis of the oscilloscope. The user moves a slide controller across the screen to select a region on the X axis that he or she wants to investigate. The amplitude of the chosen point (the Y value) is used as a control signal for the frequency of an audible tone. So a technician investigating a triangle wave, for example, may move the slider from left to right across the screen and hear the tone rise and fall repeatedly.[82] Another analogic display, the Smith Kettlewell Light Probe, extends this concept to the scanning of an arbitrary visual field. In this device, received light is coded as pitch. Users report that scanning the probe across a visual pattern can give complex and subtle cues.

Similar in design is a device that enables blind workers to analyze seismographic data. A recording of the seismic data is played back as a control signal for an oscillator. By listening to variations in the pitch of the oscillator, the analyst is able to listen to the data display at a rate five times the "real" recording time. When a significant seismic event is discerned, the tape may be slowed down and the important region may be more closely inspected.[81]

Given the pervasive use of graphs to present quantitative information, it is not surprising that some effort has been expended to make the same type of display available to blind people. Mansur developed a technique for auditory display of x–y "plots" using continuously varying pitch to represent the dependent variable (y) and time the independent variable (x).[62] Mansur found that, with only limited training (typically two or fewer sessions), the subjects were able to recognize overall qualities of the data, such as linearity, monotonicity, and symmetry, on 79 to 95 percent of the trials.

A more intricate auditory display for instrumentation may be seen in the work by Lunney and Morrison.[61] Designing an auditory interface for an infrared spectrograph, tones with pitches corresponding to the wavelength of the spectra were used to represent selected compounds. Two melodic patterns and a chord were then constructed from these notes: (i) a melody played in order of descending pitch, with the duration of the notes controlled by the intensities of the peaks; (ii) a melody played in order of decreasing peak intensities, placing the most intense peaks first and with all notes of equal duration; and (iii) a chord playing notes representing all of the peaks simultaneously. Informal testing showed that students were able to accurately identify all of the compounds for which they had previously learned the melodies and chords.

Lunney and Morrison's work represents an interesting hybrid of an analogic representation and aurally encoded symbols. The display was generated by a direct mapping of data to an auditory variable, while the presentation of the data was optimized for picking out patterns from an already learned vocabulary of chemical signatures. The display functions, perhaps, more like a tone-oriented reading machine than a data exploration or system-monitoring device. In their recent work, reported at ICAD '92, the researchers have used neural nets to perform preprocessing of the chromatograph output, further distilling the pattern recognition task.

Auditory data display provides blind people access to computer programs originally designed for sighted users. Edwards has been working since 1987[27,71] to develop auditory aids for users of general computer interfaces. Recent work by Mynatt, described in these proceedings, proposes general guidelines and particular implementations for an auditory interface that provides blind people the use of software based upon X-windows.

With the advent of optical character recognition and speech synthesis, the field of ADs for vision-impaired users, which began with analogous representation, has swung towards the symbolic end of the spectrum. With the recent development of complex graphics, the need to unlock standardized graphical user interfaces, and enhanced sound synthesis and processing technologies, analogic representations—both auditory and tactile—are once again ascending.

The discipline of designing displays for vision-impaired users has been informative for sonification designers. Ongoing progress in various areas of sound computation and signal processing, including physical modeling, spatialization, and speech recognition, should continue to increase the designer's repertoire in creating interfaces for the blind. At the same time, the sophisticated display techniques developed for complex data representation should provide useful tools in the creation of more capable general interfaces. It is likely that this cross-pollination between general interfaces for the blind and sonification will continue.

11. SONIFICATION AND MUSIC
11.1 BACKGROUND

For years data structures have been perceived in sound and these structures have become a basis for musical systems. Predating Pythagoras, who analyzed the structure of harmonics and applied them to musical scales, we see the application of natural law to human-generated sound-producing systems. Pythagoras referred to his results as "sounding numbers." Refinements and extensions of this law defined the development of the musical scales in use all over the world, from the shakuhachi (bamboo flute) of Japan to the diatonically tuned music synthesizers of global popular music.

Manipulating sound for musical ends based upon data or mathematically derived structures arises from a distinguished tradition. Early in the Christian era, the astronomer Ptolemy remarked on the elements of musical modulation and wrote widely studied books on harmonics, as did Kepler and Newton. The composer Guillaume Dufay wrote "Nuper rosarum flores" on commission to dedicate the completion of Brunelleschi's dome for the cathedral of Florence in 1438. The design of the composition mirrors the floor plan and elevation of Il Duomo.[1] Other composers employed similar devices, ranging from Mozart, who used dice, to John Cage, who used the I Ching, or Chinese Book of Changes, as the source of data to guide the details of a composition.

Natural phenomena were further harnessed by composers using electronics to extend their reach. "The Earth's Magnetic Field" by Charles Dodge, and the flood of recent compositions using data generated by chaotic systems are two examples. Sound art, the creation of sound in installations or objects, may likewise use data, but this data may be generated by an audience as they interact with their environment.

11.2 SIMILARITIES OF STRUCTURE

Assembling, finding, and manipulating sonic materials: This could describe auditory display design. It could just as well characterize music composition and performance. What, then, are the differences between the two?

In order to look at the correlation between music and sonification, we will break down the two activities schematically. As shown in Figure 3, our schematic includes a control source, the renderer, and the listener.

The act of performing on a musical instrument is a special case of data generation, where the data is usually generated by the physical movements of breath, lips, and fingers. The physical movements may be guided by a predetermined data set (a composition) or may be improvised. This data is sent to the sound-generating means, or renderer, which is in this case a musical instrument. In the case of an acoustic musical instrument, there is a very direct, tightly coupled relationship between the data source (composer/performer) and sound-generating means (the musical instrument). Since the data is the form of physical movements rather than numbers, the structural relationship music performance to sonification may not be obvious. The relationships between data generator and sound generator are more explicit in a digital musical instrument, wherein the physical manipulations of the instrument are translated into digital control signals which are then used to control sound synthesizers. The listener decodes the data that is thereby embedded in the sound, extracting from it structure, meaning, and emotion.

An Introduction to Auditory Display

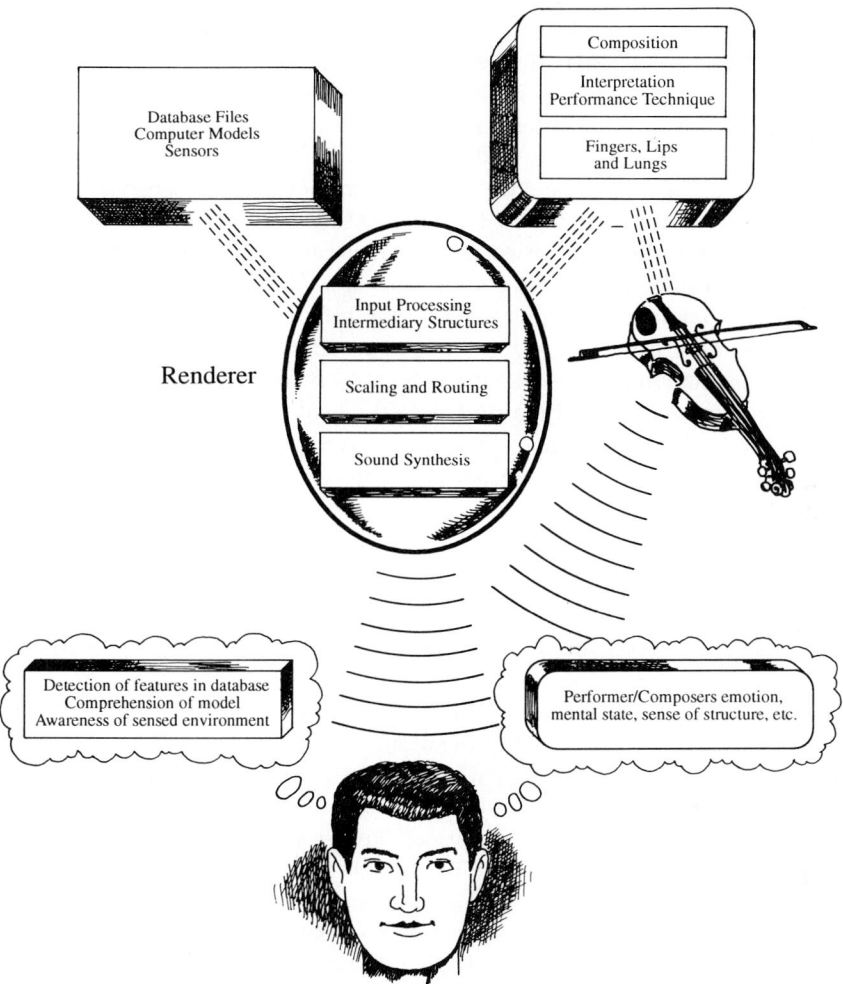

FIGURE 3 Sonification and Music.

Just as an enologist can determine much about the origin of a particular wine with a mere sip, a skilled listener can discern remarkable particulars about the performer as a result of the direct connection between the performer and his instrument. The ways and means of this conveyance constitute the study of psychomusicology and are beyond the scope of this paper. Suffice to say that a subtle and powerful array of techniques for controlling sound have been developed and the best employers of these techniques are intimately familiar (cognitively or otherwise) with what input creates a given output and where the rich ambiguities lie.

Now consider a sonification system. Schematically it is the same as a music performance system. The data sources are no longer notes on a page, fingers, and lips but computer files of numbers representing phenomena or outputs of sensors. The renderer must include a configuration of the sound-generating means, the data scaling and routing mechanisms, and other elements that represent the data with maximum clarity. In both music and sonification, the sound output is then presented to the listener for decoding.

Given the similarities between music generation and AD, we should not be surprised that the two share many concerns. Auditory perception, from psychoacoustics to psychomusicology, may inform both pursuits. The role of affect in auditory display may not be as obvious as it is in music, but display designers may well learn from the ways composers and musicians use sound to convey emotion. Bela Bartok said, "I cannot conceive of music that expresses absolutely nothing." Can one conceive of *any* type of auditory experience that does not have the potential to convey meaning, intended or unintended?

11.3 SHARED SKILLS

A host of skills are shared by practicioners of both music and AD. Included are: creating and manipulating sounds; fitting those manipulations to our perceptual systems; familiarity with data structures on multiple time scales; familiarity with the relationship of sound and affect; and, of course, familiarity with the process of careful listening. More specific crossover skills include sound computation and digital signal processing, psychomusicology, experiential (and some theoretical) knowledge of auditory perception, and, for many composers, the ability to interactively manipulate the data structures that comprise a sound-generating system.

As one example of how skills translate from music to auditory display, consider the role of aesthetics in display design. It may be fair to say, "If the sound is ugly, people won't use it." While this may not always be true, it is a familiar experience of people working in AD that a sonification will be running and it becomes sufficiently annoying that we just turn it off to take a break. Likewise, overly simple, intrusive, or simply unpleasant auditory computer interfaces are turned off, even if they have some utility.[15]

The craft of composition is important to auditory display design. For example, a composer's skills can contribute to making auditory displays more pleasant and sonically integrated and so contribute significantly to

---

[15] Gaver relates that SonicFinder was frequently disabled, Mynatt reports that poorly designed sounds degraded Mercator, and Kramer considers some of his sonification experiments downright ugly.

the acceptance of such displays. There are clear parallels between the composer's role in AD and the graphic artist's role in data visualization. Improved aesthetics will likely reduce display fatigue. Similar conclusions can be reached about the benefits of a composer's skills to making displays more integrated, varied, defined, and less prone to rhythmic or melodic irritants.

11.4 CONVEYING SUBTLETY AND MATTERS OF INTENTION

If a physical therapist were to use data from the body to control a sonification system, gauging the progress of the patient by how the sound changes, how would this differ from a musical performer turning body gestures into sound? Perhaps the difference is really one of intention. The former is driven by the intention of the therapist to extract data, the latter by the intention of the audience to extract emotion or meaning. Intention, as Jonathan Cohen points out in this volume, is in the ear of the beholder as well as the sonifier/composer. If you listen to a piece of music with the intention of understanding the data structures that generated it, whether the data were derived from molecular structures or the micro-fluctuations of someone's fingertips, perhaps you are receiving as a sonification what was intended as a piece of music.

The finger of the violinist on the string conveys a great deal of information about the player. It is at that point of contact (and at the bow) where the medium that embodies the idea (data) is transferred to the sound-making device. The data in this case are the notations of pitch, rhythm, and, to some extent, loudness markings of the composers. This is overlaid with a parallel data filter of the performer's technique, which may convey subtle elements of the performer's personality in the micro movements of the finger. The quality of the violin, the performance space, and other variables impact the sonic experience. These may be understood as characteristics of the display. Mistakes on the part of the performer may be understood as display-induced anomalies.

If this same performer is playing an electronic musical instrument, perhaps a violin with digital control outputs, the points of translation become more explicit. The idea (data) drives the muscle movements (again: data filter) which, in turn, cause the finger on the string to move. The transducers convert the finger placement or the string vibration into a digital number. This number is routed to a sound generator which is controlled in various ways by the data. The signal from the sound generator is sent to an amplifier and speaker. If the loss of fidelity in the system is not too great (which, for most performance transduction systems, it currently is), a skilled listener could discern the same aspects of the performer's abilities or mental states by listening to the resultant sound as might be gleaned via performance on a "direct" (acoustic) musical instrument.

It might even someday be possible for this same performer, were he or she to injure a finger, to go to a physical therapist and to attach a sensor to the finger. By then having the unlucky musician wiggle the finger in certain ways, a data set could be generated and injected into a diagnostic sonification system. The therapist would thereby gain the insight into the state of the finger necessary to promote healing. This concept could conceivably be extended to sports medicine, occupational rehabilitation, and general physical therapy.

If the data generated by the human subject is replaced by data generated from a computer model, it stands to reason that some sort of understanding of that model would be possible. As above, the insights into the data would be limited by the fidelity of the various translation processes and how well those particular translations matched the data under investigation.

One might speculate as to whether systems optimized for sonification will yield musical tools that embody a clarity of expression as a result of the rigorous demands of data representation. Likewise, might a music performance system of extraordinary breadth and subtlety function as a broad and subtle sonification system? There may also be important relationships between musical traning and sonification training. Teaching a listener how to appreciate a musical composition, discern temporal structures, or analyse orchestrations may correlate with how a system user is taught to most productively listen to a sonification. Cross-pollination is undoubtedly necessary and can be expected to yield benefits to both disciplines.

12. CONSIDERING THE ACOUSTIC ECOLOGY

The *acoustic ecology* of an auditory interface may have one or many elements. Speech input, speech output, auditory telepresence via microphones, spatialized and synthesized sounds, alarms, and data sonification may all be essential aspects of an interface. How these elements interrelate to form a comprehensible environment is an area that has received very little study. Our investigation of acoustic ecologies can be informed by consideration of acoustic environments occurring in natural and social environments.

Barry Truax, in his book *Acoustic Communication*, refers to an acoustic community as "a system within which acoustic information is exchanged."[96] He enumerates three characteristics which are essential for such a system to be functioning successfully. These are:

1. A *variety* of different kinds of sound.

2. A *complexity* within the sounds themselves and the types and levels of information they communicate. Listeners who are familiar with the environment are able to decode and interpret subtleties in the sound that the novice does not recognize.
3. A functional *balance* operates within the environment as a result of spatial, temporal, social and cultural constraints on the system.

Variety, complexity and balance may be useful design considerations in auditory display as well. Just now, *variety* is becoming an option in sound interfaces. Buxton made the clever observation that archaeologists of the future, upon finding a 1980s personal computer, would have concluded that the creatures living at that time had very poor hearing indeed. Why else would they design a tool with no audible output except a "beep?"[17] While the technology for providing greater variety in auditory interfaces has taken significant leaps forward, perhaps it was the conceptual limitations and visual bias of the system designers that produced this ludicrous situation. The reasoning seemed to be, "If the computer is just to display text (a dominant paradigm of the premultimedia era), then there is no need to provide any audio output other than a single alerting sound."

Complexity was likewise, hampered by inadequate technology and conceptual limitations. Indeed, while complexity is necessary for an information-rich display, it can demand more of the person using the system. People familiar with the display then are more adept at using it. Those that are unfamiliar with the acoustic ecology and the layers of meaning it contains may be intimidated by it. This sets up resistance to change.

Balance in acoustic communication, as described by Truax, is accomplished by natural and social forces. Acoustic sounds, unaided by electronic storage, transmission, and amplification, can only radiate limited distances. This assures a certain balance of how many sounds there are and how loud they are. Frequency area, another aspect of balance mentioned by Truax, is likewise guided by natural laws. Certain types of physical events are required to produce very low frequency sounds and others for high-frequency sounds.

Our auditory systems have evolved to reflect the acoustic balance between loud/soft and high/low pitched sounds. Extremely high pitched and loud sounds, which naturally occur infrequently, may strike us as unpleasant or even hurt our ears. Many sounds of nearly identical frequency range are less likely to occur than sounds with a variety of frequency ranges. The social forces in operation, such as work schedules and the general consideration people may have for one another, further inclines an acoustic ecology towards balance. (It is the imbalances, loss of definition, effects of technology, and means of becoming aware of and establishing a workable acoustic communication environment that are key concerns of Truax's excellent book.)

The designer of an auditory display is not bound by the physics of our natural acoustic environment. Sounds can be arbitrarily loud or soft and of any duration. Many sounds may occur simultaneously and in any frequency range. The spatial location of the sounds need have no reference to any other spatial scheme, and the sounds as well as what they represent may be chosen arbitrarily and changed at will.

Given this freedom and its potential for a chaotic and incomprehensible result, what then will be the forces guiding us towards balance? The laws of physics are replaced by the laws of perception. As William Bricken has said, "Psychology is the physics of virtual reality."[15] When we create our worlds (in Bricken's case, virtual environments), the rules of those worlds and any limitations to our experiences in them are dictated by what we can perceive and understand (and, of course, physically deliver). The limitations are not dictated only by what can be made to happen to stimulate our sensorium.

13. PLATFORMS

For years the computer interfaces market has been visually oriented. The reliance on early character-based CRT displays yielded to the dominance of graphical user interfaces. Now, authors of software, as well as computer designers and manufacturers, are realizing the extent to which they have neglected to exploit the possibilities afforded by the sophisticated use of sound. Only recently has the potential of audio for making products easier to use, more engaging, and more powerful been widely recognized.

Dissemination of AD-supported applications has been hampered by the erratic development of sound capabilities in computers. Sound-generating solutions have included plug-in sound boards, MIDI interfaces, special-purpose outboard processors, a variety of built-in dedicated sound chips, and software intended to drive D/A converters. None of these solutions have been standardized or made mutually compatible. Without an installed base of standardized "soundware," applications developers do not know if their potential customers will have the necessary sound-generating means to make use of the more sophisticated audio capabilities their products would demand. Typically, the result of the slow start in audio support has been products that are written to the lowest common denominator for sound on a given platform (typically low-bandwidth, low-resolution playback of a limited set of sound samples), or which ignore sound altogether (except for the beeps we have all become too familiar with).

Independent researchers and a few companies dedicated to sound-related research have been working to fill the void left by major suppliers of hardware and software. Most of this activity has been associated with music, and has resulted in software and/or hardware with which composers can experiment. Sound synthesis systems based upon specific hardware offer the advantages of speed and relative affordability when performing specific tasks; however, tying into any hardware-dependent solution carries with it problems of limited portability and extensibility, a reliance on a single company's research and development (R&D) team, and the possibility of rapid obsolescence. Alternatively, sonification and auditory interface designers have been attempting to bend general computing environments more towards their needs by integrating their display-oriented code with music-oriented sound synthesis software. These efforts are discussed below.

13.1 MIDI

Currently, the market forces of multimedia applications in general, and entertainment packages in particular, are driving a rapid trend towards increased sound capabilities in computers and computer-assisted systems. To the extent that this trend affects the market, the standards set by entertainment software will influence the diffusion of sonification into scientific, business and other application areas, at least in the short term.

As a case in point, there has been a wide acceptance of sound cards supporting GMIDI (General Musical Instrument Digital Interface).[65] GMIDI is an extension of the MIDI protocol, which specifies that certain byte streams will trigger specific sounds and/or send specific control and timing information to the sound-generating hardware (and in some cases to nonsound-related hardware as well). While MIDI has been an extremely useful tool in professional sound studios, and although it has facilitated sound tracks for games and provided a certain portability across platforms, its usefulness for general auditory interfaces and sonification is extremely limited at this time.

A key limitation of GMIDI is its specific vocabulary, while the key limitation of MIDI is its limited resolution. Vocabulary refers to the types of sounds specified for output by the sound generator. At this point, the popular music and game markets driving GMIDI fail to provide the everyday types of sounds that would be most suitable for the experiments reported in this volume by Cohen, Gaver, and Mynatt. It is worth noting, however, that because earcons use standard musical instruments for most of their notifications, the research reported here by Blattner et al. and Brewster et al. would be well served by the GMIDI specification.

The resolution problems associated with MIDI are simple yet intransigent. The two major issues are timing resolution and data resolution.

MIDI is a serial interface which specifies that you can transmit or receive a maximum of 31,250 bits (31.25 Kb) through each port. Any more bits will be buffered, causing timing delays. At 31.25 Kb/sec, large volumes of sonification data would create unacceptable timing delays in temporally sensitive data. Multiple MIDI channels or proprietary high-speed MIDI protocols could alleviate this problem; however, once one has left the most generic GMIDI area, its advantages quickly dissolve. Stripped of its portability, the deficiencies of GMIDI far outweigh its strengths.

The data resolution issue is very straightforward: if the data to be sonified has greater than 7-bit resolution (128 steps), then funneling it down to MIDI's 7 bits will be tantamount to throwing away the data's original fine grain. If, for example, one's data set has a range of 0 to 4096, then for every single increment in MIDI value the data will increment 32 steps. This is equivalent to a loss of 5 bits of information. In many cases this poor display resolution will be unacceptable. It is possible within the MIDI specification to use another 7 bits of controller resolution, for a total of 14 bits, but this foray outside of MIDI's most generic use eliminates much of the desired portability. Most synthesizers are not configured to respond to 14-bit control data for any variables other than pitchbend. If an additional control byte were used, this would also involve a degradation of timing resolution, since two bytes rather than one would be required to transmit each data value.

13.2 OTHER SOLUTIONS

As more computers are shipped with powerful, built-in digital signal processors, and as CPU's become increasingly powerful and capable of setting aside the cycles necessary for sound computation, the need for specialized sound-generating hardware may be reduced. Network servers[16] are being developed that provide a layer of software that insulates the user from specific sound-generating devices. To date, much of this work has been implemented using MIDI and thus is subject to the limitations described above. Other network-based systems provide extensive sound files along with the means of accessing them from within applications. By and large, these systems are modular and extensible and can be made to encompass other sound-generating targets.

There has been a trend towards specialized subsystems which are designed as solutions to a particular AD system requirement. These subsystems generally take the form of a hardware card that plugs into the selected computer's bus. The card is accompanied by software which is installed in the host system. The MIDI sound boards mentioned above are an example of one such subsystem. Specialized voice recognition and speech synthesis

[16] Systems such as those described in this volume by Madyastha and Reed and at ICAD by Roger Powell of SGI, along with many others being developed by hardware and software developers at most major computer companies.

An Introduction to Auditory Display

cards are another. These have both enjoyed substantial development of late. While some speech input/output systems are costly and powerful, simple systems are available for most computer platforms.

There has also been great interest in sound spatialization hardware and software. Locating sonified data in space can be a tremendously powerful tool. Since the Convolvotron first came on the market,[100] a number of manufacturers have attempted to apply spatial sound perception research to digital signal processor algorithms in order to synthesize spatial cues. Originally used Primarily for creating virtual environments and architectural simulations and now found regularly on game-oriented sound cards, these processors epitomize the modular approach to AD system design.

One design solution that has been used to great effect in the evolution of virtual reality systems is utilization of multiple computers working in parallel. In this sort of system, each of the components can perform independent tasks while coordinating with the other functioning units. Data import and scaling may be handled by one system, sound synthesis by another, and graphic display by yet another. Speech input and output are also managed by a subsystem. In addition to affordably providing substantial computing power, modular subsystems allow the researcher to take advantage of the hardware and software strengths of each component. Of course, communication between the system's components becomes an issue and portability can be a problem.

13.3 BUILDING TOOLS

Clearly, the sonification researcher is on his or her own to develop the necessary tools with which to carry out AD research. Even if one can accept the limitations of MIDI, it is still necessary to deal with input functions (data import, mapping, and scaling) and, in some cases, data gathering for assessment of the display itself. If one decides to proceed with sound synthesis software, whether running on the CPU, a built-in DSP, or a specialized outboard platform, it is still incumbent upon the AD designer to handle input functions, the precise configuration of the sound-generating software and the interface between those two elements.

The frustration expressed by numerous researchers[17] is that one has to spend much time working out the broad parameters of a system, and—while developing that system—risk becoming separated from the original sonification research issues and/or applications one set out to study. As in the early days of physics and astronomy, we all become tool builders. As pioneers in many other fields have discovered, the sophistication of our

---

[17] Private conversations with J. Ballas, S. Smith, and others. This author, too, feels these frustrations.

tools determines, to some extent, the outer boundaries of our explorable universe. Good tools do not assure good research, but good research is difficult without good tools. Furthermore, if the tools are difficult to configure and use, time and energy will be diverted from the primary task as the researcher devotes his or her energies to the tools.

The difficulties described here are associated primarily with the development of a highly flexible research and development system. The reader may find some solace in the fact that the performance requirements of R&D systems are different from those of most final applications. A powerful system may be required in order to work out the broad parameters of a sonification implementation, but then one can specify a system with just the capabilities needed. For example, analysis systems may need more flexibility while most monitoring systems can be predefined. In many cases, final sonification applications, such as manufacturing monitoring or medical display systems, will be satisfied by hardware/software systems designed to fulfill one specific purpose. In such cases, all elements of a system might be housed within a single assemblage, eliminating such issues as protocols, hardware interfacing, and availability of the selected synthesis technique.

13.4 A BRIEF LOOK AT SYSTEM POWER

General auditory interfaces have been designed to use straightforward sample playback capacity and, more recently, simplified sound synthesis capacity. Audification demands large memory and/or storage resources if large data files are to be listened to, but unless extensive filtering or compression is to be employed, the computational demands are not too great. By contrast, the demands for computer power made by sonification are substantial. Any attempt to quantify the CPU power needed to develop a sonification system must be individually tailored to the requirements of each researcher and his/her problem. Each researcher's needs will be different, and the necessary synthesis and signal processing algorithms—like hardware—are changing rapidly. The numbers offered here are only intended as a general gauge.

Informal estimates from a number of researchers working with sound suggest that, assuming 44.1-KHz sample rates, a sustained rate of 100 MIPS is about the power one would need to handle the sound computation for a reasonably sophisticated sonification system.[18] This figure assumes adequate cache memory and no serious problems with disk access and data I/O. A 100-MIPS system might be able to handle either three complex

[18] S. Smith, A. Peevers, D. Wessel.

voices with perhaps five multivariate, real-time controls for each voice. Alternatively, such a system could generate 50 to 100 simpler voices with simple parametric control.

If sophisticated spatial processing is included in the computation demands (which is not essential for all applications), the sustained MIPS figure might double to 200, especially if one considers spatializing multiple voices. This figure could increase substantially if a highly detailed rendering of room acoustics is implemented for virtual environments. If one then considers data import, scaling, mapping, and other system functions, the power needs increase further. Where integrated, simultaneous psychometric testing of the auditory display itself is to be included, yet more computation capacity would be necessary.

MIPS are only part of the problem. The size and complexity of sonification system code can be a more serious concern to the interface designer than the number of MIPS required. In lieu of an off-the-shelf sonification package, simply designing an interface can be a multi–man-year project. Pieces of a project can get out of control and beyond the designer's capabilities, as the sonification researcher attempts to take on a task normally associated with substantial commercial efforts or heavily funded scientific development.

13.5 INTERFACING WITH THE REST OF THE WORLD

Sonification displays, like other systems, must to some extent be interactive. What are the optimal hardware-based interface strategies? Are mouse- or data-glove-based interactions appropriate to this display type? Are voice command strategies well suited to sonification systems? If extensive real-time control capabilities are found to be essential, CPU power and scheduling software to support this need must be made available. If integration of sonification with graphics capabilities is a goal, then the system designer must not only calculate the raw power necessary to run concurrent visualizations and sonifications, but also cope with the complex scheduling issues that are bound to arise.

Amplifiers, speakers, and headphones may seem like too pedantic an issue to raise when considering AD systems but, as the final link to our auditory systems, their importance should not be overlooked. Any step up from the single two-inch speaker mounted somewhere near the computer's power supply may be appreciated, but many options need to be considered. What frequency range should be considered essential to a good-quality display? While 20 Hz to 20 kHz represents a full-range system by professional audio standards, perhaps 40 Hz to 16 kHz would be adequate.

Some spatialization algorithms perform best when transduced by free-field systems (speakers), while others perform optimally over headphones.[63]

Near-field monitoring (speakers placed near the listener so as to minimize the effect of listening-room acoustics) has certain advantages but results in a smaller display space. Stereo playback may be essential to some researchers, whereas monaural playback may be adequate for others. Some display designers, on the other hand, may exploit multiple channel or multiple speaker playback systems, which would provide the means for moving the sound around and over the listener. A lack of standards in this area is to be expected at this point.

13.6 TAKING DOWN THE ROADBLOCKS

The sophisticated tools that are available to producers of commercial music indicate what can happen when a well-defined market exists for sound products. Disk-based audio editing systems; a variety of music synthesizers employing advanced algorithms; sophisticated audio signal processors; automated mixing consoles; digital recorders in multiple formats ranging from 2 to 128 tracks (if modular systems are considered); MIDI sequencers with elegant editing, filtering, and quantizing capabilities; and controllers allowing breath control, foot control, keyboard aftertouch, and multiple faders are all tantalizing legacies of music-oriented research and development. The music technology market also boasts an admirable degree of data interchangibility. While the limitations of expressivity and timbral flexibility in current music and sound technology may leave something to be desired, when compared to that of AD, the latter comes in a distant second.

    This advanced state of music technology, however, points to where AD can go, given a clear agenda and a modicum of solidarity. If the variety of sound synthesis algorithms currently implemented or under study at music research centers were written so as to be portable, well-documented, with common interfaces and control hooks, they could all become available to serve as modules of a sonification system. Examples include inverse FFT,[34] physical modeling,[87] stochastic deterministic analysis/resynthesis,[83] and granular synthesis,[77] all of which offer powerful sound synthesis and manipulation capabilities. If the flexibility and power of programming and control environments were integrated into a widely available software package which possessed the capacity to readily control the synthesis modules, these tools in combination would allow a large number of researchers to do meaningful work without purchasing specialized hardware or building up their own sonification system from scratch. Some examples of higher order music-oriented programming environments which don't require specialized hardware include MAX, Fugue, HMSL, and Csound. Such tools might be combined with data import modules which then could be continually updated to handle new data base formats. Furthermore, to assess the unique psychophysical characteristics of a customized and experimental AD, user-configurable psychometric testing packages could be formulated and integrated into the entire system (see Smith et al., in this volume).

The problems posed by such a task are substantial. However, in reviewing the history of sonification's predecessor, data visualization, we may derive some encouragment from the fact that similar problems have been surmounted in the course of visual display evolution. It is not uncommon to find graphing tools in common spreadsheet software that, until a few years ago, were only available in advanced statistical analysis software. Meanwhile, the statistics software, along with packages for imaging volumetric data sets, now offer visualization capacities to the personal computer user which were once reserved only for specialized teams at supercomputing centers.

As scientists with intensive data comprehension needs team up with specialists in auditory display, it is to be expected that the possibilities and need for AD will become increasing clear and sonification techniques will be extended. This should result in an expanded AD market. As the market enlarges, tools will improve and standards will evolve, which will further accelerate the progress of fundamental research.

14. OPEN QUESTIONS

New approaches to using sound have highlighted related research topics in the areas of auditory perception, psychomusicology, cognitive science, knowledge representation, sound computation, and digital signal processing. A new look at display needs in the host of fields potentially served by AD is also in order.

14.1 AUDITORY PERCEPTION RESEARCH

Audition research has traditionally focused on elemental phenomena, such as responses to sine waves and noise bursts. This author advocates an expansion of the research to include a thorough consideration of how we hear and process complex, multivariate sounds. How we hear and use environmental sounds (see Ballas, in this volume) and music[24] provide some foundation upon which we can draw as we seek ways to assess complex listening. To adequately extend our base of knowledge, however, we still need perception studies specifically designed to investigate the important processes of hearing and comprehending complex synthetic environments such as sonifications, audifications, and auditory interfaces.

How do we design acoustic ecologies? What will the sound design practices of film sound or the fine art of orchestration have to contribute? Will studies on speech intelligibility offer guidelines to designers of busy, nonspeech auditory interfaces? How precisely will we cope with sonifications when more than one auditory stream is perceived? Will we obtain information from all streams, as our attention jumps from one to the other? Is parallel processing of multiple auditory streams too much to expect? Will the principles discovered in studying speech and music apply when the new sounds and tasks diverge so radically from such well-explored contexts?

The importance of speech as an information output channel for humans is generally considered self-evident. As a consequence, speech generation and perception have been widely studied, with much of this research being driven by the huge market for telephony. The relationship of speech to nonspeech audio for purposes of human/machine interfaces could benefit from a good deal more investigation. Might speech synthesizers be used to generate parametrically controllable abstract sound to convey numerical information or system states? How might inflection harness our affective responses to speed up or enhance comprehension?

It has yet to be seen whether a widely accepted set of acoustic variables will become the lingua franca of sonification. If so, will we be able to quantify, or at least rank order, their forcefulness, or perceptual strength, in relationship to each other (see Kramer, in this volume)? Even if such a common language is developed, how differently will diverse AD users hear the same display? To what extent will variations in hearing and auditory processing ability effect performance? Might a variety of different auditory variables (assuming they are developed) be used to compensate for the weaknesses of a selected population in effectively using an auditory display. As the field of AD develops, a standardized approach for relating display technique to auditory perception will become increasingly necessary.

In any multivariate system, as the number of independent variables increases the potential for confounding interaction effects escalates exponentially. Will interference between auditory variables determine a ceiling on display dimensionality? When multiple variables are used simultaneously, how can we accurately gauge the effectiveness of any single variable or group of variables for conveying information?

Indeed, what is the ceiling on display dimensionality? Can this ceiling be extended by designing multimodal displays? What are the synergistic effects on the efficacy of a display when information is presented visually, haptically, and audibly (see Lewandowski[58] and Fitch, in this volume)? Do multimodal displays and environments lead to the generation and perception of multimodal gestalts? If so, how do these relate to auditory gestalt formation in conveying complex states?

What consideration should be given to the emotional responses associated with auditory displays? It is argued elsewhere (see Kramer, in this volume) that affect should be considered in display design. Regrettably, hard research on affect and AD is lacking. In what ways might the time-tested affective tools of music be used and when should they be avoided? Are the answers to these questions universal or do they vary with data types and display tasks? How does familiarity with a display effect emotional responses to it? In fact, what are the long-term effects of using ADs? Will they result in pleasant experiences because of increased engagement and ease of use, or will they arouse irritation or discomfort on the part of frequent display users? Will familiarity breed affection or contempt? Will frequent use cause display fatigue and, if so, what solutions to this problem can be designed into the display (for example user-configurable options)?

It should be interesting to discover what the effects of culture will be on how any given display is perceived. Will differences in how different cultures perceive pitch,[25] harmony, and rhythm[19] affect the design of reliable, effective displays across national borders?

Precious little research has been conducted on the impact of learning on display effectiveness. Some researchers have commented on the role of training in the productiveness of auditory displays (see Fitch and Brewster, both in this volume and others[9,31,64]). Patterson's[69] exploration of the training issue was amongst the most comprehensive. As one would expect, researchers generally report that subjects improved their performance with initial training. Plateaus and other limitations of what can be learned, learning curves for different display types, retention of skills and transfer of training in using ADs, and ability to learn as a function of display complexity are all issues that warrant further attention.

In addition to the role of learning in ADs, we might also consider the role of ADs in learning. If, as has been suggested, some people learn better aurally while others learn better visually,[45,48] might not sonification and auditory cues be employed in education and training? A related series of questions might revolve around auditory information processing and creativity. If we inspect a complex system using sonification, will this lead to different observations than would result from visual inspection? Do the strengths of auditory pattern recognition offer novel insights, or simply increased efficiency and ease of use? What are the roles of cognitive and precognitive information processing in audition-based learning and in comprehending complexity? When it comes to early information preprocessing, to what extent shall we rely on the human nervous system as opposed to mathematical preprocessing and statistical analysis?

[19] For a consideration of cultural impact on music perception, see the difficult to obtain *Proceedings of the First International Conference on Music Perception and Cognition*, Kyoto, 1989.

14.2 MUSIC AND PSYCHOMUSICOLOGY

When it comes to the manipulation of nonspeech audio, it is hard to imagine a more sophisticated body of knowledge than that of music. Musical orchestration represents, perhaps, the best understood and most subtle craft for sculpting sound. While the orchestrator's collection of skills may serve as a fountain of inspiration, it is as yet unclear whether musical sounds are effective for conveying quantitative information. Do the associations we have with certain pitch combinations, or even certain instruments, render them somehow especially suitable or unsuitable for sonification or general interfaces? Might our expectations of "musicality" interfere with or enhance our interpretation of sonified data or of a system state? Whether or not musical sounds prove to be effective tools for AD, the guidelines and rules of orchestration might still come to be profitably employed in the service of designing interfaces.

Similar issues may be raised in relation to rhythm. What associations do we have with different tempi and how might those impact our reactions to an auditory display? Are some tempi agitating or irritating while others are soothing or uncompelling? Will rhythmic irregularity impact the accuracy of our judgements about other dispay variables? Will enculturated expectations cause us to perceive rhythmic "closure" that is not representative of the data? How do different harmonies effect our interpretation of auditory displays? Clearly, each musical variable presents us with a new set of issues to consider.

14.3 INTERACTIVITY

Sonification enables the user to be immersed in and interactively navigate through audible representations of a data base. How important is this type of interactivity to sonification? When one moves through a data field at his or her chosen pace, as one might in an immersive sonification, temporal relationships will be lost. Therefore, when one wants to maintain temporal relationships, as in seismic audification, this may not be the method of choice. Even when the user is not actively navigating through the data, user access to filtering, compression, and other signal-processing techniques, may be provided allowing them to optimize a display for a selected task. Given that interaction with one's display is important, which variables in an AD should be user configurable and which should be fixed by a skilled display designer? What types of hardware interfaces might be optimal for interaction with the display system? Research in virtual environments should have something to contribute here. Knowledge amassed from the design of musical instruments, bio-feedback displays and weaponry might also prove useful.

14.4 FORMATS, UNITS, AND REPRESENTATIONS

As it now stands, experimenting with sonification of different data sets usually means first wrestling with how to get that data set into a format that will be readable by one's sonification system. Even accomplishing this is no guarantee that the result will be translatable and transferable to the systems developed by other researchers. Given that data could come from a geologist, a physicist, a bank, or a network analysis tool, we can only hope that a common data format will one day be possible. In the meantime, the lack of established data formats will no doubt continue to be a problem to sonification researchers and their visual counterparts alike.

Even if a common data format is established within the sonification research community, proliferation of sonification techniques will probably lead to other difficulties in collaborative and portable research. Suppose, having detected an interesting trend or similarity, one researcher wants to share the experience of his display with another researcher, but the first researcher's system is designed around a particular sound synthesis technique that the second one does not have implemented. Since the auditory variables in one may have little or no relationship to those in another, how can they achieve their goal of a common interactive experience?

Just as there is no common verbal language for detailed descriptions of sound, so there is no common method for describing sound signals, control variables, and temporal structures for purposes of sound computation. These difficulties are shared with computer music. We can achieve some insight as to the intractability of this problem by considering the variable we refer to as timbre. Dannenberg says, only half jokingly, "...we picked out the two things we understood, pitch and amplitude, and called everything else timbre. So timbre is by definition that which we cannot explain."[23] He goes on to say that as we understand individual elements of complex acoustic phenonmena, such as spatial location and reverberation, we regard them seperately. We then give them names, algorithms, and, in many cases, discrete control variables for those algorithms.

How, then, do we develop a general language for computing sound that transcends individual algorithms, if we do not even have descriptions for what happens within a single, complex algorithm? For example, we can speak of changing the amplitude of a sinusoid which is frequency-modulating another sinusoid. The resultant sound may have more upper partials and be perceived as brighter. How can we generalize this transformation from pure to bright so that it can be represented in a system without FM synthesis capacity? Clearly this problem becomes increasingly intractable when we consider multiple levels of frequency modulation, modulation with nonsinusoid waveforms, other, possibly more complex synthesis algorithms and their component variables, and problems of representing the way that the transformations happen over time.

Even within a single sonification system, what units should be used to describe the magnitude of different auditory variables? Some variables may have units based upon the physical attributes of the sound, such as milliseconds for duration, or Hertz for pitch. But what about such percepts as brightness or roughness? What about such synthesis-specific variables as grain density and grain size? Smith et al. have suggested normalized units, or NUs, as a measure of such variables.[20] Should then pitch or duration be expressed in NUs as well? Should acoustic variables be described wherever possible in terms of physical variables and, when this is not possible, in terms of NUs? If so, then how will the system user relate physical units and normalized units? And what assures us that NUs will have any meaning to the system user? Perhaps NUs should be themselves normalized perceptually, so that each NU crosses a specified number (probably one) of JNDs.

14.5 AREAS OF APPLICATION

Will displays that are effective for one research area, say, meteorology, be effective for other disciplines, e.g., geology? What about if the data sets are different in nature, such as quantum physics and sociology? More generally, how specific are the matchings of data type and display technique?

As the use of sonification increases, questions of format, now arising in relationship to data file formats will begin to apply to whole data types. To what extent will classification of data types transcend disciplinary boundaries? Are some sonification techniques well suited to applications employing these generic data types? Will other types of data be best rendered by particular sonification techniques suited to a specific application. In more general terms, how is a researcher to ascertain if AD might provide useful tools for his or her work?

One can always analyze the benefits and deficiencies of ADs vis-à-vis the requirements of a particular display task. If sonification software is available, one can simply try to sonify his or her data and see how it works. Likewise, one can use readily available tools to add auditory icons or tone notifications to an interface and informally observe user performance. Since there is, as yet, no common wisdom as to what types of applications will benefit most from AD, each attempt to apply AD to a new area, even if it yields disappointing results, will advance the field.

[20] Private conversations with S. Smith.

14.6 QUESTIONS ABOUT MULTIMODAL DISPLAYS

The questions raised here multiply if we consider systems that may just as easily map a data variable to a visual as to an audible representation medium. How might one arrive at the most useful balance of auditory, visual, and haptic displays for any given task? What are the stengths and weaknesses of the various display modalities? How might displays that employ different senses complement one another? What guidelines will govern the combining of modalities in order to yield an effective display? Are the data format considerations similar for sonification and visualization? Is it useful, or even possible, to provide perceptual parameter weightings that are consistent across modalities? If data is mapped to display destinations that stimulate more than one sense, what rules govern the resultant perceptual weightings? Would it be desirable to employ normalized units across modalities or would doing so only generate confusion?

As in any new area of inquiry, for some time we can expect the questions to multiply as quickly as the answers.

15. CONCLUSIONS

Sonification is at the point that data visualization was perhaps ten years ago: the field is poised for serious progress towards solutions to some very difficult problems. Effective architectures for providing flexible sound synthesis and control in computer networks, the mechanisms of spatial perception, and the role of learning in the effectiveness of sonification all stand to make substantial strides. Insights into these problems will no doubt lead us to genuinely useful applications. Auditory interface designers are developing strategies that are likely to become the underpinnings of easy to use and powerful links between humans and computers. The integration of these strategies into future multimodal interfaces will help ensure that a system's capabilities are fully employed.

There is a clear need for new ways to make sense of vast quantities of data. Employing all of our sensory capacities is a reasonable, even essential approach to accomplishing this task. Audition, heretofore neglected, is a powerful input channel. The technology with which to construct this auditory information conduit is falling into place and the perception research is underway. Barriers that have hampered research between disciplines are gradually dissolving. With the newfound patterns of collaboration, there is a dawning recognition of the important work to be done. It is my hope that a context for the work presented in this volume has been established and that the reader's curiosity has been piqued by some of the salient issues.

ACKNOWLEDGMENTS

I would like to gratefully acknowledge the support I've received from colleagues in the preparation of this paper. References provided by Steve Frysinger, Meera Blattner, Beth Wenzel, and Jim Ballas were helpful in filling out the scope of the paper and grounding it in prior research. Larry Scadden's feedback on the *Blind* section, Tom Hanna's feedback on *Sonar*, Alan Peevers on *Platforms*, and Jim Ballas's input on a myriad of auditory perception issues helped me to iron out my facts. Stuart Smith and Steve Frysinger generously read and commented on several sections of this paper, providing invaluable feedback and support throughout the writing process. Finally, I would like to thank by brother, R. Jonathan Kramer, for his work and patience in editing this paper. His efforts provided an indispensable contribution to the paper's clarity and accessibility.

REFERENCES

1. Aarset, Tim. "Musical Learning as a Model for Speech Recognition Algorithms." MIT Lincoln Laboratory, unpublished proposal.
2. Astheimer, P. "Realtime Sonification to Enhance the Human-Computer Interaction in Virtual Worlds." In *Proceedings Fourth Eurographics Workshop on Visualization in Scientific Computing*, held April 1993 in Abingdon, England.
3. Baecker, R., M. Mantei, W. Buxton, and E. Fiume. "The University of Toronto Dynamic Graphics Project." In *Proceedings of CHI '91, ACM Conference on Human Factors in Software*, 467–468, 1991.
4. Ballista, A., E. Casali, J. Chareyron, and G. Haus. "A MIDI/DSP Sound Processing Environment for a Computer Music Workstation." In *Computer Music Journal*, edited by S. T. Pope, Vol. 16(3). MIT Press, 1992.
5. Begault, D. R., and E. M. Wenzel. "Techniques and Applications for Binaural Sound Manipulation in Man-Machine Interfaces." *Intl. J. Aviation Psycho.* (1992).
6. Bidlack, R. "Chaotic Systems as Simple (but Complex) Compositional Algorithms." *Comp. Music J.* **16(3)** (1992): 33–47.
7. Blattner, M. M., D. A. Sumikawa, and R. M. Greenberg. "Earcons and Icons: Their Structure and Common Design Principles." *Hum.-Comp. Inter.* **4(1)** (1989): 11–44.
8. Blattner, M. M., and R. M. Greenberg. "Communicating and Learning Through Non-Speech Audio." In *Multimedia Interface Design,*

edited by Alistair D. N. Edwards and Simon Holland. NATO ASI Series, Series F, Vol. 76. Berlin: Springer-Verlag, 1992.
9. Bly, S. "Sound and Computer Information Presentation." Unpublished Ph.D. Thesis, University of California, Davis, 1982.
10. Bly, S., S. P. Frysinger, D. Lunney, D. L. Mansur, J. J. Mezrich, and R. C. Morrison. "Communicating with Sound." In *Human Factors in Computing Systems*, edited by L. Borman and W. Curtis. Proceedings of CHI '85, 115–119. New York: ACM, 1985.
11. Bly, S, S. P. Frysinger, D. Lunney, D. L. Mansur, J. J. Mezrich, and R. C. Morrison. "Communication with Sound." In *Readings in Human-Computer Interaction: A Multidisciplinary Approach*, edited by R. Baecker and W. A. S. Buxton, 420–424. Los Altos: Morgan Kaufmann, 1987.
12. Boff. "Measurement of Visual Acuity." In *Handbook of Perception and Human Performance*, edited by K. R. Boff and J. E. Lincoln, 198–200. New York: Wiley, 1986.
13. Brabyn, J. A. "The Design of Auditory Instrument and Computer Displays for the Blind." SID '92 Digest, Society for Information Display, 1992.
14. Bregman, A. S. *Auditory Scene Analysis*. Cambridge, MA: MIT Press, 1990.
15. Bricken, W. "Progress in Virtual Reality." In *Proceedings Imagina '92*, held 1/92, Monte Carlo.
16. Brown, M. H. "An Introduction to Zeus: Audiovisualization of Some Elementary Sequential and Parallel Sorting Algorithms." In *CHI '92 Proceedings*, 663–664. New York: ACM, 1992.
17. Buxton, W. "There's More to Interaction than Meets the Eye: Some Issues in Manual Input." In *User Centered System Design: New Perspectives on Human-Computer Interaction*, edited by D. A. Norman and S. W. Draper, 319–337. Hillsdale, New Jersey: Lawrence Erlbaum, 1986.
18. Buxton, W., and T. Moran. "EuroPARC's Integrated Interactive Intermedia Facility (iiif): Early Experience." In *Multi-User Interfaces and Applications*, edited by S. Gibbs and A. A. Verrijn-Stuart, 11–34. Proceedings of the IFIP WG 8.4 Conference on Multi-User Interfaces and Applications, Heraklion, Crete. Amsterdam: Elsevier, 1990.
19. Chambers, J. M., M. V. Mathews, and F. R. Moore. "Auditory Data Inspection." Technical Memorandum no. 74-1214-20, AT&T Bell Laboratories, 1974.
20. Chareyron, J. "Digital Synthesis of Self-Modifying Waveforms by Means of Linear Automata." *Computer Music Journal*, edited by S. Pope, Vol. 14(4). MIT Press, 1990.
21. Cherry, E. C. "Some Experiments on the Recognition of Speech with One and with Two Ears." *J. Acous. Soc. Amer.* **25** (1953): 975–979.

22. Dannenberg, R. B., C. L. Fraley, and P. Velikonja. "A Functional Language for Sound Synthesis with Behavioral Abstraction and Lazy Evaluation." In *Readings in Computer-Generated Music*, edited by Denis Baggi. Los Alamitos, CA: IEEE, 1992.
23. Dannenberg, 1993, need reference
24. Deutsch, D. *The Psychology of Music*. Academic Press, 1982.
25. Deutsch, D. "The Tritone Paradox: An Influence of Language on Music Perception." *Music Perception* **8** (1991): 335–347.
26. Durlach, N. I. "Auditory Localization in Teleoperator and Virtual Environment Systems: Ideas, Issues, and Problems." *Perception* **20** (1991): 543–554.
27. Edwards, A. D. N. "Adapting User Interfaces for Visually Disabled Users." Ph.D. Thesis, The Open University, July 1987. (Available on microfiche from the British Library, Shelf number DX 80409.)
28. Edwards, A. D. N. "Soundtrack: An Auditory Interface for Blind Users." *Hum.-Comp. Inter.* **4(1)** (1989): 45–66.
29. Fish, R. M. "A New Auditory Display for the Blind." *Proceedings of the Annual Conference on Engineering in Medicine and Biology* **13** (1971): 175.
30. Fisher, R. A. "The Use of Multiple Measurements in Taxonomic Problems." *Ann. of Eugenics* **7** (1936): 179.
31. Forbes, T. W. "Auditory Signals for Instrument Flying." *J. Aeronautical Soc.* **May** (1946): 255–258.
32. Francioni, J. F., L. Albright, and J. A. Jackson. "Debugging Parallel Programs Using Sound." In Proceedings of the ACM/ONR Workshop on Parallel and Distributed Debugging, 68–73. (1991).
33. Frantii, G. E., and L. A. Leverault. "Auditory Discrimination of Seismic Signals from Earthquakes and Explosions." *Bull. Seismol. Soc. Amer.* **55(1)** (1965): 1–26.
34. Freed, A., X. Rodet, and Ph. Depalle. "Synthesis and Control of Hundreds of Sinusoidal Partials on a Desktop Computer Without Custom Hardware." In *Proceedings of the International Conference on Electronic Engineering Times*, 1024–1030. Santa Clara, CA: DSP Associates, 1993.
35. Freed, A., X. Rodet, and Ph. Depalle. "Synthesis and Control of Hundreds of Sinusoidal Partials on a Desktop Computer without Custom Hardware." In *Proceedings of the 1993 International Computer Music Conference*, 98–101. Held in Tokyo, Japan. Computer Music Association, 1993.
36. Frysinger, S. P. "Pattern Recognition in Auditory Data Representation." Unpublished Masters Thesis, Stevens Institute of Technology, Hoboken, 1988.
37. Frysinger, S. "Applied Research in Auditory Data Representation." In *Extracting Meaning from Complex Data: Processing, Display and*

Interaction, edited by E. Farrell. SPIE Proceedings. Bellingham, WA: SPIE, 1990.

38. Fubini, E, A. De Bono, and G. Ruspa. "System for Monitoring and Indicating Acoustically the Operating Conditions of a Motor Vehicle." U. S. Patent #4,785,280, U.S. Patent and Trademark Office. (An Italian patent was issued in 1986.)

39. Gaver, W. W., R. B. Smith, and T. O'Shea. "Effective Sounds in Complex Systems: The ARKola Simulation." In *Proceedings of CHI '91*, held April 28 to May 2, 1991, in New Orleans. New York: ACM, 1991.

40. Gaver, W. W. "The SonicFinder: An Interface that Uses Auditory Icons." *Hum.-Comp. Inter.* **4(1)** (1989).

41. Gaver, W. W. "Sound Support for Collaboration." *Proceedings of the Second European Conference on Computer-Supported Collaborative Work*, held September 24 to 27, 1991, in Amsterdam. Dordrecht: Kluwer, 1991.

42. Gaver, W. W., R. B. Smith, and T. O'Shea. "Effective Sounds in Complex Systems: The ARKola Simulation." In *Proceedings of CHI '91*, held April 28 to May 2, 1991, in New Orleans. New York: ACM, 1991.

43. Gaver, W. W., T. Moran, A. MacLean, L. Lvstrand, P. Dourish, K. Carter, and W. Buxton. "Realizing a Video Environment: EuroPARC's RAVE System." In *Proceedings of CHI'92*, held May 3 to 7, 1992, in Monterey, CA, 1992. New York: ACM, 1992.

44. Grinstein, G., and S. Smith. "The Perceptualization of Scientific Data." In *Proceedings SPIE/SPSE Conf. Elec. Imaging* **1259** (1990): 190–199.

45. Gardner, H. *Frames of Mind, the Theory of Multiple Intelligence.* New York: Basic Books, 1985.

46. Hayes, P. "Some Problems and Non-Problems in Representation Theory." In *Readings in Knowledge Representation*, edited by R. Brachman and H. Levesque. Los Altos, CA: Morgan Kaufmann, 1985.

47. Hirsch, H. R. "Perception of the Range of a Sound Source of Unknown Strength." *J. Acous. Soc. Amer.* **43** (1968): 373–374.

48. Hunter, W. E., and L. S. McCants. "The New Generation Gap: Involvement vs. Instant Gratification." Topical Paper No. 64, National Institute of Education, U.S. Department HEW, 1977.

49. Kaiser, W., and H. Greiner. "Warning System for Printing Presses." U.S. Patent #4,224,613, U. S. Patent and Trademark Office, 1980. (A German patent was issued in 1977.)

50. Kramer, G. "Audification of the ACOT Predator/Prey Model." Unpublished research report prepared for Apple Computer's Advanced Technology Group, Apple Classrooms of Tomorrow, 1990.

51. Kramer, G. "Audification: Using Sound to Understand Complex Systems and Navigate Large Data Sets." Proceedings of the Santa Fe Institute Science Board, Santa Fe Institute, 1990.
52. Kramer, G., and S. Ellison. "Audification: The Use of Sound to Display Multivariate Data." In *Proceedings of the International Computer Music Conference*, 214–221. Montreal: International Computer Music Assoc., 1991.
53. Kramer, G. "Sonification of Financial Data: An Overview of Spreadsheet and Database Sonification." In *The Proceedings of Virtual Reality Systems '93*. New York: SIG Advanced Applications, 1992.
54. Kramer, G. "Sound and Communication in Virtual Reality." In Communication In The Age of Virtual Reality, edited by F. Biocca. Lawrence Earlbaum Assoc., 1994.
55. Kreigyer, John B. "Sound Variables, Sound Maps, and Cartographic Visualization." Unpublished Thesis, Dept. of Geography, Penn. State University, University Park, PA, 1992.
56. Lauer, H., and L. Mowinski. "The Unknown Reading Machine for the Blind—Listening to Shapes: An Unfinished Project." Internal report, Central Rehabilitation Section for the Visually-Impaired and Blinded Veterans (124), VA Hosp, Hines, Illinois, 1989.
57. Lauer, H., and L. Mowinski. "Reading Machines for the Blind—A Multimedia Approach." Internal Report, Central Rehabilitation Section for the Visually-Impaired and Blinded Veterans (124), VA Hosp, Hines, Illinois, 1989.
58. Lewandowski, L. J., and D. A. Kobus. "Bimodal Information Processing in Sonar Performance." *Hum. Perform.* **2(1)** (1989): 73–84.
59. Little, A. D., D. H. Mershon, and P. H. Cox. "Spectral Content as a Cue to Perceived Auditory Distance." *Perception* **21** (1992): 405–416.
60. Loomis, J. M., C. Hebert, and J. G. Cicinelli. "Active Localization of Virtual Sounds." *J. Acoust. Soc. Am.* **88** (1990): 1757–1764.
61. Lunney, D., and R. Morrison. "High Technology Laboratory Aids for Visually Handicapped Chemistry Students." *J. Chem. Ed.* **58** (1981): 228.
62. Mansur, D. L. "Graphs in Sound: A Numerical Data Analysis Method for the Blind." Unpublished Master's Thesis, University of California, Davis, 1984.
63. Martens, W. L. "Demystifying Spatial Audio." In *Proceedings of the 3D Media Technology Conference*, edited by Hal Thwaiter. Held in Montreal, 1992.
64. Mezrich, J. J., S. P. Frysinger, and R. Slivjanovski. "Dynamic Representation of Multivariate Time-Series Data." *J. Amer. Stat. Assoc.* **79** (1984): 34–40.

65. *MIDI 1.0 Detailed Specification Version 4.2 (c) 1993.* Published and distributed by the International MIDI Association, Los Angeles, CA, 1993.
66. Mowbry, G. H., and J. W. Gebhard. "Man's Senses as Informational Channels." In *Human Factors in the Design and Use of Control Systems*, edited by H. W. Sinaiko, 115–149. New York: Dover, 1961.
67. Neff, W. D., and W. R. Thurwood. "Auditory Discrimination in Sonar Operation." In *A Survey Report on Human Factors in Undersea Warfare*, prepared by the Panel on Psychology and Physiology, Committee on Undersea Warfare, National Research Council, Washington, DC, 1949
68. Newman, W. R., A. Krickler, and B. M. Bove. "Television, Sound, and Viewer Perceptions." In *Proceedings Joint IEEE and Audio Engineering Society Meeting*, held in Detroit, MI, 1991.
69. Patterson, R. D. "Guidelines for Auditory Warning Systems on Civil Aircraft." Civil Aviation Authority, London, 1982.
70. Perrott, D. R., K. Saberi, K. Brown, and T. Z. Strybel. "Auditory Psychomotor Coordination and Visual Search Performance." *Percep. & Psycho.* **48** (1990): 214–226.
71. Pitt, I. J., and A. D. N. Edwards. "Navigating the Interface by Sound for Blind Users." In *People and Computers VI*, edited by D. Diaper and N. Hammond, 373–383. Proceedings of the HCI '91 Conference. Cambridge, MA: Cambridge University Press, 1991.
72. Polansky, L., and P. Ourk, eds. *HMSL (Hierarchical Music Specification Language): A Theoretical Overview*, edited by L. Polansky and P. Burk, 136–178. Perspective in New Music, 28/2, Summer, 1990.
73. Pollack, I., and L. Ficks. "Information of Elementary Multidimensional Auditory Displays." *J. Acous. Soc. Amer.* **26** (1954): 1550–158.
74. Pressing, J. "Nonlinear Maps as Generators of Musical Design." In Computer Music Journal, edited by C. Roads, Vol. 12(2). MIT Press, 1988.
75. Puckette, M. "Combining Event and Signal Processing in the MAX Graphical Programmin Environment." *Comp. Music J.* **15(3)** (1991): 68–77. Also available from Opcode, Inc., MAX Documentation, Palo Alto, CA, 1990.
76. Rabenhorst, D. A., E. J. Farrell, D. H. Jameson, T. D. Linton, and J. A. Mandelman. *Complementary Visualization and Sonification of Multi-Dimensional Data, Extracting Meaning from Complex Data: Processing, Display, Interaction*, edited by E. J. Farrell, SPIE Vol. 1259, 147–153. (1990).
77. Roads, C. "Granular Synthesis of Sound." *Comp. Music J.* **2(2)** (1978): 61–61.
78. Rogowitz, B., D. Ling, and W. Kellogg. "Task Dependence, Veridicality, and Pre-Attentive Vision: Taking Advantage of Perceptually-Rich

Computer Environments." In *Proceedings of the SPIE/IS&T Conference on Human Vision, Visual Processing and Digital Display III*, Vol. 1666, 1992.
79. Scaletti, C., and A. Craig. "Using Sound to Extract Meaning from Complex Data." In *Extracting Meaning from Complex Data: Processing, Display, Interaction II*, edited by Edward J. Farrell, SPIE 1459, 207–219. (1991).
80. Sloman, A. "Afterthoughts on Analogical Representations." In *Readings in Knowledge Representation*, edited by R. Brachman and H. Levesque. Los Altos, CA: Morgan Kaufman.
81. Scadden, L. A. "Annual Report of Progress." Rehabilitation Engineering Center of the Smith-Kettlewell Institute of Visual Sciences, San Francisco, CA, 1976-77.
82. Scadden, L. A. "Annual Report of Progress." Rehabilitation Engineering Center of the Smith-Kettlewell Institute of Visual Sciences, San Francisco, CA, 1977-78.
83. Serra, X. "A System for Sound Analysis/Transformation/Synthesis Based on a Deterministic Plus Stochastic Decomposition." Ph.D. Thesis, Stanford University, 1989. (Also CCRMA/Dept of Music Report No. STAN-M-58.)
84. Sloman, A. "Afterthoughts on Analogical Representation." In *Readings in Knowledge Representation*, edited by R. Brachman and H. Levesque. Los Altos, CA: Morgan Kaufmann, 1985.
85. Sloman, A. "Interactions Between Philosophy and A.I.—The Role of Intuition and Nonlogical Reasoning in Intelligence." In *Proceedings of the Second IJCAI*, London. Reprinted in *Artificial Intelligence* **2** (1971).
86. Smith, S. "An Auditory Display for Exploring Visualization of Multidimensional Data." In *Workstations for Experiment*, edited by G. Grinstein and J. Encarnacao. Berlin: Springer-Verlag, 1991.
87. Smith, J. O. "Techniques for Digital Filter Design and System Indentification with Application to the Violin." Doctoral Thesis, Stanford University, 1983. (Also CCRMA/Dept of Music Report No. STAN-M-14.)
88. Sorkin, R. D. "Design of Auditory and Tactile Displays." In *Handbook of Human Factors*, edited by G. Salvendy, 549–576. New York: Wiley & Sons, 1987.
89. Sorkin, R. D., D. E. Robinson, and B. G. Berg. "Detection Theory Method for Evaluating Visual and Auditory Displays." *Proc. Hum. Fact. Soc.* **2** (1987): 1184–1188.
90. Sorkin, R. D., B. H. Kantowitz, and S. C. Kantowitz. "Likelihood Alarm Displays." *Human Factors* **2** (1988): 445–459.
91. Sorkin, R. D. "Perception of Temporal Patterns Defined by Tonal Sequences." *J. Acoust. Soc. Am.* **30** (1990): 1695–1701.

92. Speeth, S. D. "Seismometer Sounds." *J. Acous. Soc. Amer.* **33** (1961): 909–916.
93. Stevens, S. S. "Matching Functions Between Loudness and Ten Other Continua." *Percep. & Psycho.* **1** (1966): 5–8.
94. Terenzi, F. "Design and Realization of an Integrated System for the Composition of Musical Scores and for the Numerical Synthesis of Sound (Special Application for Translation of Radiation from Galaxies into Sound Using Computer Music Procedures)." Physics Dept, University of Milan, 1988.
95. Triesman, A., and G. Gelade. "A Feature Integration Theory of Attention." *Cog. Psych.* **12** (1980): 97–136.
96. Truax, B. *Acoustic Communication*, 43. Norwood, NJ: Ablex Publishing Corp., 1984.
97. Urick, R. J. *Principles of Underwater Sound*. New York: McGraw-Hill, 1967.
98. Vercoe, B., and D. Ellis. "Real-Time CSOUND: Software Synthesis with Sensing and Control." In *Proc. 1990 International Computer Music Conference*, 209–211. San Francisco: Computer Music Association, 1990.
99. Weber, C. R. "Sonic Enhancement of Map Information: Experiments Using Harmonic Intervals." Unpublished dissertation, Department of Geography, State University of NYew York at Buffalo, 1993. 1993 need reference
100. Wenzel, E. M., F. L. Wightman, and S. H. Foster. "Development of a Three-Dimensional Auditory Display System." *SIGCHI Bull.* **20** (1988): 52–57.
101. Westheimer. "Spatial Configurations for Visual Hyperacuity" Gerald Westheimer and Suzanne P. McKee *Vision Research* **17** (1977): 941–947.
102. Witten, M. "Increasing Our Understanding of Biological Models Through Visual and Sonic Representations: A Cortical Case Study." *Intl. J. Supercom. Appl.* **6(3)** (Fall 1992): 257–280.
103. Wolff, R. "Sounding out Images."
104. Yeung, E. S. "Pattern Recognition by Audio Representation of Multivariate Analytical Data." *Anal. Chem.* **52** (1980): 1120–1123.
105. Zweig, G. "Auditory Speech Preprocessors." In *The Proceedings of the DARPA Speech and Natural Language Workshop*, 229–235. Los Altos, CA: Morgan Kauffman, 1989.
106. Zweig, G. "Finding The Impedance of the Organ of Corti." *J. Acous. Soc. Am.* **89** (1991): 1229–1254.

James A. Ballas
Naval Research Laboratory, Code 5535, Washington, DC 20375–5337, (202) 404-7988; e-mail: ballas@itd.nrl.navy.mil

Delivery of Information Through Sound

The potential to deliver information through sound is rapidly expanding with new technology, new techniques, and significant advances in our understanding of hearing. Although these changes raise important new issues about the design of sound delivery systems, there is already a wide range of knowledge scattered through different disciplines about communicating information through nonspeech sound such as sonification. An overview of how sound can deliver information is presented using a framework of linguistic analogies. Areas that will be discussed in some detail include contextual and expectancy effects, which operate when tonal sounds as well as realistic sounds are interpreted.

INTRODUCTION

My objective in this paper is to present some ideas about how information is effectively delivered through sound. The notions are not strikingly new. In fact, one intention is to illustrate that we do not have to invent completely new notions about how sound communicates information in order to exploit the new technology that is available. We already know much about how sound can be used effectively. In the spirit of rediscovering worthwhile ideas, I revisit a cliché:

> "If a tree falls in the forest and no one is around, does it make a sound?"

This conundrum provides interesting insights into the nature of human aural perception. It suggests that sound is present only when it is heard. Contrast this to the situation when the equivalent question is asked about vision: "If a tree falls in the forest and no one is around, does it make visible movement?" It is difficult to make sense of this question. It seems easier to separate sense data from an event in the aural domain than in the visual domain. As Kneale[15] put it, "We say that bodies make or cause sounds and smells, but never that they make or cause views." This difference also applies to inference mechanisms that are used to discover information about the world from the senses. Kneale makes this point also: "We say sometimes that the presence of a body in the neighborhood may be inferred from the occurrence of some sound or smell, but very rarely, if ever, that the presence of a body may be inferred from the occurrence of a view." The important implication for auditory display design is that listening is not necessarily the perception of events but can be the perception of stimuli caused by the event. Although there are persuasive arguments[11,13,14] that listening is intended to process events in both the speech and nonspeech domain, the idea persists that listening is a process of stimulus analysis.

For auditory display design, this conundrum is evidence of opportunities and pitfalls. The dissociation between an event and its sound means that sounds can be used to represent phenomenon other than the sound-producing event. For those interested in using sound parameters to convey information about data, the dissociation opens opportunities. But for those interested in communicating information about causal events, the possibility of a dissociation means that the listener may not perceive the sound as it was intended. For those who want to exploit the dissociation, there are complex perceptual interactions that need to be understood and managed to effectively map complex phenomena to sound parameters. For those who want to avoid the dissociation, the expectations of the listener and the context of the sound need to be considered.

The development of effective methods to deliver information through aural mechanisms should be aided by a general understanding of the kinds of information conveyed through sound and of the perceptual processes that are involved. This paper presents an overview of how information is represented through sound, provides some information about relevant perceptual processes, and provides some information about expectancy and context effects. There are different ways of characterizing the informational functions of synthetic sound. Several of these are described in this volume (Bargar, Kramer, and Scaletti) and elsewhere.[13,21] The framework I present is intended to be inclusive and readily understood. I will describe five types of functions and will express all of the functions with linguistic analogies: exclamation, deixis, simile, metaphor, and onomatopoeia. I use these linguistic analogies without taking a stand about whether nonspeech sound and music are forms of language. It is simply that many of the functions of language are found in nonspeech sound that communicates information. Essays about comparisons between speech and nonspeech sound are found elsewhere.[14,21] For each of the linguistic devices, I briefly discuss some of the perceptual issues involved in designing a sound to achieve the function. It should be noted at the outset that many applications combine several of these basic functions. For example, three-dimensional sounds often combine exclamation, deixis, and one of the representational devices. I also should mention that this classification scheme was developed primarily for purposes of presenting an overview of informational sound and is best thought of as a pedagogical device.

EXCLAMATION

A frequent function of sound is to get one's attention. In doing so, sound serves as an exclamation and is extremely effective in this function. Perception of exclamation is a process of detection and is related to the signal-to-noise ratio and the observer's response bias. Design guidelines cover how to enhance detection through appropriate signal levels and spectral composition.[20] However, increasing exclamatory effectiveness purely by raising the signal level produces only louder and louder sounds and is unsatisfactory as a general solution. An alternative that is presented in detail by Edworthy, Loxley, and Dennis[9] is to utilize temporal and spectral patterns to encode urgency. Their results are summarized in Table 1. Changing either the pulse or burst parameters in the manner indicated will change the perceived urgency of the alerting sound. Although we have little understanding of why these parameters cause a change in perceived urgency,

TABLE 1 Relationship between pulse and burst parameters and the perceived urgency of a sound, summarizing the results of Edworthy, Loxley, and Dennis (used by permission of the authors).[9]

	Perceived Urgency		
	Higher		Lower
Pulse			
Envelope			
Harmonic regularity	Random		Regular
Interpulse interval	Shorter		Longer
Burst			
Rhythm	Regular	Syncopated	Slowing
Average pitch	High		Low
Pitch range	Large		Small
Pitch contour	Random		Down/Up

and many details remain uncertain such as tradeoffs between the parameters, this does provide a method of designing different levels of urgency in alerting sounds.

DEIXIS

The ubiquitous computer and appliance "beep" is equivalent to "this" and works as a type of pointing gesture similar to the deictic pronouns used in language. For example, when a person says "this one," they are using language to point to a particular thing near them. When the computer produces a "beep," it is pointing to itself and by implication to something or some event related to its operation. In providing deictic reference, sound has an important advantage over vision: sound surrounds, where sight only provides a 2.5-dimensional view of unocculuded surfaces. The missing half-dimension and surface occlusion effect make much of the environment unavailable for sight. Sound can provide information about events and objects hidden from sight. When this added information is unavailable through sound (e.g., through deafness or masking), then part of the environment is always hidden.

It is important to realize that deictic devices cannot be understood without contextual information such as the position of the listener and the referenced object or event. When a person says "this one," we have to know where the person is standing in order to understand the reference. Likewise, in a room with several computers, a "beep" would be confusing if we did not know which computer it came from. But even knowing which computer produced the sound may not be sufficient context. The sound is a spatial gesture directing attention to a particular machine and by implication to an entity or event within the machine. The entity or event itself is often hidden. For example, many computer systems produce a beep when the user enters a response that currently is not valid. The invalid response typically is not echoed, and the user must remember it. If feedback is delayed or if the beep is given to a complex series of responses, only one of which is incorrect, then there may be a problem of inadequate contextual specification.

In complex systems, the aural alert is often used to indicate information and events that are not currently displayed visually. An interesting example of the use of aural alerts is the modern cockpit. In modern commercial aircraft, the usage of aural alerts increased beyond the point that pilots thought acceptable.[23] This increase in aural alerts occurred over a period when the number of visual displays in the cockpit increased and then decreased (Figure 1). The reversed trend in these two types of information displays after 1970 probably is not coincidental, but rather reflects the use of aural displays to refer to increasing amounts of hidden information. The decrease in visual displays occurred when multifunction computer displays began to replace the dedicated steam gauge displays (i.e., round gauges linked to a specific parameter). The multifunction displays could present different pages of information, but the particular page has to be chosen by the pilot. This replacement hid information that had been present continuously in previous cockpit designs and raised the need for alerts to inform the pilot about critical information. As the multifunction displays were being introduced, automation capability was also increasing, which meant that there were more operations (and more information) hidden within the computer. Thus Veitengruber[23] concluded that the rapid increase in alerts was due especially to the introduction of flight management software.

However, there is a limit on the number of aural alerts that can be used. Patterson and Milroy[17] found that large sets of warning signals could be learned, but the learning time and retraining requirements are considerable and current guidelines[20] suggest a maximum of four to six. One solution is to use very general aural alerts and provide detailed information about the alerting condition in a visual display listing the alerts. In this solution, the aural alert is a cross-reference device and is similar to anaphora in linguistics.

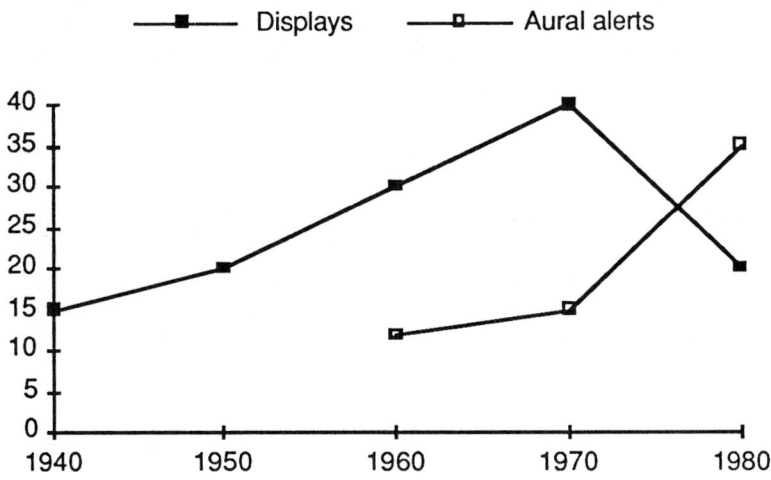

FIGURE 1 Average number of visual displays and aural alerts in commercial and military cockpits from 1940 through 1980. Based upon data from several sources.[8,19,20,23]

Deictic sounds can provide information for source localization and provide an advantage over complementary visual information.[18] The recent development of effective methods of presenting three-dimensional audio through headphones (e.g., the Convolvotron [see Durlach and Wenzel, in this volume]) has expanded the possibilities of delivering spatial deixis information. Coupled with a visual system, either a real scene or one generated through virtual reality technologies, the three-dimensional sound can indicate an external location. An interesting application of this technology in aircraft cockpits would be to indicate the position of an aircraft that is below and outside the field of view of the pilot.[4] This audio technology may eventually complement the commercial cockpit displays that provide collision avoidance warning.

SIMILE AND METAPHOR

Simile and metaphor are the core of sonification and current status in achieving these functions and outstanding issues are covered by papers in this volume which address parameter mapping techniques (see Bargar, Kramer, and Scaletti). When sounds are being used as similes, a sound property or parameter is being likened to some property or parameter in another domain. The interpretation of the sound is defined by something equivalent to the legend in a graph. In this sense, the legend acts as the adverb *like* in a linguistic simile. Ideally, the basis of the similitude should not be arbitrary. One way to improve the effectiveness of the simile is to base it on commonly understood relationships. There are research studies and a subfield of psychophysics called cross-modality matching that address this issue. As an example of research studies, Walker[24] found that the following visual/aural relationships were commonly understood: sound frequency with vertical placement, wave form with pattern, amplitude with size, and duration with horizontal length. Consistency of these matches was related to musical training and age and, lesser so, to cultural and environmental factors. Sound simile must also consider the effect of psychophysical transformations (e.g., joint effects of frequency and amplitude on loudness). Mappings between aural and visual dimensions and between aural dimensions and data should be consistent with established literature on cross-modality matching.[1] While experts would recognize the need to attend to these issues, casual users of sonification techniques may not recognize this need and thus may produce sonifications that have inadvertent perceptual effects.

Metaphor is an extension of the simile to an identity relationship. Rather than saying that one thing is like another, a metaphor states that the first thing is the other. The result is a tight coupling between the sound and its intended interpretation. The development of sonification metaphors is also an indication of the evolution of this technology, in much the same way that metaphor expands language. Dead metaphors are common in language (e.g., cold person, hot date, quick temper, lofty idea). They do not puzzle or entertain anymore; they communicate an idea very effectively and enter common language usage. A goal of sonification should be to move similes into metaphors and ultimately produce dead metaphors. The Geiger counter is one of the first dead metaphors that is commonly used in aural display design. Many signaling sounds such as sirens, telephones, and doorbells also might be considered to be dead metaphors. These signaling sounds are quickly identified[2] and convey information rapidly, sometimes even more rapidly than the equivalent speech sound. For example, Burrows[6] was perhaps the first to demonstrate an advantage of using a signaling sound to convey information when he showed that a fire-alarm bell produced quicker and more accurate responses than alerting words.

ONOMATOPOEIA

The function of this type of sound is to indicate events by imitating the sound that is typically caused by the occurrence of the event. Using onomatopoeia to describe this function is a compromise. Gaver[12,13] has written about using these sounds in computer systems suggesting the terms nomic, iconic, and everyday. Elsewhere in this volume Kramer uses the term realistic. However, here I want to emphasize that the sound is synthesized and imitates the sound caused by some real event. The linguistic term "onomatopoeia" refers to sounds that are designed to imitate another sound and represents accurately the communication function of this type of sound.

One of the advantages of this type of sound may be an achievement of high-identification accuracy without extensive training (see, Fitch and Kramer, in this volume). But this will not come with all everyday sounds. There are a variety of factors that are involved in the interpretation of this type of sound including acoustic properties, perceptual interpretations, cognitive expectancies, and ecological frequency.[2] Two types of cognitive expectancy are involved: expectancy about the usual cause associated with a sound, and expectancy about the typical sound associated with a cause. The first type of expectancy refers to the degree of causal uncertainty for a sound. Some sounds can be produced by several very different types of causes. For example, an elephant trumpeting can sound very similar to a squeaky pipe valve. The second refers to the mental concept that people have for a sound. For example, a doorbell is thought of as a high-low note pattern, rather than the reverse.

Effective delivery of information through sound should produce responses that are quick and accurate. My recent studies[2,3] provide some insight into how to insure that the cause of a synthesized sound is unambiguous. These studies examined the factors that contributed to the quick and accurate identification of a set of onomatopoeic sounds. Although generalization to other sounds that are more metaphorical may be limited, these studies provide insights into causal identification since this seems to be the immediate primary response of the listener to onomatopoeic sounds (i.e., when asked to describe these sounds, listeners mention the causal events, rather than the acoustics or other reactions such as affective response[22]).

The results of the studies showed that sound identification accuracy and speed were related to several factors including acoustic variables, ecological frequency, causal uncertainty, sound typicality, and context. Only certain complex acoustic variables were related to performance; simple acoustic variables were not. The variables tested included properties in dimensions of amplitude, frequency, and time (e.g., peak, FFT spectra, duration) and moments computed in each of these dimensions. A combination of two orthogonal acoustic variables were related to performance: the presence of harmonics in continuous sounds and similar spectral bursts in noncontinuous sounds. These two variables represent the continued delivery of similar spectral information. With continuous sounds, the harmonics persist; with noncontinuous sounds, the repeated bursts have a similar spectral content. Continuity in two dimensions, frequency and time, are present in these variables.

The identification of a sound is also related to sound typicality. Typical sounds are identified faster than nontypical sounds. An example of a typical sound given previously is the high-low note pattern of a doorbell. Although people can learn to associate another pattern or a buzzer with this type of information, responses will be quicker if the sound matches the mental expectations for a listener. Studies that examined this effect tested how quickly listeners confirmed that a sound could indicate an event. The listeners first were primed with the event and then heard the sound. They had to agree or disagree about whether the sound could cause the event. Thus the paradigm required them first to think of an event that can cause a sound, and then to evaluate a particular sound as a possible outcome of the event. This priming paradigm is used to activate a mental concept, and then force a comparison between this concept and a sound. The key results are those trials when subjects agree upon a match between the primed concept and the sound. The response time for this agreement is thought to reflect the time needed to compare the mental concept to the sound. Results in my studies showed that this time is quicker when typical versions of sounds are employed, even though listeners will agree that the nontypical versions could have been produced by the event. In other words, acceptance of a match between a sound and a causal event is quicker if the sound conforms to the stereotype.

The identifications for a sound also can be useful in understanding the causal uncertainty associated with the sound. Causal uncertainty, an entropy measure, reflects in one statistic the number of different causal attributions for a sound and the distribution of responses across these attributions. In these studies I found that causal uncertainty correlates highly with identification time and accuracy and, significantly, but weakly, with ecological frequency. The relationship between causal uncertainty and ecological frequency was interesting. It varies somewhat with the type of sound. Signal sounds (e.g., telephone ring) had lower causal uncertainty than would be estimated on the basis of ecological frequency. Impact sounds and sounds involving the modulation of noise had higher causal uncertainty than would be estimated on the basis of ecological frequency. Essentially this means that sounds used as signals are well understood probably because they are not used indiscriminately for different types of signals. Learning of these signals is enhanced by consistent, rather than extensive, experience.

In order to assure quick and accurate responses to the intended information, these studies show that effective delivery of information through sound should minimize causal uncertainty and should be consistent with the mental concepts that people have for sound. This is true for any synthetic sound and especially so for onomatopoeic sounds. The intended cause of the sound is often the information that we wish to deliver. Whether this information is data that controls the sound as in sonification, or data that is being made audible as in audification, we need to control the causal interpretation of the sound. With sonifications, the intended cause of the sound is the data controlling the synthesis. With audification, the intended cause is the data that is itself being made audible. In either case the listener should receive a sound that is unambiguous in its cause.

QUANTIFYING CAUSAL UNCERTAINTY

Causal uncertainty may be a useful measure in sonification design, because it correlates well with identification time and accuracy, it has been shown to be consistent for different listeners and different examples of a particular sound[2] and is derived from information theory. However, the usual, cumbersome method of calculating it requires obtaining and sorting a set of identification responses from listeners, tallying the number of alternative

TABLE 2 Forty-one sounds used in obtaining estimates of causal uncertainty.

1	Telephone ring	2	Clock ticking	3	Car horn
4	Doorbell	5	Automatic rifle	6	Boat whistle
7	Water drip	8	Bell buoy	9	Foghorn
10	Water bubbling	11	Bugle	12	Gunshot indoors
13	Lawn mower	14	Church bell	15	Oar rowing
16	Door knock	17	Toilet flush	18	Footsteps
19	Fireworks	20	Cigarette lighter	21	Touch tone
22	Door opened	23	Bacon frying	24	Hammering
25	Sub dive horn	26	Clog footsteps	27	Car ignition
28	Tree chop	29	Power Saw	30	Door latched
31	Cork popping	32	File drawer closed	33	Door closed
34	Car backfire	35	Jail door closed	36	Gunshot outdoors
37	Light switch	38	Stapler	39	Telephone hung up
40	Sawing	41	Electric lock		

causes and calculating an entropy statistic from this data. The method is similar to the production measures used to quantify aspects of verbal materials.[7]

An alternative to a production measure is to ask listeners to estimate the number of alternative causes, similar to the method that Brooks[5] used to obtain estimates of verbal category size. Work with verbal materials has demonstrated that measures on similar dimensions obtained by different methods (e.g., rating scales and production methods) are significantly correlated.[7,16] The correspondence of multimethod estimates in verbal research also may occur in estimating causal uncertainty. In the following experiment, subjects were asked to estimate the number of alternative causes for a sound. These estimates were compared to causal uncertainty values computed from identification responses obtained in a previous experiment.

STIMULI. The set of stimuli were the same 41 sounds as those in the first experiment reported by Ballas.[2] All the sounds were brief (less than 600 msec) and represented a variety of events including signals, explosive events, impacting objects, footsteps, and water events. Those sounds, listed in Table 2, were used because extensive data were already available.

PARTICIPANTS AND PROCEDURE. Thirty-six undergraduate students participated on a volunteer basis and were paid for their participation. None reported any hearing disorders. Twenty students were tested without an estimation anchor and sixteen were tested with the anchor. The task was to estimate the number of reasonable, potential causes of the sound. The listeners had the opportunity to hear the sound as often as they wished. No constraints were placed on the size of the number that could be entered. Listeners in the anchor condition were told that, in previous research, it had been found that the number of potential causes of the sounds ranged from one cause for some sounds to as many as 35 for other sounds. This range was presented to the listeners in the instructions and was presented as part of the prompt on the computer screen requesting the entry of an estimate of the number of causes. The value for the upper anchor came from previous studies of the sounds.

RESULTS. Geometric mean estimates of the number of possible causes, averaged across listeners, increased when an anchor was provided but both sets of estimates correlated significantly with calculated causal uncertainty[2] ($r = .74, .64, p < .05$ without and with an anchor, using Spearman rank-order correlations). Both sets of estimates were also significantly correlated with identification time ($r = .77, .65, p < .05$). Differences between the correlations were not significant. These results show that a quick approximation of the causal uncertainty of sounds can be obtained by simply asking listeners to estimate it. An anchor for the upper range of the estimate does not have to be provided. For purposes of evaluating the causal uncertainty of a sound being considered for an auditory display, these estimates can be used as an indicator of both how long it would take to identify the sound and how accurately the sound would be identified.

CAUSAL UNCERTAINTY OF REMEMBERED SOUNDS

The ability of listeners to estimate the relative number of alternative causes raises the possibility that this information might be obtained by describing the cause of the sound but not by actually producing the sound. In other words, could a verbal description of a cause produce accurate estimates of causal uncertainty? Is the memory of a sound sufficiently close to actual instances to be useful in evaluating causal uncertainty? To explore this issue, listeners were given verbal labels for the causes of the 41 sounds and asked to generate alternative causes without actually hearing the sound.

PARTICIPANTS AND PROCEDURE. Participating for class credit, 25 students were given a list of the 41 sounds—described by the cause of the sound—and asked to write down other possible causes of the sound—causes which would produce the same kind of sound. A brief explanation of alternative causation was provided with an example. The order of sounds was randomized in five sets.

RESULTS. The responses were sorted according to the guidelines and sound categories used in previous studies of the sounds,[2] adding more sound categories as necessary. Causal uncertainty values were calculated as before. The correlation of these values with those obtained when the sounds were heard was not significant ($r = .27$, $p = .08$). The mean of the values obtained from verbal labels alone was significantly larger than the mean of the values obtained when the sound was heard ($t(67) = 4.75$, $p < .05$), and the variance was significantly smaller, ($F(40, 27) = 10.38$, $p < .05$).

These results show that the causal uncertainty for a sound is higher and less variable when a person is given a label for the sound as a cue rather than the sound itself. Apparently verbal cueing for alternative causes produces alternatives that are not consistent with the causes produced upon hearing the sound. The increased average uncertainty but reduced variability in uncertainty suggests that the verbal labels produce a limited random search for alternative causes. Too many alternatives are generated for sounds that are highly identifiable, and too few for sounds that are highly ambiguous. It is likely that the cognitive process of considering alternative causes when presented with a sound does not access a linked set of verbal labels but rather a set of causes activated by the acoustic retrieval cues.

CONTEXT AND CAUSAL UNCERTAINTY

Context also can influence the identification of ambiguous everyday sounds, according to studies by Ballas and Mullin.[3] These studies employed nearly homonymous pairs of test sounds. That is, the members of a pair sounded similar but were aurally discriminable. For example, one pair of sounds was walking in leaves and a fuse burning. These test sounds were presented in isolation to get baseline identification performance and were presented within sequences of other everyday sounds to assess contextual effects. One member of each pair was included within a sequence. These sequences were: (1) semantically consistent with the included member of the pair; (2) semantically biased toward the other member of the pair, which was not in the sequence; or (3) composed of randomly arranged sounds. Two paradigms, binary choice and free identification, were used. The results indicated that context had significant negative effects and only minor positive effects. Performance was consistently poorest in biased context and best in consistent context. A signal detection analysis indicated that there is a bias against detecting an out-of-context sound, especially with a free-response paradigm. Labels added to enhance context generally had little effect.

One of the surprising findings of these studies was the lack of a positive context effect. I know of no research in language perception that would parallel the differential effects of context found in these studies. The result may be due to a negative effect of embedding the test sound within other everyday sounds. This effect is present even if all the sounds in whole sequences were consistent with the meaning of the sound. This suggests a fundamental difference between language perception and the perception of everyday sounds. In language, the perception of words is enhanced by embedding them within sentences. Encoding efficiency is sufficient to identify the words at a rate that can take advantage of the context. With everyday sounds of the type we used, encoding may be too slow to generate sufficient contextual information on the fly to offset the negative embedding effect. Thus perception of a discrete sound may not be enhanced by embedding it within sounds that could also occur within the same environment. When the context is biased or random, then added negative effects accrue in addition to the embedded effect. Response bias due to context becomes greater as the paradigm shifts to produce greater context effects. Overall, these results show that the effects of context are clearly detrimental when the task is to detect a potentially ambiguous sound embedded in a context that is biased against the meaning of the sound.

SUMMARY

These brief analogies to linguistic devices are reminders that certain types of nonspeech sound, including sonifications of data, are intended to communicate information. An extensive presentation of the communication perspective is available in Truax's paper.[21] The factors that must be considered in delivering the intended message can be extensive, as the studies of ambiguous everyday sounds demonstrate. But, as those studies also demonstrate, many of the factors are similar to those involved in the perception of language such as context and expectancy. The specific effects of these factors often are similar.

It is useful to remember that whereas the development of new technology will change the artifacts of sound production, the communication functions are less flexible since they are limited by what the person can interpret. An expanded frequency range can be produced by the technology but will have little utility if it exceeds human capability. The challenge for sonification design is to capitalize on the extensive understanding of sound production and perception in the fields of language, music, and acoustics to invent new sounds that communicate from abstract sources and domains to a listener who has complex but fixed receiving and interpretation capability.

ACKNOWLEDGMENTS

The experimental studies reported here were conducted while the author was at George Mason University and were supported by the Office of Naval Research, Perceptual Science Program. Support for preparing this paper came from the Office of Naval Research.

REFERENCES

1. Baird, J. C., and E. Noma. *Fundamentals of Scaling and Psychophysics.* New York: Wiley, 1978.

2. Ballas, J. A. "Common Factors in the Identification of an Assortment of Brief Everyday Sounds." *J. Exper. Psych.: Human Percep. & Perform.* **19** (1993): 250–267.
3. Ballas, J. A., and T. Mullin. "Effects of Context on the Identification of Everyday Sounds." *Human Performance* **4** (1991): 199–219.
4. Begault, D. R., and E. M. Wenzel. "Techniques and Applications for Binaural Sound Manipulation in Human-Machine Interfaces." Technical Memorandum 102279, NASA, 1990.
5. Brooks, J. E.. "Judgments of Category Frequency." *Amer. J. Psych.* **98** (1985): 363–372.
6. Burrows, A. A. "Choice Response in Quiet and Loaded Channels with Verbal and Non-Verbal Auditory Stimuli." *Human Factors* **4** (1962):187–192.
7. Cofer, C. N. "Properties of Verbal Materials and Verbal Learning." In *Woodworth & Schlosberg's Experimental Psychology*, edited by J. W. Kling and L. A. Riggs, 847–904. New York: Holt, Rinehart and Winston, 1971.
8. Doll, T. J., and D. J. Folds. "Auditory Signals in Military Aircraft: Ergonomic Principles Versus Practice." *Appl. Ergonomics* **17** (1986): 257–264.
9. Edworthy, J., S. Loxley, and I. Dennis. "Improving Auditory Warning Design: Relationship Between Warning Sound Parameters and Perceived Urgency." *Human Factors* **33** (1991): 205–232.
10. Fitch, W. T., and G. Kramer. "Sonifying the Body Electric: Superiority of Auditory Over A Visual Display in a Complex, Multivariate System." This volume.
11. Fowler, C. A. "Auditory Perception is Not Special: We See the World, We Feel the World, We Hear the World." *J. Acoust. Soc. Amer.* **89** (1991): 2910–2915.
12. Gaver, W. W. "The Sonicfinder, an Interface that Uses Auditory Icons." *Human-Comp. Inter.* **4** (1988): 67–94.
13. Gaver, W. W. "Auditory Icons: Using Sound in Computer Interfaces." *Human-Comp. Inter.* **2** (1986): 167–177.
14. Handel, S. *Listening*. Cambridge, MA: MIT press, 1989.
15. Kneale, W. C. "What Can We See?" In *Observation and Interpretation: A Symposium of Philosophers and Physicists*, edited by S. Korner, 151–159. New York: Academic Press, 1957.
16. Mervis, C. B., J. Catlin, and E. Rosch. "Relationships Among Goodness-of-Example, Category Norms, and Word Frequency." *Bull. Psychonomic Soc.* **7** (1976): 283–284.
17. Patterson, R. D., and R. Milroy. "Auditory Warnings on Civil Aircraft: The Learning and Retention of Warnings." Technical Report, MRC Applied Psychology Unit, Civil Aviation Authority Contract 7D/S/0142, 1980.

18. Perrott, D. R., T. Sadralodabai, K. Saberi, and T. Z. Strybel. "Aurally Aided Visual Search in the Central Visual Field: Effects of Visual Load and Visual Enhancement of the Target." *Human Factors* **33** (1991): 389–400.
19. Sexton, G. A. "Cockpit-Crew Systems Design and Integration." In *Human Factors in Aviation*, edited by E. L. Wiener and D. C. Nagel, 495–526. San Diego: Academic Press, 1988.
20. Sorkin, R. D. "Design of Auditory and Tactile Displays." In *Handbook of Human Factors*, edited by G. Salvendy, 549–576. New York: Wiley & Sons, 1987.
21. Truax, B. *Acoustic Communication.* Norwood, NJ: Ablex, 1984.
22. Vanderveer, N. J. "Ecological Acoustics: Human Perception of Environmental Sounds." *Dissertation Abstracts International* **40** (1979): 4543B. (University Microfilms No. 8004002).
23. Veitengruber, J. E. "Design Criteria for Aircraft Warning, Caution, and Advisory Alerting Systems." *J. Aircraft* **15** (1978): 574–581.
24. Walker. R. "The Effects of Culture, Environment, Age, and Musical Training on Choices of Visual Metaphors for Sound." *Percep. & Psychophys.* **42** (1987): 491–502.

Sheila M. Williams
Departments of Computer Science and Psychology, University of Sheffield, Regent Court, 211 Portobello Street, Sheffield, S1 4DP, UK;
e-mail: s.williams@dcs.shef.ac.uk

Perceptual Principles in Sound Grouping

Essential to the development of a theory of Auditory Display is a thorough understanding of the perception of complex sounds. An introduction to Auditory Grouping principles is presented here, explaining the difference between analytic and synthetic listening and introducing examples of different "gestalt" processes as they apply in the auditory mode. Sound examples are provided to demonstrate each of these examples. This is followed by a brief overview of methods of investigating sound grouping and an outline of modeling methods that have been applied. The STREAMER computational model of perceptual grouping is introduced, together with a brief explanation of some of the experimental methods employed in the acquisition of data to support the development of the model. The chapter concludes with a consideration of the implications of sound grouping for the purposes of Auditory Display.

1. INTRODUCTION

A fundamental principle of Auditory Display is that the sounds and sound changes that represent the information in the display system must be capable of being detected by the listener. Further, it is essential that particular configurations of sound parameters should convey consistent percepts to the user or, in the case of a multiuser system, to a wide range of listeners.

Auditory grouping is the perceptual process by which the listener separates out the information from an acoustic signal into individual meaningful sounds. So, it includes all the sequential and cross-spectral processes which operate to assign relevant components of the signal to perceptual objects denoted "auditory streams." Among these processes will be some that exclude part of the signal from a particular stream as well as the ones that help to bind each stream together. Each auditory object may itself have numerous attributes which may be independently described; pitch, duration, loudness and rate of vibrato are a few of these.

Certain characteristics of acoustic sources are correlated with distinct attributes of the auditory objects that result from them. A rise in the frequency of a pure-tone oscillator will normally be heard as a rise in the pitch of the resultant auditory object associated with it. Other aspects may be completely disguised or not so easily interpreted. For instance, the introduction of a new acoustic element that is harmonically related to many existing components arising from the same source may be imperceptible.

Thus, each stream may be composed of information derived from different parts of the frequency range at any given point in time and may extend over time. It could, for instance, be the percept of a voice or the sound of a series of footsteps on a hard surface. A stream is a psychological organisation with perceptual attributes that are not just the sum of the percepts of its components but are dependent upon the configuration of that stream. That is, a component such as a pure tone integrated into a simple stream may present a pitch and timbre directly related to its frequency value whereas, when integrated into a more complex stream, with other co-occurring tones, it may only add to the perceptual richness of the synchronous tone group percept. A simpler example is the contrast in loudness between a watch alarm when it sounds in a crowded train and the same alarm in the relative silence of an examination room.

Similarly, the components that normally form an independent stream can, in a different context, form part of another stream. So, the voice that is heard independently speaking within a crowded room can be completely integrated into a chorus of voices singing in unison.

It is important to distinguish between the physical sources that contributed to the formation of the acoustic signal and these perceptual streams which constitute part of the interpretation of it. For ecological purposes, the ideal situation occurs when each source is associated with a stream that represents all of the relevant information for the listener to identify the source accurately.

By analogy with the processes involved in visual perception, the acoustic pressure wave carrying the combined evidence from all of the sound sources present has been called the "Auditory Scene" and the process of decoding it, which occurs in auditory perception, is denoted Auditory Scene Analysis.[8,9]

Because a knowledge of the potential perceptual streams that may arise from a particular acoustic signal is essential in order to predict the possible interpretations for that signal, a considerable amount of research has been focused on auditory grouping. Such research encompasses a range of approaches varying from empirical studies directed at isolating primitive grouping factors to the study of listeners' responses to complex musical compositions. The former permits accurate formulations of the sensory and/or perceptual consequences of adjusting individual parameters of simple sounds whereas the latter may require complex hierarchical descriptions of the resulting percept (see Kramer in this volume) and may involve measures of the listeners' preferences[39] or emotive reactions (see Sloboda[48]). Other research investigates potential mechanisms whereby such grouping may be performed by the auditory systems of humans and animals. Recently, this research has led to the development of computational models to replicate such grouping effects.

An outline of some of this research that is likely to be important for predicting the outcome of mapping datasets to sound is included here, together with a slightly more detailed account of one attempt to devise a computational model of auditory grouping based on psychoacoustic studies of auditory streaming.

2. AUDITORY GROUPING

The term "auditory grouping" encompasses all of the sequential and cross-spectral processes that operate to assign relevant components of the signal into a single perceptual object, denoted as an auditory stream. This includes processes that exclude parts of the signal as well as those that include or even enhance other components.

Bregman[8,9] describes the concept of auditory streams as the assignment of sound to one or more sources depending on grouping processes, analogous to the Gestalt grouping principles identified in visual perception.

More formal definitions of these concepts include the following:

"A stream may be defined as a sequence of auditory events whose elements are related perceptually to one another, the stream being segregated perceptually from other co-occurring auditory events."[3]

"A source is a physical event ... A stream, on the other hand, is a psychological organization whose function is to mentally represent the acoustic activity of a single source over time."

Also:

"Timbre seems to be a perceptual description of a stream, not of an acoustic waveform."[5]

Auditory grouping effects may be sequential or cross-spectral. Many grouping factors operate concurrently in a conservative evidence-accumulating process over relatively long time periods.[6] For instance, sequential groupings of successive tones may depend on the proximity in time and frequency of the tones whereas spectral integration, of tones occurring at the same time, into complex timbres may be due to common onset/offset times of the component tones.[18]

2.1 ANALYTIC VS. SYNTHETIC LISTENING

Streaming depends on individual differences[21] and on focus of attention.[43] **Synthetic** perception takes place when the information presented is interpreted as generally as possible; for example, hearing a room full of voices or listening to the overall effect of a piece of music. **Analytic** perception takes place when the information is used to identify the components of the scene to finer levels; for instance, listening to a particular utterance in the crowded room or tracking one instrument in an orchestral piece or identifying the components of a particular musical chord.

The interpretation of environmental sounds generally involves combinations of analytic and synthetic listening. For example, hearing the message of a particular speaker requires the synthesis of many components arising from a single voice source, in order to separate the voice from other background noises. This takes place concurrently with the analysis of those components with respect to each other, in order to distinguish the contrasts that identify the phonemic structure of the utterance. The whole process is mediated by other levels of linguistic knowledge so that actual but implausible sounds are "repaired" to construct a meaningful interpretation, without the listener necessarily being aware of their occurrence.[61]

The achievement of an appropriate balance between analysis and synthesis appears to be dependent on training. Most people with unimpaired hearing are adept at tracking one voice through noise but the extent to which they can identify components of a musical chord may be related to the extent of their musical experience.

Even abstract sound is not exempt from this interaction between analysis and synthesis. Terhardt, Stoll, and Seewann[59] found that pitch perception of complex sounds depends upon competition between the effects of spectral pitch, derived from the components that have been "heard out" through analytic listening, and virtual pitch, derived from a holistic perception (synthetic listening). They noted that the balance between these factors depends on at least three variables: the listener's inherent preference for holistic or analytic listening; the listener's state of attention and consciousness; and the functional and acoustic context in which the stimulus is presented.

Thus, the way sounds are perceived may vary for every individual in a room and the acoustic signals will themselves differ in different parts of the room due to interactions with the fabric of the room such as reverberation and absorption.

2.2 AUDITORY GESTALTS

Gestalt psychology proposes that the ability to perceive form is inborn. It emphasises that the *whole is more than the sum of its parts*. The relationships between components and other attributes are significant properties of the object perceived and recognition of these attributes is an important part of perception.

Gestalt psychology is the study of the human propensity to recognise patterns and configurations which appear in the environment. "Gestalt," in German, means an independent entity which has a definite shape or form. "Gestalten" are familiar objects as they are perceived in the mind. The form of the object is independent of its substance. This implies that features have independent entity from the wholes which they are part of, even if they can have no independent existence.

Gestalt theory began in the study of perceptual wholes, especially in visual figures, but its concepts have been extended to apply to other psychological phenomena including learning and personality.

Gestalt psychology is based on the concept that all behaviour takes place in an environmental, or psychophysical, field in which interacting forces compete until a stable state is reached, as in physical fields such as electromagnetic fields. The field is subject to figure ground effects in which an organised pattern becomes a "figure" that is separated from the ground by a contour or boundary. The figure stands out from the ground and tries to maintain its structure following certain principles.

The figure-ground relationship reflects the fact that environmental objects exist in space and time. Changes to one parameter of the sound emanating from a single object (or of its size, shape, or mass) necessarily have repercussions for other parameters of the same figure and may also modify the ground. Thus, gestalt perceptual psychology investigates the relationship between the composite structure of the percept and the inherently complex environmental source from which it arises. The figure-ground relationship is exploited in Auditory Display when multiple parameters are mapped onto a single auditory stream. This enables the listener to interpret several elements of the data with respect to each other. Apart from reducing the perceptual processing load on the listener, this technique is especially beneficial when it is the interactions between the data elements rather than simple contrasts between them which provide the significant information (see Bly, Fitch and Kramer, Kramer, and Scaletti, all in this volume).

The fundamental principles of gestalt perception are "pragnanz," a term introduced by Wertheimer, and "stability."[33]

Pragnanz states that a figure always becomes as regular, symmetrical, simple, and stable as prevailing conditions permit. The following processes, which promote pragnanz, have been identified in the context of perception, mostly for the purposes of vision research and the explanation of visual illusions[33,42]:

- **Similarity**, components which share the same attributes are perceived as related.
- **Proximity**, the closer two components are, the more likely they are to belong together.
- **Good continuation**, components that display smooth transitions from one state to another are perceived as related.
- **Habit** or **Familiarity**, recognition of well-known configurations among possible subcomponents leads to these subcomponents being grouped together.
- **Belongingness**, a component can form part of only one object at a time and its percept is relative to the rest of the figure-ground organisation to which it belongs. Conflicting relationships with other components create tensions in the field which must be resolved in order for a stable state to be achieved.
- **Common fate**, components which experience the same kinds of changes at the same time are perceived as related.
- **Closure**, incomplete figures tend to be completed. An incomplete figure creates tension in the system which needs to be resolved. Discrete parts can be grouped to form part of a continuous possible object in the presence of evidence of occlusion and/or the absence of evidence to contrary.

Stability, which is also known as **set**,[42] predicts that having achieved one interpretation, that interpretation will remain fixed throughout slowly changing parameters until the original interpretation is no longer appropriate. The better the "pragnanz" or "goodness" of the object, the greater the stability that it displays under subsequent changes.

"**Articulation**,"[33] the separation of the figure from ground, requires energy; for example, lack of attention in a stuffy lecture theatre can lead to auditory and visual fields becoming more homogeneous, with little or no separation of figure from ground. The speaker's voice, for instance, begins to blend in with background noises until its meaning can no longer be interpreted. This is an example of the interactions between the principles of maximum and minimum simplicity.

A minimum simplicity is the simplicity of uniformity. This requires little energy to perceive and the perception reflects an averaging of the qualities of the perceived environment spread over the whole of its region. It corresponds to synthetic or holistic listening. By contrast, a maximum simplicity is that of perfect articulation and occurs when a high degree of energy is available. The scene is separated into the lowest level components, or figures, which it contains and each figure obeys principles of good shape and continuation. This corresponds to analytic attention.

Thus, attention can be seen as the directing of energy into the grouping processes. The energy may be focused in response to a stimulus in the environment or be consciously controlled. For instance, by making an effort, we can listen to a particular voice among many in a crowded room. Conversely, overhearing a word or two relating to a topic of particular interest to us may be sufficient to cause a conversation, which had previously been masked by the louder voices accompanying it, to suddenly stand out from the background noise. This is known as the "cocktail party effect."[14]

Attention plays an important part in perception. Broadbent modeled attention as a filter in which one channel is selected, by analogy with mechanical communication systems used in engineering. In this model, the concept of a channel not only encompasses the independent sensory pathways from the sense organs to the brain, but is defined operationally to be that which can carry any class of sensory messages which can be selectively ignored or attended to. Thus, the channel can be rejected on the basis of any features identified at the first stage of processing.[10]

The focus of energy can be divided between more than one channel at a time, not necessarily equally. For instance, while processing or interpreting one channel, other channels may be monitored for important signals.[60]

The deliberate directing, or focusing, of attention can lead to alternative interpretations of the same signal. This focusing is known as that attentional "set."[42,43] By concentrating on listening to a whole set of tones or to one or another of its possible subsets, perceptual fusion into a single sound figure or fission into two or more separate strands of sound can be induced.

The perception of a particular component can vary depending on whether it is seen as part of the figure or part of the ground. For example, in the Wertheimer-Benary effect, a small triangle, of fixed intensity intermediate between figure and ground hues, appears brighter when it is seen as belonging to the dark, cross-shaped figure than it does when perceived as part of the light ground.[32]

Moreover, the principles of gestalt suggest that it is not merely the relationships between the components which are recognised, but that once sufficient evidence is accumulated, the higher level or more complex object, or figure, defined by the relationships, replaces the components in the perception. Gestalt factors thus describe the ways we eliminate redundant information and code information more efficiently.[42] Evidence of the difficulty in switching between alternative perceptions of visually ambiguous figures once an interpretation has been achieved supports this hypothesis, for example, in the case of the Ames' room.[27] Once one interpretation of the scene has been achieved, it remains even if it is illogical, until either the relationships of the scene components have changed considerably or intense concentration of attention on other aspects of the scene shift the focus. Having achieved more than one interpretation of the same visual scene, it is then often possible to shift the focus between the interpretations, at will, e.g., in the Mach-Eden effect[42] or the Necker cube illusion.[27]

Psychoacoustic research provides evidence for the existence of many primitive grouping principles. Some, presented here, exemplify the concepts identified in vision and other gestalt research and link these to auditory grouping processes.

SIMILARITY. Components which share the same attributes are perceived as related.

For example, the decomposition of two fused sinusoidal glides into separate streams, by the inclusion of an alternating captor glide, is dependent on the similarity of both frequency and glide direction between the captor and the target glide.[49]

Some examples of primitive auditory grouping concepts which demonstrate the operation of the principle of **similarity** are: common onset, common offset, common frequency, common frequency modulation, and common amplitude modulation.

Tr 1,2   The first pair of examples (1.1 and 1.2) provided demonstrate the effects of similarity in loudness on the perceived grouping of tones. While the tones are all of the same frequency and of equal duration, in example 1.2 a distinct galloping rhythm is usually heard, as the beat of the more intense tones stands out from that of the softer tones. The separation of the single sequence of tones into two sets or streams, grouped on similarity of intensity, causes the dominant distinguishing percept to be the uneven rhythm of the more intense stream rather than the contrast in intensity directly presented to the listener.

By associating the trace of each processing unit with a particular pitch, Jackson and Francioni (in this volume) were able to model the behaviour of a multiprocessor computer system in the auditory mode. Specific events were represented by different timbres by mapping each event onto a different synthesised musical instrument voice. The similarity between the timbres caused the different pitch streams to blend, perceptually producing a much louder sound when the same event occurred on different processors at the same time. At a lower level, the consistent mapping of the same aspect of each data element over time to a single sound parameter (see Madhyastha and Reed in this volume) is an example of the use of the similarity principle to enable the progress of the data attribute to be tracked through the sound interface.

PROXIMITY. Components that are close to each other are more likely to be grouped together.

For example, frequency proximity appears to dominate over trajectory alignment in determining the perceived continuity where tone/glide conditions vary across a noise burst.[15]

Also, in a continuous cycle of alternating pure tone and two-tone simple chords, greater decomposition into a fast pure tone stream accompanied by a slower pure tone stream, was found for all the conditions of captor-target relationships, when the alternating sequence of captor and glide pair was presented more rapidly.[6]

Examples of primitive auditory grouping concepts that demonstrate the operation of the principle of **proximity** are: temporal proximity and frequency proximity.

Proximity poses extra problems in audition compared with vision analysis. At fast presentation rates, temporal proximity appears to be inversely related to frequency proximity when calculating the overall affinity between two distinct sounds. Further, in vision, within the two-dimensional or three-dimensional physical plane that defines the source object there is a common ratio scale along every axis, so distances along one axis can be compared directly with those along another axis and shortest paths throughout the object space can be calculated. In order to discuss relative distance within the frequency * time * amplitude plane, the ratio between their scales needs to be established, whether this is constant across the plane or varies with respect to local frequency.

In examples 2.1 and 2.2, the intensity and duration of each tone in the sequence is the same. The frequency relationships between the first and second and between the third and fourth tones are equivalent on a log scale. However, in the second example, 2.2, the frequency relationship between

Tr 3,4

the second and third tones is relatively greater, leading to a perception of two separate streams, a high alternating two-tone pattern and a similar but lower alternating two-tone pattern in contrast to the four-tone repeating pattern usually heard from example 2.1. Again, the most prominent distinguishing percept between the two examples is the uneven rhythm of the two streams in example 2.2.

Grouping by frequency proximity is also sensitive to the rate of presentation of the tones, resulting in apparent trade-off effects between the rate of presentation and the frequency separation. Thus, the greater the frequency difference between two groups of alternating tones, the slower the rate of presentation needs to be, to avoid perception as two independent sound sources.[44]

GOOD CONTINUATION. Components that display smooth transitions from one state to another are perceived as related.

For example, two tones separated by a noise burst are more likely to be heard as continuous if the start frequency of the second tone matches the end frequency of the first tone.[16]

Examples of primitive auditory grouping concepts that demonstrate the operation of the principle of **good continuation** are: proximity in time of offset of one component with onset of another, frequency proximity of consecutive components, constant glide trajectory of consecutive components, and smooth transitions from one state to another state for the same parameter.

Tr 5,6

The steady state portion of the tones corresponds precisely in examples 3.1 and 3.2. Only the gliding transitions in example 3.1 and the onsets and offsets in 3.2 are contrasted. However, the perceptual contrast tends to be between an alternating two-tone pattern and a single continuous note that fluctuates in frequency.

The principle of continuity is exploited in Jackson and Francioni's work (in this volume) on representing parallel processes. The note of the sending processor may be sustained until the message has been received by another processor, enabling the duration of the communication processed to be mapped auditorily.

HABIT OR FAMILIARITY. Habit, which is also known as Familiarity, refers to prior expectations about sound groupings which have been acquired through previous experience. It operates to ensure that relationships between sounds that have been attributed a particular meaning in the past will preferably be assigned the same meaning when they occur again.

Components, or relationships between components, that have attributes that can be categorised easily, due to previous experience with similar sounds, are more likely to be grouped with other items of the same category. They are also perceived or interpreted within the context of that category.

For example, while researching the effects of spectral fusion in music, McAdams[38] found that, by applying different types of modulation, subsets of a large tone mass that displays no apparent harmonic structure could be perceived as singing male voices. Both harmonicity of the frequencies of synchronous tones and relative familiarity of the spectral envelope seem to provide cues to perceptual fusion of spectral entities into independent auditory objects.[19,38]

Examples of primitive auditory grouping concepts which demonstrate the operation of the principle of **habit** are: good harmonic ratio and amplitude ratio inversely related to harmonic ratio. However, more complex sound structure relationships, known as schemas,[9] whose particular significance has been learned, are of special importance here. Schemas are stored knowledge of familiar patterns which may be used "top-down" to assist in decoding the signal. These include the complex frequency structures of music, such as the diatonic scale and particular spectral envelope characteristics, e.g., speech (either the complex frequency and amplitude relationships typical of a human voice or the specific perceptual attributes of a familiar voice). Once activated by a partial match of the perceived sound attributes, the schema focuses the matching process on the gathering of sufficient evidence to confirm the occurrence of the event (or type of event) that it represents. Activation of a particular schema will depend upon both its level of familiarity and the closeness with which it matches the new auditory evidence. Thus, an acoustic signal that partially matches a very familiar pattern may be recognised as the more familiar event in preference to a more unusual event whose pattern it more closely matches.

Examples 4.1 and 4.2 were created by the computerised signal-processing system in exactly the same way as the simpler examples demonstrating primitive grouping concepts but serve to demonstrate that once sufficient cues are present to invoke music or speech schemas, interpretation becomes a matter of recognising the meaning carried by the signal, not of diagnosing its perceptual attributes. This reflects Matlin's **cognitive** level[37] rather than the **perceptual** level. It is of importance for Auditory Display techniques as perception mediated by schemas is often categorical in nature and small deviations in signal parameters, even though easily detectable at a sensory level, may no longer be assigned any significance or recorded. That is, the signal processing by the listener seems to map directly to cognitive descriptions (what the sound waves mean) without much attention being paid to the perceptual level (what they actually sound like).

Tr 7,8

Familiarity may be exploited in Auditory Display by mapping data to sounds commonly associated with them (see Cohen, Fitch and Kramer, and Gaver, all in this volume; see also Gaver[25]) or to sounds with appropriate affective responses (see Kramer, and Madyastha and Reed, both in this volume) but it can also cause major problems due to unexpected dissonances (see Jackson and Francioni in this volume) or to the disguising of relevant events. Certain low-level percepts such as increases in pitch and loudness tend to be naturally associated with increases in quantity which may or may not be appropriate in the data interpretation.

BELONGINGNESS. A component can normally form part of only one disjunctive object at a time and its percept is relative to the rest of the figure to which it belongs.

For example, segregation of localization information affects the perceived quality of the remaining sounds.[7] The principle of disjoint allocation applies so that a sound allocated to one stream tends to be excluded from all other possible streams.[4]

Tr 9,10

In example 5.1, the repeated three-tone complex is usually heard as a single stream of tones, having a relatively rich timbre. The introduction of an intervening low captor tone in example 5.2 tends to re-group the pattern into two streams; one of pure low tones at a more rapid tempo and a higher pitched stream at the original tempo. This raising in pitch of the slower stream demonstrates that the tones captured into the faster stream no longer contribute to the pitch percept and richness of the slow pattern even though all the tones present in example 5.1 are also present in example 5.2.

Belongingness causes problems for Auditory Display when the percept related to a particular data attribute is hidden or distorted by its context or when data items that must be contrasted are attributed to different perceptual objects or streams.

COMMON FATE. Components that undergo the same kind of changes at the same time are perceived as related.

For example, comodulation acts to tie together the noise component of a complex sound formed by adding sufficient noise to a tone to mask it, releasing the tone as a separate stream. This is denoted "comodulation masking release."[28]

Examples of primitive auditory grouping concepts that demonstrate the operation of the principle of **common fate** are: common onset, common offset, common frequency modulation, and common amplitude modulation. Modulation may operate at both macro and micro levels. That is, correlated changes over relatively extended periods of time, such as frequency glides, and common variations on a very small time scale, such as the application of vibrato, may both act to group components together that would otherwise be perceived independently or to subdivide components that would otherwise all group together into a single stream.

Examples 6.1 and 6.2 contrast two patterns, in each of which three tones are presented simultaneously. The first example is generally perceived as a single stream of rich tone glides while the second example demonstrates the segregation of a stream of tones that have the same centre frequency as the middle tone glides presented in the first example but a different glide direction from the accompanying concurrent stream of higher and lower gliding tones. The perception of the second example as two rather than three separate streams is confirmed by estimating the relative pitches of the two streams. If the steady-state tones appear higher in pitch than the tone glides, then both upper and lower tones are being perceived as a glide complex for which the pitch percept is primarily dependent on the lower tone. This is due to their close harmonic relationship (the upper tone frequency is a simple multiple of that of the lower tone).

Common fate is responsible for causing sounds that start or stop in unison to be grouped together.

Tr 11,12

CLOSURE. Incomplete forms tend to be completed. This is analogous to the phenomenon of visual occlusion in which discrete fragments may be observed as components of a single object which has been partially obscured by a nearer object or objects. However, as sound is transparent (components from multiple sources are convolved into complex acoustic waves rather than hiding each other), evidence for occlusion must be local. That is, there must be sufficient energy at the right frequencies to stimulate the same parts of the auditory system as the missing components. Thus, the fragments of sound that form a perceptual auditory entity may be scattered across time and the spectrum, providing that sufficient relevant evidence for occlusion is present.[17]

Examples 7.1, 7.2, and 7.3 demonstrate the conditions under which a steady-state tone alternating with a noise burst may be heard as continuing throughout the noise. Bregman[9] has identified four rules that govern the sufficient and necessary conditions for this illusion to occur:

Tr 13,14

i. No discontinuity in A: there should be no evidence that the tone has actually stopped or started within or at the edges of the noise burst.

ii. Sufficiency of evidence: the stimulation of the peripheral units of the auditory system during the noise burst must include that which would have been provided by a continuation of the tone.

iii. A1-A2 grouping: the tones separated by the noise burst must share sufficient attributes to be grouped into a single stream of tones, if the noise burst was not present.

iv. A is not B: there should be no evidence of a transition from tone to noise and back again.

Tr 15   In the sound examples, 7.1 presents the repeating pure tone alternating with a short period of silence. The onsets and offsets are abrupt but the tones generally group together into a single stream. This establishes that rule iii applies. Example 7.2 presents the same tones with the gaps filled by random noise of greater amplitude (sufficient to conform to rule ii). The pure tone can usually be heard to continue throughout the noise bursts, although this effect is cumulative so may be heard more clearly if the example is played several times. The third example presented here, 7.3, demonstrates the effect of varying the amplitude of both tone and noise bursts at their onsets and offsets. This may cause the illusion to fail and the sound pattern to be heard as either alternating tone and noise or as a single stream of sound that is transformed over time from pure tone to noise and back. Example 7.3 violates both rules i and iv above.

This illusion is also subject to attention effects, analytic listening being sufficient to induce a perception of alternating tones and noise bursts, even in example 7.2.

Closure can be a major problem for Auditory Display as it induces the listener to hear the percept as more regular and complete than it really is, thus disguising subtle variations in the data. This becomes more pronounced if the data is mapped to an acoustic signal which would normally be perceived categorically. Pitches close to musical tones may be assigned to pitch classes, phoneme-like sounds may be interpreted linguistically, hiding variations between similar sounds.

STABILITY. Having achieved one interpretation of an acoustic signal, that interpretation will remain fixed through slowly changing parameters until the original interpretation is no longer appropriate. The sound is perceived or interpreted in the context of what preceded it, not just its current form.

For example, "effect propagation" appears to apply, "forces of grouping" interacting iteratively to achieve groupings with least violation of grouping rules.[8]

The figure-ground effect has also been demonstrated in auditory perception:

> Artists and composers exploit the relationships between spectral fusion and auditory streaming in music in order to provoke conscious transformations of perception by directing attentional focus to different levels of form and structure.[38]

However, auditory and visual modes differ greatly in their dependence on temporal aspects and it would be a mistake to assume that the perceptual processes that apply are directly analogous.

Tr 16,17   In examples 8.1 and 8.2, the effects of stability on auditory grouping are demonstrated by alternating tone patterns in which the intertone intervals are progressively decreased (example 8.1) or increased (example 8.2) throughout the pattern. At the slowest speed of presentation, the tone may be perceived as a single stream where high and low tones alternate at equal temporal intervals. At the fastest rate, the tones separate into a high stream in which the same tone repeats at regular intervals, and a similar low stream. It should be noted that the actual intervals between the tones correspond inversely for pattern 8.1 and 8.2. That is, there are exactly the same number of intervals of each size in each example but the examples are contrasted by reversing the order of the intervals. Perceptual effects whereby either the single stream or the two-stream percept persists for longer in one pattern are solely due to the attentional set of the listener.

The cumulative nature of auditory streaming is also demonstrated here. The auditory system seems to be biased towards interpreting the whole acoustic signal as derived from a single source; the more similar the sound elements are, the more evidence seems to be required before the existence of multiple streams is established.[6] In example 8.2 which begins with the fastest rate of presentation, the first few tones may be perceived as coherent before the pattern splits into two streams. This is due to the time taken to accumulate sufficient evidence for the separate high and low streams.

Stability may cause problems for auditory monitoring. Frequency or intensity changing gradually may exceed prescribed data limits by a considerable amount before the listener detects a change of state. If such state changes are important, such as safety limits, it may be necessary to monitor them automatically at the interface, deliberately introducing a new "warning" element into the Auditory Display as soon as the limit is exceeded.

In practice, it may be difficult to ascribe particular grouping processes to the specific gestalt principles listed here. Many of the factors identified in auditory grouping appear under more than one of the above categories. Models of timbre space[39] and tonal hierarchies,[34] for instance, map *similarity* into a spatial domain in which the operator appears to be that of *proximity*.

3. TECHNIQUES FOR INVESTIGATING AUDITORY PERCEPTION

In order to discuss perceptual processes, we need to agree about what we mean by perception, as opposed to say, sensation, as these terms are frequently used interchangeably in the literature. Throughout this chapter, I adopt Matlin's[37] distinctions: **sensation** refers to immediate and basic experiences generated by isolated, simple stimuli; **perception** involves the interpretation of those sensations, giving them meaning and organisation;

and **cognition** involves the acquisition, storage, retrieval, and use of knowledge.

Different approaches to perception have focused on each of these aspects of the perceptual process. For example, early attempts to design machine perception systems began from the structuralist viewpoint that if you could capture all of the responses to the separate components that were presented to the sensors, this would result in a complete percept of the scene. However, a description of a speech signal in terms of its component frequency/intensity characteristics fails to address the perceptual problem of the communicative purpose of speech. It is also extremely difficult for a human listener to achieve, requiring intensive analytic attention.

3.1 TECHNIQUES FOR EVALUATING SENSORY PROCESSES

The view that the perception of an object was simply the sum of all the sensations received from interacting with it in some way was put forward by the British School of Psychology in the nineteenth century. It led to a focus of research on the development of techniques for evaluating sensations.

The only way sensations can be investigated directly is by introspection, the self-observance and description of the stimulus event by the person who has been subjected to it. Comparison between the responses of several subjects to the same stimulus is problematic because of the difficulty of distinguishing between the variations in the perception of the stimulus and the variations in the way subjects report their perceptions.[42]

The search for more principled methods of studying sensations led to the development of Sensory Psychology which is concerned with identifying measurable relationships between physical stimuli and behavioural responses and with devising techniques for accurate measurement of them.

One of the earliest and most common introspective techniques used in psychophysical experimentation is that of measuring a threshold, or "limen." A limen is not a real threshold in the sense that values below it are not perceived and those above it are, but an average value of the stimulus at which it will be perceived a certain percentage of the time.[42] Values of the stimulus that fall below the limen are referred to as subliminal and those above it as supraliminal. Some stimuli presented at subliminal values may be interpreted without being consciously perceived. The limen thus represents a hypothetical milepost on the curve of statistical likelihood that a given intensity of stimulus will be perceived. An absolute threshold is usually defined as the intensity at which a particular stimulus will be recognized as being present 50% of the time. Similarly, a differential threshold, or difference limen, is the amount of variation required in a given aspect of a stimulus before a difference is perceived in 50% of cases. The difference limen is also known as the "just noticeable difference" (jnd).

Early in the nineteenth century, Ernst Heinrich Weber discovered that the ratio of the jnd to the original stimulus is constant within each mode of perception. This rule was later formalised as Weber's Law by Gustav Fechner, who applied mathematical principles to formulate the relationship as a logarithmic expression.[47] Although the rule has been since been found to be inaccurate, it is sufficiently precise providing that extremes of ranges are not considered, and so is still commonly applied.

Thresholds were originally found by the method of adjustment. This involves asking the subject to adjust the stimulus to a level where it (or the difference between it and a reference signal) can just be heard. This method is subject to personality, training, and instructional biases. For instance, the threshold is higher or lower according to whether subject is directed to respond whenever the stimulus is suspected or only when certain that the stimulus has occurred. A more recent method, based on Signal Detection theory, uses constant stimuli and allows for the contribution made by response criteria to be calculated independently.[55]

By asking subjects to estimate the intensity of various stimuli relative to given standards, Stevens[52] established that people are remarkably consistent in their value judgements about sensations. He hypothesised that the relationship between stimulus and sensation conforms to a power law (i.e. $\mathbf{S} = \mathbf{kI}^n$, where $\mathbf{S}$ is sensation, $\mathbf{k}$ is a constant related to the standard, $\mathbf{I}$ is intensity of stimulus, and $\mathbf{n}$ is constant for any particular attribute of the stimulus).

So, the principle aim of psychophysics is to establish the correlation between physical stimuli and physiological and behavioural responses. Traditionally this was investigated through comparison of stimulus data with the introspective reporting of perceived sensations, although, recently some investigators have preferred to concentrate on neurophysiological measurement techniques. Direct measurement of the neural firing rate corresponding to the intensity of a given stimulus allows low-level perception mechanisms to be studied without influence from the subjects' interpretation of the event. This provides an accurate means of mapping the relationship between a stimulus and the physiological response to it. However, these methods are intrusive, involving the use of fine electrodes implanted in neurons, and so cannot be applied to human subjects.

Recent research into the operation of gestalt grouping principles in auditory perception has been mainly concerned with investigating the availability of physical mechanisms to support such processes. Neurophysiological data derived from experimentation with animals, such as cats, are compared with estimates of human absolute and difference limens for the perception of events in which the interaction of competing grouping principles is predicted. The estimates are obtained by applying signal detection methodology. The information derived in this way is used to design models of the human auditory system.

In summary, psychophysical measurements are measurements of indicants of perception, not measurements of the perception itself which can only be described qualitatively, through introspection.[42] However, to develop a qualitative model of audition, we need to understand the relationships between the stimuli and the sensations and percepts that are a consequence of them and must resort to psychoacoustic data, supported by the results of physiological studies, in order to do so.

3.2 TECHNIQUES FOR EVALUATING PERCEPTUAL PROCESSES

Investigations into the perceptual processes of hearing generally focus on the aspects of pitch, timbre, rhythm, and localisation in space percepts arising from the frequency, duration, and intensity attributes of the acoustic signal.

Stimulus comparison methods are used to devise metrics for simple percepts such as the pitch and loudness of pure tones or simple tone groups. This involves specifying a standard tone of a fixed frequency and intensity and relating the perceived pitch and loudness of other tones to it in order to develop a scale. The Mel[50] scale for frequency and the Sone[51] and Phon (equal loudness contours)[26] scales for loudness were derived in this way.

However, metrics devised for simple percepts do not necessarily apply directly in calculating the predicted perception of complex acoustic waveforms. Masking effects occur between frequency components presented simultaneously or close to each other in time and frequency.[12] That is, the perceived prominence of a particular pitch component is dependent on the context in which it is heard, the complete auditory scene.[59]

More complex groups of simultaneous components are distinguished by differences in their holistic percept, described as a texture or timbre. Models of mutlidimensional "timbre space" have been developed for the purposes of musical research. These are perceptual models that represent the basic perceptual attributes of sound texture. The number of different dimensions portrayed and the relative distances along those dimensions are derived from empirical studies using analogy techniques.

Sounds are attributed to spatial locations depending on the disparities between the percepts at each ear (see Wenzel, in this volume).[22,30,62]

Sounds are allocated to perceptual groups, or streams, depending on their perceived attributes rather than as a direct result of the attributes of the acoustic signal, so the resulting percept may depend on attentional factors or previous training or familiarity with similar sounds (see Bregman,[9] and others). The component sounds may be grouped according to many different attributes, such as differences in modulation rates between components, an example of "common fate" applying in the auditory mode.[24]

The percept of a stream also depends on the relationships between the components assigned to it and is not simply an accumulation of the percepts of each component.

3.3 TECHNIQUES FOR EVALUATING COGNITIVE PROCESSES

While investigation into the cognitive aspects of auditory perception is less well developed, attempts have been made to study the effects of higher level structure on auditory stream analysis. For instance, the phonemic restoration effect on speech signals[61] shows that linguistic knowledge is used in the closure of fragmented signals providing there is suitable evidence of occlusion. That is, the listener tends to hear the signal as if it were complete, presumably using a combination of evidence derived from the signal and prior knowledge of phonetic and linguistic constraints. Further, sounds with only minimal speech characteristics tend to get interpreted as speech even when no linguistic meaning can be derived.[53]

Research into musical sounds shows that expectancies can be set up through musical structure (e.g., at a local level, imperfect cadences, broken rhythms) which must be satisfied if the piece is to sound complete. These may be socially acquired.

4. MODELING AUDITORY GROUPING

Theoretical and computational models have been devised to demonstrate sensory, perceptual, and cognitive aspects of auditory grouping. Computational representations formalise the descriptions derived through empirical studies and enable theoretical descriptions to be tested systematically. They may be used predictively to identify acoustic signals that cannot be distinguished perceptually and are potentially capable of deducing the most likely interpretation of any given auditory scene, although this has yet to be achieved.

4.1 MODELS OF THE AUDITORY PERIPHERY

Computational models of the auditory periphery have been available for some years now (e.g., see Carlson and Granstrom[13]) but more recent developments include models of mechanisms that promote auditory grouping in the early stages of auditory processing such as harmonic grouping and common onset/offset times,[11,16] promote musical structuring of the acoustic signal,[46] and mathematical methods such as autocorrelograms that enable dynamic displays of auditory processing to demonstrate coincident effects arising from harmonically related components in the acoustic signal.[23,54]

However, the descriptions that arise from physiology-based studies fail to explain why perfect pitch is such a rare phenomenon, given that frequency appears to map fairly consistently into the peripheral auditory system. For human listeners, pitch intervals seem to be of much greater significance than specific pitch values.

4.2 MODELS OF PERCEPTUAL EFFECTS

In contrast, although theoretical models of perceptual functions of the auditory system have been available for some time,[8,9,43] few computational models seem to have addressed this level. Exceptions are the STREAMER model[63] which is directed towards modeling perceptual effects rather than physiological mechanisms and the Beauvois and Meddis model[2] in which a physiology-based model of the auditory periphery is extended by the incorporation of attentional mechanisms to account for simple streaming effects in two tone data.

Tr 18-20   Examples 9.1, 9.2, and 9.3 present alternating two-tone patterns such as those used by van Noorden[43,44] and Beauvois and Meddis.[2] At relatively fast presentation rates, tones close in frequency value can still be tracked in a single coherent stream (9.1). At wider frequency disparities and the same rate of presentation, two distinct streams are heard (9.2). However, the larger frequency disparity does not prevent a single coherent stream from forming if the rate of presentation of the tones is slower (9.3).

These studies are important for Auditory Display techniques which model continuous data rather than discrete events as they investigate the effects on perception over relatively extended time periods. They contrast with signal-processing approaches to complex sound pressure waves where a "bacon slicer" technique is frequently adopted in which only the instantaneous data collected by sampling the signal as specific time points is considered.

4.3 MODELS OF MUSICAL STRUCTURE

Models of musical structure include gestalt-based models such as Tenney and Polansky's computational model[56] which derives the hierarchical structure of simple pieces of music from an analysis of their frequency and temporal components, and the theoretical model of Terhardt[58] based on secondary pitch contours derived from a sound spectrum that is continuous in both the temporal and frequency dimensions. Deutsch,[20] Parncutt,[45] and Krumhansl[34] have all proposed comprehensive theories of music perception and cognition involving complex pitch-processing strategies while

McAdams[39] has developed a multidimensional model of musical timbre based on perceptual "similarity," aspects of which have been implemented computationally.

However, many of these models are based on instantaneous perceptual descriptions,[59] or restrict themselves to the cognitive level, assuming that the mapping from acoustic signal to pitch classes is achieved elsewhere.[45]

5. STREAMER: MAPPING THE AUDITORY SCENE

Work at Sheffield, currently funded by the British Cognitive Science Initiative, seeks to discover how the components of an acoustic signal are grouped together into perceptual objects that can be interpreted by the hearer. Combining computational modeling with psychoacoustic experimentation, we aim to trace out the interactions between the frequency, time, and amplitude aspects of the sound waves and the percepts to which they lead, with particular reference to those attributes that cause sounds to be grouped together.[64]

Although it is well known that a pure tone captor tends to stream with a potential target tone which is near in frequency, the metric by which such nearness is measured had not hitherto been explored. We had predicted that the metric would be the same as that found in other perceptual studies, such as the Bark[41] or the Mel[50] scale, but this is clearly not the case. Attempts to explore this further have demonstrated that different scales operate when the target tones are presented simultaneously than when they are presented at discrete points in time (no overlap) and that the metric varies across different parts of the frequency range. Although music perception theory[31,57] suggests that a pitch-weighting factor may be involved here, our results do not entirely support Terhardt's[57] model. Further investigation into this effect indicates that temporal factors also play a role in the operation of frequency proximity, different metrics applying at different rates of presentation of the components of the acoustic signal.

5.1 THE COMPUTATIONAL MODEL

The STREAMER prototype[63] is a partially developed tool to be used in the search for principled "level-1" solutions to the problems of auditory perception, analogous to those being developed for visual processing. Level-1 is the computational theory of a process, which is independent of the algorithm and implementation levels, since it is determined solely by the information-processing task to be solved (see Marr,[36] p. 337). That is, the primary aim of the STREAMER model is to represent the functions of human auditory grouping rather than the mechanisms that may underlie it.

Although it is shaped by the results of neurophysiological and other psychoacoustic research, the model is essentially a qualitative framework in which to test interactive grouping principles to determine the nature of these principles and their interactions. Quantitative elements in the early version of the prototype were chosen with regard to the psychoacoustic literature only to set the model within a "natural" acoustic environment and to be able to monitor the effects of changes in the relationships of its parameters. More recently, we have directed our attention towards ascertaining perceptual "metrics" that might be applied.

Following the example of Marr and Nishihara,[35] the models have been based on an object-centred, modular, hierarchical structure. Within this structure, representations describing the location in time, frequency, and magnitude of intensity of contiguous segments of simple sounds are grouped, through the application of grouping principles, to form progressively higher levels of organisation. Evidence from previously reported empirical research is being used to suggest suitable grouping concepts to be applied in the auditory domain.

The first STREAMER prototype represented the grouping processes which operate to form the auditory streams, as explicit conceptual objects, connected by an inheritance lattice derived from the major gestalt principles described earlier.[63] However, it is becoming evident that the gestalts identified as being of primary importance in vision research may not be directly reflected in auditory perception. For example, in the auditory mode, *proximity* operates in different ways in the frequency dimension from its function in the temporal domain and the interaction between the two cannot be ascertained as a simple summation of these effects.[2,43,65] Hence, the latest STREAMER model currently avoids commitment to prespecified representations of the relationships between the auditory grouping processes. It is intended that correspondences between the grouping processes operating in the auditory mode will be identified as more detailed representations of the individual processes emerge. These will then be incorporated in the model at a later date.

The current model provides a framework in which to test out theories about individual grouping processes and the interactions between them, upon data representing the simple auditory scenes which result from the stimuli employed in our experiments. The model can also be applied, in the same form, to more complex representations, allowing us to make predictions as to the effects of such processes on more ecologically realistic sound environments.

Perceptual Principles in Sound Grouping

5.2 THE EXPERIMENTAL PROGRAM

As some of the relationships that pertain in the perceptual grouping of sounds are formed cumulatively over relatively extended periods of time,[6] it is essential that models of continuous data are based upon empirical studies using such data rather than being pieced together from studies of discrete stimulus comparisons.

The empirical research to support the STREAMER model is based upon the streaming methodology by Bregman and Pinker.[5] In order to study the operation of frequency proximity in competitive grouping, we varied the captor in frequency between that of the two target tones and presented the stimuli at such a rate that streaming always occurs for most subjects.[1] From the responses of our human subjects we were able to estimate the point at which the captor stopped streaming with the upper tone and began to stream with lower tone. By determining the midpoint of the frequency proximity metric that applies in streaming in this way, we hoped to be able to select the most appropriate metric to employ in our model.

We used two sorts of stimuli: a sequential pattern, in which each of the tones is presented at a different time, and a synchrony pattern, in which the captor tone was presented alternately with the tone pair B and C synchronous. These are shown in Figure 1.

The streaming midpoint for the synchrony stimulus pattern was significantly less than that for the sequential stimulus pattern. Further, neither of the results was particularly close to the values we had estimated using frequency distance metrics derived elsewhere. In each case our results showed that the streaming midpoint for simple competitive data was substantially lower than log, bark, or mel midpoints.

The repeating three-tone patterns presented in examples 10.1, 10.2, and 10.3 are are similar to the sequential patterns presented to our subjects during the experimental trials and consist of a pattern with a relatively high captor value (10.1), one with a low captor value (10.2), both of which are selected from actual stimuli used in the experiments, and one whose value is around the average midpoint found for our subjects.

Tr 21-23

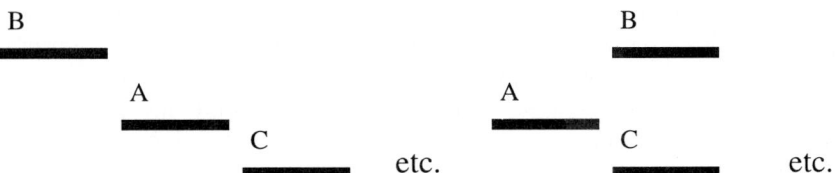

FIGURE 1 Schematic diagram of the tonal patterns used in the experiments: the sequential pattern (left) and the synchrony pattern (right).

More recent research on ERB scales has suggested alternative methods by which the ERB can be used to predict the likelihood of multiple streaming versus temporal coherence and it is possible that these methods could be applied to predict a competitive streaming midpoint. Beauvois and Meddis[2] have produced an auditory grouping model based on the ERB mechanism which accounts for both temporal and frequency proximity effects in predicting the assignment of alternating tones to one or two streams. The output from the model reflects the experimental data from the human subjects which demonstrates that temporal coherence occurs if the tones are within a certain frequency disparity *or* if their onset times are sufficiently far apart.

The form of the outputs from the Beauvois and Meddis[2] model differs in both these cases due to the presence of an attentional mechanism that selects only the output channel with the largest excitation rate. In the case of tones presented very rapidly, the two sequential tones have to be sufficiently close to produce greatest excitation at a channel intermediate between the tones in order for them to be assigned to the same auditory stream. For tones presented more slowly, the attentional mechanism can switch between two output channels centred on the frequency of each of the tones and so the limits on their frequency disparity are less constrained.

Three-tone streaming midpoints obtained at fast data presentation rates[65] show that a metric close to the log scale operates at intertone onset intervals of 50–70 ms. This different midpoint adds support to the hypothesis that a different mechanism is arbitrating the grouping process at data presentation speeds of less than around 100 ms intertone onset interval than that which applies above.

Returning to the perceptual level, music theorists have long recognised that the timbre or perceptual quality of a note is dependent, not just on the number of harmonics present in the note but also on the frequency range which its components occupy. Even pure sine-wave tones vary in their perceptual clarity, sounding most clear at around 700 Hz and becoming progressively more rough sounding as they tend towards the upper and lower extremes of the frequency range.

Terhardt's spectral dominance theory[57] proposes that there is a region of the spectrum from around 300 to 2000 Hz in which these perceptual changes happen relatively rapidly and suggests a weighting factor by which the amplitudes of tones in this region should be adjusted in order to predict their spectral dominance over and above the adjustment applied to equate loudness variation attributable to the equal loudness curve. Terhardt's pitch-weighting factor is also known as **tone salience**. There are other music theory models available that offer alternative explanations for this phenomenon.

6. CONCLUSIONS

All these different approaches provide us with more knowledge as to how any given acoustic signal is likely to be perceived by the human listener but we do not yet have an accurate model of the whole process. Physiology-based models demonstrate how information is passed from the auditory periphery to higher levels of the auditory system. Perception- and cognition-based models tend to focus on particular aspects and so do not give us a general picture of auditory processing.

Experimentation in support of the development of a computational model of the perceptual processes employed in human auditory grouping demonstrates that, even for simple auditory scenes containing pure tones, the principles that operate in competitive grouping are more complex than could be represented by any of the previously existing metrics.

It is possible that by combining information from a number of the existing models of perception and perceptual grouping, we might get a much closer approximation to a model of competitive streaming than the existing models currently provide. It may also be possible to develop a perceptually derived model from further experimentation. Both these possibilities are being investigated at Sheffield.

7. IMPLICATIONS FOR AUDITORY DISPLAY

The lack of a complete theory of human auditory perception and the intrinsic variability between human listeners makes it impossible to predict the interpretation of any given acoustic signal on the basis of existing knowledge. This implies that the development of Auditory Display techniques is necessarily experimental, requiring validation through user evaluation.

From the auditory grouping examples presented here, it can be seen that changes in the physical parameters of the auditory signal may not lead to corresponding changes in its associated perceptual attributes. Intensity variations may lead to rhythm changes, differences in the signal, detectable by the auditory periphery, may be masked by mapping to categorical interpretations where the contrastive data falls within the same perceptual category. The specific frequency values of certain components may be obscured by their integration into complex sound objects due to harmonicity or common motion with synchronous components. Such masking may be exploited in Auditory Display in situations where the actual value of each of many parameters is not significant but deviations by one or few parameters, from the general trend, must be identified or the trend itself is to be monitored. Multiple parameters may then be mapped to adjacent or overlapping frequency ranges, for instance. While all are changing at a similar rate only the overall direction of the change will be easily perceptible, but one item which changes against the trend or remains static while other values change may become prominent enabling the listener to identify the deviant cue.

Direct temporal comparisons may be obscured by the allocation of the items to be compared to separate streams. This latter problem is of particular significance in the choice of marker signals intended to set reference points in the acoustic data, (e.g., beacons (see Kramer, in this volume; also see Hayward, in this volume)). If the reference marker is too similar it may obscure the data or be obscured by it. On the other hand, if it is too different, it will be streamed out, allocated to another source that is not perceptually related to the data pattern so that the temporal relationships between the two patterns will be discarded as irrelevant. The effect of all these factors is to add an extra level of indirection between the data source and the perception of the sound. Even the most direct translation from data waveform to the audible domain (e.g., audification (see Kramer, in this volume); 0th-order mapping (see Scaletti, in this volume)) does not necessarily imply a direct relationship between data and auditory percept.

The experimental studies presented here demonstrate that far from being simple, the primitive grouping principles that apply to segregate the components of the acoustic signal into perceptual structures such as streams are highly context-dependent. Frequency relationships that pertain at one rate of data presentation may differ at another. Schema-driven auditory grouping is dependent on the previous experience of the listener and therefore must differ for each hearer.

Much basic research into the processes of auditory perception has already been undertaken. While many principles which promote segregation into auditory streams have been established, the interactions between them are highly complex and need considerable further investigation. However, in recent times this research has been led more by the need to understand processes that are schema-driven such as the recognition of meaningful structure in speech and music. It is anticipated that the needs of Auditory Display interfaces will lead to an increasing interest in the investigation of primitive auditory grouping principles and the relationships between the different factors that affect the construction of the auditory signal and the resulting perceptual changes which arise therefrom.

ACKNOWLEDGMENTS

The STREAMER project is currently supported by the UK MRC/ESRC/SERC Joint Council's Initiative in Cognitive Science/HCI, Grant No. SPG8921799. The sound examples were all created using the MIT-SYN sound synthesis system.[29] I would like to thank Kevin Baker and the late Rob Waltham for their help in preparing and recording the sound examples, and Greg Kramer for inviting me to the ICAD workshop and for his endless patience and advice throughout the editing process.

REFERENCES

1. Baker, K. L., S. M. Williams, and R. I. Nicolson. "Evaluating Frequency Proximity in Stream Segregation." Mapping the Audiroty Scene: Working Paper I, Department of Computer Science of Psychology, University of Sheffield, UK, 1992.
2. Beauvois, M. W., and R. Meddis. "A Computer Model of Auditory Stream Segregation." *Qtr. J. Exp. Psych.* **43** (1991): 517–41.
3. Bregman, A. S., and J. Campbell. "Primary Auditory Stream Segregation and Perception of Order in Rapid Sequences of Tones." *J. Exp. Psych.* **89(2)** (1971): 244–249.
4. Bregman, A. S., and A. Rudnicky. "Auditory Segregation: Stream or Streams?" *J. Exp. Psych.: Hum. Percep. & Perf.* **1** (1975): 263–267.
5. Bregman, A. S., and S. Pinker. "Auditory Streaming and the Building of Timbre." *Canad. J. Psych.* **32** (1978): 19–31.
6. Bregman, A. S. "Auditory Streaming is Cumulative." *J. Exp. Psych: Hum. Percep. & Perform.* **4** (1978): 380–387.
7. Bregman, A. S., and H. Steiger. "Auditory Streaming and Vertical Localization: Interdependence of 'What' and 'Where' Decisions in Audition." *Percep. & Psychophys.* **28(6)** (1980): 539–546.
8. Bregman, A. S. "Auditory Scene Analysis." In *Proceedings of the 7th International Conference on Pattern Recognition*, 168–175. Silver Spring, MD: Computer Society Press, 1984.
9. Bregman, A. S. *Auditory Scene Analysis.* Cambridge: MIT Press, 1990.
10. Broadbent, D. E. *Perception and Communication.* London: Pergamon, 1958.
11. Brown, G. J. "Computational Auditory Scene Analysis: A Representational Approach." Ph.D. thesis, University of Sheffield, UK, 1992.

12. Buser, P., and M. Imbert. *Audition*. Translated by R. H. Kay. Cambridge, MA: MIT/Bradford Books, 1992.
13. Carlson, R., and B. Granstrom, eds. *The Representation of Speech in the Peripheral Auditory System*. Amsterdam: Elsevier, 1982.
14. Cherry, E. C. "Some Experiments on the Recognition of Speech with One and with Two Ears." *J. Acous. Soc. Am.* **25** (1953): 975–979.
15. Ciocca, V., and A. S. Bregman. "Perceived Continuity of Gliding and Steady-State Tones Through Interrupting Noise." *Percep. & Psychophys.* **42(5)** (1987): 476–484.
16. Cooke, M. P. *Modelling Auditory Processing and Organisation*. Cambridge, MA: Cambridge University Press, 1993.
17. Dannenbring, G. L. "Perceived Auditory Continuity with Alternately Rising and Falling Frequency Transitions." *Canad. J. Psych.* **30** (1976): 99–114.
18. Dannenbring, G. L., and A. S. Bregman. "Streaming vs. Fusion of Sinusoidal Components of Complex Tones." *Percep. & Psychophys.* **24(4)** (1978): 369–376.
19. Darwin, C. J. "Perceiving Vowels in the Presence of Another Sound: Constraints on Formant Perception." *J. Acous. Soc. Am.* **76** (1984): 1636–1647.
20. Deutsch, D. "Grouping Mechanisms in Music." In *The Psychology of Music*, edited by D. Deutsch. New York: Academic Press, 1982.
21. Deutsch, D. "The Tritone Paradox: An influence of Language on Music Perception." *Music Percept.* **8** (1991): 335–347.
22. Divenyi, P. L., and S. K. Oliver. "Resolution of Steady-State Sounds in Simulated Auditory Space." *J. Acous. Soc. Am.* **85** (1988): 2042–2052.
23. Duda, R. O., R. F. Lyon, and M. Slaney. "Correlograms and the Separation of Sounds." *Proceedings of the Twenty-Fourth Asilomar Annual Conference on Signals, Systems and Computers*, 457–461. Pacific Grove, CA: Maple Press, 1990.
24. Gardner, R. B., S. A. Gaskill, and C. J. Darwin. "Perceptual Grouping of Formants with Static and Dynamic Differences in Fundamental Frequency." *J. Acous. Soc. Am.* **85** (1988): 1329–1337.
25. Gaver, W. W. "The Sonic Finder: An Interface that Uses Auditory Icons." *Hum.-Comp. Inter.* **4** (1989): 67–94.
26. Gelfand, S. A. *Hearing*. New York: Marcel Dekker, 1981.
27. Gregory, R. L. "Visual Illusions." In *New Horizons in Psychology 1*, edited by B. M. Foss. London: Penguin, 1966.
28. Hall, J. W., M. P. Haggard, and M. A. Fernandes. "Detection in Noise by Spectro-Temporal Pattern Analysis." *J. Acous. Soc. Am.* **76(1)** (1984): 50–56.

29. Henke, W. L. *MITSYN: A Synergistic Family of High-Level Languages for Time Signal Processing* (Version 8.1). Computer Software. 1990. Mailing Address: 133 Bright Rd., Belmont, MA 02178.
30. Hilkhuysen, G. "The Influence of Harmonicity on the Lateralization of Harmonics." Masters dissertation, University of Leiden, NL, 1990.
31. Huron, David. Research note received in personal communication, 1991.
32. Kanizsa, G. *Organization in Vision: Essays on Gestalt Perception.* New York: Praeger, 1979.
33. Koffka, K. *Principles of Gestalt Psychology.* London: Kegan Paul, 1936.
34. Krumhansl, C. L. *Cognitive Foundations of Musical Pitch.* Oxford Psychology Series, Vol. 17. Oxford: Oxford University Press, 1990.
35. Marr, D., and H. K. Nishihara. "Representation and Recognition of the Spatial Organisation of Three-Dimensional Shapes." *Proc. Roy. Soc. Lond.* **B200** (1978): 269–294.
36. Marr, D. *Vision.* San Francisco: Freeman, 1982.
37. Matlin, M. W. *Sensation and Perception,* 2nd ed. Massachusetts: Allyn & Bacon, 1988.
38. McAdams, S. "Spectral Fusion and the Creation of Auditory Images." In *Music, Mind and Brain,* chapt. XV. New York: Plenum Press, 1982.
39. McAdams, S. "Perception and Memory of Musical Timbre." Paper presented at the XXV International Congress of Psychology, Brussels. Abstract appears in *Intl. J. Psych.* **27 (3&4)** (1992): 146.
40. Meyer, L. B. *Emotion and Meaning in Music.* Chicago: University of Chicago Press, 1956.
41. Moore, B. C. J., and B. R. Glasberg. "Suggested Formulae for Calculating Auditory-Filter Bandwidths and Excitation Patterns." *J. Acous. Soc. Am.* **74** (1983): 750–753.
42. Miller, G. A. *Psychology, the Science of Mental Life.* England: Penguin, 1962.
43. Noorden, L. P. A. S. van. "Temporal Coherence in the Perception of Tone Sequences." Ph.D. dissertation, Eindhoven University of Technology, NL, 1975.
44. Noorden, L. P. A. S. van. "Minimum Differences of Level and Frequency for Perceptual Fission of Tone Sequences ABAB." *J. Acous. Soc. Am.* **61** (1977): 1041–1045.
45. Parncutt, R. *Harmony: A Psychoacoustical Approach.* Berlin: Springer-Verlag, 1989.
46. Patterson, R. D. "A Pulse Ribbon Model of the Peripheral Auditory System." In *Auditory Processing of Complex Sounds,* edited by W. A. Yost and C. S. Watson. Hillsdale, NJ: Lawrence Erlbaum, 1987.

47. Plutchik, R. *Foundations of Experimental Research*, 2nd ed. New York: Harper, 1974.
48. Sloboda, J. A. "Music Structure and Emotional Response: Some Empirical Findings." *Psych. Music* **19** (1991): 110–120.
49. Steiger, H., and A. S. Bregman. "Capturing Frequency Components of Glided Tones: Frequency Separation, Orientation, and Alignment." *Percep. & Psychophys.* **30(5)** (1981): 425–435.
50. Stevens, S. S., and J. Volkmann. "The Relation of Pitch to Frequency: A Revised Scale." *Am. J. Psych.* **53** (1940): 329–353.
51. Stevens, S. S. "The Measurement of Loudness." *J. Acous. Soc. Am.* **27** (1955): 815–829.
52. Stevens, S. S. "On the Psychophysical Law." 1957. Cited by: R. Plutchik, In *Foundations of Experimental Research*, 2nd ed. New York: Harper, 1974.
53. Summerfield, Q., A. Sidwell, and T. Nelson. "Auditory Enhancement of Changes in Spectral Amplitude." *J. Acous. Soc. Am.* **81(3)** (1987): 700–708.
54. Summerfield, Q., A. Lea, and D. Marshall. "Modelling Auditory Scene Analysis: Strategies for Source Segregation Using Autocorrelograms." *Proc. Inst. Acous.* **12(10)** (1990): 507–514.
55. Swets, J. A. "Is there a Sensory Threshold?" 1961. Cited by: R. Plutchik, In *Foundations of Experimental Research* 2nd ed. New York: Harper, 1974.
56. Tenney, J., and L. Polansky. "Temporal Gestalt Perception in Music." *J. Music Theor.* **24(2)** (1979): 205–241.
57. Terhardt, E. "Toward Understanding Pitch Perception: Problems, Concepts and Solutions." In *Psychophysical, Physiological and Behavioural Studies in Hearing*, edited by G Van den Brink and F A Bilsen. Netherlands: Delft University Press, 1980.
58. Terhardt, E. "Gestalt Principles and Music Perception." In *Auditory Processing of Complex Sounds*, edited by W. A. Yost and C. S. Watson. Hillsdale, NJ: Lawrence Erlbaum, 1987.
59. Terhardt, E., G. Stoll, and M. Seewann. "Algorithm for Extraction of Pitch and Pitch Salience from Complex Tonal Signals." *J. Acous. Soc. Am.* **71(3)** (1982): 679–688.
60. Treisman, A. "Human Attention." In *New Horizons in Psychology 1*, edited by B. M. Foss. London: Penguin, 1966.
61. Warren, R. M. *Auditory Perception: A New Synthesis*. New York: Pergamon, 1982.
62. Wightman, F. L., and D. J. Kistler. "Headphone Simulation of Free-Field Listening II: Pychophysical Validation." *J. Acous. Soc. Am.* **85** (1989): 868–878.

63. Williams, S. M. "STREAMER: A Prototype Tool for Computational Modelling of Auditory Grouping Effects." Research Report No. CS-89-31, Department of Computer Science, University of Sheffield, UK, 1989.
64. Williams, S. M., R. I. Nicolson, and P. D. Green. "STREAMER: Mapping the Auditory Scene." *Proc. Inst. Acous.* **12(10)** (1990): 567–575.
65. Williams, S. M., K. L. Baker, and R. I. Nicolson. "Mapping the Auditory Scene: Temporal Proximity and 3-Tone Streaming." *Proc. Inst. Acous.* **14(6)** (1992): 621–628.

Elizabeth M. Wenzel
Aerospace Human Factors Research Division, NASA Ames Research Center, Mail Stop 262-2, Moffett Field, CA 94035-1000; 415-604-6290; FAX: 415-604-3729; e-mail: beth@eos.arc.nasa.gov

Spatial Sound and Sonification

Immersive or artificially generated, three-dimensional environments are increasingly becoming a goal of advanced human-machine interfaces. While the technology for achieving truly useful multisensory environments is still in its early developmental stages, techniques for generating three-dimensional sound are now both sophisticated and practical enough to be applied to acoustic displays. This paper provides a brief description of three-dimensional sound synthesis and describes the performance advantages that can be expected when these techniques are applied to sound streams in sonification displays. Specific examples, and the lessons learned from each, are discussed for applications in telerobotic control, aeronautical displays, and shuttle launch communications.

INTRODUCTION

Virtual acoustics, also known as spatial sound, three-dimensional sound, and auralization, is the simulation of the complex acoustic field experienced by a listener within an environment. Going beyond the simple intensity panning of normal stereo techniques, the goal is to process sounds so that they appear to come from particular locations in three-dimensional space. Although loudspeaker systems are being developed, much of the recent work focuses on using headphones for playback and is the outgrowth of earlier analog techniques. For example, in binaural recording, the sound of an orchestra playing classical music is recorded through small microphones in the two ear canals of an anthropomorphic artificial or dummy head placed in the audience of a concert hall. When the recorded piece is played back over headphones, the listener passively experiences the illusion of hearing the violins on the left and the cellos on the right, along with all the associated echoes, resonances, and ambience of the original environment. Current techniques use digital signal processing to synthesize the acoustical properties that people use to localize a sound source in space. These techniques provide the flexibility of a kind of digital dummy head, allowing a more active experience in which a listener can both design and move around or interact with a simulated acoustic environment in real time. Such simulations are being developed for a variety of application areas including advanced communications and information display systems, telepresence and virtual reality, navigation aids for the visually impaired, architectural acoustics, and as a test bed for psychoacoustical investigations of complex spatial cues.

TECHNIQUES FOR CREATING SPATIAL DISPLAYS

The success of spatial synthesis relies on understanding the acoustical cues that are used by human listeners to locate sound sources in space (Figure 1). In the original duplex theory of sound localization based on experiments with pure tones (sine waves), interaural intensity differences (IIDs) were thought to determine localization at high frequencies because wavelengths smaller than the human head create an intensity loss or head shadow at the ear farthest from the sound source.[42] Conversely, interaural time differences (ITDs) were thought to be important for low frequencies since interaural phase (delay) relationships are nonambiguous only for frequencies below about 1500 Hz (wavelengths larger than the head). The duplex theory, however, cannot account for the ability to localize sounds on the vertical median plane where interaural cues are minimal. Also, when subjects listen to sounds over headphones, they usually appear to be inside the head even though ITDs and IIDs appropriate to an external source location are present. Many studies now indicate that these deficiencies of the duplex theory reflect the important contribution to localization of the direction-dependent filtering that occurs when incoming sound waves interact with the outer ears. Experiments have shown that spectral shaping by the pinnae is highly direction-dependent,[46] that the absence of pinna cues degrades localization accuracy,[21] and that pinna cues are at least partially responsible for externalization or the "outside-the-head" sensation.[40]

The synthesis technique typically used in creating a spatial auditory display involves the digital generation of stimuli using location-dependent filters constructed from acoustical measurements made with small probe microphones placed in the ear canals of individual subjects or artificial

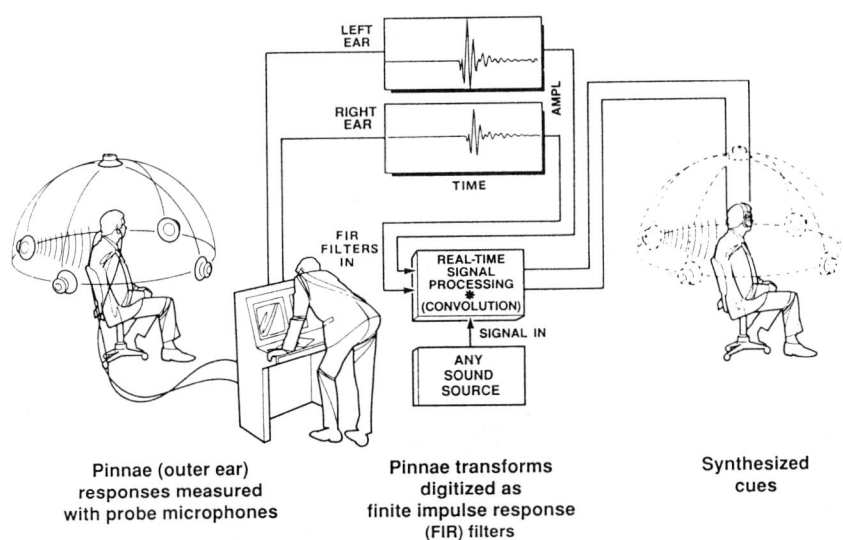

FIGURE 1 Illustration of the technique for synthesizing virtual acoustic sources with measurements of the head-related transfer function. An example of a pair of finite impulse responses measured for a source location at 90° to the left and 0° elevation (at ear level) is shown in the insets for the left and right ears. The placement of the loudspeakers in the drawing is illustrative only.

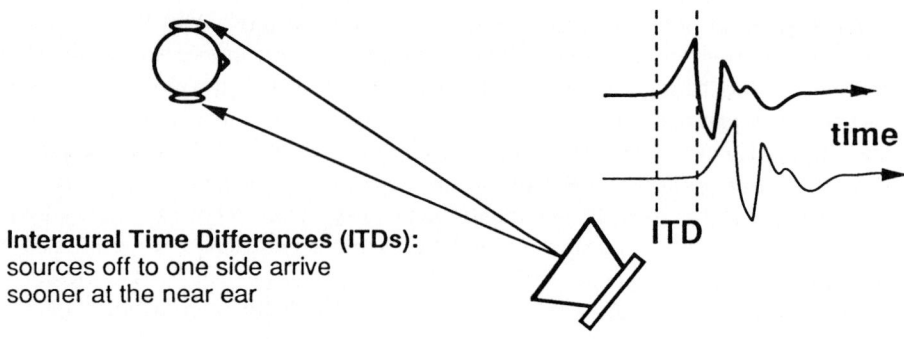

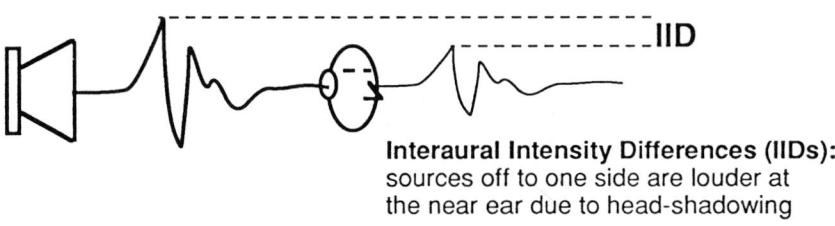

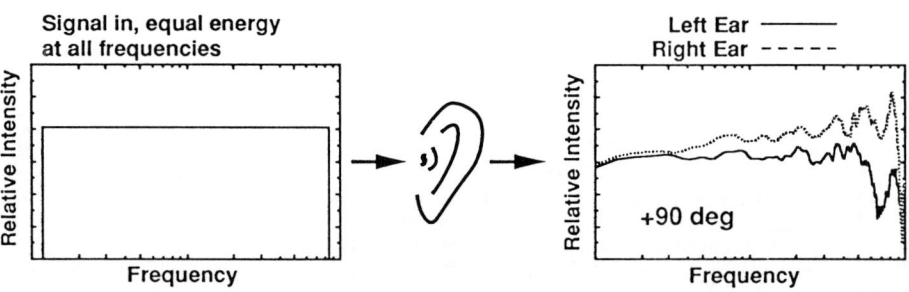

FIGURE 2 Illustration of the primary cues for sound localization: interaural intensity differences (IIDs), interaural time differences (ITDs), and the spectral coloration produced by the pinnae. Adapted from Wenzel with permission from MIT Press.[49]

Spatial Sound and Sonification

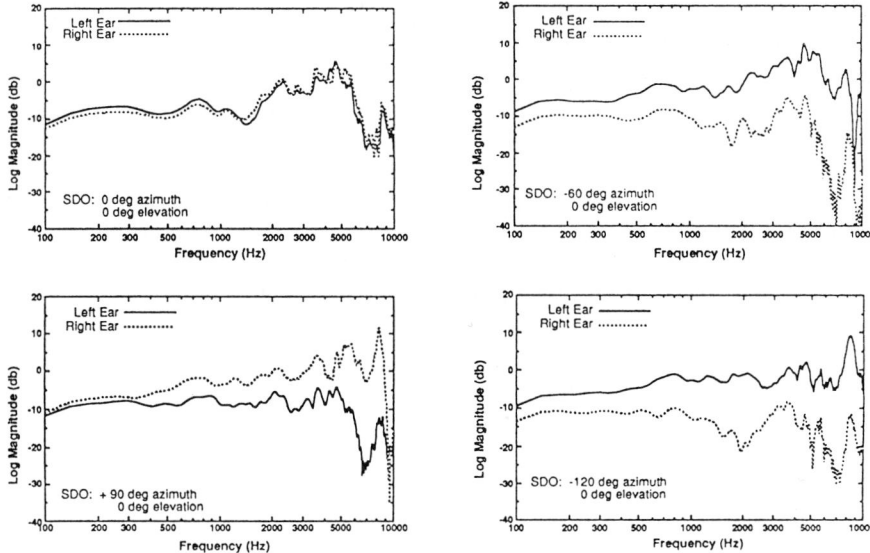

FIGURE 3 Illustration of the effects of spectral shaping by the pinnae. The panels show representations in the frequency domain (intensity only shown here) of a broadband acoustic signal after interaction with the outer ear (and other body) structures. Magnitudes are plotted for the left (solid line) and right (dashed line) ear canals of a single individual for stimuli delivered from loudspeakers at four different locations: directly in front (0° azimuth, 0° elevation), directly to the right (90° azimuth), −60° (on the left), and −120°.

heads for a large number of different source (loudspeaker) locations[54] (Figure 2). The filters constructed from these ear-dependent characteristics are examples of finite impulse response (FIR) filters and are often referred to as head-related transfer functions (HRTFs). Acting something like a pair of graphic equalizers, the HRTFs capture the essential cues needed for localizing a sound source: the ITDs, IIDs, and the spectral coloration produced by a sound's interaction with the outer ears. Figure 3 illustrates these complex spectral effects. The panels, which plot representations in the frequency domain, show what happens to a broadband sound source (e.g., a train of noise bursts) delivered from four different locations after interaction with the outer ears. The differences between the left and right intensity curves are the IIDs at each frequency. Spectral phase effects (frequency-dependent phase, or time, delays) are also present in the measurements, but are not shown here for clarity. While the interaural phase differences are somewhat frequency-dependent, in practice they are sometimes approximated by inserting a single delay between the two ears that increases in size with increasing displacement from the median plane (e.g., a delay estimated from the peak of the cross-correlation function of the left and right HRTFs[28]).

Using these HRTF-based filters, it is possible to impose spatial characteristics on a signal such that it apparently emanates from the originally measured location. Spatial synthesis can be achieved either by filtering in the frequency domain, a point-by-point multiplication of an input signal with the left and right HRTFs, or by filtering in the time domain, using the FIR representation and a somewhat more computationally intensive, multiply-and-add operation known as convolution. Of course, the localization of the sound will also depend on other factors such as its original spectral content; narrowband sounds (sine waves) are generally very difficult to localize while broadband, impulsive sounds are the easiest to locate. Filtering with HRTF-based filters cannot increase the bandwidth of the original signal, it merely transforms the energy and phase of the frequency components that are initially present.

In most current systems (Table 1), from one to four moving or static sources can be simulated (with varying degrees of fidelity) in an anechoic (free-field or echoless) environment by time-domain convolution of incoming signals with HRTF-based filters chosen according to the output of a head-tracking device. The head-tracking device allows the display to update the directional filters in real time to compensate for a listener's head motion so that virtual sources remain stable within the simulated environment. Motion trajectories and static locations at finer resolutions than the empirical data are generally simulated either by switching or, more preferably, by interpolating between the measured HRTFs.[51] Also, in some systems, a simple distance cue can be provided via real-time scaling of amplitude.

It should be noted that the spatial cues provided by HRTFs, especially those measured in free-field environments, are not the only cues likely to be necessary to achieve accurate localization.[7,16,49] Research indicates that synthesis of purely anechoic sounds can result in perceptual errors; in particular, increases in apparent front-back reversals, decreased elevation accuracy, and failures of externalization. These errors tend to be exacerbated when virtual sources are generated from nonpersonalized HRTFs, a common circumstance for most virtual displays. Other research suggests that such errors may be mitigated by providing more complex acoustic cues from reverberant environments. For example, acoustic features such as the ratio of direct to reflected energy in a reverberant field can provide a cue to distance (closer sources correspond to larger ratios) as well as enhance the sensation of externalization. See work by Blauert, Wenzel, Wightman, and Durlach for more complete discussions of the theoretical basis, synthesis, and psychophysical validation of spatial cues.[8,16,49,50,51,55]

TABLE 1 Examples of current three-dimensional sound systems and their performance characteristics.

AKG Creative Audio Processor[39]
    headphone presentation, 32 sources/reflections, room modeling, large stand-alone hardware system.
    HRTFs: individualized and artificial head.

Beachtron, Convolvotron, and Acoustetron, Crystal River Eng.[19]
    headphone presentation, 2 anechoic sources to 16 anechoic sources or 4 sources plus 6 early reflections each, Doppler effects, variable source radiation, interactive head-tracking supported, PC host.
    HRTFs: several individuals based on published behavioral data.

Focal Point, Gehring Research[23]
    headphone presentation, 2 anechoic sources per card, interactive head-tracking supported, PC.
    HRTFs: unknown source.

HEAD Acoustics (Sonic Perceptions[24])
    headphone presentation, 4 anechoic sources or 1 source plus 3 early reflections, stand-alone hardware system.
    HRTFs: based on a structural model.

McKinley and Ericson,[34] Wright-Patterson AFB
    headphone presentation, 2 to 4 anechoic sources, interactive head-tracking supported, stand-alone hardware (lab and flightworthy systems).
    HRTFs: individualized artificial head.

Roland Spatial Sound Processor[13]
    headphone and loudspeaker presentation, 4 anechoic sources, stand-alone hardware system.
    HRTFs: single individual.

While noninteractive room modeling or auralization, as it is known in Europe, has been implemented for some time,[24,25,27,30,41,43] recently some progress has been made toward interactively synthesizing reverberant cues. For example, in one system (the Convolvotron), the walls, floor, and ceiling in an environment are simulated by using HRTF-based filters to place the "mirror image" of a sound source behind each surface to account for the specular reflection of the source signal.[20] The filtering effect of surfaces such as wood or drapery can also be modeled with a separate filter whose output is delayed by the time required for the sound to

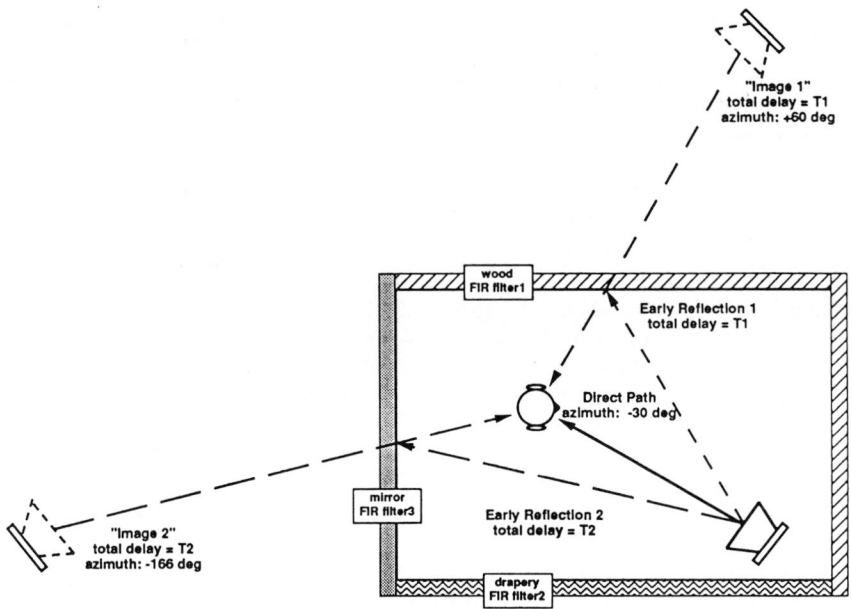

FIGURE 4 Illustration of the image model used for simulating reflections in real time in the Convolvotron. Only the direct path plus two first-order reflections are shown here for clarity. Currently, the Convolvotron simulates the direct path plus six first-order reflections.

propagate from each reflection being represented (Figure 4). Such dynamic modeling requires enormous computational resources for real-time implementation in a truly interactive (head-tracked) display. Currently it is not practical to render more than the first one or two reflections from a small number of surfaces. For example, the Convolvotron requires approximately 200 million operations per second to render the direct path plus six first-order[1] reflections for a single sound source, or a total of seven acoustic

[1] First-order reflections refer to the fact that only the first bounce off each surface is being modeled. Obviously, in a real environment, there are essentially an infinite number of possible angles of reflection incidence (not just one per room surface) and each reflection can bounce off several surfaces in succession, forming first-order, second-order, third-order bounces, and so on. Directional characteristics (e.g., simulated by HRTF-based filters), however, are thought to be important only for some number of the early reflections. In the limit, the later-order reflections can be modeled by an environmentally specific, but essentially nondirectional, noise process known as late reverberation. See Allen and Berkley,[1] Kendall and Martens,[27] and Schroeder.[45]

images that can be interactively updated in real time (up to 4 sources and 28 images in an Acoustetron). Future work in this area will examine the perceptual consequences of using dynamic reflection models and eventually extend the approach to more realistic models of acoustic environments.

INTEGRATING SPATIAL SYNTHESIS AND SONIFICATION TECHNIQUES

Although simulation of the spatial cues used by the listener has received the most attention in recent work, it represents only one component of a virtual acoustic display. The development and validation of HRTF-based processing and the environmental modeling techniques summarized in Figure 5 could clearly encompass decades of research, yet in many ways the problems that need to be solved in spatial synthesis are at least fairly well understood. The same cannot be said for the real-time generation of other qualities of acoustic sources that are also critical for simulating acoustic objects in a virtual display. As discussed in many of the chapters in this book, in addition to spatial location various acoustic features, such as temporal onsets and offsets, timbre, pitch, intensity, and rhythm, can specify the identities of individual objects and convey meaning about discrete events or ongoing actions in the world and their relationships to one another. One can systematically manipulate these features, effectively creating an auditory symbology for computer interfaces that operates on a continuum from literal everyday sounds, such as the clunk of a file being thrown into the trash can, to a completely abstract mapping of statistical data into multidimensional sound parameters. Unfortunately, both current synthesis technology and understanding of the perceptual consequences of simulating such complex, multidimensional stimuli are not yet well developed.

The ideal synthesis device would be able to flexibly generate the entire continuum of nonspeech sounds described above as well as be able to continuously modulate various acoustic parameters associated with these sounds in real time. As implied by Figure 5, such a device or devices would act as the generator of acoustic source characteristics, which would then serve as the inputs to a sound spatialization system. Thus, initially at least, source generation and spatial synthesis would remain as functionally separate components of an integrated acoustic display system. While there would necessarily be some overhead cost in controlling separate devices, the advantage is that each component can be developed, upgraded, and utilized as stand-alone components so that systems are not locked into an outmoded technology.

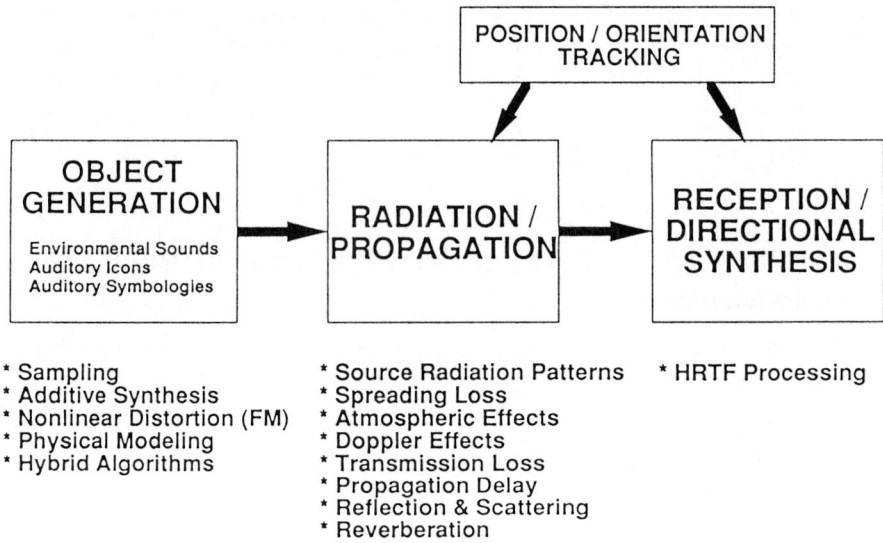

FIGURE 5 The primary components required for simulation of acoustic objects in a virtual acoustic display.

Current devices available for generating nonspeech sound sources tend to fall into two general categories: "samplers," which digitally store sounds for later real-time playback, and "synthesizers," which rely on analytical or algorithmically based sound generation techniques originally developed for imitating musical instruments (see Scaletti, in this volume). With samplers, many different sounds can be reproduced (nearly) exactly, but substantial effort and storage media are required for accurately prerecording sounds and there is usually limited real-time control of acoustic parameters. Synthesizers, on the other hand, afford a fair degree of real-time, computer-driven control.

Most widely available synthesizers and samplers are based on MIDI (musical instrument digital interface) technology. The baud rate of such devices (31.25 Kbs), especially when connected to standard serial computer lines (19.2 Kbs), is still low enough that continuous real-time control of multiple sources/voices will frequently "choke" the system. In general, synthesis-based MIDI devices, such as those which use frequency modulation (FM), are more flexible than samplers in the type of real-time control available but less general in terms of the variety of sound qualities that can be reproduced. For example, it is difficult to generate environmental sounds such as breaking or bouncing objects from an FM synthesizer (see Gaver, in this volume).

Large-scale systems designed for sound production and control in the entertainment industry or in music composition incorporate both sampling and digital synthesis techniques and are much more powerful. However, they are also quite expensive, require specialized knowledge for their use, and are primarily designed for off-line sound design and postproduction. A potential disadvantage of both types of devices is that they are primarily designed with musical performance and/or sound effects in mind. This design emphasis is not necessarily well suited to the generation and control of sounds for the display of information, and again, tends to require that the user/designer have specialized knowledge of musical/production techniques.

A few cue-generation systems have been integrated for virtual environments and data sonification using currently available devices,[47,52] while a few designers are developing special-purpose hardware and software systems for acoustic displays (see Madyastha and Reed, in this volume).[29,44,53] However, far more effort needs to be devoted to the development of sound-generation technology specifically aimed at information display. Perhaps even more critical is the need for further research into lower level sensory and higher level cognitive determinants of acoustic perceptual organization[9] (see Williams, in this volume), since these results should serve to guide technology development.

PERFORMANCE ADVANTAGES OF VIRTUAL ACOUSTIC DISPLAYS

Useful features of acoustic signals in general include that they tend to produce an alerting or orienting response and that they can be detected more quickly than visual signals.[37,38] Such characteristics are probably responsible for the most prevalent use of nonspeech audio in simple warning systems, such as the malfunction alarms used in aircraft cockpits or the siren of an ambulance. Another advantage of audition is that it is primarily a temporal sense and we are extremely sensitive to changes in an acoustic signal over time.[37] This feature tends to bring a new acoustical event to our attention and, conversely, allows us to relegate sustained or uninformative sounds to the background. Consequently, audio is particularly suited to monitoring state changes over time, for example, when the hard drive of your computer suddenly begins to malfunction (see Cohen, in this volume).

Acoustic displays can be further enhanced by taking advantage of the auditory system's ability to segregate, monitor, and switch attention among simultaneous streams of sound, particularly when such acoustic objects are distinguished by different locations in space.[9,37] Another advantage of the binaural system, related to the so-called "cocktail party effect," is that the spatial separation of sounds improves the intelligibility of signals in a background of noise or other voices.[10,14] Thus, the two primary performance advantages that can be expected from using spatial sound are enhanced situational awareness, or the direct comprehension of object relationships in a three-dimensional task space, and enhanced comprehension of multiple, simultaneous sound streams or voices (Table 2).

Some of the kinds of tasks that can benefit from spatial presentation of information are given in Table 3. In each case, the various roles that spatial cues could play—direct representation of spatial information, spatial metaphors for displaying nonspatial information, and enhanced stream segregation—are outlined.

For example, the direct representation of spatial information could be used in architectural design, where the ability to interactively simulate, or "auralize," room acoustics would be of great use in exploring and avoiding possible acoustically undesirable effects that may not be obvious from a visual design.[20,39] Similarly, when coupled with some type of range-finding device, artificial acoustical environments could be used as aids for the visually impaired by audibly representing the surfaces and obstacles through which a blind person must navigate.[32]

TABLE 2 Performance advantages of the spatial presentation of sound.

Enhanced Situational Awareness
    Direct representation of spatial information.
        the eyes."
    Omnidirectional Monitoring: "The function of the ears is to point
    Reinforces (or replaces) information in other modalities.
    Enhances the sense of presence or realism.
Enhanced Multiple-Channel Presentation
    The "Cocktail Party Effect": Improves intelligibility of voices in a
        background of noise or other voices ("natural" noise cancellation).
    The "Cocktail Party Effect": Improves discrimination and selective
        attention among multiple sources.
    Enhanced Stream Segregation: Allows separation of multiple sounds
        intodistinct "objects."

TABLE 3 Performance advantages associated with the spatial presentation of sound in various application areas.

Architectural Acoustics: Acoustical CAD/CAM Systems
Direct representation of spatial information: "auralization" of models of room acoustics
Enhanced source intelligibility/separation: simultaneous sources

Data Spaces: Large-Scale Databases/Information Systems
Symbolic representation via spatial location: database navigation: an architectural/spatial metaphor for database organization
Enhanced intelligibility/separation: simultaneous icons/symbologies

Data Visualization: Computational Fluid Dynamics, Virtual Wind Tunnel
Direct representation: airflow/noise patterns produced by an aircraft engine
Symbolic representation: localized intensities → size of error measures dispersed over the sample grid of a flow model
Enhanced intelligibility/separation: simultaneous icons/symbologies

Aeronautics: ATC Displays, Cockpit Warning Systems
Direct representation: incoming aircraft locations; left vs. right engine malfunctions
Symbolic representation: different aircraft systems mapped to different locations (a cockpit data space)
Enhanced intelligibility/separation: simultaneous radio communications

Telerobotic Control: Space Station Construction and Repair
Direct representation: contact cues; range finding
Enhanced intelligibility/separation: simultaneous icons/symbologies

Spatial metaphors for displaying the organization of information could also be quite useful in navigating through large-scale databases, as in spatial mnemonic devices in which related topics are located in adjacent rooms of a metaphorical building. Similarly, complex, multidimensional variables from models in computational fluid dynamics could be represented spatially (e.g., representing the dispersion of errors for modeled vs. actual turbulence data via localized intensity cues; modeling aerodynamic flow patterns using a "virtual wind tunnel"[12]).

Tr 24 Segregation enhancement can be critical in applications involving both simultaneous speech channels, as in aviation communication systems, and the kind of encoded nonspeech cues proposed for scientific "visualization" or sonification. Examples include aeronautical displays for air traffic control in which the controller hears communications from incoming traffic in positions that correspond to their actual location in the terminal area (see CD, sound example 1),[6,7] sonification displays for the acoustic representation of multi-dimensional, statistical data (see Bly; Blattner; Kramer; Madyastha and Reed; and Smith, Pickett and Williams, all in this volume), and alternative computer interfaces for the visually impaired[17] (see Mynatt, in this volume).

Another aspect of auditory spatial cues is that, in conjunction with the other senses, they can reinforce the information content of a display and provide a greater sense of presence or realism in a manner not readily achieved by a single (usually visual) modality.[15,48] Similarly, in direct-manipulation tasks, auditory cues can provide an alternative medium for the representation of tactile or force-feedback cues.[52] This is a quite difficult interface problem for multimodal displays, which is only beginning to be solved.[35] Intersensory synergism will be particularly important in applications involving telepresence, including advanced teleconferencing,[33] shared electronic workspaces,[18,22] and monitoring telerobotic activities in remote or hazardous situations.[52] Similarly, the interaction of the senses will be critical in purely virtual environments for visualization, large-scale data management and systems control,[11,18] and entertainment.[27]

APPLICATIONS CASE STUDIES

Much of the work to date in virtual acoustic displays has been concerned with developing the technology for real-time rendering of acoustic sources and perceptually validating the spatial synthesis techniques in basic psychophysical studies.[7,49,50,51,55] There have actually been very few formal, or even informal, studies of the consequences of using such techniques in real-world applications. This section reviews a few of the applications case studies conducted at NASA Ames and discusses some of the lessons that can be learned from each.

TELEROBOTIC CONTROL

Wenzel, Stone, Fisher, and Foster described a system to provide auditory feedback for telerobotic control[52] in the NASA Ames' VIEW system.[18] In this system, the acoustic display is capable of generating localized acoustic cues in real time over headphones using a Convolvotron. An auditory symbology, a related collection of representational auditory "objects" or "icons," can be designed using ACE, the auditory cue editor, which links both discrete and continuously varying acoustic parameters with information or events in the display. During a given display scenario, the symbology can be dynamically coordinated in real time with three-dimensional visual objects, speech, and gestural displays. The types of displays feasible with the system range from simple warnings and alarms to the acoustic representation of multidimensional data or events.

The VIEW telerobotic scenario was designed to illustrate the capability of telepresence, i.e., the manipulation of objects or interaction with persons or objects remote from one's location. The participant, upon donning a stereoscopic head-mounted display, could "couple" his or her arm with that of a life-size model of a Puma robotic arm so that the robotic arm moved, to the extent of its kinematic capabilities, in correspondence to the movement of the user's arm. An end effector with a viselike gripping apparatus was opened and closed by opening and closing the hand.

The efficacy of the control mapping between human and machine was informally tested by requiring operators to remove and replace a "circuit card" in a slot on a "task board." With perfect telepresence, this task could be accomplished with little more difficulty than if one were using one's own hand and a real task board and circuit card. However, factors such as slower-than-ideal graphic refresh rates, lower-than-ideal contrast and focus, etc., made the precision manipulation required somewhat difficult. In such a situation, auditory feedback can make an important difference, particularly with the current paucity of good haptic or force feedback displays.

At the simplest level, auditory feedback was used to indicate the occurrence of discrete events in the scenario. For example, many commands and actions in VIEW are initiated by hand gesture. A VPL "data glove" reports finger positions to the host computer, which examines those positions for correspondence to any of several predefined gestures, such as "single-finger point" or "fist." When one of these gestures is detected by the host, a sound is made by the auditory display to indicate gesture recognition. Other simple auditory cues fall into the category of environmental sound effects; bumping an end effector into a "solid" object in the virtual (or teleoperated) world caused a "bump" sound to be produced. Since direct force feedback was not available, this form of audio display was particularly critical, as it warned of a situation that could cause damage to a real-world robotic arm or to objects with which it is colliding. At a more mundane level, this sort of sound effect enhanced the sense of presence; objects tend to make a sound when they collide in the real world, so it is reasonable to expect them to do so in a virtual environment.

Audio feedback that supplemented or replaced force feedback could also be represented as a continuum by changing one or more sound parameters in correspondence to the force's intensity. This type of display was utilized for a special circumstance in the telerobotic scenario. If the user attempted to force the replacement circuit card into the task board without orienting it correctly, a force-reflection display called "push-through" was initiated. This started as a soft, steady tone but the harder the participant pushed on the misaligned card, the louder, brighter (higher harmonics are let through the filter), and more frequency-modulated it became. In this way, a potentially damaging increase in user input was signalled by an increasingly harsh and strident auditory warning.

Taking this idea one step further, not only force, but any arbitrary continuation of data could be displayed. Perhaps the most successful use of auditory feedback came into play when the participant attempted to guide the replacement circuit board into the target socket. As the board reached a certain proximity to the socket, a cue was initiated consisting of two sustained tones; the pitch of one of the tones was deflected with respect to the other by an amount proportional to the distance between the circuit card and its slot. As the card neared the slot, the two pitches came closer together (resulting in an obvious decrease in the beat frequency produced by the increasingly adjacent pitches); at distance zero, the tones were in unison. The cue functioned as an auditory "range finder" and greatly facilitated the proper positioning of the card in its slot.

While formal experimentation was not done, the telerobotic scenario provided a useful exercise for formulating some basic guidelines for the design of auditory icons and the development of an auditory symbology in a virtual environment. First, the most effective cues are simple cues, particularly when simple events are being represented. Long sequences or elaborate clusters of tones not only tend to clutter the auditory display, they can increase the load on cognitive processing and memory required to interpret the information, and in the long run, become downright annoying. Similarly, a "thud" sound suffices to signal a bump in a virtual world; it is not necessary to have a speech synthesizer say "You have bumped into something" at each and every collision. While these may be extreme examples, the basic principle holds: an auditory icon should be as simple as possible.

The need for simplicity is even more critical when several cues occur in close proximity to each other. For instance, in the telerobotic scenario, a gesture-recognition cue might be immediately followed by a sound that indicates movement of the jaws of the end effector. If the jaws were then to close over the circuit board, a "board-grasped" cue would result. These three cues can occur in rapid succession, so they must be of short duration for the correct sequence of events to be properly represented.

This situation also points out the need for carefully choosing the sound signatures or synthesizer patches (configurations) that form the fundamental units of an auditory symbology. Patch design, including spectral content, amplitude and filter envelopes, and various special effects, is the chief distinguishing feature of a simple icon. The best way to make an icon recognizable is to give it a distinctive sound. Much effort in the design of auditory icons is therefore concentrated in selecting or building an appropriate synthesizer patch. It is also probably the area in which art, rather than science, is currently most operative in acoustic display design and where Gestalt principles of perceptual organization can be of most assistance.

As noted before, guidelines can also be derived from the fields of psychoacoustics, music, and perceptual psychology. As illustrated by the auditory range finder, the close tuning of two pitches is a continuous parameter to which the human ear is very sensitive. However, the amplitude modulation or beat frequency that signals the change in proximity will only occur for a limited range of frequencies which must be considered when mapping the distance data to the difference in pitch. In developing a symbology, one can also take advantage of what one might think of as "natural" or metaphorical mappings even if a literal sound, such as a "bump," is not possible. For example, the "push-through" cue described above clearly signals an increasing violation of the allowable forward movement when inserting the task board at an incorrect orientation by a harsh sound which increasingly "violates" the ears (see also Kramer's discussion of affective association, in this volume).

To minimize cognitive effort, it is also important to build meaning by the relationships between icons. In the telerobotic scenario, icons that provide feedback for related gestures have similar timbres that are distinguished by their temporal structure. For example, larger changes in pitch, at the level of short sequences, are used (much like the familiar two-chime doorbell). This type of icon has the virtue of being reversible like a short musical motive; a "grasp" gesture is represented by a high note followed by a slightly lower note. The complementary "release" gesture is the same two notes, only in reverse order (see also Blattner et al., in this volume). As much as possible, this relationship between sound and meaning should remain consistent throughout a display system. Thus, in VIEW, the cues which provide feedback for the various gestures remain the same across the different types of display scenarios that have been developed.

AERONAUTICAL DISPLAYS

AUDITORY FEEDBACK. One very simple but useful example of the use of auditory icons for situational awareness in aeronautics is as a substitute for tactile feedback in a cockpit touch-panel display. In a recent flight simulator experiment designed to evaluate automated touch-panel checklists, pilots were unable to tell when a virtual switch was actually "engaged" or merely touched due to the absence of feedback. Specifically, there was a lack of a perceptible difference between merely "sliding across" several switches to an intended target switch and actually engaging the switch. With a real aircraft switch, tactile feedback is naturally available to cue the difference between these states; on menu-driven computer operating systems, such as an Apple Macintosh, there are also visual cues. Begault, Stein, and Loesche used auditory cues to suggest the result of engaging a virtual switch that worked in response to tactile activity.[5] Recordings of an aircraft-quality switch were made and loaded into a sampler to be activated in response to the virtual switches. The switch had a different sound when pushed in than when released; the "push in" sound was used for making finger contact on the virtual switch, while the "release" sound was used when the switch was actually engaged. Reports from pilots who have used the system so far have been very favorable. The additional feedback provided by the sound also allows one pilot to know if the other pilot is engaged in menu selection activity, without requiring the use of an already overcrowded visual channel.

Tr 25

TRAFFIC COLLISION AVOIDANCE SYSTEM. Recently, in an example of the direct use of spatial information, Begault also evaluated a simple, noninteractive "head up" spatial auditory display for the recently developed TCAS (traffic collision avoidance system) being studied in the flight simulator of the NASA Ames Man-Vehicle Systems Research Facility.[3] TCAS is intended to supply flight crews with real-time information about aircraft in their vicinity that may represent potential collision threats. Currently, the TCAS traffic advisory consists of the spoken word "traffic" and is nondirectional: the actual position of the traffic is usually obtained visually, through instrument monitoring and out the window acquisition. Previous work on TCAS displays had found no difference in performance between current visual-only or visual-plus-auditory (nonspatial) versions of a display. However, in Begault's study, the out-the-window position of the traffic was linked to the virtual auditory position of the word "traffic" heard through headphones without the use of a head-down display (see CD, sound example 2). He found that the time interval for traffic acquisition was reduced by approximately 2.2 sec when spatial sound was used to suggest the direction for head-up visual search of the target, compared to monotic (single earpiece), normal-practice conditions. Such improvements in reaction time can be extremely critical in emergency situations where appropriate evasive actions must be made quickly and correctly.

SHUTTLE LAUNCH COMMUNICATIONS

Begault has also recently developed a noninteractive, four-channel display device for the purpose of separating simultaneous voice channels in real time. The system is intended for use in mobile communications vehicles at the NASA Kennedy Space Center to aid voice intelligibility during shuttle launch procedures, since currently launch personnel must frequently monitor several voice channels at once over a single earphone. Thus, while up to four static locations can be represented by this system, it is primarily intended as a device for channel segregation rather than the display of position information per se.

Although the prototype has not yet been evaluated at KSC, a preliminary study was recently conducted at NASA Ames to determine intelligibility levels as a function of different horizontal positions (every 30 degrees) for different call signs (stereotypical identifying letters used by KSC personnel during launch control, e.g., "NTOC") in a background of nonspatialized speech babble.[4] The stimuli were synthesized by a Convolvotron using reduced-length filters that had been reconstructed from minimum-phase approximations of HRTF measurements from a single subject (SDO; e.g., see the control condition described by Kistler and Wightman[28]). Begault found that the intelligibility of call signs was improved by about 6 dB (compared to a nonspatialized call sign condition) for the most lateral locations in the front (60 and 90 degrees on both the left and right sides). That is, speech at these lateral positions required about half as much signal strength in order to be as intelligible as the nonspatialized speech. While the generalizability of these data is limited by the fact that only one speaking voice was tested for a limited set of positions, the results do suggest that intelligibility in a communications system can benefit from the use of spatial synthesis techniques.

CONCLUSIONS

While much work remains to be done in the area of auditory spatial synthesis, the basic technology needed for adding at least minimal spatial cues to sonification displays is now available. Admittedly, this currently comes at a fairly high cost, although this is bound to become less of a factor as digital signal-processing technology becomes cheaper in the near future. What is more critical for the future success of sonification displays is a systematic approach to understanding the perceptual and practical constraints involved in developing the real-time technology needed for generating source characteristics in as general a manner as possible. In the near term, this probably means continuing to use approaches based on standard synthesis and sampling techniques, or some hybrid version of both. In the long term, researchers must begin to think about instantiating more complex approaches to simulation; for example, using techniques based on algorithmic or structural modeling (see Gaver, in this volume).[26,31,36] In parallel with the development of a sonification infrastructure, it also behooves us to do some formal validation studies of the systems we develop in the context of real-world tasks. While we all may be personally convinced of their efficacy, we need to start providing concrete evidence of when and how sonification displays can help users get the job done.

ACKNOWLEDGMENTS

Work supported by NASA and by the Naval Command, Control and Ocean Surveillance Center, San Diego. Thanks to my colleague, Durand Begault, for his valuable input to the paper.

REFERENCES

1. Allen, J. B., and D. A. Berkley. "Image Model for Efficiently Modeling Small-Room Acoustics." *J. Acous. Soc. Am.* **65** (1979): 943–950.
2. Begault, D. R. "Perceptual Effects of Synthetic Reverberation on Three-Dimensional Audio Systems." *J. Audio Engr. Soc.* **40** (1992): 895–904.
3. Begault, D. R. "Head-Up Auditory Displays for TCAS Advisories: A Preliminary Investigation." *Human Factors*: in press.
4. Begault, D. R. "Call Sign Intelligibility Improvement Using a Spatial Auditory Display." Technical Memorandum TM104014, NASA, 1993.
5. Begault, D. R., N. Stein, and V. Loesche. "Advanced Audio Applications in the NASA-Ames Advanced Flight Simulator." Unpublished manuscript. Contact: D. R. Begault, NASA-Ames Research Center, Mail Stop 262-2, Moffett Field, CA 94035-1000.
6. Begault, D. R., and E. M. Wenzel "Technical Aspects of a Demonstration Tape for Three-Dimensional Sound Displays." Technical Memorandum TM102826, NASA, 1990.

7. Begault, D. R., and E. M. Wenzel. "Techniques and Applications for Binaural Sound Manipulation in Human-Machine Interfaces." *Intl. J. Aviation Psych.* **2** (1992): 1–22.
8. Blauert, J. *Spatial Hearing: The Psychophysics of Human Sound Localization.* Cambridge, MA: MIT Press, 1983.
9. Bregman, A. S. *Auditory Scene Analysis.* Cambridge, MA: MIT Press, 1990.
10. Bronkhorst, A. W., and R. Plomp. "The Effect of Head-Induced Interaural Time and Level Differences on Speech Intelligibility in Noise." *J. Acous. Soc. Am.* **83** (1988): 1508–1516.
11. Brooks, F. P. "Grasping Reality Through Illusion—Interactive Graphics Serving Science." In *Proceedings of CHI, 1988*, 1–11, held in Washington, May 15–19. New York: ACM, 1988.
12. Bryson, S., and C. Levit. "The Virtual Wind Tunne." *IEEE Comp. Grap. & App.* **12** (1992): 25–34.
13. Chan, C. J. "Sound Localization and Spatial Enhancement with the Roland Sound Space Processor." In *Cyberarts: Exploring Art and Technology*, edited by L. Jacobson, 95–104. San Francisco, CA: Miller Freeman, 1992.
14. Cherry, E. C. "Some Experiments on the Recognition of Speech with One and Two Ears." *J. Acous. Soc. Am.* **22** (1953): 61–62.
15. Colquhoun, W. P. "Evaluation of Auditory, Visual, and Dual-Mode Displays for Prolonged Sonar Monitoring in Repeated Sessions." *Human Factor* **17** (1975): 425–437.
16. Durlach, N. I., A. Rigopulos, X. D. Pang, W. S. Woods, A. Kulkarni, H. S. Colburn, and E. M. Wenzel. "On the Externalization of Auditory Images." *Presence* **1** (1992): 251–257.
17. Edwards, A. D. N. "Soundtrack: An Auditory Interface for Blind Users." *Hum.-Comp. Interaction* **4** (1989): 45–66.
18. Fisher, S. S., E. M. Wenzel, C. Coler, and M. W. McGreevy. "Virtual Interface Environment Workstations." *Proc. Human Fac. Soc.* **32** (1988): 91–95.
19. Foster, S. H. *Convolvotron™ User's Manual.* Crystal River Engineering, Inc., 12350 Wards Ferry Road, Groveland, CA 95321, 1988.
20. Foster, S. H., E. M. Wenzel, and R. M. Taylor. "Real-Time Synthesis of Complex Acoustic Environments." In *ASSP (IEEE) Workshop on Applications of Signal Processing to Audio and Acoustics.* New York: New Paltz, 1991.
21. Gardner, M. B., and R. S. Gardner. "Problem of Localization in the Median Plane: Effect of Pinnae Cavity Occlusion." *J. Acous. Soc. Am.* **53** (1973): 400–408.
22. Gaver, W. W., R. B. Smith, and R. O'Shea. "Effective Sounds in Complex Systems: The ARKola Simulation." *Proceedings of CHI, 1988*, 85–90, held in Washington, May 15–19. New York: ACM, 1988.

23. Gehring, B. *Focal Point™ 3-D Sound User's Manual.* Gehring Research Corporation, 189 Madison Avenue, Toronto, Canada, M5R 2S6, 1990.
24. Gierlich, H. W. "The Application of Binaural Technology." *Appl. Acous.* **36** (1992): 219–244.
25. HEAD Acoustics. *Binaural Mixing Console.* [Product literature.] Contact: Sonic Perceptions, 114A Washington St., Norwalk, CT 06854
26. Jaffe, D., and J. Smith. "Extensions of the Karplus-Strong Plucked String Algorithm." *Comp. Music J.* **7** (1983): 43–55.
27. Kendall, G. S., and W. L. Martens. "Simulating the Cues of Spatial Hearing in Natural Environments." In *Proceedings of the International Computer Music Conference,* 1984.
28. Kistler, D. K., and F. L. Wightman. "A Model of Head-Related Transfer Functions Based on Principal Components Analysis and Minimum-Phase Reconstruction." *J. Acous. Soc. Am.* **91** (1992): 1637–1647.
29. Kramer, G., and S. Ellison. "Audification: The Use of Sound to Display Multivariate Data." In *Proceedings of the International Computer Music Conference,* 214–221, 1991.
30. Lehnart, H., and J. Blauert. "Principles of Binaural Room Simulation." *Appl. Acous.* **36** (1992): 259–292.
31. Li, X., R. J. Logan, and R. E. Pastore. "Perception of Acoustic Source Characteristics." *J. Acous. Soc. Am.* **90** (1991): 3036–3049.
32. Loomis, J. M., C. Hebert, and J. G. Cicinelli. "Active Localization of Virtual Sounds." *J. Acous. Soc. Am.* **88** (1990): 1757–1764.
33. Ludwig, L., N. Pincever, and M. Cohen. "Extending the Notion of a Window System to Audio." *Computer* **23** (1990): 66–72.
34. McKinley, R. L., and M. A. Ericson. "Digital Synthesis of Binaural Auditory Localization Azimuth Cues Using Headphones." *J. Acous. Soc. Am.* **88** (1988): S18.
35. Minsky, M., O. Ming, O. Steele, F. P. Brooks, and M. Behensky. "Feeling and Seeing: Issues in Force Display." *Comp. Graph.* **24** (1990): 235–243.
36. Morse, P. M., and K. U. Ingard. *Theoretical Acoustics.* New York: McGraw-Hill, 1968.
37. Mowbray, G. H., and J. W. Gebhard. "Man's Senses as Informational Channels." In *Human Factors in the Design and Use of Control Systems,* edited by H. W. Sinaiko, 115–149. New York: Dover, 1961.
38. Patterson, R. R. "Guidelines for Auditory Warning Systems on Civil Aircraft." Paper No. 82017, Civil Aviation Authority, London, U.K., 1982.

39. Persterer, A. "A Very High Performance Digital Audio Signal Processing System." Paper prepared for ASSP (IEEE) Workshop on Applications of Signal Processing to Audio and Acoustics, New Paltz, NY, 1989.
40. Plenge, G. "On the Difference Between Localization and Lateralization." *J. Acous. Soc. Am.* **56** (1974): 944–951.
41. Poesselt, C., J. Schroeter, M. Opitz, P. Divenyi, and J. Blauert. "Generation of Binaural Signals for Research and Home Entertainment." Paper B1-6, Proceedings of the 12th International Congress on Acoustics, Toronto, 1986.
42. Lord Rayleigh [Strutt, J. W.] "On Our Perception of Sound Direction." *Philo. Mag.* **13** (1907): 214–232.
43. Richter, F., and A. Persterer. "Design and Applications of a Creative Audio Processor." Preprint 2782 (U-4), 86th Convention of the Audio Engineering Society, Hamburg, 1989.
44. Scaletti, C., and A. B. Craig. "Using Sound to Extract Meaning from Complex Data." *Proceedings of the SPIE* **1459** (1991): 207–219.
45. Schroeder, M. R. "Digital Simulation of Sound Transmission in Reverberant Spaces." *J. Acous. Soc. Am.* **47** (1970): 424–431.
46. Shaw, E. A. G. "The External Ear." In *Handbook of Sensory Physiology, Vol. V/1, Auditory System*, edited by W. D. Keidel and W. D. Neff, 455–490. New York: Springer-Verlag, 1974.
47. Smith, S., R. D. Bergeron, and G. G. Grinstein. "Stereophonic and Surface Sound Generation for Exploratory Data Analysis." *Proceedings of CHI, 1988*, 125–132, held in Washington, May 15–19. New York: ACM, 1988.
48. Warren, D. H., R. B. Welch, and T. J. McCarthy. "The Role of Visual-Auditory 'Compellingness' in the Ventriloquism Effect: Implications for Transitivity Among the Spatial Senses." *Percep. & Psychophys.* **30** (1981): 557–564.
49. Wenzel, E. M. "Localization in Virtual Acoustic Displays." *Presence: Teleoperators & Virtual Env.* **1** (1992): 80–107.
50. Wenzel, E. M., M. Arruda, D. J. Kistler, and F. L. Wightman. "Localization of Non-Individualized Head-Related Transfer Functions." *J. Acous. Soc. Am.*: in press.
51. Wenzel, E. M., and S. H. Foster. "Perceptual Consequences of Interpolating Head-Related Transfer Functions During Spatial Synthesis." *Proceedings of the ASSP (IEEE) Workshop on Applications of Signal Processing to Audio and Acoustics*, held in New Platz, NY, OCt. 17–20, 1993. Piscataway, NH: IEEE, 1993.
52. Wenzel, E. M., P. K. Stone, S. S. Fisher, and S. H. Foster,. "A System for Three-Dimensional Acoustic 'Visualization' in a Virtual Environment Workstation." *Proceedings of the IEEE Visualization '90 Conference, San Francisco* **1** (1990): 329–337.

53. Wenzel, E. M., W. W. Gaver, S. H. Foster, H. Levkowitz, and R. Powell. "Perceptual vs. Hardware Performance in Advanced Acoustic Interface Design." *Proceedings of CHI, 1988*, held in Washington, May 15–19. New York: ACM, 1988.
54. Wightman, F. L., and D. J. Kistler. "Headphone Simulation of Free-Field Listening I: Stimulus Synthesis." *J. Acous. Soc. Am.* **85** (1989): 858–867.
55. Wightman, F. L., and D. J. Kistler. "Headphone Simulation of Free-Field Listening II: Psychophysical Validation." *J. Acous. Soc. Am.* **85** (1989): 868–878.

Robin Bargar
National Center for Supercomputing Applications, and School of Music, University of Illinois at Urbana-Champaign, 152 Computing Applications Building, 605 East Springfield Avenue, Champaign, IL 61820; (217)244-4692; e-mail: rbargar@ncsa.uiuc.edu

Pattern and Reference in Auditory Display

1. INTRODUCTION

This paper addresses the potential for identifying common concerns and collaborative potentials linking scientific research methods with the field of auditory display, a field that is closely related to music composition. The capability of listeners to differentiate sounds meaningfully is a complex construct that involves a system having a sound-producing potential and an organized observation of that system by a sound designer who may be considered a composer. By describing the application of music composition techniques to the auditory display of scientific data, a connection can be established between compositional thought processes and scientific observation.

1.1 ACTIVE LISTENING

The design of sound to help comprehend data will be aided by descriptions of the listeners' comprehension of sound. Active listening means comprehension by purposeful interaction with the environment[15] through sound.

This is to say that a sound conveys information when a listener is capable of retrieving that information from the sound. This capability rests upon the actions and decisions of the listener and upon the design of the sound to be differentiated by a listener's actions and decisions. Auditory display may be approached as a collaboration between sound designer and listener, where sounds inform a listener to extract information from them. The listener determines *how* to extract meaning based upon previous encounters with sounds carrying information. This model depicts listeners extracting meaning by reformulating a sound, and constructing references between the perceived signal and previous experience.

1.2 WHAT'S IN A MESSAGE?

Cognitive research can evaluate a listener's potential ability to comprehend sounds. This research is valuable and necessary; at the same time there is a complementary body of empirical research regarding listeners' receptions of complex messages, and that is the practice of constructing messages. Message structures contain clues to ways in which those messages are received. Asking what a message may reveal about it's receiver, we find a designer of an auditory message creates a description of the person(s) for whom the message is intended. For example, the designer must think of the repertoire of symbols that are known by the likely listener; a *repertoire* consists of a number of message elements and their possible combinations.[16] By using these combinations, meaning is constructed consisting of shared cultural designations that de Saussure refers to as la langue, and individual articulations that he names *la parole*.[6] The structure of a message anticipates a listener who is capable of decoding it, exchanging *langue* and *parole* during individual formulations of meaning. This decoding process will not work unless the receiving repertoire includes a greater or equal number of elemental components as the sending repertoire. By observing the ways a message predicts the presence of listeners with a certain repertoire, we can note the repertoire (references to ideas) and the structures in the message that denote the repertoire (references among sounds). Observing references among sounds that allow listeners' exchanges of sounds and meanings, we may establish principles to support a repertoire for nonspeech auditory display.

1.3 COMPOSITION[1]

A close relationship between composition and nonspeech auditory display can be found in structures for sound production and articulation of sound combinations. Composers are message designers predicting the potential presence of listeners capable of formulating meanings from a repertoire of auditory signals. In compositions, composers provide descriptions of listeners by offering auditory clues how to gain meaning from complex nonspeech auditory events. Structures used in music composition can inform auditory display designers of the descriptions of listeners and sound-production techniques. This paper introduces auditory display techniques based upon composition techniques, providing auditory examples of their construction and discussing several display paradigms—auditory scenes, frames of reference, iteration, sequence, and fidelity—observing differences between composition techniques and auditory practices currently supported by computer hardware and software manufacturers.

Reformulating the question "how to display this data?" we may ask how sound and composition function and how they configure data for display. Responses to this question include suggestions for new contexts for computational research: changes to the structure of human-computer interaction to utilize multimodal feedback and display, and changes to the structure of information to better correspond to listening and sound.

2. AUDITORY DISPLAY PARADIGMS
2.1 AUDITORY SCENES

An auditory scene describes listening to an environment and may provide a model for auditory display. In machine hearing research, an auditory scene describes a natural acoustic environment which includes perceivable as well as physical properties; auditory scene analysis of acoustic environments requires models of auditory organization in human perception.[5] Auditory scenes also appear as proposed display spaces for complex messages about data or computation; their design has been referred to as Acoustic Ecology.[4,13] Characteristics of auditory scenes have been discussed by Bregman,[3] McAdams,[11] Mont-Reynaud,[12] Gaver,[9] and others.

The auditory scene concept is closely aligned to the visual perception of three-dimensional objects and spatial locations. The use of the term "scene"

---

[1] Composition refers to the organization of sound to make music and other auditory presentations in which a natural language is not the primary organizing structure, including sound conjoining visual media and/or performance.

reinforces visual models of three-dimensional space. While this is appropriate for auditory localization, the n-dimensional space within acoustic events is not spatial. Boundaries are unclear for establishing perceptually orthogonal auditory dimensions. The visual perception of orthogonal spatial dimensions is relational rather than orthogonal: one compares dimensions to distinguish breadth, depth, and height. Likewise auditory perception of complex dimensions arises from perceptible intervals between sound characteristics, not merely from the quantitative separation of characteristics. In other words, differences are more perceivable if they are systematic, quantized to a specific set of intervals. Analogies drawn between traditional display parameters (pitch, loudness, timbre) and distinct perceptual dimensions are questionable. A one-to-one correspondence between physical parameters and perceptual parameters cannot be assumed.

Music compositions provide design examples of complex auditory scenes. Sound is the primary source of information in music, and is alleged to be the sole source of information; visual information and written text play a supporting role that is often not acknowledged (see Section 2.3). In a composition, sound and potential sound are continuously present, representing a contract between the presenter and the listener, an understanding that sound comprises a seamless artificial space; experimental compositions notwithstanding, the standard musical experience is designed to offer a diegesis—a fictional world an observer takes as seamless[2]—similar to that of novels and movies that encourage a "willing suspension of disbelief," in this case, a listener giving full attention to unfolding auditory events. This can be observed, for example, in popular music where the listener is encouraged to identify with the character the singer portrays (a similar function to that of actors in narrative cinema). Events in musical space are of varying duration and are often overlapping; their design is intended to direct the listener's attention, providing simultaneous events which demand different degrees of attention. Composition describes listening as consciously seeking the message potential of sounds, engaged in a continuous auditory diagetic space, capable of following transformations of sounds, and attributing meaning to differences among sounds.

Since "auditory scene" is a term with several uses, here *auditory space* is used to denote the information region of auditory display. A designer modeling an auditory space specifies a number of sound types with variable acoustic characteristics; a sound type produces one or more instances of a sound that can be transformed, can appear simultaneously with other sounds, and can be used to construct sound sequences. Auditory space includes the range of transformations a sound may undergo and the sources of data that provide transformation values.

---

[2] "Diegesis" is used as it is used in film theory and criticism to denote the novelistic, fictional world of a narrative.

Music provides an exclusively auditory space for auditory display. Implementing auditory display in a desktop environment suggests examining the compatibility between workstation paradigms and principles of music composition. There are many differences.

Regarding the computer interface, note that certain computer games utilize audio extensively, often emulating leitmotivs[3] and sound effects used in movie soundtracks. Attending to these auditory cues gives advanced knowledge of game conditions. These sounds constitute a rather limited auditory space; for example, the sounds rarely undergo transformations. At best these games present multiple simultaneous events of mostly invariant sounds. A highly redundant scenelike texture results.

Traditional computer interfaces are marginally multimodal. The desktop is an uncomposed environment where sounds among applications occur without a common protocol for their design. The user's capability of interpreting computational events using audio remains largely untapped. Applications are first designed to function visually, in silence; auditory messages are intermittent, not continuous. This type of environment describes a hypothetical listener who has little expectation for auditory information and is not able to follow sound transformations, nor to attribute significance to changes among sounds. In addition to constraining a useful modality, these interfaces teach users to disregard listening skills. Users adopting a description of themselves as uninformed listeners can experience difficulty when a wider palette of auditory signals is introduced into the environment.

2.2 FRAMES OF REFERENCE

A frame of reference is a symbol used to measure one or more other symbols or potential symbols. The measurement context is hierarchical: perceptions are referenced to symbols, which are referenced to an even smaller class of still more prevalent symbols, and so on. In music a sound provides a frame of reference for other sounds; a sound that is most often heard, or most easily remembered among a collection of sounds, will tend to become a frame of reference for the others in the collection. A large portion of composition technique is devoted to establishing frames of reference in multiple layers. Recurrence, including intermittent iteration and self-similarity, provides temporal orientation toward intended messages in sound (see Mayer-Kress et al., in this volume). By constructing and altering frames of reference a composer establishes guidelines for listeners, who are formulating descriptions of symbols based upon the sounds they hear.

---

[3] A leitmotiv is a short musical theme associated with a particular character or narrative event; Wagner used leitmotivs as the basis for associating musical and dramatic development in opera.

Learning from this practice, a designer of an auditory display can notice that in music, separate sounds do not necessarily represent separate musical ideas; some sounds are serving as references for others. If an auditory display uses analogous hierarchical organization, the number of sounds will be greater than the number of data dimensions being represented. In music, reference sounds are associated with a background layer that changes less often than the musical foreground. At the same time, the background is often perceived through the foreground rather than as a separate musical idea (rhythm, melody, and ornament are examples of foreground while harmony and tonal center are identified as background materials). Adopting foreground-background relations in auditory display may allow for rapid changes of display context by altering the background in which an auditory pattern appears. Further research may indicate that these models are most effective when the references and data changes are bound into a comprehensive musiclike fabric.

2.3 MULTIMODEL FRAMES OF REFERENCE

At the computer interface there are three frames of reference that, in most cases, take precedence over auditory information: texts, images, and the user's actions. Auditory display can accommodate these by making explicit reference to them or by deactivating them: in the first case by indicating that the nonauditory references will provide information utilized in tandem with the auditory display, and in the second case by providing information that supersedes unintended combinations of auditory and nonauditory information. For example, NCSA Audible Image Software[18] utilizes visually defined regions of a data set to give meaning to sound sequences—the shape and size of the region, the sonification method specified for the region, and the mapping of color to pitch; these regions assist the user in establishing a bandlimited association between sound and data. The sound patterns can enhance the visual change across the display of numerical data by providing more noticeable auditory change; without the screen and mouse the same sound patterns would not acquire their data-correlated relevance, though other relevances might occur.

The triangulation of action, image, and sound provides correlations that can override unintended messages more effectively than image and sound without interaction. Users making specifications and observations among images and sounds can detect covariations emerging across the three modalities (vision, hearing, action). In ecological psychology there are called *intermodal invariants*. By returning meaningful information to more than one perceptual modality, intermodal invariants reveal consistencies in an environment or a system. When users are able to formulate increasingly consistent descriptions of the environment, they are encouraged to distinguish structural features from local coincidences that are not relevant to the underlying data.

Multimedia presentations often display simultaneous yet uncorrelated representations. Correlation is not a natural result of simultaneous multimodal perceptions. Correlation can occur when a media designer (composer) specifies relationships among the display rates and symbolic units of each medium. Relationships may be thought of as relative capabilities for change among modalities; change potentials may be specified as ratios between system parameters (for example, "parameter a covaries inversely with parameter b at a rate of $2:1$"). Observers can extract correlations from covariance that consistently manifests this type of relationship.

Nonauditory frames of reference include the presence of performers, texts, and other listeners. These are often present during music listening, and composers have developed extensive techniques for accommodating them, either by incorporating them into a composition or by deactivating them with other references such that unintended messages are minimized.

2.4 UNINTENDED MESSAGES

Beyond the composer's action to make sounds, yet within the composer's design, are references to sounds from outside of the composition which may reside in a listener's memory. The composer creates sounds that potentially evoke a reference to memorable sounds from other music (which outnumber the sounds that occur within a single composition). A composer who generates a significant symbolic construct will also articulate a relationship to the history of sounds in her or his environment. This articulation includes techniques for avoiding unintended references.

Mapping numerical data to a display medium establishes a frame of reference, yet it does not necessarily generate greater significance than the observation of other unintended frames of reference. Mapping associates a numerical value with more than a display parameter; the mapped variable becomes correlated with the entire display context. Meaning accumulates from each frame of reference involved in the display hierarchy. For example, interpretation of a standard scientific visualization integrates two-dimensional position, color, text, image-processing, and display functions, and the shape and size of patterns emerging from the display.

The accumulation of references can occur in unexpected ways unless specifically counteracted; for example, mapping a numerical value from an image to a sound will acquire meaning in a videotape display based upon the history of the motion picture sound track (including the tendency to trivialize sound via synchronization[1]) and the passive role video assigns to viewers[2,14]; the quantitatively identical mapping will provide different meaning in an interactive computer display, providing a greater chance for the user to discover correlations between sounds and images, as well as other potentially unintended associations that may emerge from the workstation context (such as references to computer games).

2.5 ITERATION AND SEQUENCE

A display iterates a field of information; an observer iterates an observation. That is, observation is not continuous; attention is directed and acquired information is formulated into discretized impressions.[15] Patterns in auditory messages anticipate iteration and grouping of perceptions. Iteration in music tends to have chaotic properties[4]; beginnings and endings of events overlap in time, and repetitions are not periodic. The size of a meaningful unit can change from one to many notes. Musical time is specified by sequential structures that present the listener with multiple frames of reference for length of event and rate of change. Changes of pitch and duration in sequences are measured against previous sequences within a composition, and constitute music's temporal-symbolic context. Brief and long-term events occur simultaneously, and a listener's formulation of duration and rate of change depend upon the events that the listener identifies as the perceptual foreground.[5]

Computer interfaces define auditory experience in terms of text-based or icon-based tasks. In these interfaces, auditory events are not designed to represent multiple time scales; apart from tasks there is no concept of duration. Interface tasks tend to trivialize the notion of sequence: an event consists merely of the actions that the user performs to complete a computational task. Once underway, a task-based event offers little symbolic information other than its status of completion. The potential use of sequence as a display space for complex symbolic events is subordinated to the need for task completion. For example, if every time a user makes a selection the computer issues a sound confirming that selection, the sequence of sounds is entirely redundant to the users' actions and cannot convey other information.

The SonicFinder,[8] an interface tool which produces sounds for mouse-driven desktop tasks, reveals the symbolic shortcoming of the monotonic-task-based workstation. The sounds are informative of actions; however, most of the actions are too trivial to merit this information. This triviality is magnified by the sounds, due partly to their unchanging auditory

[4] Chaotic behavior includes periodic, quasi-periodic, intermittent, and seemingly random sequences, in local and broad regions of an attractor.[19]

[5] Sources and elaboration of these ideas may be found in the Preliminary Examination papers for the Doctor of Musical Arts degree, by Insook Choi, University of Illinois at Urbana-Champaign.[10]

characteristics and partly to the low resolution of sound reproduction. In a sense the idea is too rich for the environment: sounds are channeled into a narrow representation space where they are not given the capacity to undergo transformations or form symbolic groups (hierarchical references are not established). As an alternative, a multitasking workstation environment could allow users to apply varying degrees of attention and interactive steering to a number of simultaneous processes with different processing rates; here the need to represent layers of computational events could be met by composition techniques for articulating multiple auditory channels at multiple time scales. The availability of such an interface might stimulate the development of advanced interactive multiprocess computing.

2.6 FIDELITY AND AMBIGUITY

When sound can be reproduced with high fidelity, the differentiation of similar sounds can be used for complex representations. When sounds are reproduced with low fidelity, the listener's task changes from differentiation to categorization. Sounds in a low-resolution display can only be used for messages which convey information regarding membership in established categories. The standard example is the "sound byte,"[6] a brief sound effect used to communicate the status of a process. Sound bytes have static properties and are rarely used in sequence; they are essentially a class of alarms: they merely require recognition. Auditory display in most computational environments identifies recognition as the outer limit of its functionality. Interpretation of changes in dynamical sequences requires a sensitivity to acoustic ambiguity which most computer audio systems do not support.

In high-resolution audio environments, sound synthesis techniques can provide a continuum from recognizable sounds at one end to abstract sounds at the other. Traversing this space allows a wider range of meanings by introducing ambiguity into the auditory language. Ambiguity in a composition is the intended reference of a sound pattern to more than one established and recognized interpretation.

When is ambiguity valuable for auditory display? First, when it is the property of an intended message and not of a poor display system (one needs a high-fidelity sound system to experience this distinction). In a high-fidelity environment, the creation of a space between two or more recognizable sounds can be experienced as an intended message. For example, an unfamiliar sound produced by interpolating between two known sounds can indicate a system is in an unstable state between two potentialities. When the data being represented has important ambiguous properties, or has not yet been classified (fuzzy matching), or awaits user input to assist

[6]This term is commonly used in media production.

its further clarification, in these cases a symbolic continuum between known sounds could assume meaningful properties by its capability to maintain ambiguity.

3. AUDITORY DISPLAY TECHNIQUES
3.1 THE DISPLAY SPACE

There is a great deal of research currently devoted to virtual three-dimensional auditory display. Issues of spatial location and externalization (see Wenzel, in this volume, and Durlach et al.[7]) are closely correlated to visual display of a virtual space. Spatial localization has long proved valuable in electro-acoustic music and is a standard audio recording technique. For example, most popular music albums contain selections where various instruments appear to be positioned both in a stereo field and at a distance from the listener. This distribution in artificial space (it is unlikely that the ensemble was recorded all at once) is intended not so much for reproducing a realistic three-dimensional performance as for the clarification of the sound of each instrument in the mix. Live performances rarely offer the spatial resolution of a studio mix; in a live environment the listener obtains spatial information by means not available in a recording (head movement, for example, and sound reflections from the sides and the rear of the room if one is indoors).

Localization research is devoted to supporting the relevance of these natural perceptual mechanisms (such as head movement) in a virtual environment. Localization and externalization add spatial dimensions to n-dimensional auditory space. Wenzel has demonstrated that head-related transfer functions (HRTFs) can be used to reproduce fully dimensional sound localization[17]; digital multichannel real-time applications of HRTFs may be anticipated in the near future.

As localization becomes a real-time functionality there will be a demand to design sounds that render well when localized. The organization of the n-dimensional symbolic space is a separate task from the display of 3-space; one can think of constructing sounds that are to appear in localized space. A number of composition and studio recording techniques offer organizational tools for articulating sound as an n-dimensional display space. These dimensions include synthesis parameters, auditory parameters, and symbolic structures that a listener formulates from psychoacoustic dimensions. It will be important to understand the auditory side effects that localization may cause in a sound, so that sounds may be designed to avoid undue alteration and to maintain their relevant characteristics in different locations.

3.2 TECHNICAL RESOURCES

The material properties of mechanically reproduced sound can help determine the symbolic potential of an auditory display. Auditory display theory and practice will follow from a designer's capability to generate acoustic material that is intelligible.

The sound examples accompanying this text systematically explore a number of techniques absent from many auditory displays. Included are music composition techniques which are supported by aspects of sound synthesis and by engineering techniques used for audio recording. These examples were produced at various facilities at the University of Illinois at Urbana-Champaign. Sound synthesis software was developed in the Computer Music Project; techniques for data sonification were implemented in the Software Development Group facilities of the National Center for Supercomputing Applications; production engineering, audio recording and mixing of the completed examples occurred in the Experimental Music Studios.

Tr 25

3.3 AUDITORY EXAMPLES

In the accompanying auditory examples, the techniques demonstrated are organized around symbolic constructs, including sequence, inflection, meter, rhythm, register, tonal center, polyphony, spatial location, and environmental ambiance. These are presented in musical excerpts.

A sequence may be defined by the physical parameters of the start time and duration of its constituent sounds. A listener is likely to define sequences in terms of pattern and repetition, which originate perceptually. Acoustic display depends upon the length and complexity of sequences; it is helpful to correlate the synthesis description of a sequence with characteristics that a listener might describe. This can be done by grouping synthesis parameters into syntactic units, that is, by designating along the synthesis continuum particular parameter combinations that are acoustically distinct, and locating these within the minimum and maximum values to be synthesized.

Inflection may be synthesized by altering loudness and timbre. This is demonstrated in example one, where inflection creates tone color changes that allow listeners to parse the sound stream into phrases that have irregular, speechlike timing.

Tr 26

Inflection is derived from vocal articulation, and is usually perceived as a component of speech, where spoken text may be thought of as the carrier signal and inflection as a modulating signal that adds significance by indicating which of several possible meanings a word is intended to convey. Inflection is a binding context for messages at a level above lexical meaning and syntax: inflections indicate which sounds among a sequence have the greatest influence on content. Inflections may be transferred to nonspeech auditory materials and are commonly used in music performance to enhance sequences, where notes are given emphasis in proportion to their importance in the tonal and metric hierarchy. Inflection may be synthesized by altering loudness, timbre, and envelope.

Tr 27   Meter is a regular time reference which underlies duration; meter is implied but not continuously instanced; in this regard it can be distinguished from rhythm, which is a specific pattern of durations. Meter is a reference across events at regular time intervals; even when a meter is asymmetrical, its elementary units are regular. Meter summarizes local time into groups, allowing listeners to predict the onset of the next group and thus to reflect upon the current group as it becomes the past. This contributes to the memory of events in reference to their order. Examples one and two are not organized to convey a regular meter, though it is possible to describe the timings of patterns as rhythms. In examples three and four, some individual sequences are organized around a meter, however the overall structure is not tied to a single meter.

Tr 28   Rhythm is a sequence of durations measured in micro and macro units. Durations are divisible into elementary units; at the same time this quantization allows regular groupings of elementary units to display a longer period, which is perceived in addition to the rhythmic pattern. This implied longer period allows a rhythm to convey the meter by which it is measured.

Register (frequency range interpreted as pitch) and tonal center express relative pitch; relative pitch range is displayed hierarchically. The octave is a natural perceptual boundary; pitch range can be described as either "within one octave" or "across octaves." To reinforce this hierarchy most pitch sequences unfold within an octave. Transposition (relocation) of pitch by one or more octaves is considered an identity operation (pitch has not changed, but octave has); if the primary pitch display parameter is kept within an octave, then register change can convey an additional parameter. In each audio example, register is used to create one or more *voices*, sound streams that display variation within a pitch bandwidth. Distinct registers can improve the display of simultaneous sequences.

Pitch is the dominant variable for organizing traditional musical sequences. In the first audio example, sequences are constructed by variations in loudness and ttone color (brightness), with limited pitch variance. In example two, a repeated pitch pattern with periodic timing (called *ostinato* in music) imposes a sequence that is me recognizable (predictable) upon the timbre-based sequences. Examples three and four include transitions between pitch-based and timbre-based sequences, creating ambiguities for listeners.

Tonal center expresses a much smaller frequency range; the division of the octave into discrete steps allows the perception of the frequency continuum as a class of frequency ranges around each step. This cannot be perceived unless the quantization value is regularly reiterated. In other words, without a reference tone or reference interval, it may be difficult to hear small pitch shifts that are *out of tune*. Tonal centering and tonal deviation can be symbolically nested to make a three-tiered hierarchy: deviation around a tonal value, within an octave, at a certain register. In the audio examples, tonal centers appear whenever a limited pitch set is the primary organizing method for a sequence. Example one has no tonal center. In example two the ostinato creates a pitch-related tension that is resolved, demonstrating that a tonal center is present. In examples three and four, brief melodies (with tonal centers) are created when pitch sequences are reinforced by regular rhythms and narrow pitch registers.

Polyphony, the presence of multiple simultaneous parts, depends upon the impression that a sequential stream can be distinguished from another simultaneously played sequence. This distinction is established by register, timbre, and characteristic rhythmic and pitch changes. Equivalent perception of two ongoing events is difficult to perceive quantitatively for an extended time; rate and bandwidth influence the perception of the number of streams.[11] Traditional musical polyphony tends to depend on each stream manifesting unique pattern characteristics which are quickly summarized in a listener's memory. Examples two through four provide multiple streams by layering sequences organized primarily by pitch, with others organized by timbre, further distinguished by register, rhythm, and inflection.

Tr 29

Spatial distance and ambiance characteristics are similar to those identified by localization research. These examples utilize traditional stereo display techniques. Stereo position is synthesized by loudness differences, interaural filtering, and interaural delay; depth cues are synthesized by loudness, filtering, and reverberation (involving multiple time delays to iterations of a signal, with attenuation and filtering increased dynamically). In example four many sounds are located in a simulated middle ground or background depth, also in a stereo field. For example, a rhythmic percussion voice is set in the far background and travels back and forth in the stereo field. Its repetitive characteristics help to keep it identifiable, so that it can be used to establish a perceptual cue for distance. These techniques are used to expand the virtual auditory space as an n-dimensional space and to distinguish signals, and are not intended to represent a visual field auditorially (see Section 3.1).

4. CONCLUSIONS

There are rich aspects of auditory display related to musical structure. Composition techniques may be utilized to increase the complexity and the intelligibility of auditory display. The technical fidelity and the conceptual framework of most computer-based audio displays are incapable of supporting complex display techniques. In addition to improved sound quality, the reformulation of the monotonic task-based workstation, eliminating its constraints upon sequential display, is a prerequisite to empowering auditory display with the design aspects that composition techniques can provide.

REFERENCES

1. Bargar, Robin. "Composition and Synchronized Sound, from Opera and Cinema to the Computer." In *Proceedings of the New Music and Art Festival*. Held at Bowling Green State University, Bowling Green, OH. Forthcoming.
2. Berger, John. *Ways of Seeing*. New York: British Broadcasting Corporation and Penguin Books, 1972.
3. Bregman, A. S. *Auditory Scene Analysis: The Perceptual Organization of Sound*. Cambridge, MA: MIT Press, 1990.
4. Buxton, Bill. Reference made during a presentation at the International Conference on Auditory Display. Held at the Santa Fe Institute, in Santa Fe, NM, October 1992.
5. Chowning, John, and Bernard Mont-Reynaud. "Intelligent Analysis of Composite Acoustic Signals." Technical Report STAN-M-36, Department of Music, Stanford University, Stanford, CA, May 1986.
6. de Saussure, Ferdinand. *Course in General Linguistics*. Translation by W. Baskin, edited by C. Bally and A. Secheyae. New York: McGraw-Hill, 1966. With permission from The Philosophical Library, New York, 1959.
7. Durlach, N. I., et al. "On the Externalization of Auditory Images." *Presence* **1(2)** (Spring 1992): 251–257.
8. Gaver, William. "The SonicFinder: An Interface that Uses Auditory Icons." *Hum.-Comp. Inter.* **4(1)** (1989): 67–94.
9. Gaver, W., R. B. Smith, and T. O'Shea. "Effective Sounds in Complex Systems: The ARKola Simulation." In *Reaching Through Technology*. CHI '91 Conference Proceedings, Human Factors in Computing Systems, 85–90, Held in New Orleans, LA, April 1991. ACM Press, 1991.
10. Insook, Choi. "Preliminary Examination Papers for the Doctoral of Musical Arts Degree." Unpublished paper, School of Music, University of Illinois at Urbana-Champaign, Urbana, IL, 1991.

11. McAdams, Steve, and Albert Bregman. "Hearing Musical Streams." *Comp. Music J.* **3(4)** (1979): 26–43.
12. Mont-Reynaud, Bernard. "Machine Hearing Research at CCRMA: An Overview." In *Research Overview, Center for Computer Research in Music and Acoustics*, edited by Patte Wood, 24–32. Stanford, CA: Stanford University Press, 1992.
13. Schafer, R. Murray. *The Tuning of the World*. Philadelphia, PA: University of Pennsylvania Press, 1980.
14. Sekula, Alan. *Photography Against the Grain*. Halifax: Press of the Nova Scotia College of Art and Design (no date).
15. Stoffregen, Thomas A., and Gary E. Riccio. "An Ecological Theory of Orientation and the Vestibular System." *Psych. Rev.* **95(1)** (1988): 3–14.
16. Umberto, Eco. *A Theory of Semiotics*. Bloomington: Indiana University Press, 1976.
17. Wenzel, E. M. "Localization in Virtual Acoustic Displays." *Presence* **1** (1992): 80–107.
18. Wilson, Chris, and Robin Bargar. *NCSA Audible Image User's Manual*. Urbana, IL: National Center for Supercomputing Applications, University of Illinois at Urbana-Champaign, 1992.

Stuart Smith,* Ronald M. Pickett,† and Marian G. Williams‡
*Computer Science Department, University of Massachusetts, Lowell, MA 01854, e-mail: stu@cs.ulowell.edu
†Psychology Department, University of Massachusetts, Lowell, MA 01854, e-mail: pickett@cs.ulowell.edu
‡Center for Productivity Enhancement, University of Massachusetts, Lowell, MA 01854, e-mail: mwilliam@cs.ulowell.edu

Environments for Exploring Auditory Representations of Multidimensional Data

The field of auditory data representation has produced several intriguing proof-of-concept systems, but there has been little formal research to measure the effectiveness of auditory data displays or to increase our understanding of how they work and how to improve them. We argue that formal assessment is necessary throughout the process of developing new auditory display technologies in order to learn how to restrict the universe of possible sound attributes to those that are most effective for data representation. The ability to run quick psychometric tests to obtain quantitative figures of merit for alternative auditory representations is a requirement for auditory display, researchers engaged in the development of new technologies. This capability can be realized with a special-purpose workstation designed to generate and administer psychometric tests automatically using test patterns generated from statistically well-specified synthetic data. We outline the requirements for such a workstation and describe a testing method for the development of a new type of auditory data displays that we have been working with for the last few years.

1. INTRODUCTION

In the last decade much has been made of the burgeoning flow of scientific data and of the need for greatly improved capabilities for exploratory data analysis. Techniques for data visualization offer enormous promise but, while many advances are being made in that direction, the limitations of visualization in dealing with data of high dimensionality have become increasingly apparent.

Investigators seeking alternatives to visual display or ways to raise the dimensionality of visual data representations have been attempting to develop effective methods for encoding quantitative information in sound.[3,20,21,24,38,41,50] In experiments comparing subjects' performance on a variety of data analysis tasks using visual data representations alone and combined visual and auditory data representations, subjects have shown modestly improved performance when using the combined representations (see, for example, Kramer and Fitch (in this volume) and also Bly,[3] Mezrich et al.,[24] and Williams et al.[49]). While these studies have proved the concept of auditory data representation, much work remains to be done to establish the value of auditory data representation as a tool for analyzing and exploring data.

There has been little formal research, either to measure the effectiveness of auditory data displays in real-world applications or to increase our understanding of how they work and how to improve them. Frysinger[9] has given a concise statement of the overall task that must be accomplished:

> We must discover the set of truly useful auditory parameters and understand their perceptual transfer functions so that displays can be designed to take advantage of them. Likewise, we need to understand which data analysis tasks can most benefit from auditory data representation, and what types of displays to apply to them. Finally, the interaction between visual and auditory data representation should be understood so that the best combination of the two can be chosen for a given analysis task. (p. 136)

Research in this area has been severely hampered by the kinds of sound generation devices available. These devices, mostly Musical Instrument Digital Interface (MIDI) units, provide the capability to play notes in a variety of familiar sonorities, but they offer only limited control over the detailed structure of sound. While all MIDI sound generators provide some control over pitch and loudness, most offer little or no control over such useful attributes as the rates of vibrato and tremolo, the rates of attack and decay, and changes in the spectral content of sounds over time. The low resolution of MIDI data—seven bits maximum in most implementations—restricts the degree to which subtle shifts in data can be represented in sound.

Most of the useful sound attributes available in a given MIDI sound generator must be controlled via the MIDI "System Exclusive" command, a device-dependent bypass of the MIDI standard. The values of the sound attributes controllable via System Exclusive are specified with device-dependent binary codes rather than in conventional units such as Hertz, milliseconds, decibels, etc., and manufacturers typically give only vague indications of the relationships between MIDI data values and standard units. Thus, it is often difficult to know precisely the characteristics of a signal generated by a MIDI unit.

MIDI is adequate for the development of toy or demonstration systems, but it is clearly inadequate for the development of production-quality systems to be used for scientific work. Systematic and quantitative studies of auditory data representation require much more capability for precise specification and manipulation of sounds. In particular it is necessary to have sound generation techniques that offer rich sets of predictable, perceptually relevant transformations. These transformations provide the essential "hooks" that allow data dimensions to be associated with perceived attributes of sound. Among the many potentially useful sound synthesis methods are frequency modulation,[5,6] nonlinear distortion or "waveshaping,"[16] and granular synthesis.[33,34]

A sound facility for auditory data representation should offer, as a minimum, the ability to synthesize fairly complex sounds in real time and the ability to construct arbitrary sound-synthesis software modules in a convenient way. The facility will need to have both special-purpose sound-generating hardware and special-purpose software that offers (1) powerful sound description, (2) the control and timing functions necessary to cause each precisely specified sound to happen at a precisely specified time, and (3) a graphical user interface. Recently, two systems offering such capabilities have appeared: the IRCAM Music Workstation,[17,18,30,31,46] which is built around the NeXT computer, and the Kyma system,[36,37] which uses a Macintosh II or 486 PC and a special, attached sound processor. While neither of these systems was designed specifically for auditory data representation, both are adaptable to this purpose.

2. A STARTING APPROACH TO REPRESENTING DATA IN SOUND

Recent developments in sound generation systems, described above, along with a steady trend toward higher performance computing, open up a world of possibilities for auditory data representation. We will sketch some of the available alternatives and comment on their complexity.

Consider a large collection of multidimensional data, in which each datum comprises r individual values, called its dimensions. We find it helpful to consider the creation of an auditory representation of the data at two levels. On the first level, an auditory element is created from each datum. On the second, a display is created from a whole collection of such elements.

There are many design options at the first level. A fundamental option is whether to create natural-sounding elements or to create arbitrary or abstract ones. While natural sounds like those emitted by real physical objects and events are generally more difficult to implement computationally, such sounds may evoke more effective perceptual analysis of the data than abstract sounds (the key issues concerning the synthesis of natural sounds for computer interfaces are discussed in Gaver's paper "Synthesizing Auditory Icons," in this volume).

Each auditory element has distinctive audible attributes that are controlled by the datum's values. A multidimensional datum can be mapped onto such an element in a great many different ways. Its dimensions can be mapped, for example, to pitch, loudness, duration, attack and decay rates, the parameters of amplitude and frequency modulation, the parameters of various kinds of filters, rates of pitch slide (glissando or portamento), and changes in waveform. For n sound attributes and r data dimensions ($r \leq n$), the number of different ways to pair data dimensions with sound attributes is the number of permutations $P_r^n = n!/(n-r)!$. Even for n and r as small as 10, there are well over three *million* possible ways to pair sound attributes and data dimensions. The many different mapping functions that can be used for each pair increase the number of possibilities even further.

Current knowledge of human auditory perception does not tell us which combinations of sound attributes are the right ones to use for the representation of multidimensional data. Most basic research on audition has focused on the perception of single sound attributes[39,40] or interactions between two attributes.[7,8,11,29,34,45-47,53] Investigations of the perception of musical timbre[2,11,12,28] have attempted to find a small number of auditory dimensions on which differences in the tonal qualities of sounds can be represented; however, these efforts have not produced a satisfying comprehensive theory of timbre perception. There are no timbre models corresponding to the various color models in vision, whose dimensions behave in psychologically simple ways and have straightforward physical definitions.

Because of the limitations of the available sound generation devices, we have, until now, been spared the necessity of dealing with the combinatorial explosion of possible representations. The size of the universe of sound attributes, as well as the user's control over those attributes, has been artificially restricted by the limited capabilities of MIDI sound generation devices. The default model of auditory data representation has been simply to map data dimensions arbitrarily to available sound attributes. When high-performance, general-purpose, sound generation hardware becomes widely available, the multitude of possible representations in this large, poorly understood domain will have to be faced directly. In Section 3, we consider the need to provide efficient evaluation procedures for comparing alternative auditory representations; in Section 4, we discuss the need for tools for managing the exploration of the world of possible representations.

At the second level of designing an auditory representation, we are concerned with how to display a collection of auditory elements in a sufficiently compact way to evoke perception of the statistics of the data. One basic approach is to present the elements in rapid sequence; another is to present them in parallel, as a kind of ensemble; and a third is to create ensemble sequences. Until the feasibility of computational implementation intercedes, perceptual considerations should be paramount. What kind of display would evoke rich and discriminating perceptions? Again, as with the design of individual sound elements, we believe it is important to think about natural displays and the natural processing of such displays.

As a starting approach to auditory data representation we suggest a method that is an extension of an "iconographic" approach we have been developing for data visualization.[13,26] In our visual iconographic approach, each datum is represented by an icon whose visual attributes are controlled by the data. In Figure 1 is shows an icon consisting of five line segments joined end-to-end.

Data can be mapped to the angles, lengths, or intensities of the line segments, or to any combination of these attributes. Two of the data variables, not necessarily independent of those controlling the icon attributes but hopefully largely independent of each other, control the position of each icon on the display surface. With sufficient density, the icons form a surface texture display, and structures in the data are revealed as streaks, gradients, or islands of contrasting texture.

Figure 2 is a typical iconographic picture. It depicts the solution of a partial differential equation that describes a reaction inside an inductively coupled, plasma chemical reactor.

Five of the 22 parameters at each solution point are mapped redundantly to the lengths and angles of the five line segments from which the icons in this picture are formed. The principal feature in the picture is the contour of the plasma, which enters the picture near the upper left and then loops over to the center of the right side, where the plasma is drawn by an electromagnet. The information in this picture is conveyed by the textures formed through the orientations and overlappings of the icon elements.

FIGURE 1 Five-limbed stick-figure icon.

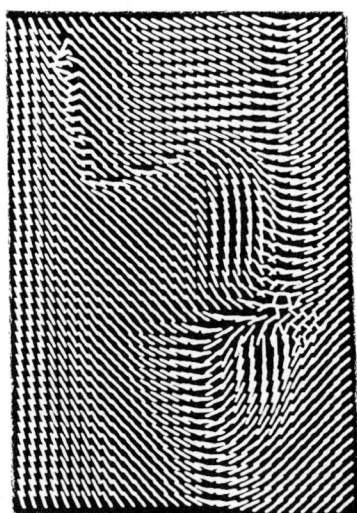

FIGURE 2 Iconographic composite of partial differential equation data.

We have used the success of displays based on visual texture perception as a lead for the development of data representations based on the perception of auditory texture. A second example of iconographic display illustrates how we use sound textures to assist visual data analysis. In Figure 3 we show three gray-scale images with normally distributed random intensities.

Exploring Auditory Representations of Multidimensional Data   **173**

These images were generated from vectors of normally distributed deviates having a mean of 0 and a standard deviation of 1. There is a correlation of approximately +0.5 between each pair in the top halves of these images and approximately -0.5 between each pair in the bottom halves. These relationships are not obvious from casual inspection. In Figure 4, an iconographic composite of these three images, a subtle textural difference between the upper and lower halves of the picture is visible.

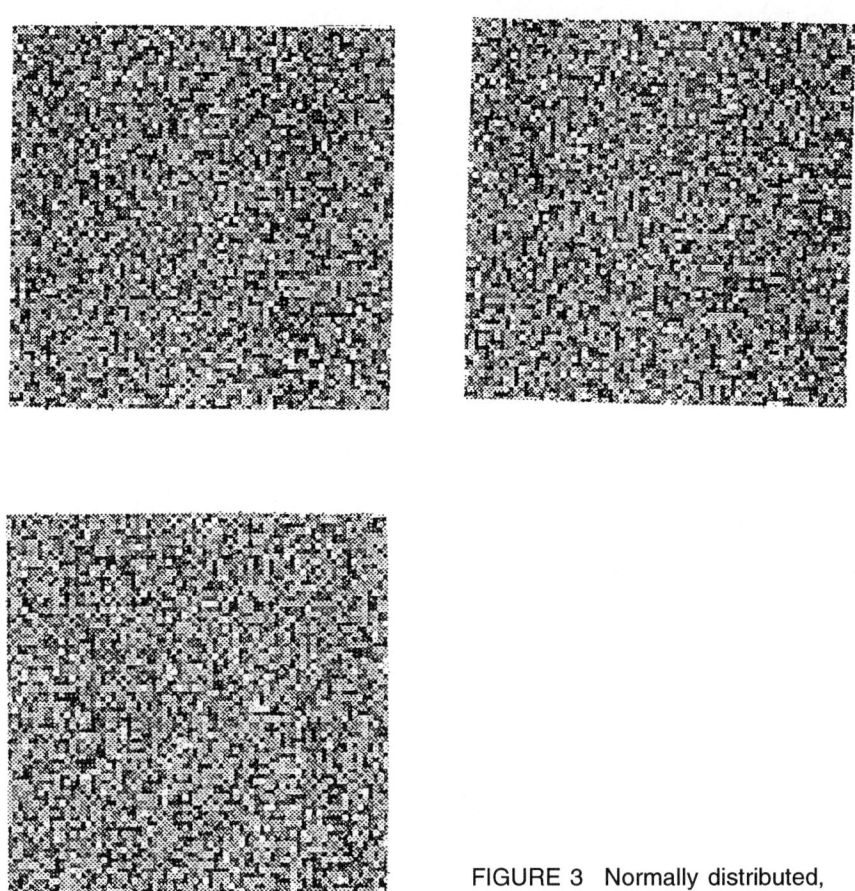

FIGURE 3 Normally distributed, random-intensity, gray-scale images.

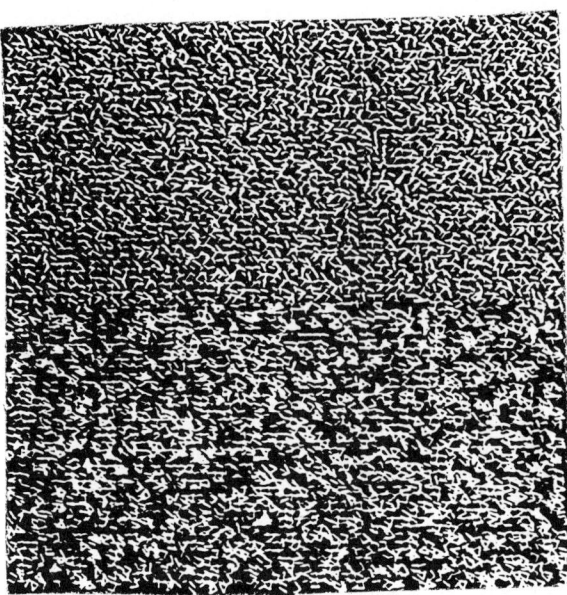

FIGURE 4
Iconographic
composite of the
random-intensity
gray-scale images.

By sweeping the mouse cursor over an iconographic picture like Figure 4, the user can trigger an auditory data representation consisting of multiple small and partially overlapping sonic events, such as tones or noise bursts. The textural character of the sound produced can change audibly when the statistics of the data controlling the sounds change, thereby indicating the presence of a boundary or contour like that in Figure 4.

The idea for using sound in this way finds some support in the work of Bregman. Bregman[4] has speculated on the existence of *auditory* grain analyzers capable of segregating "streams" of sound on the basis of texture differences. Bregman defines an auditory stream as the perceptual unit that represents a single happening (see Bregman,[4] p. 10). The stream plays the same role in auditory mental experience that the object plays in visual experience. Bregman cites an experiment by Warren and Verbrugge[47] which shows that the auditory system, essentially on the basis of texture differences, can easily distinguish the sound of a bottle bouncing on a resilient surface from that of a bottle breaking against a hard surface. Bregman (see Bregman,[4] p. 121) concludes that "a granular analysis might provide a perceptually relevant description of the sounds of events that can be considered iterations of multiple small events."

Bregman cautions (see Bregman,[4] p. 121) that it may not be practical to use granularity statistics to create a *sharp* boundary between streams because the auditory system might need a temporally extended sample to assess the statistical properties of each granularity. We do not consider this a serious problem for our approach because we use sound not alone but in support of visual analysis. Therefore, it is not necessary that sound establish a sharp boundary between regions. It is sufficient if sound can call attention to a boundary and indicate its approximate location for subsequent visual and conventional mathematical analysis.

3. A STARTING APPROACH TO EVALUATION

At this early stage in the development of the field of auditory data representation, we are concerned not only with determining how well particular representations reveal structures of particular kinds in data but also with understanding *how* these representations work. We need this knowledge to steer our explorations of the vast space of possible representations towards the ones that work best. Gathering this knowledge requires systematic and quantitative testing of auditory representations, using test patterns that contain well-specified and systematically manipulable data structures.

Like designing auditory representations of data, evaluating representations confronts the investigator with many options. There are two general considerations: the type of data structure to study and the kind of auditory task to use to study it. The data structure question will be examined first, followed by a brief discussion auditory tasks.

In choosing data structures, one direction to look is toward the real world settings in which the representations are expected to be employed; however, the common problem in exploring real data is that more often than not the structures in such data are not well understood. If they were well understood, it would not be important to study them with the techniques discussed here. Also, there is usually little available truth data (i.e., information independent of the database that establishes the presence and location of a known structure) on which to select reliable test patterns. Ultimately, a representation should be validated with real data, but it is reasonable to begin by looking at fully specified and systematically controlled data structures embedded in *synthetic* data.

A synthetic database is one created by the operation of some random process. The output of the process is mapped to the spatial or temporal interval of the display and onto the attributes of the elements to be displayed at each locus within the interval. Data structures are created by altering one or more parameters of the process as it is mapped over the display interval. In this way, one can create sets of test patterns in which the type of structure, its location in the database, and its degree of sharpness or clarity can be precisely specified and controlled. The challenge for the researcher is to make good choices of the random process, the parameters to be altered, and the patterns of alteration over the display interval.

It is reasonable to work first with a familiar and well-understood random process, like a multidimensional Gaussian, and to work with shifts in a simple statistic like the mean or variance on one or more dimensions. One might graduate then to manipulating correlations among the dimensions. The random-process samples from a Gaussian-shaped cloud in n-space. Shifts in the mean on one or more dimensions move the cloud around in n-space, while shifts in the variance on one or more dimensions elongate or compress the cloud along those dimensions. Shifts in correlation change the cloud's shape and orientation. Working within this concept of sampling clouds of points in n-space, Bly[3] created test patterns by sampling points from cylindrical structures. But where do we go from here? The world of possibilities is mostly beyond our current understanding. There is a critical need for help, possibly from specialists in statistical pattern recognition.

Three main types of tasks are used in studies of this kind: detection, recognition, and scaling. Detection, the simplest task, is the one we recommend using in the starting phase of research. The listener is given a series of test patterns that contain a shift in the structure of interest, randomly interleaved with test patterns that do not contain a shift. The listener's task is to identify the patterns that contain a shift. The listener's performance provides objective and quantifiable evidence of how effective a given representation is in making the shift audible.

Detection studies are a very powerful form of evaluation; however, they provide information only about whether the listener hears a shift, not about the nature of the shift itself. That richer kind of information has to come from studies using recognition and scaling. A recognition task is one in which the listener is asked to demonstrate not just that the shift is audible, but also that it is audibly discriminable from one or more other kinds of shifts. A scaling task is one in which the listener provides information that characterizes the shift.

A full program of research, aimed not just at measuring effectiveness but also at understanding how the representation works perceptually and therefore how to improve it, would require studies using all three tasks. For simplicity in this presentation, we will briefly describe a starting approach to evaluation limited to a particular type of detection task and to simple data structures embedded in synthetic data.

The general approach is to create test patterns consisting of streams of auditory events in which each event is an auditory rendition of a datum from a random normally distributed statistical generator. A two-alternative, forced-choice testing procedure is employed in which, for each test trial, two such streams are presented in succession. One of the streams, chosen at random, contains a midway shift in its statistical characteristics.

The shifts can be either in the mean of one or more of the dimensions controlling the attributes of the events or in the degree of correlation among two or more of the dimensions. The listener is required to identify which of the two streams contains the shift.

The test trials follow an adaptive up-and-down testing procedure[15] in which the size of the shift in statistics is varied from trial to trial depending on the listener's performance. On the first trial, the size of the shift is set to be so large that the listener cannot fail to hear it. In subsequent trials, the size of the shift is incrementally decreased until the listener begins to make mistakes in choosing which of the streams contains the shift. The size of the shift then is incremented on subsequent trials until the listener demonstrates reliably correct choices. From an averaging of the degree of shift at which each reversal occurs, a threshold value can be computed. These threshold values then can serve as quantitative figures of merit for comparing discrimination performance across different data representations and across individual attributes or sets of attributes within a single data representation.

4. DEVELOPMENT ENVIRONMENTS

An environment for inventing and studying auditory representations of data must provide tools for running formal and informal experimental studies, like those described in the preceding section. The environment must automate routine activities, such as creating test patterns, presenting the test patterns to human test subjects, recording the test subjects' responses, and analyzing the recorded data. It must also provide support for nonroutine activities, such as designing the parameters for an experiment. In this section, we describe the model of experiment design developed from our observations of experiment designers at work. We also describe a development environment based on that model.

A variety of workstations that embody well-known experimental paradigms for psychometric studies already exist (see, for example, Bartram et al.[1]). Such workstations automate parts of the experimentation process, such as creating test cases and scoring the performance of test subjects. However, they do not assist the user in designing the parameters of the experiment. Assisting with the design of an experiment requires not only knowledge of an experimental paradigm, but also knowledge of the domain (for example, auditory display) being studied.

We have examined the process by which experienced and novice designers create psychophysical experiments.[48] We observed the designers while they designed experiments to study the auditory extensions of the iconographic display technique described in Section 2. From these studies, we have been able to characterize experiment design in the domain of auditory display. We have identified eight primary steps in designing an experiment in this domain:

1. stating a hypothesis,
2. specifying the parameters of the data that must be manufactured,
3. specifying what will vary from test case to test case,
4. selecting or designing one or more auditory attributes to use,
5. specifying how data fields will be associated with the auditory attributes,
6. selecting an experimental protocol for presenting the test cases,
7. specifying what will be measured and how the measurements will be recorded, and
8. selecting test subjects.

These eight steps are not entirely sequential because of interdependencies. For instance, the number of data sets required for an experiment (specified in step 2) may depend upon how many test cases there are (specified in step 6) and how the data sets are associated with the test cases (specified in step 5). We view the introduction of experimental controls as a "guardian angel" task that occurs concurrently with the steps listed above, because we observed that scientists do not design experimental controls as a separate step but evaluate the need for internal and external controls throughout the process of experiment design.

Although we state the eight steps in terms of actions that the designer takes, each action is preceded by decision making. Five of the steps (specifically, steps 1, 2, 3, 7, and 8) require decision making that is standard for any perception experiment. The other three (steps 4, 5, and 6) involve decisions about the auditory representation of data. A development environment for auditory display technologies must support this kind of domain-specific decision making. The following discussion concerns the nature of a development environment that addresses the needs implied by this model.

In order to select auditory attributes and to design a mapping of data dimensions onto them, the researcher must be able to listen to sample test patterns, just as a visualization researcher must look at sample graphical patterns. A test pattern will always be in the form of one or more streams of events. The component events of a stream may be sequential or overlapping, and two or more streams may exist simultaneously. The researcher must design both the characteristics of the individual sounds and the characteristics of the stream and must be able to produce a series of streams that demonstrate the range of variation in one or more parameters.

How the streams of events are presented to, or tapped by, a test subject during an experiment should be a designable feature of the display. The system should provide two display modes—one in which the test subject is a passive listener to streams, where the flow is controlled by the system, and another where the test subject can control the flow, including backward and forward replaying of selected portions of the stream. End users of an auditory display technology will also need to have these two display modes available.

At the heart of the development environment is a toolkit for creating test data. As we pointed out in Section 3, test patterns need to be generated primarily from synthetic data. This toolkit includes tools for specifying and editing synthetic data generators, for generating data, for storing and loading data sets, and for customizing the data generation facilities according to the personal preferences of an experimenter. Tools are also provided for selecting, storing, and loading real data.

The development environment needs to make available a library of previous results and previous experience. In particular, since not all parameters are necessarily data-driven, it needs to provide suggested defaults for various auditory parameters. One of the designers in our studies knew from experience that if all of the sound elements in an auditory display had the same pitch, the display might annoy listeners rather than convey information to them. Consequently, he chose to vary pitch at random when it was not data-driven. Such expertise needs to be embedded into the development environment.

To specify the parameters of an experiment, the researcher must have tools for selecting auditory attributes onto which to map data, specifying the mappings of data fields onto auditory attributes, specifying the testing protocol (including the task that the listener will perform), specifying the scoring of the experiment, and customizing the experiment facilities according to personal preferences. Moreover, the researcher must be able to save an entire experiment and reload it at a later time.

All of the tools in the development environment should be available within a visual programming environment that allows the designer to build a graphical representation of the parts of an auditory display and the relationships between those parts. For example, the designer will graphically link data dimensions to sound attributes. The visual interface will invoke the various toolkits, which, in turn, invoke the workstation's sound generation software.

SUMMARY AND CONCLUSIONS

Once powerful sound generation and manipulation systems become widely available, auditory-display researchers will have to consider a vastly expanded universe of possible auditory data representations. This universe is so large that researchers will need tools that allow quick pruning of unpromising representations and permit the exploration of alternative representations to converge rapidly and efficiently on those that are the most effective. We believe that the best way to attack this problem is to provide researchers with an environment that supports systematic and quantitative testing of a wide range of auditory data representations. The heart of this environment would be a toolkit for creating test patterns that contain well-specified and systematically manipulable data structures. This environment would provide tools for running formal and informal experimental studies, and support all of the other activities of the psychometric experimenter as well. The environment would also provide tools to keep track of the path and results of exploration so that it will be possible to build systematic knowledge and theory in this new, poorly understood area.

REFERENCES

1. Bartram, L., K. Booth, and W. Cowan. "Issues in the Design of Workstations for Psychology Experimentation." In *Workstations for Experiments*, edited by J. Encarnacao and G. Grinstein. Berlin: Springer-Verlag, 1991.
2. von Bismarck, G. "Timbre of Steady-State Sounds: A Factorial Investigation of Its Verbal Attributes." *Acustica* **30** (1974): 146–159.
3. Bly, S. "Presenting Information in Sound." *Proceedings of the CHI '82 Conference on Human Factors in Computer Systems*, 371–375. New York: ACM, 1982.
4. Bregman, A. *Auditory Scene Analysis.* Cambridge, MA: MIT Press, 1990.
5. Chowning, J. "The Synthesis of Complex Audio Spectra by Means of Frequency Modulation." *J. Audio Engr. Soc.* **21(7)** (1973): 526–534.
6. Chowning, J., and D. Bristow. *FM Theory and Applications.* Tokyo: Yamaha Music Foundation, 1986.
7. Doughty, J., and W. Garner. "Pitch Characteristics of Short Tones II: Pitch as a Function of Duration." *J. Exper. Psych.* **38** (1948): 478–494.

8. Fletcher, H., and W. Munson. "Loudness, Its Definition, Measurement and Calculation." *J. Acoust. Soc. Amer.* **5** (1933): 82–108.
9. Frysinger, S. "Applied Research in Auditory Data Representation." *Proceedings of the SPIE/SPSE Conference on Electronic Imaging* **1259** (1990): 130–139.
10. Green, D., T. G. Birdsall, and W. P. Tanner. "Signal Detection as a Function of Intensity and Duration." *J. Acoust. Soc. Amer.* **29** (1957): 523–531.
11. Grey, J. *An Exploration of Musical Timbre*. Technical Report No. STAN-M-2, Department of Music, Stanford, CA, 1975.
12. Grey, J. "Multidimensional Perceptual Scaling of Musical Timbres." *J. Acoust. Soc. Amer.* **61** (1977): 1270–1277.
13. Grinstein, G., and S. Smith. "The Perceptualization of Scientific Data." *Proceedings of the SPIE/SPSE Conference on Electronic Imaging* **1259** (1990): 190–199.
14. Howe, H. *Electronic Music Synthesis*. New York: W. W. Norton, 1975.
15. Levitt, H. "Transformed Up-Down Methods in Psychoacoustics." *J. Acoust. Soc. Amer.* **49** (1972).
16. Le Brun, M. "Digital Waveshaping Synthesis." *J. Audio Engr. Soc.* **27(4)** (1979): 250–26.
17. Lindemann, E., F. Dechelle, B. Smith, and M. Starkier. "The Architecture of the IRCAM Musical Workstation." *Comp. Music J.* **15(3)** (1991): 41–49.
18. Lindemann, E., and M. de Cecco. "Animal: Graphical Data Definition and Manipulation in Real Time." *Comp. Music J.* **15(3)** (1991): 78–100.
19. Lunney, D., and R. Morrison. "High Technology Laboratory Aids for Visually Handicapped Chemistry Students." *J. Chem. Ed.* **58(3)** (1981): 228–231.
20. Lunney, D., and R. Morrison. "Auditory Presentation of Experimental Data." *Proceedings of the SPIE/SPSE Conference on Electronic Imaging* **1259** (1990): 140–146.
21. Mansur, D., M. Blattner, and K. Joy. "Sound Graphs: A Numerical Data Analysis Method for the Blind." *J. Med. Sys.* **9(3)** (1985): 163–174.
22. Mathews, M. *The Technology of Computer Music*. Cambridge, MA: MIT Press, 1969.
23. Mathews, M., and J. Pierce, eds. *Current Directions in Computer Music Research*. Cambridge, MA: MIT Press, 1989.
24. Mezrich, J., S. Frysinger, and R. Slivjanovski. "Dynamic Representation of Multivariate Time Series Data." *J. Amer. Stat. Assoc.* **79(385)** (1984): 34–40.

25. Moore, F. *Elements of Computer Music.* Englewood Cliffs, NJ: Prentice-Hall, 1990.
26. Pickett, R., and G. Grinstein. "Iconographic Displays for Visualizing Multidimensional Data." *Proceedings of the 1988 IEEE Conference on Systems, Man and Cybernetics.* Beijing and Shenyang, People's Republic of China, 1988.
27. Plomp, R., and M. Bouman. "Relation Between Hearing Threshold and Duration of Pulses." *J. Acoust. Soc. Amer.* **31** (1959): 749–758.
28. Plomp, R. "Timbre as a Multidimensional Attribute of Complex Tones." In *Frequency Analysis and Periodicity Detection in Hearing,* edited by R. Plomp and G. F. Smoorenburg. Leiden, Netherlands: Sijthoff, 1970.
29. Pollack, I., and L. Ficks. "Information of Elementary Multidimensional Auditory Displays." *J. Acoust. Soc. Amer.* **26** (1954): 155–158.
30. Puckette, M. "FTS: A Real-Time Monitor for Multiprocessor Music Synthesis." *Comp. Music J.* **15(3)** (1991): 58–67.
31. Puckette, M. "Combining Event and Signal Processing in the MAX Graphical Programming Environment. *Comp. Music J.* **15(3)** (1991): 68–77.
32. Reichardt, W., and H. Niese. "Choice of Sound Duration and Silent Interval for Test and Comparison Signals in the Subjective Measurement of Loudness." *J. Acoust. Soc. Amer.* **47** (1970): 1083–1090.
33. Roads, C. "Granular Synthesis of Sound." *Comp. Music J.* **2(2)** (1978): 61–61.
34. Roads, C. "Asynchronous Granular Synthesis." In *Representations of Musical Signals,* edited by G. De Poli, A. Piccialli, and C. Roads. Cambridge: MIT Press, 1991.
35. Roads, C., and J. Strawn, eds. *Foundations of Computer Music.* Cambridge, MA: MIT Press, 1985.
36. Scaletti, C. "The Kyma/Platypus Computer Music Workstation." In *The Well-Tempered Object: Musical Applications of Object-Oriented Software Technology,* edited by S. Pope. Cambridge: MIT Press, 1991.
37. Scaletti, C., and K. Hebel "An Object-Based Representation for Digital Audio Signals." In *Representations of Musical Signals,* edited by G. De Poli, A. Piccialli, and C. Roads. Cambridge: MIT Press, 1991.
38. Scaletti, C. "Using Sound to Extract Meaning from Complex Data." *Proceedings of the SPIE/SPSE Conference on Electronic Imaging* **1459** (1991): 207–219.
39. Scharf, B., and S. Buus. "Audition I: Stimulus, Physiology, Thresholds." In *Handbook of Perception and Human Performance,* edited by K. R. Boff, L. Kaufman, and J. P. Thomas, Vol. 1, 14.1–14.71. New York: Wiley, 1986.

40. Scharf, B., and A. Houtsma. "Audition II: Loudness, Pitch, Localization, Distortion, Pathology." In *Handbook of Perception and Human Performance*, edited by K. R. Boff, L. Kaufman, and J. P. Thomas, Vol. 1, 15.1–15.60. New York: Wiley, 1986.
41. Smith, S., R. Bergeron, and G. Grinstein. "Stereophonic and Surface Sound Generation for Exploratory Data Analysis." *Proceedings of CHI '90*. Seattle, WA, 1990.
42. Speeth, S. "Seismometer Sounds." *J. Acoust. Soc. Amer.* **33** (1961): 909–916.
43. Stevens, S. "The Relation of Pitch to Intensity." *J. Acoust. Soc. Amer.* **(6)** (1935): 150–154.
44. Terhardt, E. "Pitch of Pure of Tones: Its Relation to Intensity." In *Facts and Models in Hearing*, edited by E. Zwicker and E. Terhardt. New York: Springer-Verlag, 1974.
45. Verschuure, J., and A. A. van Meeteren. "The Effect of Intensity on Pitch." *Acustica* **32** (1975): 33–44.
46. Viara, E. "CPOS: A Real-Time Operating System for the IRCAM Musical Workstation." *Comp. Music J.* **15(3)** (1991): 50–57.
47. Warren, W., and R. Verbrugge. "Auditory Perception of Bouncing and Breaking Events: A Case Study in Ecological Acoustics." *J. Exper. Psych.* **(10)5** (1984): 704–712.
48. Williams, M. "Interactive Assistance for Experimentation on the Visual and Auditory Properties of Iconographic Data Displays." Dissertation, University of Massachusetts Lowell, 1992.
49. Williams, M., S. Smith, and G. Pecelli. "Experimentally Driven Visual Language Design: Texture Perception Experiments for Iconographic Displays." *Proceedings of the IEEE 1989 Visual Languages Workshop*, 62–67. Rome, Italy.
50. Yeung, E. "Pattern Recognition by Audio Representation of Multivariate Analytical Data." *Anal. Chem.* **(52)7** (1980): 1120–1123.
51. Zwislocki, J. "Temporal Summation of Loudness." *J. Acoust. Soci. Amer.* **46** (1969): 413–441.

Gregory Kramer
Clarity, Nelson Lane, Garrison, NY 10524, e-mail: kramer@santafe.edu and Santa
Fe Institute, 1399 Hyde Park Road, Santa Fe, NM 87501

Some Organizing Principles for Representing Data with Sound

Techniques for auditory data representation, and the perceptual issues they raise, are discussed. Sonification, audification, and audiation are defined in terms of mediating structures between the data and the listener. A software system for sonification research is described and parameter nesting, the control of a single auditory variable on several time scales simultaneously, is suggested as a technique for achieving high-dimensional displays. The use of both realistic and abstract sounds for auditory display is discussed in the context of parameter nesting. Techniques for the weighting and balancing of attentionally compelling display components are discussed and it is suggested that a 100% balanced display can only be approximated. The technique of using "beacons" for orienting oneself within an auditory display is discussed and examples of applications are suggested. Gestalt formation is recognized as an operant factor for auditory display in general and beacons in particular. The techniques of data family/stream association, data type/parameter association, global, inter-stream and per stream linking, and metaphorical and affective association are described and suggested as the means of making sonification displays more intuitive, comprehensible, and easier to use.

1. INTRODUCTION

This paper presents a number of means for linking perceptual issues in auditory display with techniques for their practical implementation. Some of these approaches could be the subject matter for entire papers and many of the techniques are in need of perceptual testing to confirm their usefulness. The author's intention here is to introduce the reader to some organizing principles, particularly as related to the use of nonrepresentational sound for data display, and in the course of doing so raise the perceptual questions they imply. In the discussion below it is not the least of our objectives to explore the problems that these techniques may address and the complications that may arise as a consequence of auditory data representation.

2. DIFFERENT LEVELS OF DIRECTNESS

It is assumed that when creating an auditory display there is, at the very least, a transducing mechanism between the data and the listener. We define "directness" as the degree to which the original data input is "removed" from the final sound. As shall become evident, the translation between data and user breaks up into natural divisions for the purposes of categorizing display techniques.

2.1 AUDIFICATION

A direct translation of a data waveform to the audible domain for purposes of monitoring and comprehension will be referred to as *audification*.[16][1][2] In audification, the data itself is shifted into the audible domain (multiplied by a time constant to fall between 20 Hz and 20 KHz), looped (if necessary), converted to the analog domain (if necessary), and amplified. Other than frequency shifting, there needs be no intermediary element. Filters and other signal processing techniques may be used to assist the user of the display in isolating certain elements, but there are no sound-*generating* elements introduced. Electroencephalography, seismogram (see Hayward, in this volume) and radio telescope data, and equation-generated waveforms (see Mayer-Kress et al., in this volume) are all examples of data types that may benefit from audification.[24]

---

[1] In these earlier papers I used the word audification for what is now being referred to as sonification.

[2] This is what Scaletti refers to as 0th-order sonification (in this volume).

2.2 SONIFICATION

The use of data to control a sound generator for the purpose of monitoring and analysis of the data will be referred to as sonification. In sonification, there are substantial mediating factors, as the sound generation technique need not have any direct relationship to the data being generated. The simplest sonifications include a direct mapping of the data to a simple auditory parameter, such as pitch or loudness.

Due to the limited number of low-level variables available for manipulation, simple sonification has limits as to the number of data variables that can be displayed. In order to create higher dimensional displays, we have been working with intermediary structures, such as complex control waveforms, envelope generators, pulse streams, and so on. Parameter nesting, described below, is one example of employing higher level structures to increase display dimensionality.

While intermediary structures do, in one sense, remove the display one more step from the data, they may also be used to make the display richer and more coherent. As an analogy, a mechanic listening to an automobile engine may be able to tell many things about the engine based upon such things as pitch (rpm), rhythmic regularity (are all cylinders firing?), tapping sounds (knocking), and so on. These sounds are comprehensible for two basic reasons: experience and their relationship to Newtonian laws.

The "Virtual Engine" (see Figure 1) is a hypothetical means of using a model to engage our everyday listening skills to understand aurally represented data. Such data may or may not have an underlying relationship with the model being used.

When a sound is louder, we assume that it is because an object was hit harder. When a sound is higher in pitch, the machine is running faster, and so on. It stands to reason that if an intermediary structure, such as an engine simulation, is introduced into a sonification architecture, with different data inputs causing one or more auditory variables to change in "natural" (or what is often referred to in perception research as "ecologically valid")[12] ways, then the complex structure stands to increase the comprehensibility of the display further.

Other structures, simulations or otherwise, may also be of considerable use when inserted into the sonification architecture. Examples of such higher-level structures from the Sonification Toolkit are described below. They include tools for using incoming data to manipulate complex sound fields in a variety of ways.

An intermediary structure such as the Virtual Engine described above may be designed to have any number of impacts on the auditory output of the sonification system. Some of these may be enhancing relevant indices while de-emphasizing extraneous ones, fabricating a Newtonian reality to produce a more comprehensible display, and translating relationships in data to relationships in the sonic output. There is always the possibility, it should be noted, that more complex the intermediary structures will result in a greater likelihood of artifacts, which may then interfere with clear comprehension of the data itself.

2.3 AUDIATION

When we form an imagined auditory image from the sonified output, additional recall is added as a further mediating structure. In those cases where recall of the sonic experience is necessary, I will call the resultant process *audiation*.[3]

In mentally reviewing sonic experiences with the auditory display, we are audiating. When, in designing a new auditory display, we consider the sound component of ideas, we are audiating. When we consider what might happen if we performed a certain action upon the data or the display system, we are audiating. Precious little research has been done on our capacity to audiate and to manipulate our audiations. A compilation of work can be found in a paper by Reisberg.[22] The importance of audiation should be recognized. Audiation, the level of least directness by our definition above, may, ironically, be the most "direct" or "immediate" in terms of the user's subjective experience.

2.4 CATEGORIES ARE NOT DISCREET

In Figure 1 I provide a schematization of the varying levels of directness. While the categories in the above illustration imply hard and fast boundaries, the reality is much more of a continuum. There are some structures inherent within any synthesis system. In order to create a data-controlled sound, for example, an oscillator may introduce a cosine function between

---

[3] The word audiation was inspired by its proximity to *ideation*. In earlier drafts of this paper I had been using the term *auralization*,[19] which was intended to correlate with how psychologists, as opposed to computer scientists, use the word visualization. After discovering that the term *auralization* is sometimes used to refer to the realistic "spatialization" of sound and, of course, knowing that it is occasionally used to refer to what is frequently called sonification, I decided to drop the term lest I add to the confusion on that front. I first suggested the term audiation in a 1990 paper prepared for the Santa Fe Institute Science Board. While I do not expect my usage of the word *audiation* to be universally accepted as synonymous with auditory imagery, perhaps by naming it we take a step towards recognizing its importance in auditory experience in general and auditory display design in particular.

Some Organizing Principles for Representing Data with Sound

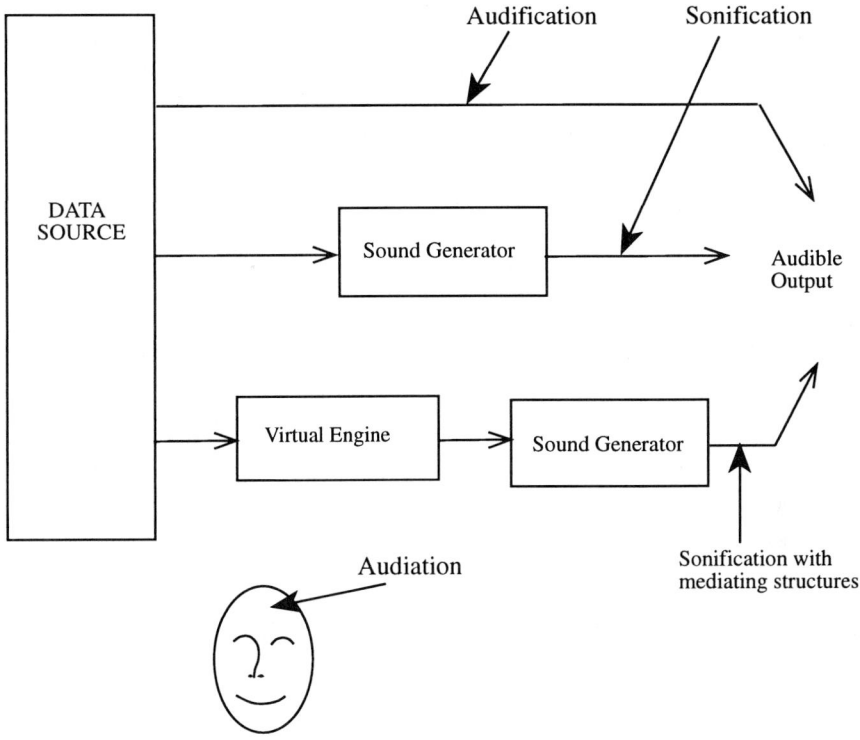

FIGURE 1 Varying Levels of Directness.

the data and the listener. Controlling this tone with another cosine function (frequency modulation) will still only create one tone, but another structure has been introduced. It soon becomes clear that the designation of intermediary structure (or lack of same) is necessarily somewhat arbitrary.

In considering the two approaches to comprehending data (audification and sonification), we should not overlook the possibility that the first deviation from the data marks a significant departure, even in the case where the mathematical distinction between the two appears trivial. From a practical standpoint, listening to the data and listening to a sound controlled by the data may be substantially different.

To illustrate the significance of the distinction between listening to data and data-controlled sound, consider the case of seismic data: For seismic data, the samples must obey the elastic wave equation. In this case audification may work better than sonification because the direct translation preserves the underlying relationships in the data (nearly the same for seismic as for audio data; see Hayward, in this volume).

In audification, there exists for each sample a specific underlying relationship between itself and the surrounding samples. Sonification, on the other hand, is inherently less direct and is typically nonlinear. Therefore, unless executed with great care, sonification may destroy a necessary relationship by virtue of this nonlinearity.

Another way to look at this question is to consider whether the physics of the sound or the perceptual capacities of the listener are the primary consideration. Processes that maintain the underlying physical relationships in the data, such as linear filtering and expansion or compression (multiplying the data by its amplitude envelope or the inverse of that envelope), would be used in audification. Processes that disrupt the relationships of successive samples in favor of simplifying and enhancing features of the data, such as multiplying the data by a cosine wave, would be classified as sonification.

This paper will deal primarily with the techniques and perception issues related to sonification and sonification using intermediary structures. Simple, direct conversion of data to sound will receive little attention here. Fundamentally the discussion will concern the creation and manipulation of structures which, in turn, control a sound generator (as opposed to more direct, low-level control over sound synthesizers). Perceptual issues associated with these and other techniques will also be considered.

3. ABOUT THE SYSTEM USED FOR THIS RESEARCH

The system used for this research was specified by the author and designed and coded by Stephen Ellison. Currently referred to as the *Sonification Toolkit*, the system is centered around flexible, object-oriented software for receiving and mapping data to a number of sound-generating and signal-processing targets. It is coded in the C programming language and MAX.[21,20] We have created a palette of sonification tools which are configured into systems. The data that drives the system is either generated within MAX, read in from a text file, or received in real time via serial communications or concurrent applications such as HyperCard. User control of the system is provided by mouse, keyboard, MIDI, PowerGlove, and voice. The size and complexity of these systems are restricted by memory and CPU type and speed of the Macintosh. Simple systems have been run on the Macintosh SE; however, most useful systems require a Macintosh II or later series computer. A version of the Sonification Toolkit is currently being developed for the UNIX operating system, and an object-oriented synthesis system is being integrated into the toolkit.

Some Organizing Principles for Representing Data with Sound

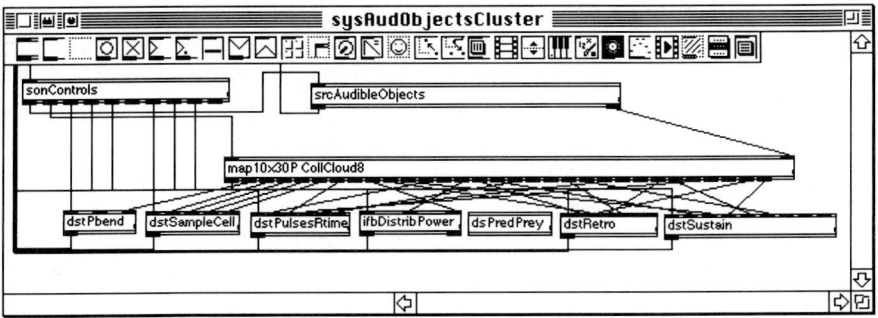

FIGURE 2 Sonification Toolkit.

Since we are interested in developing a large number of sonification techniques and applying these to many different types of data sources, we needed to design a system with sufficient flexibility and expansion potential. The tools we created consist of data sources, maps, destinations, intelligent feedback modules (ifb), and the central control tool sonControls. This last tool is used to channel messages between other tools and serve as a control center for the system.

In general the system functions as follows:

Sources provide the data that drives the sound; maps route and scale the information; and destinations directly control the sound-generating means. By using internally consistent naming conventions and intertool protocol, we have been able to achieve a high degree of flexibility in configuring new sonification systems. Intermediary structures such as the hypothetical Virtual Engine (described above) may be designed and inserted anywhere in the system architecture. Such structures may provide the display balancing, ecologically valid behavior, and multivariate sound parameter relationships discussed above.

The selection and placement of sonification tools provides each specific system with its unique characteristics. New systems are created by connecting tools together systematically in a new MAX "patcher." A variety of high-level structures have been devised which can be inserted into this sonification architecture.

One such tool enables the creation of a sound field of discrete percussive sonic events, with the input data controlling randomness, density, volume, and timbre of the sound. (This is conceptually similar to granular synthesis; see Roads[27] and Scaletti, in this volume.) Other tools use incoming data to control algorithms whose output is fed back into the map to control one or more sonic parameters. In some cases, these feedback tools use one input stream of data to generate numerous output streams. For instance, one of the tools in the Sonification Toolkit creates a set of consecutive amplitudes that roughly corresponds to a normal bell curve whose center frequency and shape (Q) may be controlled. The resulting set of amplitudes may be mapped to a series of sustaining tones of various frequencies in order to produce one smoothly changing sound mass whose harmonic content changes in response to the incoming data. Alternatively, the changing amplitudes may be mapped to a series of sound pulses, creating a sound field that changes in density and definition with changes in the incoming data. Other high-level tools that employ feedback include a control wave, whose shape, frequency, and amplitude may be changed with incoming data. The result is a single control stream, which may then be mapped to any sonic parameter.

New objects for the manipulation of control data and for additional sound-generating and processing targets are continually under development. For a more in-depth discussion of the Sonification Toolkit, please see Kramer.[16]

4. PARAMETER NESTING

A basic premise of this paper is that in order to aurally represent systems of higher dimensionality, more complexity is required of the sound and that this complexity can be obtained by creating parallel auditory streams (polyphony) and by generating a single sound stream with many levels and types of parameter variability. The author developed the concept of *parameter nesting*[16] in order to achieve (in an orderly manner) the higher dimensions required for representation of complex multivariate information via sonification.

Parameter nesting is the use of basic sonic variables on different time scales simultaneously. It enables us to wring 4, 5, or more dimensions out of the most basic variables, such as loudness and pitch. We will briefly describe loudness nesting, pitch nesting, and brightness nesting. The basic concepts may be applied to other auditory variables.

4.1 LOUDNESS NESTING

There are five levels of loudness nesting described. They may be summarized as speed, duration, envelope (instantaneous loudness), cluster speed, and master loudness.

1. *Pulse Speed*: If the signal is broken into discrete "packages" (pulses) by periodically reducing the amplitude to zero, a pulsing sound is the result.

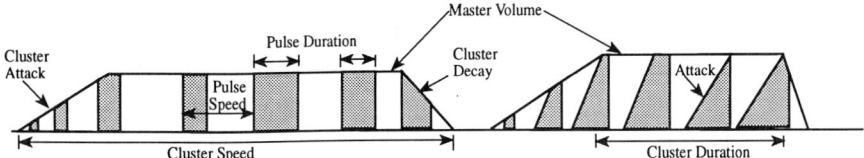

FIGURE 3 Loudness Nesting Parameters.

2. *Duration*: The length of time the sound package is sustained at full volume will be referred to as the duration. In the case of minimum attack and decay (below) this may be understood as the pulse width of the controlling waveform.
3. *Envelope* (instantaneous loudness): The attack and decay times of the pulsed signal are referred to as the envelope. By manipulating these rates, one or two more dimensions can be represented.[4][5]
4. *Cluster Speed*: If a second loudness waveform is imposed upon the first, a series of fade-ins and fade-outs of the pulsed sound will be heard. The period from the onset of one "packet" to the onset of the next will be referred to as the cluster speed. The useful range for cluster speed is approximately 1 Hz to 1 cycle every 4 seconds (.25 Hz). The attack and decay shape of these clusters have also been employed as sonification variables and found to be of some use.
5. *Master Loudness*: The peak RMS amplitude of the clusters will be referred to as master loudness.

[Examples of pulse speed and envelope can be heard on sound examples Tr 30-35 GK 1-6]

4.2 PITCH NESTING

Pitch nesting, where in pitch is controlled within several ranges and time scales, may be understood in a like manner. Musical descriptions and implementation techniques of these variables will give a summary idea of their sound. Examples include: master pitch, vibrato depth, vibrato speed, vibrato waveshape, degree of quantization of pitch changes, etc.

---

[4] Instantaneous loudness may also be employed for a tremolo effect, where a low-frequency waveform is used to control the amplitude of the sound.

[5] Instantaneous amplitude may technically refer to the amplitude of the waveform at the sample level. We refer here to the instantaneous RMS amplitude which is perceived as loudness. This should also be differentiated from very fast amplitude changes (greater than 20 Hz) which produce a timbral change (amplitude modulation).

Two levels of pitch nesting are described, vis. master pitch and instantaneous pitch. Each level may enable additional variables. When considering pitch we will find some differences in implementation and effectiveness between sounds with a well-defined pitch, for example, an oboe playing a note, and a sound with a high noise content or high-amplitude nonharmonic partials, such as the sound of rain or a bell. In the first example, we can speak in terms of explicit pitches, designating these in terms of either notes or cycles per second. In the second we can speak most productively in terms of the frequency area of the sound, or average pitch.

Master Pitch (Pitch Area). The overall pitch area of a sound will be referred to as master pitch. There may be pitch changes within this area, but only a shift in the average pitch and/or the upper and lower frequency limits will be categorized as master pitch.

The center frequency or, where applicable, pitch of the sound will be referred to as *pitch*. The pitch of a sound may change within a certain frequency area (master pitch).[6]

The changes in pitch which occur over a very small range and somewhat rapidly will be referred to as instantaneous pitch. The musical definition for this is vibrato, and the change is typically less than a whole tone in standard musical usage. The rate of this fluctuation from the (center) pitch would typically fall into the 10 Hz to .5 Hz range. The depth of the change of instantaneous pitch, its rate of change, and the waveshape of the change are all perceivable variables and could be used to represent three dimensions rather than the one simple dimension of instantaneous pitch.

If the sound has a clearly defined pitch, the additional parameter of *musical scale* becomes available. Note that while *scale* is not a continuous variable, it may be useful for indicating broad system state information. Perhaps four discreet states may be represented by *scale* without excessive training. If we also allow that instantaneous pitch may be broken down into depth, rate, and waveshape variables, this one variable actually yields three. Thus, a total of at least six dimensions may be sonified by using pitch variables. These are:

---

[6] If a sound with a clear and unambiguously perceived pitch is used, frequency will have several subsets which may represent additional dimensions. These include culturally conditioned recognition, such as the musical scales used and pitch patterns such as melody fragments, how "in-tune" pitches are, and more general perceptions such as the size of the interval jumps and the slew rate between pitches (portamento). See Deutsch,[5] Cuddy,[4] and other studies on the psychoacoustics of music for possible insights on how pitch may be used (or where it should be avoided) for data display.

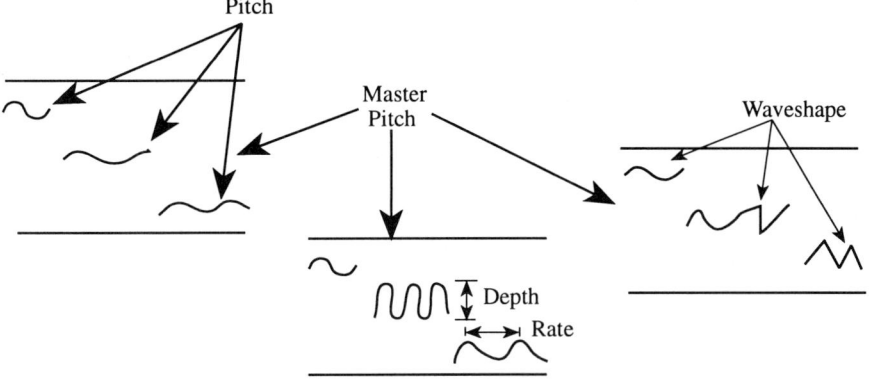

FIGURE 4 Pitch Nesting.

1. master pitch (pitch area),
2. pitch,
3. pitch scale (not a continuous variable): instantaneous pitch,
4. depth,
5. rate, and
6. waveshape.

It is important to keep in mind that we are describing a technique for working with sound which we hope will yield perceivable changes, and that the transforms we make may not always be useful for display of discreet data streams. Even though pitch perception is a copiously studied area, how numerical information might be extracted from sound using this variable is not well understood. Our experience with these techniques has led us to believe that many of them, particularly pitch nesting, are still of limited value given the current "state of the art."

4.3 BRIGHTNESS NESTING

Brightness may be defined as the high frequency content of the sound, or midpoint of the spectral energy distribution.[25] This is one aspect of the spectrum or timbre of the sound. Sounds may have natural differences in brightness, as heard when comparing a french horn and a bassoon playing the same note or comparing the sound of striking a cardboard box and a metal container.

Brightness changes may be imposed on any set of sounds by filtering the sound or implementing a variable additive process such as waveshaping,[18] thereby reducing or adding high or low frequencies. When using subtractive synthesis techniques, exploiting the brightness variable requires the use of sounds with sufficient energy in the higher partials for filtering to be effective. If there are few higher components to the sound, filtering out these frequencies will make very little difference in the timbre of the sound. Conversely, very dense sounds, with a wide frequency spectrum, will not only make the brightness variable highly audible but will facilitate more extensive brightness nesting. Most effective are sounds with a high noise content and sounds with a lot of high-frequency nonharmonic partials. Note that the dense sound that is ideal for brightness nesting is precisely the type of sound that can be problematic with pitch nesting.

While brightness and pitch nesting are not mutually exclusive, they are most perceivable on different types of sounds. We will see that where extensive brightness nesting is possible with denser sounds, the more subtle frequency nesting is not.

> The overall high-frequency content of a sound will be referred to as *master brightness*.
>
> Small-scale brightness-related variations within this timbre will be referred to as *instantaneous brightness*. This is the timbral equivalent of instantaneous pitch and it lends itself well to the same subsets as that parameter. Instantaneous brightness variations include depth, rate, and waveshape of an oscillating control signal.

In sum, four variables are described using brightness nesting. These are:

1. master brightness: instantaneous brightness,
2. depth,
3. rate, and
4. waveshape.

Additionally, the types of parameters accessible in commercially available digital signal processors have proven to be of some use in implementing "brightness" parameter nesting. These might include reverberation time, "flange" depth, speed and waveshape, and resonance. Note that the parameter-nesting technique may be applied to other timbre variables (which are all poorly described in words and little understood perceptually) such as roughness, noisiness, and so on.

5. REALISTIC VOICES AND PARAMETER NESTING

The types of sound used at the computer-human interface could be categorized as *realistic* and *abstract*. These two families of sounds are not mutually exclusive.

Realistic voices are those that are either recordings (samples) of real-world sounds or passable synthesized imitations of these sounds (see Gaver, in this volume). Examples of realistic voices might include rain or other water sounds, animal or insect sounds, speaking voices, machines, and so on. Realistic voices have mnemonic qualities that can be of great value (see below). Realistic sounds, as may be employed in what Gaver refers to as "everyday listening" and described in Gaver's SonicFinder,[8] offer distinctly different opportunities and problems from the use of what will be referred to as abstract sounds.

Abstract voices are sounds whose qualities are perceived without obvious associations to real-world sounds.[7] Blattner's earcons (see Blattner[2] and Blattner, in this volume), along with most sonification work, offer examples of the use of abstract voices. Abstract voices include the universe of acoustic instrumental sounds as well as nonimitative synthesized sounds.[8] Abstract parameter manipulations can be imposed upon realistic voices (see Fitch and Kramer and Gaver, in this volume). These voices may thus be able to include a subset of the abstract parameters discussed above (master loudness, duration, envelope, master brightness, etc.). Usually, however, the clarity of their pitch content is limited, and so the nesting of frequency parameters is somewhat limited.

Realistic and abstract voices both have a short history in the design of general user interfaces and new techniques in both areas are under development. These approaches are not mutually exclusive.

Using the parameters listed above, we may expect to achieve six dimensions using realistic voices.[9] These are:

1. speed (or density),
2. duration,
3. envelope (which may interfere with critical onset cues),
4. master loudness,
5. master pitch (pitch area), and
6. master brightness.

If the realistic sounds are synthesized rather than sampled, these basic variables may be supplemented or supplanted by more complex timbral variables. Numerous realistic voices can be used simultaneously, yielding a

---

[7] In my discussions with Gaver, different words were suggested for abstract and realistic, all with inherent problems. Representational and literal were suggested for realistic, and derivative was suggested for abstract. Realistic and abstract seem to have the least undesired connotations.

[8] In this paper we are not addressing the use of speech for data display.

[9] Note that envelope attack and decay and master amplitude attack and decay have not been designated, nor have numerous frequency variables of only occasional value in minimally pitched sounds.

complex and rich display. If the voices used have great similarities in timbre or inner movement, however, one runs the risk of the sound environment getting cluttered. Furthermore, in sounds that overlap frequency areas, masking can become a significant problem.

It is noteworthy that abstract voices are not limited by the stipulation that they remain literally identifiable within their range of manipulation. Realistic voices, on the other hand, will lose their referential quality if the parameter manipulation is too extreme.

6. USEFULNESS OF THE PARAMETERS—PARAMETER OVERLAP AND ORTHOGONALITY

We have found that while certain parameters may be audible, from a practical standpoint they can either directly interfere with the perception of other variables or simply distract from this perception. The case of reverberation provides one example of a parameter that can mask the detail of many others. When reverb time is high, attack and decay times employed in loudness nesting are smeared, as is brightness. In extreme cases many other parameters are obscured. An instance of distraction was found unexpectedly with our use of loudness nesting clusters. While the clustering technique did not disrupt perception of other variables, interruptions caused by the cluster envelopes did interfere with the subject's concentration on other, shorter time-scale variables.

It will be interesting to perform perceptual tests on our various parameters in order to determine their usefulness and interactions. If multiple parameters are tested simultaneously, however, the complexity of possible interactions may necessitate some compromises in the testing techniques.

Since we are using the same audible variable on different time scales, when these time scales are overlapped there is likely to be a loss of clarity in one or both of the nested parameters. We will refer to this phenomenon as *parameter overlap*. If, for example, pulse speed becomes too slow, then cluster speed looses definition. Conversely, as cluster speed approaches pulse speed, definition is reduced. Such interferences also occur between parameters. For example, if pulse speed becomes slower than vibrato speed, perception of the vibrato is impaired.

While parameter overlap will often be an impediment to the design of an effective display, a carefully designed sonification system may take advantage of parameter overlap to highlight relationships within a model. For instance, consider an environmental model in which independent variables describe the density of vegetation and the percentage of oxygen in the atmosphere. In such a system, when the amount of vegetation goes to 0, the amount of oxygen will also go to 0. In this case, oxygen can be mapped to cluster speed and vegetation can be mapped to pulse speed. When the amount of vegetation is small, pulse speed will decrease, as will cluster speed. Similarly, when there is a large amount of vegetation, fast, dense clusters of pulses will result. Thus, related variables can be profitably mapped to overlapping parameters.

6.1 ORTHOGONALITY AND CROSS-VARIABLE PARAMETER OVERLAP

Parameter interactions may also occur between apparently distinct auditory variables. For example, perceived brightness may be a function of more high-frequency components in the sound. However, if any sound is given a sharp attack, the sudden onset of the sound causes harmonics to be added to the sound, as the initial energy burst has a square-wave-like form. In this way, the variable *attack time* may interact with the variable *brightness* in unintended ways. In another example from the psychoacoustic literature, perceived pitch and spectrum frequently interact, such that changes in spectrum may be perceived as changes in pitch. Conversely, certain spectra will be perceived as changing when played at different pitches.[23]

When sound is being used to represent data, changes in the parameter controlled by one data stream may *appear* to cause a change in a parameter controlled by the second data stream. We may want an auditory display designed such that each of the acoustic variables results in a perceptual experience wherein each auditory variable changes independently of the others. Independence, or orthogonality, may prevent the user from confusing a change in one data stream with changes in another, in which case maximal separation of data streams will represent a useful rendering. On the other hand, in some cases we may appreciate the ways in which auditory interaction effects render the data.

While orthogonality may or may not be desirable, the author believes it is rarely obtainable. It may be possible, however, to design effective displays that employ heuristics to compensate for the perceived interactions. Clearly an up-to-date understanding of psychoacoustics is essential to auditory display design.

7. TOWARDS THE DESIGN OF A BALANCED DISPLAY

Problems related to parameter overlap and distraction are amongst the pitfalls one encounters in constructing a multivariate auditory display. Even if sound parameters that do not directly interfere with each other are used, the fact that our attention is drawn more to certain variables than others makes the design of a balanced, or unbiased, auditory display virtually impossible. For example, changes in *master pitch* generally draw a listeners attention more forcefully than the loudness variable *cluster speed*. Therefore, a display employing these parameters is influenced by the perceptual impact of the auditory variable, which in turn skew the observer's perception of the relative importance attributed to other data variables.

Three techniques are suggested that may be used to compensate for discrepancies in what we refer to as the "forcefulness" of different variables. They are:

- scaling down the range of the auditory variable;
- multiple mapping of the data to more than one variable; and
- employing a variety of mappings for the data set.

Of course, the user may take advantage of the variability of perceived forcefulness and load the display such that more important data streams are mapped to more compelling sound parameters. However, where a more equalized display is the objective, one or more of the below-described means may be employed.

Tr 30,31

7.1 SCALING

Scaling the range of auditory variables may be accomplished by narrowing the upper and lower limits of that variable in the target sound generator or, in the case of mediating structures, narrowing the range at the input to those structures. For example, since pitch is such a forceful parameter, one could limit the highest and lowest pitch that the full range of data may produce (say, from seven octaves to one octave), thereby reducing the variability induced by all the intermediary data points. The result may produce lower resolution from the pitch variable. At the same time, it may also serve to balance the display somewhat. This scaling technique can be thought of as effecting the number of just-noticeable differences (JNDs) that are traversed by the auditory variable per change in the controlling data value.[10] [CD Tracks 30 and 31]

---

[10] Rescaling may also prove to produce other effects, such as creating a change in the "contrast" of the display.

7.2 MULTIPLE MAPPING

Multiple mapping, i.e., the routing of input data to more than one auditory variable, may serve to balance the display or highlight the contour of a chosen data variable. Some parameters, such as attack time, are difficult to perceive. Because of this characteristic, such parameters may be used to incrementally adjust the perceptual impact of input data. With multiple mapping, as with the scaling technique, parameter interactions can result in unintended perceptual effects. [CD Track 32]

Tr 32

Tr 33

7.3 VARIETY OF MAPPINGS

Perhaps the most egalitarian approach to achieving a more balanced display is to simply construct a variety of mappings such that each data stream input to the map is sequenced through the selection of auditory variables.[11] In this way the system user may listen to the same data a number of times with different mappings and decide which mapping most satisfactorily displays the meaningful contours he is looking for in the data. Because the sequence of mappings can be made egalitarian as regards favoring any one data stream, the overall effect might be used for equalizing of any skew resulting from a given mapping. The author has found this technique to be promising for exploring data sets and is undertaking additional research into its application. [CD Track 33]

7.4 MAP SEQUENCING AND INTERPOLATION

Map sequencing is the technique of invoking different maps sequentially, while the data sets remain the same, (or alternates between a small set of data points; see *Beacon Sequencing* below). It facilitates the process of comparing different mappings of the same data sets by automating the comparison process. This allows the system user to evaluate different data renderings while keeping a constant data source or configuration.

There may be cases where sequencing between mappings is complicated by the use of different sound synthesis techniques. A new synthesis technique may be implemented and the new sound parameter file may have a greater or lesser number of target parameters to control with the data streams. It may, therefore, be desirable to effect an interpolation scheme whereby each data stream is gradually shifted to control of one or more variables according to a predetermined scheme. The predetermined scheme

---

[11]This may be considered analogous to remapping colors or rotating an object in a three-dimensional visual data display.

might include rules to specify which target parameters are to be given priority for use in tandem with other target parameters and what kinds of grouping of parameters will be made for subsequent control by a given data stream.

Some obvious questions arise when considering the weighting issue: What are the more (or less) compelling variables? How might one construct a psychometric test to determine these weightings[11] or will we discover better techniques for arriving at a satisfactory display? Will weightings change when more variables are used simultaneously? Are weightings the same, or even similar, between subjects? What is the effect of culture on the perceived importance of a variable?[6] How are the dynamics governing gestalt formation and parameter interaction related to weighting changes?

Clearly, these questions have yet to be thoroughly answered. Presented here are heuristics that, it is hoped, will prove useful and that may be suggestive of further experimentation. We are working from the premise that the techniques will only gain in efficiency as we attain a more thorough understanding of the complex attending issues.

7.5 NON-NEUTRALITY

While we have discussed a number of techniques for balancing the effect of auditory variables when they differ in perceptual strength or impact, a neutral, unbiased, auditory display is, in our experience, just not achievable. Ultimately, I believe that an auditory display in which different data sets are represented equally by different auditory parameters is not 100% obtainable. We have, therefore, set our sights on working with whatever artful means we can muster in order to create the most comprehensible output possible.

8. BEACONS

In the course of attempting to recognize patterns and orient ourselves within a complex, multivariate data set, we have found a need for representing some sort of fixed point within the sonified data as a basis for comparison. With this objective in mind, I have devised a structure, called a *beacon*, to assist in the important function of establishing auditory orientation. A beacon provides a means by which one can identify particular states of a system and, by referring to those states, grasp the overall status of the system.

A beacon is a distinct auditory event which represents a discrete data point or event. It is identifiable by the state of its component parameters which, in turn, determine its auditory characteristics. A beacon can be constructed that does not represent a system state but serves as an absolute reference within a data set or between different data sets. This auditory reference, called an absolute reference beacon, can be derived mathematically or empirically and can help the system user know if the system in question has achieved a given state or if the sound has changed over time.

There are two primary components to a beacon. The first is the data component and the second is the data-to-sound parameter map described earlier in this chapter. The data component of a beacon is a stored set of data points which are used to control a sound within a sonification system. The map component of a beacon is the means by which the data are routed to selected auditory parameters in order to generate the sound. The sound generator audibly represents the data values via the map.

Tr 34

8.1 THE DATA COMPONENT—A CLOSER LOOK

The data component of a beacon may be stored and retrieved independently of sound synthesis techniques and data-to-sound parameter mappings. The data component, then, corresponds to different "snapshots" of the data (and/or a reference data set), and represents different system states. By manually or automatically sequencing through these data snapshots, different system states can be compared in a common auditory framework. [CD Track 34]

When constructing an absolute reference beacon, the data values need not originate in the data set under investigation. A reference data set may be fabricated (mathematically or empirically) to serve as a cross reference between different sessions or data sets. Alternatively, a set of average data values can be generated from the set under investigation. Then, so long as the researcher is working within that data set, he can know if the different parameters are above or below average. Of course, this would not work when multiple data sets are being explored.

STATIC AND DYNAMIC BEACONS. The beacon data values may be fixed at a given point or they may have a well-defined time-varying shape. A beacon using a single data point (however many dimensions define that point) will not vary over time. This I will simply refer to as a beacon, or a *static beacon*. A beacon using time-varying data will be referred to as a *dynamic beacon*. Each controlling data stream of the dynamic beacon changes over the course of the recorded beacon interval, typically a time span of .5 to 3 seconds.

Dynamic beacon data may be stored as either a stream of data values or a beginning and end point of an index that points to the addresses of the stored data. The dynamic beacon may represent, for example, a two-second segment of a simulation. Alternatively, one may represent two seconds of sequential spreadsheet data via the dynamic beacon. In this latter case the user must specify the playback rate of the data.

Tr 35   In either case, a two-second sound "phrase" results once the data is mapped to the sound-generating means. The system user can then change the mappings and replay the same data segment, replay different dynamic beacons sequentially, compare dynamic beacons from different points in a procedure, or perform other operations with the data. Due to the animated nature of the dynamic beacon technique, features of the system may be highlighted that might otherwise be overlooked by the static beacon. [CD Track 36]

8.2 SONIC MAPS AND ALTERNATE AUDITORY VIEWS

Once the data component of a beacon is selected, it is employed as a means of controlling the sonic qualities of a particular sound generation scheme. As described above, the means by which the data are routed to selected auditory parameters is known as a map. Various states of the system in question can then be compared by injecting beacon data into the map. Likewise, a set of known reference values can be injected to ascertain how the values under consideration may have changed. If a new sonic map, perhaps referencing a different sound generation method, is implemented, the same data points will then be represented by a new auditory beacon.

Tr 33   There is a relationship between sonic maps and sound synthesis techniques. Changing from one sonic map to another may involve simply a re-routing of data variables to sound generator variables. In this case, there may be no change whatsoever in the synthesis technique implemented in the sound generator. For example, one may route data stream 1 to pitch and data stream 2 to brightness, or one may map stream 2 to pitch and stream 1 to brightness. [CD Track 33]

Changing a map may also involve invoking a configuration in which entirely different sound parameters are possible destinations for the controlling data. For example, data stream 1 may control onset time and data stream 2 may control vibrato of a sound which is made up entirely of harmonically related partials and is pulsed in nature. An example would be a cellolike sound repeatedly playing short notes.

A change in the mapping which includes a change in synthesis technique may create a sound similar to ocean waves. Since onset time and vibrato would not readily apply to a continuous and noisy sound, these variables would no longer be available in the map. In this case, data streams 1 and 2 might be used to control the noise content and rapidity of the sound.

Since the map, by definition, encompasses the parameters of the sound generator, we refer to changes in the routings as well as changes in the routings plus the synthesis technique as changes in the map. When we refer to changes in the map, it is understood that this implies a compatibility with the existent synthesis technique and its associated available sound parameters.

8.3 COMBINING DATA MANIPULATION WITH MAP MANIPULATION

The manipulation of the data and map components of beacons together constitutes a versatile means of investigating a data set via the use of sound. By changing the data set while maintaining the mappings (and vice versa), the data can be flexibly inspected.

For instance, one might use a data file to control a sound. One might then go on to save a few subsets of this data that describe states which seem interesting when sonified. One could then maintain the sonic map and employ different beacon data, thereby developing a stable auditory reference and comparing different data sets within that reference. An obvious extension is to change mappings (and possibly the associated sound generation techniques) and compare different "views" (beacons) of the same data set.

The data component is not yet a beacon until it is fed through the system to create an auditory rendering. The data component of a beacon directly represents system parameters at a point (static) or region (dynamic) in time, while a beacon is an auditory rendering of the system at that point (or region). Consider the following application: The data component of a beacon represents a critical event in a simulation, and several different auditory beacons are made from it, each assigning different variables to different sonic parameters. Then, by listening to these different auditory beacons, the most salient features of the event are represented.

On the other hand, different beacons may be compared, each of which refers to different data sets and all having the same data-to-sound-parameter map. In this way, the important variables from the separate data sets may be compared using a consistent auditory framework. The separate data sets may be derived from distinct runs of a simulation/measurement, from different points within a simulation/measurement or simply from an absolute reference data set.

Tr 34

8.4 BEACON SEQUENCING

To facilitate comparisons between different beacons, they may be sequentially played back while the system user listens for correlations or anomalies. In the simplest case, where the user is comparing two states, the sonification system can be programmed to switch back and forth between the two beacons. [CD Track 34]

A system user may become familiar with a given set of beacons and compare the running status of the system, for example, a manufacturing control system, to the auditory beacon reference. This reference may represent the ideal status of the system. This technique would enable the system user to quickly identify system problems and make adjustments based upon auditory, visual, or other display feedback.

Using dynamic beacons, a person can be trained to recognize certain sound phrases that represent desirable (or undesirable) system states. For example, a stock market analyst could become familiar with the sound of a favorable trend and make purchasing decisions based upon hearing that trend.

8.5 BEACONS AND GESTALT

When a complex multivariate auditory stream is used to convey data, an important perceptual process comes into play. In addition to the system user's ability to scan his or her attention through the sound, relationships between variables and entire system states are perceived "at a glance." Which is to say, without attention-directed effort, all of the auditory variables are perceived as a whole, or a "*gestalt*." In addition to a sound being, for example, bright in timbre, high in pitch, pulsing quickly, and loud, it is all at once recognizable as a whole entity. Our perceptual tendency to construct gestalts underlies the concept of beacons (see Williams, in this volume, for a discussion of gestalts).

8.6 EXAMPLES OF BEACON APPLICATIONS

Two examples of how beacons may be used follow. The first shows how beacons would be used in a manufacturing system, where the system user has little control over the beacons. The second describes a data analysis task where extensive control is provided to the system user.

PROCESS MONITORING. Using beacons, a sonification system user can become familiar with a given set of beacons and compare, for example, the running status of a manufacturing control system to a reference beacon representing the ideal status of the system. This enables the system user to quickly identify system problems and make adjustments based upon auditory feedback. Note that in this example, the system user does not control the beacon data or the sonic maps, but only accesses them as points of reference.

For example, if a worker in an injection molding factory is required to visually monitor the quality of the product emerging from the injection molding machinery, it may be difficult for him to also monitor the various parameters of the machine that effects the molding process. Such variables may include input temperature, output temperature, pressure, and viscosity. In order to allow the worker to continuously monitor these variables, one could provide a sonification wherein the input temperature is converted to a control signal for the pitch of a sound generator, output temperature is converted to a control signal for the vibrato of that sound, the pressure signal is used to control brightness, and the viscosity signal is used to control roughness of the sound.

By listening to the changing sound, the system user is able to continuously monitor the status of the molding machinery without taking his eyes off of the machine's output. However, if he loses track of the sound that represents the normal state, how is he to know if the current state is normal or abnormal? By pressing a preconfigured "normal beacon" button, he causes a sound to be played back that represents the normal state. By quickly comparing the resultant sound to the sound being produced by the machine, he is able to tell if one is rougher, higher, brighter, and so on than the other. Likewise, if he believes that the sound indicates that the system is headed towards a certain malfunction, he can press a button representing that malfunction and hear what sound would be produced by the sonification system when that state is reached. By comparison he can then determine if the machine is headed in that direction or not.

DATA ANALYSIS. In this second example, the system user, a financial data analyst not only records and accesses different beacons (static and dynamic), but sequences these beacons to compare different system states. Using dynamic beacons, a person can be trained to recognize certain sound phrases that represent desirable (or undesirable) system states. For example, a stock market analyst can become familiar with the sound of a favorable trend and make purchasing decisions based upon hearing that trend.

A data analyst using a sonification system to spot trends in the stock market may begin by listening to the sonification of data representing one year of daily closing values of five target stocks. Let us say that these values for companies 1–5 are being used to control pulsing speed, brightness, loudness, pitch, and onset time of a sound. For each set of different closing values, then, a different sound results. While playing back the data file, the analyst hears a point of interest and presses the "record beacon" button on the sonification system. At the point that she presses the button, the current data values are stored in a memory location and given a label for future recall (e.g., "beacon 1"). When she hears another state, she may do so again, and another beacon data set is stored, and so on.

Now, wishing to compare, let us say, the August 4 beacon with the October 14 beacon, she stops the data flow from the stock market data file to the sonification system. Now she presses the beacon 1 button, then the beacon 2 button on her computer screen. Pressing each beacon button causes the corresponding data set to be injected into the sonic map and thence to the sound generator, causing two distinct sounds representing those data sets to be played. By comparing the two sounds resulting from the two beacons, she gains insight into their similarities and differences. Now, she changes the sonic map and plays the same beacon data. Different sounds result, and so different aspects of the two data sets are highlighted. (see "Sonic Maps and Alternate Auditory Views," above.)

Now, wishing to hear these beacons in a context, she specifies that the week proceeding and following each data point also be played back. The result is a one-second phrase for each beacon, a dynamic beacon. Now, seeing that the August 4 beacon was at the beginning of a certain change and the October 14 beacon was in the middle of that "phrase," she ascertains that a trend she first expected is not likely to materialize. So she selects another beacon from November 22 that had similarities to the August beacon and plays it back as both a static beacon and a dynamic beacon. Now sequentially playing back August and November beacons adjacent to each other, she hears expected similarities. Based upon these perceptions, she decides to purchase one stock and sell another.

9. CORRELATING DATA HIERARCHIES WITH DISPLAY HIERARCHIES

When the sonification system user is trying to remember which data variable maps to which auditory variable, he may give more attention to recalling the mapping than to trends in the data itself. Even if the user is listening to the sonification in a global, overview way, it stands to reason that important features will be more discernible if the display is structured to reflect structures in the data.

Some data sets inherently lends themselves to a hierarchical grouping of the data. In order to describe these inherent groupings, we will use the terms *data families* and *data types*. A data type exists when data from multiple data sets share traits or characteristics that distinguish them as an identifiable class. A data family is a collection of data types all derived from or associated with the same source. We use the word *type* for data rather than *parameter* because we use the word parameter in referring to auditory variables.

One example of *data families* is the grouping of ecosystem data by air variables (temperature, pressure, average wind speed, etc.), water variables (rainfall, sea temperature and level, salinity, pH, etc.), and flora (biomass and diversity). Another example is the grouping of census data by geopolitical entities, such as data associated with three different countries. An example of a *data type* may be data subsets associated with each of these countries, e.g., the GNP, infant mortality, and arms expenditures.

9.1 DATA FAMILY/STREAM AND DATA TYPE/PARAMETER ASSOCIATIONS

The comprehension of the sonification can be enhanced by maintaining these data family and data type relationships in the display. *Family* relationships in data may be represented by employing a group of distinctly different timbres, which are perceived as distinct auditory streams, and associating each stream with a different data family. We refer to this as *data family/stream associations*. Common data *types* may be represented by common auditory parameters. We refer to this as *data type/parameter associations*.

To exemplify the concept described above, let's say our data set encompasses three data families representing measurements associated with three different plant species. Each of these data families (plant species) is assigned a timbre, or voice. Variables associated with plant 1 may all be represented by a pure synthesized tone modulated with noise; plant 2 variables may be represented by an active, metallic sound; and plant 3 variables may be represented by a rich complex tone.[12] Changes in the data variable effecting one of these streams may then be immediately associated with a specific data family.

Now, suppose that there are data types common to all of the data families, such as population and health. Population of a single family may be represented by pitch area of the associated auditory stream, while health is represented by brightness of that stream. Changes in pitch or brightness of any one of the auditory streams may be immediately associated with population and health, respectively, via the data type/parameter associations.

In Figure 5, Data Families 1, 2, and 3 would correspond with the three plants under study. Auditory Streams 1, 2, and 3 are different timbres associated with each data family (plant). Data Types 1, 2, and n would correspond with the population, health, and other variables of each plant. Parameters $1-n$ ($P1-Pn$) would correspond with auditory variables which are common to all three streams and are controlled by the associated data types.

---

[12] I apologize for the vagueness of these descriptions. Clearly, one of the difficulties facing auditory display theory and design is the lack of a common language to describe complex sounds.

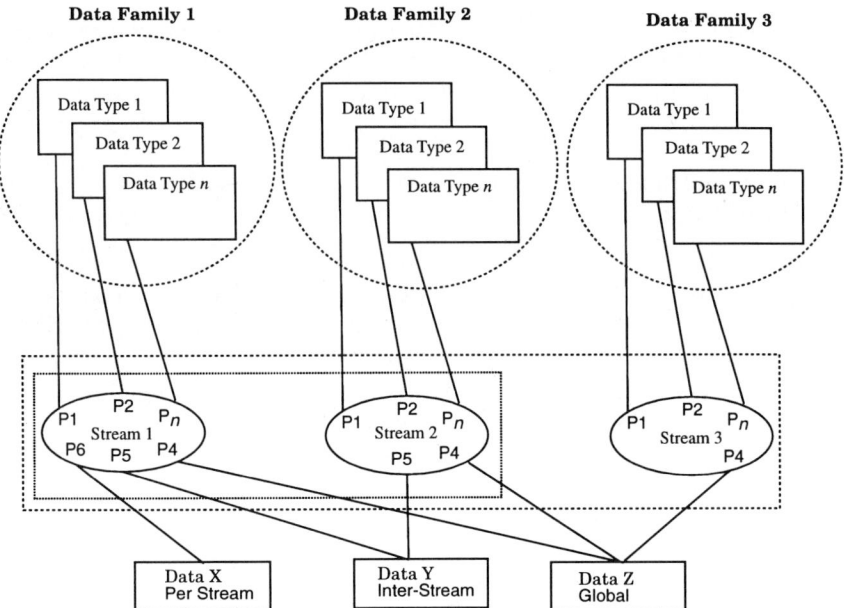

FIGURE 5 Data Family/Stream Associations, Data Type/Parameter Associations and Global, Inter-Stream and Per Stream Linking.

9.2 PER STREAM, INTER-STREAM AND GLOBAL VARIABLES

Some data may be extrinsic to the data families but affect them individually, severally, or globally. We will refer to the association of extrinsic data with auditory parameters as per stream, inter-stream, and global variables.

PER STREAM. A data variable that affects or is associated with only one data family may be used to control a sound parameter only for the auditory steam associated with that family. This will be referred to as a *per stream* variable.

> In Figure 5, Data X may represent, for example, the population of a predator which only attacks a certain plant. Such a data variable may be used to control a unique auditory variable in the appropriate auditory stream, for example, the attack times within Stream 1.

INTER-STREAM. Some data may be associated with more than one of the data families, but not all of them. This association may be in the form of a force that acts upon two or more data families. This will be referred to as an *inter-stream* variable, and connecting voices in this way will be referred to as *inter-stream linking*.

> *Data Y may represent the population of a fungus that only attacks the plants that live in fully shaded areas. As such, it will only affect plants 1 and 2 but not plant 3, because the latter is a nonshade plant. Such a data type may be used to control an auditory variable in only the streams representing the affected families, in this case Stream 1 and Stream 2. For example, the brightness of the sound in two of the streams may both be controlled by one data source, i.e., the fungus population.*

GLOBAL. Some data may represent variables that affect all of the data families. Such a data type may be used to simultaneously control an auditory variable that is common to all of the streams. This will be referred to as a *global* variable.

> *Data Z affects all of the data families and so will control a sound parameter common to all three sound streams. For example, rainfall globally affects all plants and rainfall data may be used to control the reverberant field of all three auditory streams. (In this example, just as the plants live in a changing environment, the sounds exist in a changing environment. However, an auditory variable integral to the sound and common to all streams, such as loudness, could also have been used.)*

10. DIFFICULTIES WITH SIMULTANEOUS AUDITORY STREAMS

Using the above techniques, it may be possible to bring some order to an otherwise arbitrary sonification. Still, shifts in attention from one auditory stream to another may be an impediment to an integrated understanding of certain data sets.

The gestalt concept described above in reference to beacons contributes to the comprehensibility by enabling the sonification system user to hear multiple variables as an integrated entity, with attention focused upon that single, multivariate entity. Where the sonification task is primarily the display of whether an element is on or off, as in the software performance analysis work by Francioni and Jackson (in this volume), polyphony (multiple concurrent voices) may be useful. In high-dimensional displays of continuous data variables, polyphony should be used with great care. In addition to the effect of polyphony on gestalt formation, the number of voices that are effectively discernible may be quite limited.[13] Also, the need to separate out independent streams from the "auditory scene"[3] can make the display less comprehensible. It has been noted that in certain instances, multiple voices will create problems of distraction, as the display user's attention is drawn towards one voice and away from other, possibly important, voices.[7]

Cases will arise wherein multiple abstract voices create a gestalt. For example, if multiple independent sound streams are used to represent a large number of parallel events, these may be perceived as a single "sea" of sound. Representations of an artificial life computer model or other massively parallel process may, for example, demonstrate this kind of grouping behavior. Likewise, multiple realistic voices used to mimic natural processes, such as Kramer's predator/prey model[14] or Gaver and Smith's factory floor,[10] may lend themselves well to gestalt formation. Clearly, the effectiveness of any given technique will be determined by the task undertaken and many other variables. In this domain we should be prepared for unanticipated, if not surprising, results.

11. METAPHORICAL ASSOCIATION AND AFFECTIVE ASSOCIATION

Using abstract sounds to represent data can pose obstacles to comprehensible display design. For example, since we do not associate a stock market index with a particular sound, how do we decide which auditory variable should be controlled with the stock data? How do we remember the mapping once it is made? How do we decide in which direction the auditory variable should change in order to be associated with an increase in the market data?

It may be possible in some cases to maximize comprehensibility by using *metaphorical association* and/or *affective association*. *Metaphorical association* is the association of a physical world variable with a metaphorically related change in an auditory variable. *Affective association* concerns the linking of a user's subjective affect elicited by the emitted sound with related modifications in affect aroused by meaningful changes in the data.

11.1 METAPHORICAL ASSOCIATION

The basic concepts of *more* and *less* may be found in such widely divergent sets of data as: yearly rainfall, GNP, or electrical charge. In representing these data with sound, we can harness the metaphor of "more" sound to indicate more of what is being represented by the data. Thus, a change in the quantity of these measures may be represented by a change in *master loudness* (volume) or by changing the magnitude of another auditory variable. In another instance, quantity data may be linked to changes in the *speed* of the sound. In this latter case, the parameter of *speed* might be used to designate such model variables as population size, handguns per capita, or the number of holes in a semiconductor.

The use of metaphor in auditory display has been discussed by other researchers, including Ballas,[1][13] Gaver,[8] and others. I will attempt to briefly relate the metaphor concept to auditory variables we have discussed.

A partial list of metaphorical associations includes:

- louder = more
- brighter = more
- faster = more
- higher pitch = more

- higher pitch = up

- higher pitch = faster

A brief investigation of the basis of these metaphors may help us to better understand and employ them.

- *Louder is more*: Based upon the physical reality that large objects usually make louder sounds than small objects (bells, animals, etc.). Also, louder sounds move more air and so have greater sensory impact.
- *Brighter is more*: Based upon the physical properties of sound, a sound that we perceive to be bright typically has more high harmonics, or more energy in the upper partials of the sound. This can give the sound more presence and even make it seem louder.
- *Faster is more*: Based upon the self-evident observation that more of a certain event occurs within any given time frame, all else being equal.
- *Higher pitch is more*: This could be based upon the physical fact that higher pitched sounds have more vibrations per second. It is more likely that the *higher pitch is up* metaphor (see below) combines with the *up is more* metaphor to produce the net association. The *up is more* metaphor is grounded in the physical experience of adding more to a pile and thus making it higher.

[13] What I refer to as metaphor, Ballas divides into metaphor and simile.

- *Higher pitch is up*: There are several possible sources of this metaphor. It may be related to the perceptual phenomenon of higher pitched tones being perceived as originating from higher in space.[26] It may be related to the physiology of using the voice to produce a higher pitch, i.e., humans (and some animals) often look up or stretch their necks upward when vocalizing high-pitched sounds. It may also be derived from our deep cultural imprinting of the graphical representations of pitch, although this could be a cause or an effect. In musical notation, a higher note is higher up on the page.
- *Higher pitch is faster*: Based upon the physical fact that higher pitches are associated with faster vibrations, or more cycles per second. In our everyday experience, the faster a machine goes, the higher the pitch area of the sound that it emits. The faster we scrape two rough objects together, the higher the pitch of the resultant sound.

Obviously there are many more metaphors worthy of investigation, but for purposes of understanding the issues at work and designing auditory displays employing some of the sound manipulation techniques discussed in this paper, this discussion will suffice. Needless to say, metaphorical association is a rich area for further inquiry.

11.2 AFFECTIVE ASSOCIATION

As described above, affective association is the association of feelings about the data (if such feelings exist) with feelings aroused by changes in the sound. By way of example, all of the following may be described generally as undesirable changes:

- increases in the interest rate to a construction industry analyst;
- decreases in rain pH to an ecologist; and
- increases in population to an analyst concerned with malnutrition.

As undesirable changes, these system states may cause a subtle sense of emotional discomfort. An auditory variable that causes a similar discomfort, such as the detuning of one pitch in relation to another or the addition of unpleasant upper partials to the sound, may then be used to more effectively represent these disconcerting changes in the data. Similarly, desirable changes could be represented by changes in the sound that make it more pleasing.

A partial list of affective associations includes:

- Increase in "ugliness" = increase in an undesirable data variable (or corresponding decrease in a desirable variable).
- Decrease in "richness" of sound = increase in undesirable data variable (or corresponding decrease in a desirable variable).

- Increase in "unsettling" quality = increase in undesirable variable (or corresponding decrease in a desirable variable).
- Increase in predictability = increase in desirable variable.

Examples of these affective associations include:

- *Ugliness*: A sound mutates from smooth to harsh, as high, non-harmonic partials are added.
  An increase in harshness might be associated with an increase in pollution of an ecosystem. (Or, conversely, an agent that is beneficial to the ecosystem is mapped inversely to the "ugliness" variable, such that a decrease in ugliness corresponds with an increase in a desirable variable.)
- *Richness*: A sound mutates from a full frequency spectrum (what may be called a "rich" sound) to a sound with highs and lows, but no mid-spectrum components. (Hollowness can also be evoked by creating resonances at selected frequencies. The discrepancy of power content between adjacent frequency areas can create this perception of hollowness.)
  Hollowness might be associated with an increase in infertility in a segment of a population. Desirable or not, the association with nonfullness and the lack of capacity to give birth is affectively coherent. (On the other hand, fertility might be associated with "too much," a busy, unsettling stimulus signaling global overpopulation. Paradoxical effects are further discussed below.)
- *Unsettling*: A pitched sound is created by synchronizing two tone generators. When these tone generators are detuned in relationship to each other, the resultant affect is one of discomfort.
  An increase in detuning is associated with an increase in a chemical component that destabilizes a compound. Conversely, a decrease in detuning may correspond with an increase in a chemical that stabilizes a compound.

Correlation of abstract sounds with emotional states may be more easily accomplished with some display techniques than with others. One can tap the subtle emotional associations we have with vocal formants and vocal phrasings to arrive at an effective auditory display.[14]

Creating emotionally effective auditory displays promises to be a rich area for future sonification research. Gaver's work with parameterized everyday sounds (see Gaver, in this volume) may provide a promising approach for matching affect (how we feel about a particular sound) and

---

[14] In private conversations with A. Bregman, he suggested investigating the use of the f1/f2 space for data display. This may be independent of or integrated with affective associations.

auditory data display (how changes in that sound represents changes in the data).

Because affective association involves the association of an affect with both changes in the sound and the changes in the data, it may not only be useful in cases where such associations can be made but actually impose an inadvertent detrimental effect in cases where these associations should not be made. For example, the technique may be applicable to designing a process control system (where it is assumed that certain monitored states are desirable and others are out of the desired range). At the same time the technique may be of little use (indeed, many would say, should be assiduously avoided) in a physics laboratory, where the variable being studied (e.g., the scattering of particles) has no associated affect. Objective science seems to have little tolerance for affect amidst the empirical.

11.3 INTERACTIONS BETWEEN METAPHORICAL AND AFFECTIVE ASSOCIATION

Overlap between metaphorical association and affective association is inevitable. As noted above, at times affective association will work to the system user's advantage while, at other times, it may present problems that we need to be alert to.

For example, the *brighter is more* metaphor may coexist with the *brighter is desirable* affective association. If the "more" is associated with "better," there is no conflict. If "more" is associated with "undesirable," there may be conflict.

Conversely, if the *brighter = less desirable* metaphor (a negative association; the sound is designed so that brighter is more shrill and so less desirable) is correlated with a decrease in a desirable variable, such as oxygen in the blood, we could have a conflict between the *brighter is more* metaphor and the *brighter is worse* affective association.

In the above examples we can see that it is possible in different uses to associate *brighter is better* and *brighter is worse* (brighter is correlated with a strong dollar and brighter is associated with less oxygen) while maintaining affective coherence. But in one case we remain in concordance with the *brighter is more* metaphor and in the other we conflict with it.

As this example clearly demonstrates, there is a strong subjective nature to the values that come into play when we design and use an auditory display. Any data/sound parameter associations that may be made with one sound may not apply if a different sound is used or if a different person is using the display.

11.4 CONGRUENCE, INTUITION, AND EASE OF USE

Both metaphorical and affective association are inherently ambiguous with respect to meaning. Careful examination of our attitudes towards the data in question will frequently yield far more metaphor than we may recognize at first blush (this observation has been well described by Lackoff and Johnson[17]). There is a requirement on the part of any observer to recognize inherent qualifications to the objectivity with which a data set is approached. If the listener fails to perceive a certain congruence[15] in the connection between the auditory display and data source, a loss of comprehension can readily result.

An important goal of sonification is to yield an auditory display that will be intuitively maximal in meaning to the observer. These associative sonification techniques seek to harness subjectivity in the service of mnemonics and comprehensibility.

Display design requires an artful balance of the associative effects inevitably present in the auditory output. This balance may be guided by a combination of criteria. The criteria may include objective findings, (e.g., most people in this country find an increase in a certain variable annoying), as well as the designer's (or user's) subjective reaction to the generated sounds and their component variables.

Once mastered, association techniques can become second nature and thus prove useful in designing displays that are less demanding of rememorization. Furthermore, the metaphorically or affectively associated map may be generalizable to different sets of data and various sound generation techniques, flattening the learning curve and making the sonification process more efficient.

12. SUMMARY AND CONCLUSIONS

We have seen how different auditory display techniques may be described in terms of the number and nature of intermediary structures (i.e., levels of directness). When using abstract sound for data display, the structures can include high-level structures, such as physical models and other simulations, to add ecological validity. The basic categories of *audification, sonification,* and *audiation* have been described.

The use of MIDI synthesizers to do sonification work has distinct limitations, but many concepts can be profitably investigated using these readily available devices. Our Sonification Toolkit improves upon simple control of

---

[15] For example, a certain "immediacy" or phenomenal isomorphism.

MIDI devices in that it is object-oriented with modular, high-level structures, allowing the researcher to investigate a variety of techniques with efficiency. Better still would be a sonification system with object-oriented software synthesis integrated with the modular system of high-level control structures.

Parameter nesting has the capacity to yield high-dimensional displays. Also, it is a technique that can work with variety of acoustic parameters, such as loudness, pitch, brightness, roughness, and so on. Furthermore, parameter nesting can be used with both realistic and abstract sounds. However, parameter overlap can be a problem and careful psychoacoustic testing should be done when these nesting techniques are used.

Since different acoustic parameters have different levels of forcefulness (attention-compelling characteristics for the listener), the design of a sonification system must consider the effect of these imbalances on the system user. Failure to do so may result in display-induced anomalies. Three techniques for designing a balanced display are: (1) scaling the range of the auditory variable; (2) multiple mapping of a single data stream to multiple auditory variables; and (3) using a variety of mappings. Map sequencing and interpolation can be used to effectively compare different mappings. Still in all, I believe that a 100% balanced display is not a possibility in the real world.

Beacons are a tool for establishing reference points in an auditory display system. They are useful for orienting a user within a data set. Static beacons display single, multidimensional points while dynamic beacons display regions or short time slices of the data. Both may be used to compare one point or region to another or compare a selected data point with an idealized reference. Beacons have a data component and a map component. Manipulating the data snapshots and maps, separately and together, makes for a flexible investigative technique. Underlying the beacons concept is the human ability, and tendency, to form auditory images as holistic events (without attention-directed effort to the acoustic parameters of sound). This capacity has far-reaching implications for auditory display.

Techniques to increase the orderliness and intuitive coherence of sonification displays are important to effective system design. Correlating data hierarchies with display hierarchies can add coherence, but the polyphony necessitated by this technique should be handled with great care. Polyphony, or the use of simultaneous auditory streams, can reduce the effectiveness of attempts to generate auditory gestalts. The formation of multiple auditory streams into higher level gestalts is sometimes desirable. Displaying data sets and systems that have multiple parallel events, such as genetic algorithms or massively parallel computers, are two conceivable instances of the effective use of gestalt formation with multiple voices. No doubt unexpected results will be discovered as these multidimensional techniques are explored.

Metaphorical and affective association can add coherence to an auditory display. However, they can also create conflict and should be carefully considered in display design. Whether or not one is using these associative techniques, it is important to be aware of their possible impact.

13. THE FUTURE

The design of a more flexible sonification system, one that integrates control and synthesis structures, will be one of my major efforts over the next couple of years. Increased flexibility will facilitate the development of sonification techniques. New techniques will no doubt arouse new perceptual issues, much as the beacons and parameter-nesting techniques have done. There is much perceptual testing to be done on the basic tenets set out in this paper. In particular, it is my intention to test the effects of learning on the effectiveness of sonification systems, to investigate the development of a "forcefulness" rating system for different acoustic variables, to test the efficacy of auditory beacons in a variety of data-monitoring and comprehension tasks, and to further develop the concept of, and gauge the effectiveness of, affective association. I will also continue to explore the further reaches of high-dimensional displays, in particular, the auditory representation of complex adaptive systems and massively parallel processes such as artificial life and genetic algorithms.

Ultimately, sonification is a real tool for the real world. Sonification system design should be heavily task dependent. Techniques that are applicable to one task, e.g., real-time monitoring, may not be as effective on other tasks, e.g., exploration of a high-dimensional data set. Practical use of the techniques described in this paper (and of sonification in general) will prove or disprove their utility.

ACKNOWLEDGMENTS

I would like to thank Stephen Ellison, who coded the Sonification Toolkit and followed new sonification ideas with creative design and efficient implementation. I am also indebted to my brother, R. Jonathan Kramer, for his assistance in editing this document and helping me to frame the sonification concepts in the context of psychological perception theory and arrive at an increasingly clear expression of these ideas. Apple Computer's Advanced Technology Group *Apple Classrooms of Tomorrow* provided financial and technical support for my early research.

REFERENCES

1. Ballas, J. "Interpreting the Language of Informational Sound." *J. Wash. Acad. Sci.* (1992).
2. Blattner, M., D. Sumikawa, and R. Greenberg. "Earcons and Icons: Their Structure and Common Design Principles." *Hum.-Comp. Inter.* **4(1)** 1989: 11–44.
3. Bregman, A. *Auditory Scene Analysis: The Perceptual Organization of Sound.* Cambridge, MA: MIT Press, 1989.
4. Cuddy, L., A. Cohen, and J. Miller. "Melody Recognition: The Experimental Application of Musical Rules." *Can. J. Psych.* **33** (1979): 148–157.
5. Deutsch, D. "The Processing of Pitch Combinations." In *The Psychology of Music*, edited by D. Deutsch. New York: Academic Press, 1982.
6. Deutsch, D. "Paradoxes of Musical Pitch." *Sci. Am.* **267(2)** (1992): 88–95.
7. Forbes, T. W. "Auditory Signals for Instrument Flying." *J. Aeronautical Soc.* **May** (1946): 255–258.
8. Gaver, W. "The Sonic Finder: An Interface that Uses Auditory Icons." *Hum.-Comp. Inter.* **4(1)** (1989): 67–94.
9. Gaver, W. W., and R. B. Smith. "Auditory Icons in Large-Scale Collaborative Environments." In *Proceedings of Human-Computer Interaction—Interact'90*, 735–740. Held in Cambridge, UK, August 27–31, 1990. AMsterdam: North Holland, 1990.
10. Gaver, W. W., R. B. Smith, and T. O'Shea. "Effective Sounds in Complex Systems: The ARKola Simulation." In *Proceedings of CHI 1991.* Held in New Orleans, April 28–May 2, 1991. New York: ACM, 1991.
11. Getty, D. J., J. A. Swets, and J. B. Swets. "Multidimensional Perceptual Spaces: Similarity Judgment and Identification." In *Auditory and Visual Pattern Recognition*, edited by Getty and Howard. Hillsdale, NJ: Erlbaum, 1981.
12. Gibson, J. *The Senses Considered as Perceptual Systems.* Boston: Houghton Mifflin, 1966.
13. Huron (1989) [need reference]
14. Kramer, G. "Audification of the ACOT Predator/Prey Model." Unpublished research report prepared for Apple Computer's Advanced Technology Group (*Apple Classrooms of Tomorrow*), 1990.
15. Kramer, G. "Audification: Using Sound to Understand Complex Systems and Navigate Large Data Sets." In the Proceedings of the Santa Fe Institute Science Board, Santa Fe Institute, 1991.

16. Kramer, G., and S. Ellison. "Audification: The Use of Sound to Display Multivariate Data." In *Proceedings of the International Computer Music Conference*, 214–221. San Francisco, CA: ICMA, 1991.
17. Lackoff, G., and M. Johnson. *Metaphors We Live By*. Chicago: University of Chicago Press, 1980.
18. LeBrun, M., and D. Arfib. *J. Audio Engr. Soc.* **27(4)** (1952).
19. Martin, D. Letter to the Editor: "Do You Auralize?" *JASA* **24(4)** (1952): 416.
20. Opcode Systems. *MAX User's Manual*. Menlo Park, CA: Opcode, 1991.
21. Puckette, M. "Combining Event and Signal Processing in the MAX Graphical Programming Environment." *Comp. Music J.* **15** (1991): 3.
22. Reisberg, D., ed. *Auditory Imagery*. Hillsdale, NJ: Lawrence Erlbaum, 1992.
23. Singh, P. "Interaction of Timbre and Pitch in Spectral Discrimination Tasks Using Complex Tones." *J. Acoust. Soc. Am.* (Supp. 1) **86** (1989): S58.
24. Speeth, Sheridan Dauster. "Seismometer Sounds." *J. Acous. Soc. Am.* **33(7)** (1961): 909–916.
25. von Bismark, G. "Timbre of Steady Sounds: A Factorial Investigation of Its Verbal Attributes." *Acustica* **30** (1974): 146.
26. Pratt, C.C. "Spatial Character of High and Low Tones." *J. Exp. Psych.* **13** (1930): 278–285.
27. Roads, C. "Granular Synthesis of Sound." *Comp. Music J.* **2(2)** (1978): 61–61.

Carla Scaletti
Symbolic Sound Corporation and University of Illinois, P.O. Box 2530, Champaign, IL 61825-2530, USA; (217)355-6273; e-mail: c-scaletti@uiuc.edu

Sound Synthesis Algorithms for Auditory Data Representations

A working definition of sonification is presented, and previous work is categorized as sound at the user interface, applications in specific domains, studies of sonification itself, or general tools and systems development. A brief history is given for the technology of sound synthesis; the author's sound specification language Kyma is defined and its application in data-as-event, data-as-signal, real-time, and interactive approaches to sonification is explained. Several sound synthesis techniques—data as samples, multiplication, addition, granular synthesis, nonlinear distortion techniques, filtering, physical modeling, and sampled sounds—are described and examples are given of their application in sonification. Some of the open questions in sonification are outlined, and some predictions are made on the future of sonification.

1. SONIFICATION

1.1 A WORKING DEFINITION

Sonification is such a new field that its scope and even its name are still being debated. For the purposes of this paper, the following, working definition is proposed:

> *a mapping of numerically represented relations in some domain under study to relations in an acoustic domain for the purposes of interpreting, understanding, or communicating relations in the domain under study.*

There are two parts to this definition: a technique and an intent. The technique is to map numerical data, presumably embodying some relationships in the physical world or a model world, to sound. The intent is to understand or communicate something about that world. Both halves of the definition are necessary in order to differentiate the field of sonification from other fields that involve sound computation. That the sound be data-driven is necessary but not sufficient justification for calling it sonification; it must also have been done with the intent of understanding or communicating something about the original domain.

EXAMPLES. Sound in virtual reality qualifies as sonification under this definition in that it is a mapping of virtual spatial coordinates to parameters of sound or sound processing. There is no reason to limit the sonic representation of relations in the virtual world to that of spatial relations; sonification may aid in navigating and understanding virtual worlds as it aids in the understanding of any other data. Sound at the user interface aids in understanding the state of the computer and any number of virtual machines built atop the hardware. Data-driven sound used in conjunction with a data-driven visualization is also a mapping of relations in a model world to relations in the acoustic domain.

Other examples of sound computation are less clearly classifiable as sonification. The mapping of arbitrary data points would not qualify as sonification in that it would not represent *relations* in the model world. An overly complex mapping never intended to be decipherable would not satisfy the criterion of sonification as communication. Algorithmic music composition may qualify as the mapping of numerical relations to acoustic relations; however, in many cases the primary intent of music composition is something other than the analysis and study of the domain represented by the numerical relations.

1.2 BACKGROUND

Although the desire to represent numerical relationships in sound can be traced at least as far back as Pythagoras, recent advances in computer technology have made it possible to experiment with very accurate mappings and complex sound synthesis in real time. Ironically, it is the very size and complexity of some of these computer systems and the data they can generate that have made it necessary to seek out alternative data representations and additional data presentation bandwidth.

Areas of research in sonification include sonification science—including the perceptual and cognitive validation of sonification (see Ballas, Bargar, Bly, Fitch and Kramer, Smith, and Williams, all in this volume; Malouf et al.,[16] and Wenzel et al.[33]), the use of sound at the user interface (see Blattner, Brewster et al., Cohen, Gaver, and Mynatt, all in this volume; Chamberlin[3]; and Gaver[9])—and data analysis/interpretation—including domain specific applications (see Jackson and Francioni, in this volume; Jameson, in this volume; Kramer, in this volume; Francioni et al.[8]; and Mezrich et al.[18]), virtual acoustic environments,[8,33] and systems/tools development (see Jameson, and Madhyastha and Reed, all in this volume; Scaletti and Craig[25]; Smith et al.[30]; and Wenzel[33]).

CERL/NCSA PROJECT. In the CERL/NCSA Data Sonification Project[25] our goal was to develop some prototypes for data sonification tools that could be applied to a variety of time-dependent data streams. These tools included:

- DataStream
- Mappers (DataToAmplitude, DataToFrequency, DataToStereoPosition, and others)
- Combiners
- Markers
- Analyzers
- Comparators
- SonicHistogram

To illustrate the use of these tools, we generated data-driven sound tracks for data-driven animation videos produced at NCSA.[26]

C. S. Peirce defines three ways in which a representation may relate to an object (where a representation may also be some combination of all three)[31]:

- An *icon* is typically mimetic and relates to an object that may or may not exist.
- A *symbol* has an association with a general class of objects, although it appears as a specific instance of that class.
- An *index* is causal and is affected by the object it represents.

Our sonic representations were primarily indexical; that is, streams of data were directly tied to the parameters of a sound. However, many of our indexical representations had symbolic or iconic meanings as well.

Because sound generation was done using software synthesis on a digital signal processor (rather than in the hardware of a synthesizer), we developed both the mappings and the sound synthesis algorithms. We first designed a sound, then decided which parameters of the sound should be controlled by the data, and, finally, decided on a mapping of data to sound parameters. Where the 1991 paper[27] focuses on the final stage, describing a set of "tools" or classifications for indexical data mappings, this paper focuses on the first step: the design of sound synthesis algorithms.

2. SOUND AND SOUND COMPUTATION

Sound waves are air pressure oscillations caused by the propagation of an initial deformation or displacement of the air. Common representations of sound include:

- Time-domain representations in which the change in air pressure at a single point is displayed with respect to time.
- Frequency-domain representations displaying the average strength of each frequency component over all time.
- Time-frequency representations that show the frequency content of a signal as it evolves over time.

For rigorous definitions of the time and frequency domain, see Proakis and Manolakis[21] or Oppenheim and Schaeffer.[19] For an introduction to audio applications of time-frequency representations, see De Poli, Piccialli, and Roads.[5]

If air pressure variations are translated to voltage variations by means of a microphone and then periodically sampled by an analog-to-digital converter, these samples or instantaneous air pressure values can be stored, manipulated, or simply converted back to continuous voltage variations by a digital-to-analog converter (see Figure 1).

Every step in this chain has some relevance to sonification. This discussion concentrates on the right-center block, sonification as the manipulation of streams of numbers representing digital audio signals.

Once encoded as a stream of numbers, a digitally recorded sound becomes an object in its own right, separate from the original acoustic source; it can be modified in various ways and accessed out of time order. Once sound is represented as a stream of numbers independent of its source,

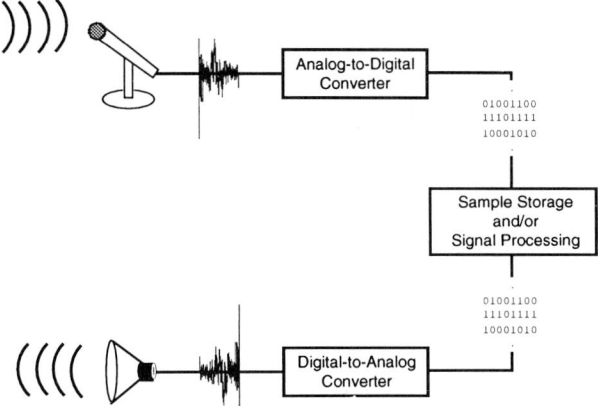

FIGURE 1 Sound computation.

it becomes clear that the stream of numbers need not have a source at all; the stream could have been generated algorithmically, or it could have come from an entirely different domain.

2.1 THE RECENT HISTORY OF SOUND SYNTHESIS

The earliest use of a digital computer to produce sound was by a group headed by Max Mathews at Bell Laboratories in the late 1950s and early 1960s.[15] Mathew's *acoustic compilers* (better known as the Music N languages since they are named Music 1, Music 2, ..., Music 5[17]) are languages that bear some resemblance to the simulation languages of the time (like GPSS). The Music N languages provide software simulations of analog circuits that can be connected together in various ways to generate and modify digital audio signals. Software sound synthesis is extremely flexible but, on most general-purpose processors, it is not possible to generate sound in real time.

By implementing a set of synthesis algorithms in hardware rather than software, researchers in the 1970s built digital synthesizers that could generate sound in real time, albeit using only a restricted set of synthesis algorithms. The adoption of the MIDI interface standard around 1983 paved the way for the explosive growth in the use of digital synthesizers controlled by computers.

The real-time response of MIDI synthesizers has made them popular with sonification researchers who control them with custom software (see Smith[30] and Kramer and Bargar, in this volume) or the MAX language.[12]

Their intended use as performance instruments in commercial music, however, can make them somewhat opaque and difficult to control; the very "hardwiring" that makes them so fast also makes synthesizers less flexible than sound synthesis in software.[29]

With the growing availability of digital signal processors and of faster general-purpose computers, there has been a renewed interest in regaining the flexibility of software sound synthesis. Languages like Kyma and Dannenberg's Nyquist,[22,23,24] embody new paradigms for software sound synthesis. There are also several implementations of the Music N paradigm targeted for digital signal processors (e.g., the NeXTStep Music Kit and the Accelerando Project[20]).

For more information on interesting languages for controlling MIDI devices and new languages for software synthesis, see back issues of *Computer Music Journal* published by MIT Press (55 Hayward Street, Cambridge, MA 02142-9902) and the *Proceedings of the International Computer Music Conference* published by the International Computer Music Association (Suite 330, 2040 Polk Street, San Francisco, CA 94109).

2.2 SONIFICATION IN THE KYMA LANGUAGE

Kyma is a visual sound-specification language that runs on a Macintosh or 386/486-based host computer and a multiple-DSP engine called the Capybara that computes digital audio samples in real time. What follows are the basic definitions needed for understanding Kyma in the context of this paper; for a more detailed description of the Kyma language, see the references.[22,23,24,27]

DEFINITIONS. The basic unit in Kyma, called a Sound, represents a stream of digital audio samples. A Sound is either atomic, or it is a function of one or of several other Sounds. A Sound is represented graphically as an icon with any argument Sounds displayed below it (see Figure 2).

TIME. What makes computation in Kyma different from computation in many other languages is the existence of temporal operators. Atomic Sounds have finite durations; Sounds that are functions of one or more Sounds derive their durations from the durations of their argument Sounds. Sounds begin at time zero unless they are descendents of a **TimeOffset** Sound. A **Concatenation** is a shorthand for a **Sum** of Sounds whose time offsets make them sequential and nonoverlapping in time.

Sound Synthesis Algorithms for Auditory Data Representations

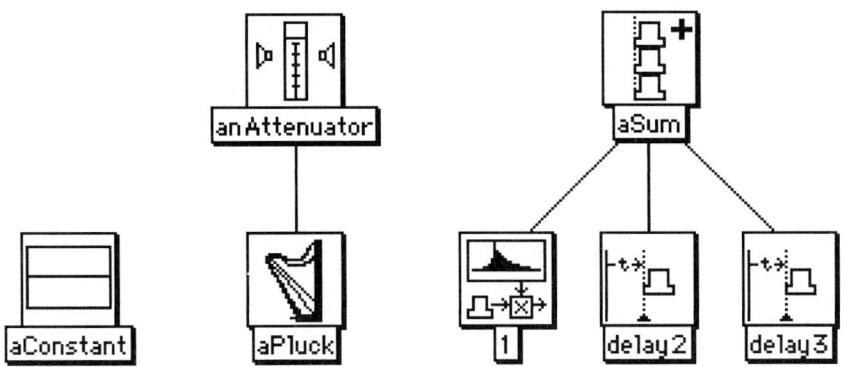

FIGURE 2 Examples of Atomic, Unary, and n-ary Sounds in Kyma.

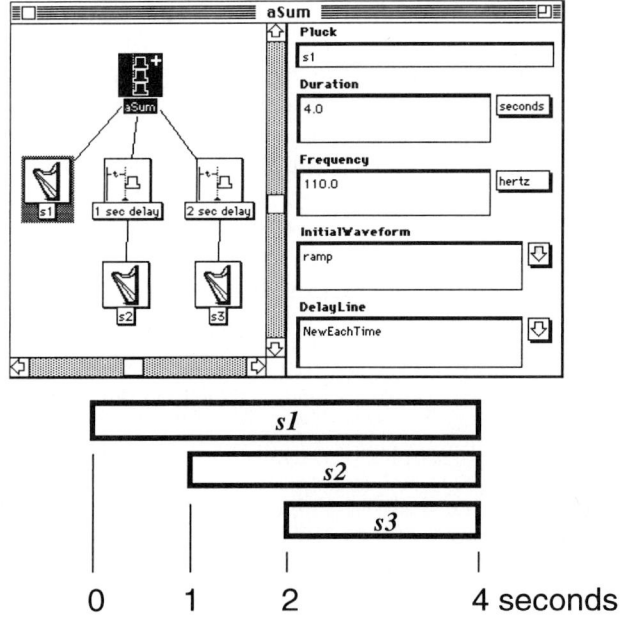

FIGURE 3 Kyma Sounds derive their start times from their context. The structure (above) results in the duration and timing shown by the horizontal bars (below).

SIGNALS VERSUS EVENTS. When data represent discrete events or elements, possibly ordered but not time dependent, then they can be thought of in the traditional way: as values stored in a file accessed as needed by the program. If, instead, the data represent a continuous, time-dependent variable (even though that variable may have been sampled so it could be represented on the computer), then the data can be thought of as a signal, on a par with sampled or synthesized digital audio signals. For example, Bly's Scenario 1 data (in this volume) represent soil samples that may or may not contain gold. A sonification of this data might map each six-dimensional data point to a single sonic event, but the timing and the order of these events is not particularly relevant; each soil sample is independent of the others. Scenario 2, on the other hand, is a representation of six time-dependent variables; each one of these variables is a function of time, just as an audio signal is a function of time.

The choice of whether to treat the data points as independent events or as continuous time-dependent streams determines how the sound synthesis software/hardware is set up and whether the mapping of data to sound parameters is done in a procedural or a functional way. In the first case, each new data point corresponds to a new event, a new instance of a sound generator with its own parameter values; data are mapped to sound parameters by a program that can have state, side effects, and random access to any of the data points. In the second case, a single, persistent instantiation of the sound generator is influenced by a continuous stream of data; the mapping of data to sound parameters is functional, without state or side effects.

DATA AS EVENTS: THE FILEINTERPRETER. In a *data-as-events* sonification, a symbolic encoding of rules and procedures maps each data point to a parameter or parameters of a sound. The sound is some fixed synthesis algorithm; a program creates instantiations of the algorithm and determines timing.

The following are examples of data-as-events sonifications:

- A C program reads data from a file; each data point is scaled and offset to lie within the range of 0 to 127, and is then sent to a synthesizer as part of a new MIDI NoteOn event.
- The data are formatted as a Music N score file, where each column supplies a parameter to a synthesis algorithm.

In both of the above examples, a symbolic program is "above" a sound generator, controlling the timing, the mapping, and thus the final result. A program arranges and organizes several *instances* of a single sound synthesis *template*.

In Kyma, the **FileInterpreter** Sound performs data-as-events sonification, reading data from a file and using a Smalltalk-80 program to map

Sound Synthesis Algorithms for Auditory Data Representations

those data to parameters of a template Sound. For example, the following mapping could be used to map Sara Bly's six-dimensional Scenario 1 data to simultaneous pitch values:

```
| paramArray t |
t := 0.
[file atEnd] whileFalse: [
    paramArray := file nextLine.
    1 to: 6 do: [ :i |
        element start: t seconds
        freq: (4 c + (paramArray at: i)) pitch
        dur: 0.25 second
        amp: 1].
    t := t + 1].
```

In Figure 4 I show the Sound template called "element"; its amplitude, frequency, and duration are set to the variables ?amp (not shown), ?freq, and ?dur.

DATA AS SIGNAL. In *data-as-signal* sonification, the sound structure emerges from a specific arrangement of functions or data streams. A single instantiation of a synthesis algorithm receives a continuous stream of input values. In Kyma, data streams can be used interchangeably with the digital audio sample streams that represent sound.

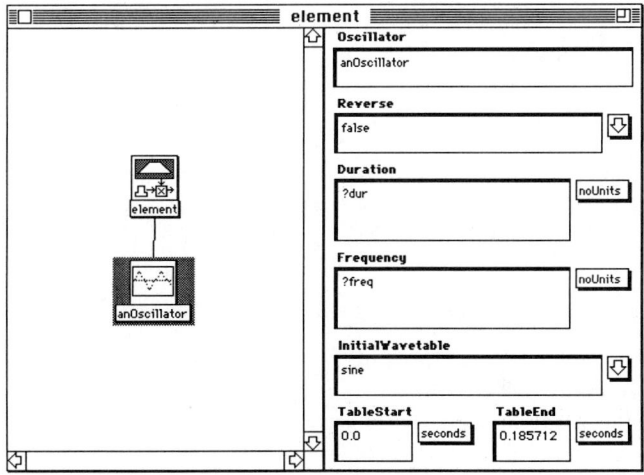

FIGURE 4   A template Sound with variables ?dur and ?freq.

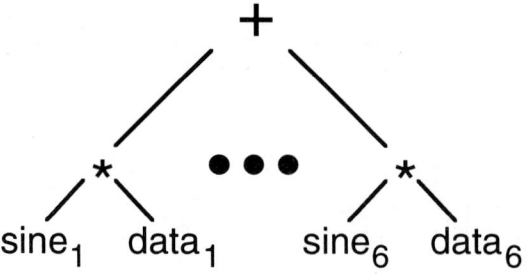

FIGURE 5 The sum of six sinusoids, the amplitude of each controlled by one of six data streams.

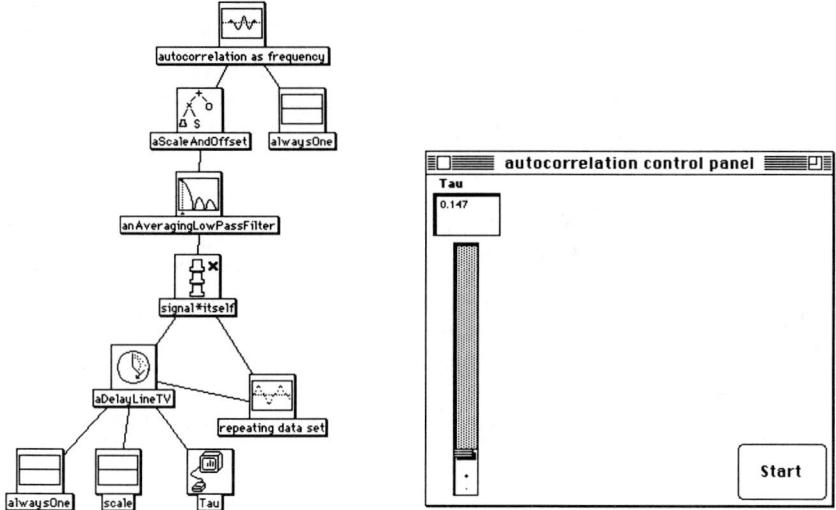

FIGURE 6 On the left is the structure for mapping an autocorrelation function to frequency. A slider on the right supplies the value of **Tau**.

Other examples of data-as-signal sonifications include:

- A continuous voltage output from a lab instrument is applied to a voltage-controlled oscillator.
- A single MIDI NoteOn event is followed by a stream of data representing continuous pitch bend control for that single note.

In these examples, the data stream "flows into" the sound generator. A single sound source varies according to its time-varying input(s).

Kyma provides prototype Sounds that read or cycle through functions stored in tables. For example, the **ScaledInterpolatingOneShot** might be used to generate an amplitude or pitch envelope by looking up the values of a function stored in a table, and an **Oscillator** might be used to generate a sine wave by repeatedly cycling through one period of the sine function stored in a table.

For example, each dimension of Bly's Scenario 2 data can be stored in a table as a 100-point function. Six oscillators, each with its own nonharmonic frequency, can each be multiplied by one of the 100-point functions (i.e., each data stream controls the amplitude of one oscillator) as shown in Figure 5. The sum of all six would be a time-varying timbre that would emphasize different frequency components at different times.

REAL-TIME DATA STREAMS—EXTERNAL VALUES, CONTROL PANELS, AND CAPYBARA DRIVER. Data-as-signal sonification can also be done interactively and/or under the control of other programming languages. **ExternalValues** are Sounds representing named data streams that can be read and written from virtual control panels on the computer display or from programs outside the Kyma environment.

For example, in the Kyma Sound shown in Figure 6, the frequency of an oscillator is controlled by the time-averaged product of a periodic data stream and the same data stream delayed by **Tau**. **Tau** is an external value that can be controlled from a virtual slider drawn on the display. The user can adjust **Tau**, listening for maxima in the correlation function.

External values can also be read and written by programs other than Kyma. At the NCSA Virtual Reality Lab, for example, a Kyma Sound implementing the sound placement algorithm outlined by Loomis,[14] is controlled by a data stream from a head-mounted Polhemus. In that environment, a C program takes Polhemus position data and maps it to parameters controlling interaural delay, head shadow, atmospheric absorption, and other factors used by the human auditory system to localize sound sources.

Other sonification programs can also access Kyma Sounds through the Capybara Driver; for example, there are MAX objects for controlling Kyma

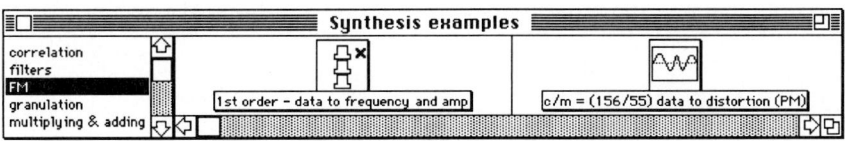

FIGURE 7 Tools developed in Kyma can be categorized and placed in a palette.

Sounds on the Capybara, as well as objects for controlling the Capybara via HMSL.

TOOL BUILDING: ABSTRACTION AND ENCAPSULATION. There are two ways to build reusable tools in the Kyma language. The first is to name a Sound and use it as a prototype in the construction of new Sounds. You can categorize these prototypes and construct palettes that make them easily accessible (see Figure 7).

The other is the specification of a new class of Sound with its own icon and parameter list. In this level of abstraction, parameters of complex Sounds are set to variables or to expressions relating parameter values to each other,[27] and a class editor is used to design the icon and editor for the new class of Sounds.

MULTIPLE AUDITORY VIEWS. Since Kyma is interactive and supports a variety of ways to "sound the data," it encourages exploration and the formation of multiple auditory "views" on the same data set. Listening to the data in multiple ways is essential if one is to minimize the distortion and bias inherent in any one choice of representation.

3. SOUND SYNTHESIS TECHNIQUES
3.1 SOUND SYNTHESIS AND TIMBRE

The ANSI standard definition for timbre says more about what it is not than what it is. Research on multidimensional timbre spaces (see Krumhansl[13] and Wessel[34]) suggests that the physical components of timbre can be separated and identified, and that timbre should be treated as an emergent property of sound rather than as a single parameter. Thus this discussion of sound synthesis techniques is only indirectly a discussion of timbre; timbre is not a goal, though it may emerge as a consequence of controlling the various parameters of a sound synthesis algorithm.

3.2 PARAMETER MAPPING

Data streams can be mapped directly or indirectly to the parameters of the sound synthesis techniques outlined here:

- In a **0th-order mapping**, the data stream itself is listened to as a stream of digital audio samples.

- In a **1st-order mapping**, the data stream controls a parameter or parameters of a synthesis model (e.g., the data controls the amplitude of an oscillator).
- In a **2nd-order mapping**, the data stream controls parameters of a synthesis model that controls the parameters of another synthesis model (e.g., the data controls the amplitude of an oscillator that, in turn, controls the frequency deviation of another oscillator).[1]

And so on. See Kramer (in this volume) for another characterization of the degree of indirection in a mapping.

3.3 DATA AS SAMPLES

A 0th-order mapping of data can be done by sending each sample in the data stream to a digital-to-analog converter. The data must be scaled to the word size of the converter and must be supplied at a regular rate (the sample rate); the data stream can be generated in real time (probably requiring some FIFO buffers in order to maintain the regular sample rate), or stored on disk or in the memory of a DSP or sampling synthesizer.

If the data are stored in a table in memory, the following algorithm is used to step through the table:

```
index := 0.
while index ≤ tableLength
    output := table at: (index rounded).
    index := index + sampleIncrement
```

To go through the data at a slower rate, each sample is sent to the converter two or more times in succession by setting the sample increment to a number less than 1; to go through the data at a faster rate, a sample increment larger than 1 is used.

This same algorithm can be used to generate envelopes (for example, if the table contains an exponential or linear function), or oscillators (for example, if the table contains one cycle of a periodic function such as a sine). In the oscillator case, however, the index into the table should be modulo the table length.

Noninteger sample increments introduce quantization noise that can be ameliorated by interpolating between successive data points. A sampled signal represents the signal exactly at the points where it was sampled. Between the sampled values, however, the value of the original signal is unknown. Linear interpolation is a computationally inexpensive method that

---

[1] Note from Editor: This differs from McCabe's use of the term **nth-order mapping** (in this volume).

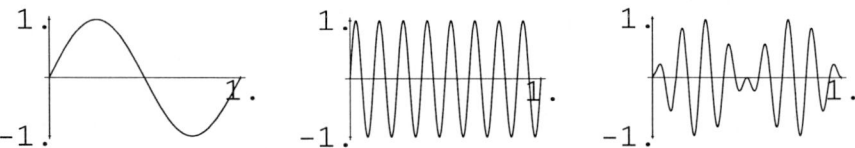

FIGURE 8 Two sine functions of frequency 1 and 8, and their product.

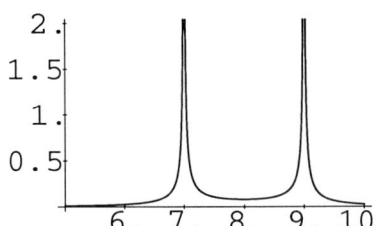

FIGURE 9 The spectrum of the product signal shown in Figure 8.

assumes that the value between two sampled values lies on a straight line drawn from one sampled value to the next. Better interpolation methods exist that are more computationally expensive.[19,21]

3.4 MULTIPLYING SIGNALS: $F_1(T) \times F_2(T)$

Multiplying a signal by another signal is one way to scale its amplitude. It is roughly analogous to a volume control; when you twist the volume knob (or move the sliders up and down), you are multiplying the audio signal by the knob position (another time-varying signal). If the knob position is replaced by a data stream, then you have a 1st-order mapping of data to amplitude.

By multiplying the left and right channels by two different signals, you can get some "stereo" or spatial effects. For example, if the output of the right channel is $f(t)g(t)$ and the output of the left channel is $f(t)\sqrt{1-g(t)^2}$, then the data stream $g(t)$ controls the perceived position of $f(t)$ between the two speakers.

When two equal amplitude cosines of frequency w_1 and w_2 are multiplied, the result is mathematically identical to the sum of two half-amplitude cosines, one whose frequency is the sum of w_1 and w_2 and one whose frequency is the difference of the two:

$$\cos w_1(t) \times \cos w_2(t) = 0.5 \cos(w_1 + w_2)t + 0.5 \cos(w_1 - w_2)t. \quad (1)$$

For example the sine waves shown in Figure 8 have frequencies of 1 and 8; the resulting spectrum of the product has components at 7 and 9 as shown in Figure 9.

For more complex signals, each spectral component (frequency) of one input generates two "sidebands" (i.e., components at the sum and difference frequencies) for each spectral component of the other input.

In electroacoustic music, the product of two signals is called *ring modulation*. A minor variant (called *amplitude modulation*)

$$\cos t\omega_1(t) \times (1 - A\cos\omega_2 t) = 0.5A\cos(\omega_1+\omega_2)t + 0.5A\cos(\omega_1-\omega_2)t + \cos\omega_1 t \quad (2)$$

produces sinusoids at the sum frequency, the difference frequency, and the "carrier" frequency.

3.5 ADDING SIGNALS: $C_1 F_1(T) + C_2 F_2(T) + \ldots + C_N F_N(T)$

A sum is the most direct way to specify simultaneous parallel sample streams. In the free field, sound pressure waves emanating from independent sources effectively add in the air. An audio mixer performs an analog or digital summation of all its inputs, and the output of most hardwaresynthesizers and software synthesis programs is, transparently to the user, the sum of several intermediate sound-generating modules (e.g., the output of a 16-voice polyphonic synthesizer is the sum of the sixteen voices.).

Except for possible phase cancellations, the spectrum of the sum of two signals is the same as the sum of the spectra of the two signals. Thus, a desired spectrum can be constructed by adding signals whose spectra form portions of the complete spectrum.

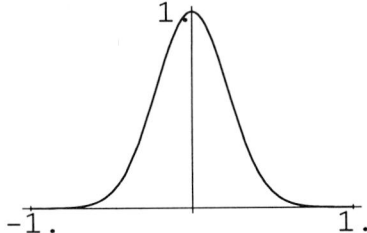
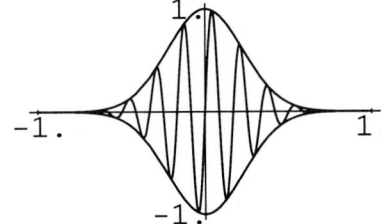

FIGURE 10 A Gaussian envelope on the left, and on the right, a grain: a Gaussian envelope multiplied by a sine wave.

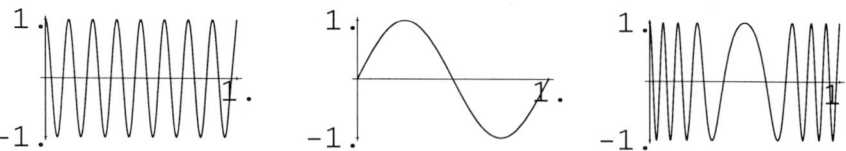

FIGURE 11 Example of frequency modulation: the carrier, the modulator, and the frequency-modulated carrier.

GRANULAR SYNTHESIS. In *granular synthesis*,[5] sound is formed by combining thousands of elementary sonic "grains," where a grain is a brief acoustical event with a duration at or near the threshold of perception. Grains are constructed by multiplying a sine wave by a Gaussian envelope of only a few milliseconds duration. Data streams can control any of several parameters including grain duration and spacing, grain frequency, and grain amplitude. Granular synthesis is a special case of additive synthesis: each grain contributes frequency content in a narrow slice of time. . .

3.6 FREQUENCY OR PHASE MODULATION: $G(\omega T + F(T))$

One way to map data to frequency deviation is to use the values from the data stream as deviations in the sample increment through a cyclic table lookup. As the step size through the table changes, the period of repetition changes with it (see Figure 11).

If the original signal (the carrier) and the signal controlling the frequency deviation (the modulator) are both sinusoids, the spectrum of the result can be described in terms of the amplitude of the modulator and the ratio of the carrier to the modulator frequencies:

$$A\sin(\omega_c t + B\sin\omega_m t) = A \begin{cases} J_0(B)\sin\omega_c t + \\ J_1(B)[\sin(\omega_c + \omega_m)t - \sin(\omega_c - \omega_m)t] + \\ J_2(B)[\sin(\omega_c + 2\omega_m)t + \sin(\omega_c - 2\omega_m)t] + \\ J_3(B)[\sin(\omega_c + 3\omega_m)t - \sin(\omega_c - 3\omega_m)t] + \ldots \end{cases} \quad (3)$$

where J_n is an nth-order Bessel function of the first kind.

In other words, the spectrum contains the sum and difference frequencies of the carrier frequency and all integral multiples of the modulator frequency; however the amplitude of each sum and difference pair depends on a Bessel function. In Figure 12 are shown the amplitude of each pair of sum and difference frequencies as a function of B, usually termed the modulation index. It is sometimes given as a rule-of-thumb that the modulation index is approximately equal to the number of sum and difference pairs; in practice the situation is complicated by negative frequencies that constructively or destructively add to positive frequencies.

3.7 TRANSFER FUNCTIONS: $F(G(T))$

Each element in a data stream g could be interpreted as an argument to a function f, where only the output of f is actually heard (or used to control the parameter of another sound). For example, if the transfer function $f(x) = e^x$ and the input function $g(x) = x$, then a stream of numbers in log space would be mapped to a stream of numbers in linear space.

In the case where $g(t)$ is a cosine and $f(x)$ is a polynomial, the spectrum of the output can be specified as a sum of polynomials. Each harmonic of the cosine $g(t)$ is generated by a polynomial T_n where:

$$T_0 = 1$$
$$T_1 = x$$
$$T_2 = 2x^2 - 1$$
$$\vdots$$
$$T_n = 2xT_{n-1} - T_{n-2}.$$

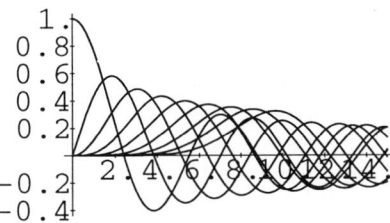 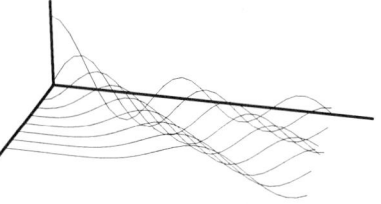

FIGURE 12 The Bessel functions J_n as a function of modulation index. Notice that, in general, the larger the modulation index, the greater the number of sum and difference frequencies at nonzero amplitudes and the lower the amplitude of the carrier. The graph on the right (in which the order of the Bessel function is plotted on the axis perpendicular to the page) shows that the number of nonzero amplitude sidebands is nearly linear with modulation index.[4]

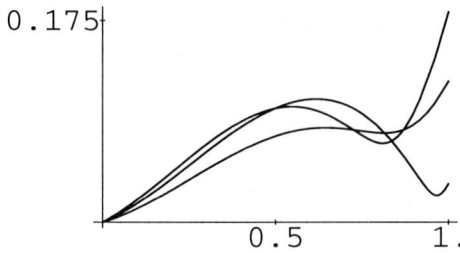

FIGURE 13 The evolution of the first three harmonic's amplitudes as a function of the input sine wave amplitude. The transfer function is $1.5x^3 - x^2 - 0.25x = .375\,T_3 - .5\,T_2 + .875\,T_1 - .5\,T_0$.

Thus, to construct a transfer function yielding some desired spectrum, add the polynomials corresponding to each component of the desired spectrum, scaling each polynomial by the corresponding component amplitude.

Most higher order polynomials are nearly linear close to zero; at larger values their nonlinearity is more apparent. Thus, a low amplitude sinusoidal input produces a nearly sinusoidal output, whereas a unit amplitude input results in harmonic distortion. An input with time-varying amplitude results in an output with a time-varying spectrum and amplitude (see Figure 13). In electroacoustic music, this technique is also known as *waveshaping*.[6]

3.8 FILTERING

A filter attenuates or amplifies specific frequency bands present in its input. For example, a low pass filter allows low frequencies to pass through it unattenuated, but attenuates higher frequencies. A band pass filter only allows frequency components falling within a specified band of frequencies to pass through unattenuated.

All filters combine scaled input with scaled and delayed versions of the input and, optionally, the output. If the output is fed back into the filter, it is an infinite impulse response (IIR) filter, since a single value fed into such a system could recirculate forever. If the output is not fed back, then any value fed into the system will eventually die out, so the filter is said to be a finite impulse response (FIR) filter.

Design and analysis techniques for digital filters constitute a study unto themselves and are a prime focus of both the basic texts[19,21] and the current research literature in digital signal processing.

3.9 PHYSICAL MODELS

Rather than modeling a one-dimensional sound pressure wave, a physical model represents a two- or more dimensional sound-producing object.[5,8,28] Physical modeling concentrates on the cause of a sound by simulating wave propagation in resonant structures. A model must describe the vibrational characteristics of the resonators, the couplings between resonators, and the forcing function that initiates the vibrations.

Simple, computationally inexpensive models can be constructed using combinations of delay lines, filters, and nonlinear transfer functions[16,28]; more complex, non–real-time models are described using numerical approximations of differential equations.[2]

3.10 SAMPLED SOUNDS

Sampled sounds (i.e., digital recordings, or streams of samples from an analog-to-digital converter) can be handled using the same algorithms described in the section on data as audio samples.

Sampled sounds have only two controllable parameters—amplitude and the rate at which the samples are presented—since the generating algorithm or the acoustic source is no longer available for alteration. A digital recording has a flat structure divorced from its original context; it is a collection of samples with no history of how it was produced and no meaning assigned to it.

By performing a frequency analysis of the recording (using Fourier analysis programs like Lemur[7]), we can "re-parameterize" the recording. Such a resynthesis provides independent control over different frequency regions in the spectrum, and the duration of these resynthesized sounds can be altered without introducing as much noise as is introduced by changing the sample increment.

3.11 FOR MORE INFORMATION

Of necessity, only terse descriptions have been given for each of these synthesis algorithms. For more detailed explanations and examples, see the texts by Dodge and Jerse,[6] Wells,[32] Chamberlin,[1] and DePoli, Picciali, and Roads.[2]

4. EXAMPLE APPLICATIONS

There are countless ways in which these synthesis algorithms can be used in data sonification. This section suggests starting points for further exploration.

4.1 DATA AS SAMPLES

Ironically, the most direct, 0th-order mapping—listening to the data stream directly as a stream of instantaneous amplitudes—has the fewest applications in auditory data representation. Most data sets do not lend themselves to direct representation as air pressure waves.

Examples of data sets that could be interesting as direct samples include: temperature measurements made at a single location several times a day over several years, deformation of some material measured at a single point for several milliseconds after the material has been excited by an ultrasonic pulse, measurements of the change in water level made several times a day at a single position over several years, or a single channel of an electrocardiogram recorded over several minutes. Each of these example data sets meets the following criteria:

- It represents a single time-dependent phenomenon (or a phenomenon that can be decomposed into several one-dimensional time-dependent processes).
- It is likely to be periodic or quasi-periodic.
- The data set is relatively large.

Periodic or quasi-periodic data sets are likely to sound more interesting than nonperiodic data sets. If the data are periodic, it should be possible to hear one or more frequencies, whereas nonperiodic data are likely to be heard as noise. In almost all cases, however, the same information could be extracted, and in more detail, by using a first-order mapping.

In order to be perceived as sound, the data should contain periods of at least 25 to 50 milliseconds at the chosen sample rate; nonperiodic changes (e.g., clicks, impulses, isolated transitions) must have about the same duration or they will not be audible. Data sets must be at least 20,000 samples long if the sound is to last for more than a second or so.

A shorter data set can be treated as a waveform, by entering it into a table and using a cyclic table lookup to cycle through it repeatedly. This is not recommended, however, as the artificial periodicity of cycling through the data is likely to overshadow any short-term structure audible in the data itself.

4.2 MOVING BETWEEN TWO OR MORE STATES

If the data set represents transitions between two or more states, the states can be associated with identifiable sounds and the data stream can control the transition between the sounds.

For example, Richard Misenheimer of the North Carolina Supecomputing Center has designed a mapping that, at one extreme, sounds like the sung vowel |u| and, at the other extreme, sounds like the vowel |a|. The vowels are generated by filtering a synthetic glottal pulse wave; each filter models a vocal formant with a center frequency, an amplitude, and a bandwidth. The data stream controls the interpolation between the amplitude value, center frequency value, and bandwidth value of corresponding formants in the two vowel sounds.

Lippold Haken's SOS system[11] allows interpolation between as many as eight different states. Each state is identified with the vertex of a cube in a "timbre space." From one to three dimensions of time-dependent data can be mapped to positions within this cube. Each vertex of the cube is associated with a sinusoidal resynthesis of a sampled sound; for positions between the vertices, the amplitudes of corresponding partials are linearly interpolated.

4.3 AXES AND GRIDS

A sonic analogy to axes and grids can provide orientation in **DataToFrequency** mappings. A constant frequency "axis" tone added to a **DataToFrequency** mapping allows the listener to detect exactly when the data stream crosses the axis; when the data points are close to but not exactly on the axis, beats are heard between the two frequencies.

If the **DataToFrequency** mapping and the constant frequency are multiplied, the sum and difference frequencies are heard. When the data stream is on the axis, the difference frequency goes to zero; at other times, the magnitude of the difference is represented by the frequency of the difference tone.

A filter or a bank of filters can be tuned to ring whenever the input hits one of the harmonics of the filter bank's fundamental frequency. The result is an emphasis and slight prolongation of harmonic frequencies, creating a kind of sonic grid for a **DataToFrequency** mapping.

Tr 41

Tr 36-38

4.4 COMPARING NEARLY IDENTICAL DATA SETS

Addition is the most transparent way to compare two or more data mappings: the mappings are sounding simultaneously, and their instantaneous differences can be compared. Placing one mapping in the left channel and the other in the right assists the listener in keeping the two streams separate. Noise values as small as one part in 1,000 can be detected as beats between two **DataToFrequency** mappings; noise values of 1 in 10,000 manifest themselves as phase cancellation (and thus amplitude changes).

Tr 36,37 When you multiply two nearly identical **DataToFrequency** mappings, the differences between them manifest themselves as difference tones. The frequency of the tones corresponds to the magnitude of the difference between the two streams, and their timing tells you exactly when those differences occur. By using amplitude modulation (Eq. (2)) rather than ring modulation (Eq. (1)), you can hear the sum, the difference, *and* one of the original mappings for reference.

Tr 39,40 By taking the difference of two data streams and mapping that difference to frequency, you can monitor the absolute value of the difference. The difference between two **DataToFrequency** mappings is extremely sensitive; when the streams are identical, the output is zero, but noise as small as one part in a million added to one of the streams is easily detectable.

To detect periodicities in a data stream, multiply the data stream by a delayed version of itself, and perform a running average of the result; this performs a sort of autocorrelation on the data stream. If you map the output of the autocorrelation to frequency and then interactively control the delay time, delays corresponding to periodicities in the data will produce higher frequencies than delays at which there is little or no correlation.

4.5 TEXTURES AND TENDENCIES

In situations where tendencies and textures are more important than precise values, distortion synthesis, the sonic histogram, and granular synthesis work well.

If the data are mapped to the amplitude of the modulator in frequency modulation, larger data values result in a "brighter timbre"; this is neither exact nor linear, but it does give a general impression of the magnitude of the data stream. A similar result can be obtained by using a **DataToAmplitude** mapping as the input to a polynomial transfer function; the higher the amplitude of the input, the more upper partials in the resulting sound.

SonicHistogram was one of the tools we developed in the CERL/NCSA project; it was used to sonify measurements taken on the age of the forest in Yellowstone National Park. In our mapping, each range of years was represented by an oscillator at a fixed frequency; each oscillator was multiplied by a function representing how much of the forest was that age. For example, if half the area of the forest was 100–200 years old, the frequency representing that age was multiplied by 0.5; if the proportion of the forest 100–200 years old grew to 75% of the total area covered by the forest, that frequency was multiplied by 0.75. Continuous, gradual changes in timbre over time showed the tendency of the forest to age during periods when there was a policy of fire suppression.

Stuart Smith is exploring the use of granular synthesis as the sonic analogy to the line segment icons of Exvis.[30] In the visual part of Exvis, one perceives not so much the individual line segments or icons, but textures, statistical tendencies, and edges. Similarly, with granular synthesis, one does not perceive the grains individually, but is aware of tendencies, textures, and abrupt changes.

4.6 SONIFICATION IN VIRTUAL REALITY

Any techniques used in sonification can be transferred to sonification in virtual reality. Physical models, because of their basis in spatial structures rather than time-based signals, hold a special appeal. If a three-dimensional object is visually simulated in virtual space, then its acoustic properties could be simulated at the same time; changes in the physical model would result in changes to both the visual appearance and the acoustic properties.

4.7 MULTIVARIATE DATA

Mapping multiple variables to a single sound presents special difficulties: not only are the perceptual parameters of sound weighted differently, but they influence each other and can confound two data streams that ought to be independent of one another. Probably the most linear and democratic way to present several dimensions of data is as the sum of multiple identical mappings (though even this method runs into problems when two components fall within the same critical band, or when the number of data streams becomes too large).

Mapping each component of a multidimensional data point to a coefficient of a polynomial and then using that polynomial as the transfer function for a sinusoidal input results in a unique timbre for each data sample; however, differences between these timbres can be subtle, and each of the dimensions will have a different perceptual weight.

4.8 OPINIONATED SONIFICATIONS

Sampled sounds offer rich evocative timbres for use in sonification; however, a sampled sound, unlike a synthetic timbre, has no parameters related to its original production and is thus difficult to modify and control in meaningful ways. For this reason, sampled sounds are often simply triggered by specified conditions.

TABLE 1 Summary

Data characteristics	Suggested synthesis techniques
Oscillating between states	timbral interpolation or morphing
Axes & grids	resonators, fixed tones
Comparison	sums, products, differences, correlation
Textures & tendencies	granular synthesis, FM, waveshaping, sonic histogram
Periodicity detection	data as samples, autocorrelation
Virtual objects in VR space	physical models, sampled sounds
Data with an attitude	instrumental sounds, musical scales, sampled sounds

As an example of sampled sound triggering, we used digital recordings of coughs in a mapping of ozone levels in the Los Angeles basin. The cough recordings were not altered in any way but, as the ozone levels rose, so did the rate of triggering; thus, the higher the ozone level, the more coughs heard per second (and the more coughs overlapped). This technique is often used in virtual reality: an action or condition triggers a sampled sound.

Of all the synthesis techniques, the use of concrete sounds is both the most powerful and the most dangerous: powerful in the sense that concrete sounds can rely on the listener's lifelong experiences,[15] and dangerous in that the overlays of semantic meaning may interfere with or even contradict the meaning of the indexical mapping of data.

As innocuous as it might seem, the use of instrumental timbres or musical scales in a sonification may convey unintended cultural and historical meaning; the researcher should be aware of these additional meanings and only use these kinds of sounds when the symbolic meanings do not contradict or confuse the indexical meaning.

5. OPEN QUESTIONS IN SONIFICATION

The current work in sonification provides hints of what may be possible and makes it clear that there is much left to be done. Broad categories for further study include:

- Applications: further and more sophisticated examples of sonification applied to specific problems

- Sonification science: studies in the perception, cognition, and neurophysiology of data-driven sound
- Systems: software and hardware for sonification

While most of the work done so far has been in the area of applications, there is still need for further and more sophisticated examples applied to actual data. It may be that the real validation of sonification will come when it is no longer noticed: when scientific papers routinely list sonification as just another one of many analytical tools and when it is an expected component of the computers and special-purpose equipment in everyday use in industry and in the professions.

The study of sonification itself will not only help to validate and improve the way we do sonification, but will also shed light on brain function, cognition, and the understanding of language. For example, studies indicate that different parts of the brain are active during speech perception than are active during pitch perception.[35] But what parts of the brain participate in nonsymbolic analytical listening like sonification? Why does music have strong associative memories and can this be used in sonification? Similarly, why do many people experience an emotional reaction to music and could this be used in sonification? Is it easier to remember things when they are grouped in regular metrical structures? Is music a sonification of something? When does sonification provide something more than could be gotten from simple statistical analysis?

Not only will systems development be shaped by results from applications research and the scientific studies of sonification, but hardware and software development will also feed back into the process by providing the essential tools for this research. Needs for future hardware and software include: integrated sonification/visualization languages, tools for getting from an imagined sound to a realized sound, the integration of sonification tools into mass market software like spreadsheets or statistical analysis packages, and more and better tools for exploratory sonification.

6. PROGNOSIS

The study of sonification forces us to ask so many intriguing questions about sound, language, meaning, perception, and the basic functioning of the human brain that it seems worth pursuing if only for the light that it might shed on other related fields.

In the most optimistic scenario, sonification will be nothing less than a new form of language. We may all be participating in the definition of a new communications path, one that could not have been possible before the technology was in place to make the connections. At the very least, it is already clear that it will provide researchers with additional tools for understanding and communicating numerical relationships.

The complexity of the topic, the number of open questions, and the vast amount of work that still needs to be done stand as an open invitation to those who enjoy leaping into uncharted, and not-yet-validated areas.

ACKNOWLEDGMENTS

To my friends in the former Visualization Group at NCSA, thank you for the discussions, the data sets, and the data-driven animations. Thanks to Kurt Hebel who read and commented on several drafts of this paper and generated many of the graphs. To everyone who participated in the Sound Computation Workshop in Urbana, thanks for the stimulating discussions on these topics.

REFERENCES

1. Blattner, M. M., D. A. Sumikawa, and R. M. Greenberg. "Earcons and Icons: Their Structure and Common Design Principles." *Hum.-Comp. Inter.* **4** (1989): 11–44.
2. Bly, S. A. "Communicating with Sound." In *Proceedings of CHI '85 Conference on Human Factors in Computer Systems*, edited by S. A. Bly, 155–119. ACM: Association for Computing Machinery, 1985.
3. Chamberlin, H. *Musical Applications of Microprocessors*, 2nd ed. Indianapolis: Hayden Books, 1985.
4. Chowning, J. "The Synthesis of Complex Audio Spectra by Means of Frequency Modulation." *J. Audio Engr. Soc.* **21** (1973): 526–534.
5. De Poli, G., A. Picialli, and C. Roads. *Representations of Musical Signals.* Cambridge, MA: MIT Press, 1991.
6. Dodge, C., and T. A. Jerse. *Computer Music: Synthesis, Composition, and Performance.* New York: Schirmer Books, 1985.
7. Fitz, K., W. Walker, and L. Haken. "Lemur: Extensions to the MacQualey-Quatieri Analysis/Synthesis Technique." In *Proceedings of the International Computer Music Conference.* International Computer Music Association, 1992.
8. Francioni, J., L. Albright, and J. Jackson. "Debugging Parallel Programs with Sound." In the Proceedings of the ACM/ONR Workshop on Parallel and Distributed Debugging. Printed as *SIGPLAN Notices*, vol. 26, 1992.
9. Gaver, W. W. "The SonicFinder: An Interface that Uses Auditory Icons." *Hum.-Comp. Inter.* **4** (1989): 67–94.
10. Grey, J. M. "Multidimensional Perceptual Scaling of Musical Timbres." *J. Acous. Soc. Am.* **61** (1977): 146–159.
11. Haken, L. "Computational Methods for Real-Time Fourier Synthesis." *IEEE Trans. Sig. Proc.* **40** (1992): 2327–2329.
12. Kramer, G., and S. Ellison. "Audification: The Use of Sound to Display Multivariate Data." In *Proceedings of the International Computer Music Conference*, 214–221. International Computer Music Association, 1991.
13. Krumhansl, C. L. "Why is Musical Timbre so Hard to Understand?" In *Structure and Perception of Electroacoustic Sound and Music*, edited by S. Nielzen and O. Olsson, 43–54. Amsterdam: Excerpta Medica, 1989.
14. Loomis, J. M., C. Hebert, and J. G. Cicinelli. "Active Localization of Virtual Sounds." *J. Acous. Soc. Am.* **88** (1990): 1757–1761.
15. Loy, G., and C. Abbott. "Programming Languages for Computer Music Synthesis, Performance, and Composition." *ACM Comp. Surv.* **17** (1985): 235–2675.

16. Malouf, F. L., M. Lentczner, and C. Chafe. "A Real-time Implementation of Physical Models." *Proceedings of the International Computer Music Conference* (1990): 191–193. International Computer Music Association, 1990.
17. Mathews, M. V. *The Technology of Computer Music.* Cambridge: MIT Press, 1969.
18. Mezrich, J. J., S. Frysinger, and R. Slivjankovski. "Dynamic Representation of Multivariate Time Series Data." *J. Am. Stat. Assoc.* **79** (1984): 34–40.
19. Oppenheim, A. V., and R. W. Schafer. *Digital Signal Processing.* Englewood Cliffs, NJ: Prentice Hall, 1975.
20. Pinkston, R. F. "The Accelerando Project." *Proceedings of the International Computer Music Conference* (1989): 242–245. International Computer Music Association, 1989.
21. Proakis, J. G., and D. G. Manolakis. *Introduction to Digital Signal Processing.* New York: Macmillan, 1988.
22. Scaletti, C. "Kyma: An Object-Oriented Language for Music Composition." *Proceedings of the International Computer Music Conference* (1987):49–56. International Computer Music Association, 1987.
23. Scaletti, C., and R. E. Johnson. "An Interactive Graphic Environment for Object-Oriented Music Composition and Sound Synthesis." *Proceedings of the 1988 Conference on Object-oriented Programming, Languages, and Systems* (1988).
24. Scaletti, C. *The Kyma Language for Sound Specification.* Champaign: Symbolic Sound Corp., 1990.
25. Scaletti, C., and A. Craig. "Using Sound to Extract Meaning from Complex Data." In *SPIE Volume 1459 Extracting Meaning from Complex Data: Processing, Display, Interaction II*, edited by E. J. Farrell, 207–219. San Jose: SPIE—The International Society for Optical Engineering, 1991.
26. Scaletti, C., and A. Craig. "Using Sound to Extract Meaning from Complex Data." In *SPIE Video Supplement Volume 1459-V Extracting Meaning from Complex Data: Processing, Display, Interaction II*, edited by E. J. Farrell. San Jose: SPIE—The International Society for Optical Engineering, 1991.
27. Scaletti, C. "Lightweight Classes Without Programming." *Proceedings of the International Computer Music Conference* (1991): 505–508.
28. Smith, J. "Waveguide Filter Tutorial." *Proceedings of the International Computer Music Conference* (1987): 9–15. International Computer Music Association, 1987.
29. Smith, S., G. Grinstein, and R. Pickett. "Global Geometric, Sound, and Color Controls for Iconographic Displays of Scientific Data."

In *SPIE Volume 1459 Extracting Meaning from Complex Data: Processing, Display, Interaction II*, edited by E. J. Farrell, 192–206. San Jose: SPIE—The International Society for Optical Engineering, 1991.
30. Smith, S., R. D. Bergeron, and G. G. Grinstein. "Stereophonic and Surface Sound Generation for Exploratory Data Analysis." *Proceedings of the Conference on Computer Human Interface* (1990).
31. Tejara, V. *Semiotics from Peirce to Barthes: A Conceptual Introduction to the Study of Communication, Interpretation, and Expression.* Leiden, The Netherlands: E. J. Brill, 1988.
32. Wells, T. *The Technique of Electronic Music.* New York: Schirmer, 1981.
33. Wenzel, E. M., S. S. Fisher, P. K. Stone, and S. H. Foster. "A System for Three-Dimensional Acoustic 'Visualization' in a Virtual Environment Workstation." *Proceedings of Visualization 90* (1990): 329–337.
34. Wessel, D. L. "Timbre Space as a Musical Control Structure." *Comp. Music J.* **3** (1979): 45–52.
35. Zatorre, R. J., A. C. Evans, E. Meyer, and A. Gjedde. "Lateralization of Phonetic and Pitch Discrimination in Speech Processing." *Science* **256** (1992): 846–848.

David H. Jameson
IBM T. J. Watson Research Center, Yorktown Heights, NY 10598

Sonnet: Audio-Enhanced Monitoring and Debugging

Sonnet is an audio-enhanced debugger under development in the Mathematical Sciences Department at IBM Research. The issues in which we are interested include the use of sophisticated yet easy-to-understand sounds to aid in understanding program execution, how to shift the programmer's focus from the narrow line-oriented view of a program to a global gestalt or holistic view, and finally how to provide a lightweight graphical user interface that allows run-time interaction rather than postmortem analysis. An important goal of Sonnet is that it should be easy to predict what sounds will be generated in advance of execution. Should an unexpected sound be produced, the user may then investigate more closely in the hope of finding the anomaly. A prototype was built on top of an internal debugger for the IBM RS/6000 workstation to evaluate the feasibility and usefulness of the sounds. A new system is now in operation and incorporates the features recognized as important from our original experiments. This paper describes the current work in progress.

1. INTRODUCTION

Unfortunately, the laws of debugging are not like the laws of Nature. Programs that are broken seem to do their best to confuse, hide, and otherwise prevent the programmer from finding the problem. One of the most ironic and frustrating aspects of the debugging process is the time required to find a bug compared with the time needed to correct the problem. It is often the case that hours, days, or even weeks are wasted trying to determine the cause of a problem but once discovered only a few minutes are needed to correct the program.

A key point to note is the ease with which a bug is often quickly isolated once the approximate area containing the error is known. Although I am not aware of any formal study of this phenomenon, there is plenty of anecdotal evidence to suggest its truth. There are countless occasions when a programmer will sit in front of his terminal for hours trying in vain to locate a bug only to have a colleague lean over his shoulder and either immediately point out the problem or make some comments that cause the programmer to realize instantly the problem.

Locating the approximate area containing the error can be exceedingly difficult because of the deceptive behavior of the broken program. Traditional methods of debugging are very time consuming mainly due to the necessity to inspect closely the state at various times during execution of a program. This forced narrow focus is a major impediment to problem solving, essentially being equivalent to the "can't see the forest for the trees" syndrome.

2. SOME SOUND JUSTIFICATION

Sound is a very underutilized medium in the computing world. Even within the multimedia explosion, the main focal points have been graphics, animation and video, with sound, and music in particular, sometimes added as an afterthought. Yet people in fields outside of computer science use sounds in many ways to help diagnose problems. A driver knows when to shift gears by listening to the pitch of a car engine. A mechanic can diagnose problems with that car engine by the sounds that it makes. A doctor listens to sounds from your heart and lungs and even to the way you cough to determine how some of the body's organs are functioning.

Sound has another very useful property compared to visual systems. Sound waves travel around corners! In other words, sound can be used to draw your attention to some event when you are not looking or paying attention. A car driver waiting at a red traffic light might not notice the light turned green until the car behind beeps the horn.[1] Many multimeters generate a beep when a circuit has no resistance, a condition known as a short circuit. An engineer checking for short-circuits on a printed circuit board can concentrate on the board and have his ears tell him a short-circuit was found.

I believe there is plenty of evidence to demonstrate that sound is indeed useful. Accepting sound is useful, we need to find out how it can be used appropriately. It seems to me that there are two important issues. The first is the question of what sounds to use, a topic currently attracting much focus. The second is a user-interface problem: how to control the sounds conveniently in an interactive environment?

3. OTHER SOUND-BASED SYSTEMS

Researchers thinking about sound look for mappings that somehow give intuitive hints as to what is happening. A trivial example might be to attach the value of an integer variable to the pitch of a note so that higher values produce higher pitches. However, I am not so much interested in intuitive hints as in using easy-to-recognize sound patterns to follow the progress of a running program.

Hotchkiss and Wampler have used sound to represent functions in scientific data[6] and also proposed using sound to represent running code. They do not discuss how to map the sounds in any way that might actually help understand the code. They do point out that one could "auralize" code execution and listen to one function calling another. They propose time stamping the beginning of each function and storing the results in a file. Afterwards, they suggest a simple mapping where notes are sounded with durations proportional to the amount of time spent in each function.

Contrast this with the approach taken by Sonnet, the audio debugger described in this paper, where instead of different notes being used for different functions, the note sounded by the first function is modulated in interesting ways by subsequent function calls. This has the effect of reducing drastically the complexity of the sound, making it easier to understand what is going on.

Francioni and her colleagues have gone a little further and discussed some mappings of sounds to execution behavior.[4,5] However, their system also does not work in real time. They collect traces from which mappings are produced later.

---

[1] In New York, the horn is beeped before the light turns green. I have not yet figured out how to apply this technique!

The InfoSound system[12] uses recognizable sound effects to monitor a telephone network simulation. For example, when the state of a phone changes from "not busy" to "busy," InfoSound generates a telephone ringing sound. The same system was also used to monitor synchronization of multiple processors. The idea was to assign different parts of a song to each processor such that when the processors synchronize, the music would also. I am not convinced that this approach is applicable in general. Simply because one **can** play music does not mean that one **should** play music! A related system, InfoProbe,[2] provided visual meters such as gauges and dials along with audio information. As with InfoSound, the audio consisted of sound effects and musical melodies. The authors have admitted that they were never too happy with the audio side of this system.[2]

In some work on sonification of multivariable data carried out a few years ago, I proposed[9] that smoothly varying sounds and timbres would be a more effective way to track data than the usual pitch changes, sound effects, and melodies traditionally employed. Sonnet extends some of these concepts so as to be helpful in the context of debugging.

Sonnet is not built around any particular debugging model. Currently, its main domain is the sequential program where we still have a lot to learn in terms of debugging methodology. The techniques and sounds should be easily applicable to parallel and distributed programs.

4. LISTENING TO CODE EXECUTION

Rather than intuitive sound mappings, I am more interested in convenient sounds that can be used to inform the programmer of interesting events. For example, I might like to know that a large **while** loop exited properly. A simple way to do this is to use a single note that starts when the loop starts and ends when the loop terminates normally. A nonterminating note is a clue that something is wrong with the loop. That is enough to at least cause the programmer to investigate that loop more closely.

One important goal of Sonnet is that it should be easy to predict what sound you expect to hear. Consider a musician about to play some notes on his instrument. The musician has a pretty good idea of what he expects to hear and will be quite disturbed if the results are not what was expected. In this sense, a Sonnet user is like the musician. Certain results are expected and only the unexpected is of concern. This is not the case with most of the existing systems where much more interpretation is needed to figure out what is happening.

[2] Conversation between Victor Miller and the InfoProbe authors.

With the primitive version of Sonnet using hard-coded sounds, one nontrivial bug was found because of this phenomenon. During a demonstration of Sonnet to a skeptical colleague, I wanted Sonnet to follow a loop in another program. Within that loop there were two function calls, named **A** and **B**. Sonnet was set up in the following way:

1. Whenever the loop was active, a single note would sound.
2. Cumulative calls to **A** would cause the volume of the note to be changed smoothly over time
3. Cumulative calls to **B** would cause the timbre of the note to be changed smoothly over time

When we ran the program, the note and continuous timbre changes were evident throughout the execution. However, the volume changes occurred for a while and then stopped. This was odd because both **A** and **B** were supposed to be called all the time under normal circumstances. A quick look at the loop showed that **A** was called conditionally based on the value of a variable that was not initialized properly. After correcting the problem, a second run produced the expected volume and timbre changes.

5. VISUAL PROGRAMMING INTERFACE

To experiment with sound, I needed a way to manipulate sounds based on the program. There are two issues here. The first is the basic question of how to capture what is happening in a running program. The second issue is how to apply the information captured so as to produce useful sounds.

The traditional way to capture what a program is doing is to collect trace information and store it for later processing. From a debugging perspective, this approach bothers me because it requires the program to be run to completion (or until it crashes) each time the source is changed so that an updated event history can be produced. Also, I really wanted a more lightweight approach based on the fact that a programmer is less likely to use a tool that requires a lot of setting up each time it is used.

Secondly, I needed a convenient tool that would facilitate the mapping of information. Such a tool would have to be open ended so that new transformations or mappings could be added easily. For example, if I wanted to listen to the difference of two variables (perhaps to understand how they

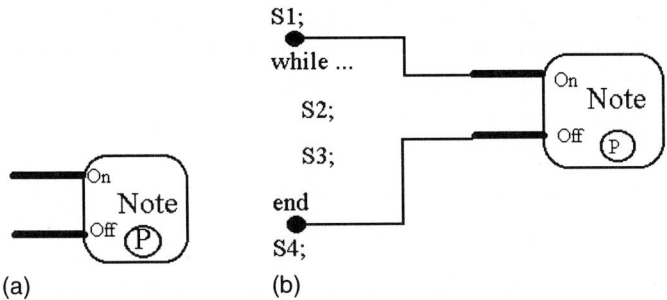

FIGURE 1 (a) A simple component that allows a single note to be turned on or off. (b) An example of some source code in a debugger window connected to a component.

were related to each other) and also to be informed if the difference became greater than some value, then arithmetic and logical operations would be required.

Eventually I decided to implement a visual programming language (VPL) that would allow me to construct operations by connecting little components together. In particular, I could connect a line of source code to a component and have the component trigger when the line executed. A description of the VPL itself is beyond the scope of this paper but the reader should get a feel for its facilities from the examples that follow.

The VPL provides sound components to setup connections between executing code and sounds that are to be heard during such execution. In its simplest form, a component triggers when an impulse arrives on any of its clocked inputs.

Figure 1(a) shows a simple component that allows a single note to be turned on or off. The "P" button allows static properties of the component to be set up. In this example, static properties include the pitch and volume of the note.

Figure 1(b) shows an example of some source code in a debugger window connected to a component. Observe that the connections are between statements rather than on them. Unlike most debuggers that allow breakpoints to be set just before a line is executed, Sonnet requires finer grained breakpoint locations. We will see why this is necessary in a moment. In Figure 1(b), an impulse is generated after S1 ends but before the **while** loop starts. This impulse is connected to the NoteOn input of the component and, hence, causes the note to start sounding. A second impulse is generated between the **end** and S4. This impulse is connected to the NoteOff input and the note is stopped at this point.

6. LOOP FOLLOWING

Suppose we find that the note does not end after some "expected" time. We would like to know more about what is going on inside the loop. There are two possible reasons why the sound never ends. The first possibility is that the test at the start of the loop always succeeds and, hence, the loop just iterates forever. The second possibility is that the program is "stuck" in one of the statements inside the loop, perhaps in some subroutine.

The obvious solution might be to cause a note to be sounded each time the loop iterates. However, this solution has several drawbacks. First of all, the high retrigger rate of the note that will occur in all but the slowest of loops produces a rather unpleasant gurgling effect. In particular, using this sound on multiple loops simultaneously creates unfathomable cacophony! Scaling the number of repeats by some factor will clean up the gurgling but again, with multiple loops, it is hard to understand what is happening because there is too much unrelated activity.

Consider instead the example in shown in Figure 2. The component here has an extra input called a volume delta (VolΔ) whose purpose is to change the volume of the note in a very special way. Each time an impulse arrives, the VolΔ takes the next point of a function and outputs the value at that point. The VolΔ input line is an example of a named Impulse Processor (IP) and it has the structure shown in Figure 3. A named IP is designed to be a shortcut so that the user does not need to fill the screen with lots of boxes and draw all the lines to connect them and so forth. The static properties of the VolΔ IP include the function (a sine wave, triangle wave, square wave, etc.) and minimum and maximum scaling values. The scaling values are useful for mapping the real-value results of the function into valid numbers in MIDI numbers.

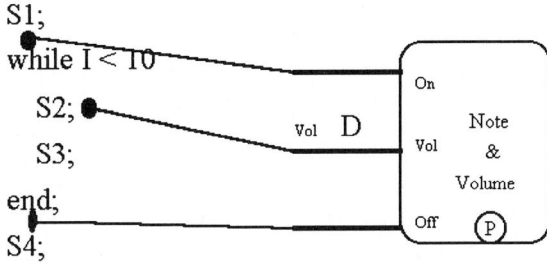

FIGURE 2 A component with an extra input called a volume delta (VolΔ) whose purpose is to change the volume of the note in a very special way.

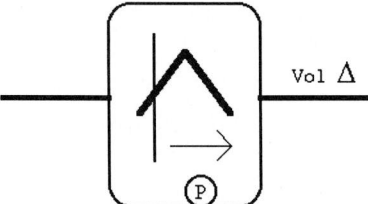

FIGURE 3 The structure of VolΔ.

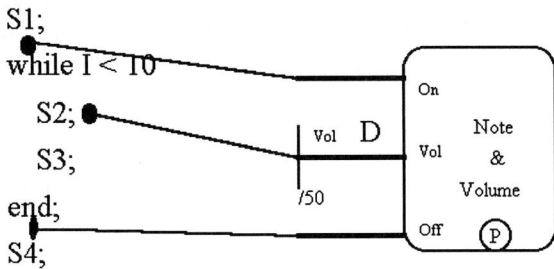

FIGURE 4 Decimating the volume changes.

Finally, to cut down on the number of impulses arriving at the component, it is useful to provide a countdown IP where an output impulse is generated once for every n input impulses. The final version is shown in Figure 4. The vertical bar at the start of the VolΔ input represents the countdown IP and the /50 indicates that only every 50th arriving impulse should be passed through to the component.

On a Yamaha TG77 using minimum and maximum volume values between 70 and 110, a pleasing tremolo effect is produced while the loop is active. Using volume changes alone we can, in fact, add yet another layer. Remembering that the tremolo effect is caused by a reasonably rapidly changing volume (≈ 4–8 Hz, say) with a small peak-to-peak amplitude (depth), we can add a slow-changing volume effect where the period is measured in tens of seconds. Kramer (in this volume) refers to this concept of incorporating multiple parameters on one sound as parameter nesting.[7]

The intent here is to represent as much information as possible using only one note. This is easier to follow than a sequence of notes which severely tests the abilities of even the most accomplished of musicians. Yet another layer of complexity can be added by smooth modulation of the timbre of a single note. Many synthesizers (including the TG77) provide low-pass filters where the cutoff frequency can be changed under program control. This can be used in the same manner as the VolΔ processor.

It is feasible to follow several loops simultaneously by using a different note for each loop where each note is modulated as described above. This is particularly useful when each note is connected to a different process.

There is a sense in which the sounds generated by Sonnet are symptoms of a running program. We expect programs to have certain symptoms and are therefore suspicious when different symptoms are produced.

7. LISTENING TO DATA

As well as following code execution, Sonnet is intended to be useful for following data within a program. Interesting issues include

- Tracking trends in the value of a variable,
- Detecting read/write access to a variable, and
- Playing data structures.

It is probably not appropriate to use pitch or the usual modulations (volume, timbre, tempo change, etc.) to indicate the precise values of variables except in the case where the variable can only take on a very small number of values. For example, Boolean variables can be only **true** or **false** and it is easy to produce an appropriate aural representation. On the other hand, if the situation is such that we only need to sample the value from time to time, then it might be appropriate to use speech synthesis to announce the value. Announcing the values of multiple variables simultaneously may be feasible by using a different (preferably harmonious) pitch for each variable. Although Sonnet has a component that allows text to be sent to a speech synthesizer, I have not yet had the opportunity to experiment with this seriously.

However, audio is very useful to indicate trends in the value of a variable   **Tr 42-46**
and comparing one variable with another on an on-going basis. For some examples of how I used Sonnet to debug and monitor sorting algorithms and loop iterations, please listen to CD tracks 42–46.

Figure 5 shows a way to keep an ear on a variable. The MS is a named IP that generates periodic impulses, and in this example an impulse is generated every 1000 milliseconds. Notice that the component used here has only an **on** connection. One of the static properties of this component is a duration that can be set appropriately.

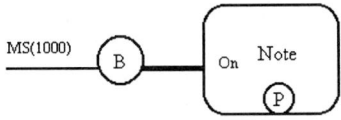

FIGURE 5   Listening to a variable.

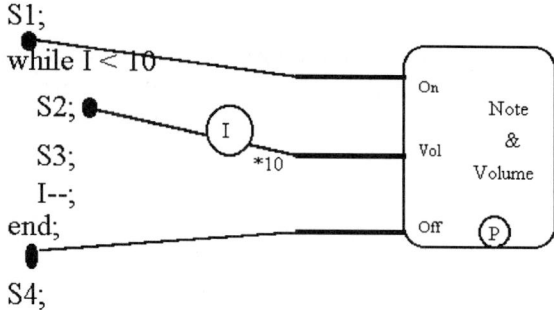

FIGURE 6 How to monitor a variable at a specific address during program execution.

Figure 6 shows how to monitor a variable at a specific address during program execution.

The circle contains an expression—in this example it is the variable I—that is evaluated each time it receives an impulse. The variable is multiplied by 10 before arriving at the component. This scaling assures that the value of I is in the allowable range for MIDI volume. Each iteration around the loop takes about 2 seconds due to the work being done. The result is a note whose volume decreases over a 20-second period. In fact, due to the nonlinearity of the synthesizer response to the volume MIDI event, no differences were heard between the values of 2 and 0.

As with following code execution, if the loop is too tight, you can chose to receive impulses less often. There is, of course, a tradeoff here since it may be important not to miss any values. One alternative is to cause the system to slow down temporarily so that there is time to hear the note. This will not change anything on a sequential program except to slow its execution time down. Unfortunately, on a parallel or distributed environment, not to mention a real-time application, slowing down a process is far too intrusive and will most probably change the execution behavior of the application.

8. DETECTING ACCESS TO A VARIABLE

A very useful debugging technique is to be able to track when a specified variable is accessed. Some processors have built-in hardware to allow watch points that detect such accesses. Unfortunately, only a few watch points are allowed, typically between two and six. I want to be able to monitor an arbitrary number of variables so that continuous sounds can be set up and modified depending on whether a particular variable is read or assigned. The actual value may or may not be relevant. To support this mechanism, we are building a special card that will allow the debugger to watch any number of variables.

9. FUTURE DIRECTIONS: PLAYING DATA STRUCTURES

Another interesting area to explore with Sonnet is playing data structures. Based on several experiences using a logic analyzer to help debug code that was part of a device driver, I found that if I could determine that there was something suspicious about the data, that would be enough information to encourage me to look at the code that was managing that data. Without such hints, it often does not occur to the programmer to look in a particular place because he "knows" that there is nothing wrong in that area or he has forgotten about that area and so forth.

Although there are many examples of interfaces to look at a data structure graphically (for example, see Myers[8] and Busaclacchi and Slagle[1]), it is not clear that they are useful when the structure becomes large. For example, it is difficult and time-consuming to examine a tree with only 100 nodes to see if it is built correctly. For one thing, the whole tree will probably not fit on the screen. Secondly, it would be useful to be able to check it quite often, something not really practical with a visual system, particularly since it would be very easy to miss important information.

Consider, for example, a binary tree that is growing over time. It might be useful to know that the tree is reasonably well balanced and that it is not degenerating into a linear list. One way to do this is to use a smoothly changing sound as we descend the tree. For example, it is possible to generate a sequence of pitch tones such that they appear to be either ascending or descending forever. This paradoxical effect was first demonstrated by Roger Shepard[10] and an excellent description of similar effects is provided by Deutsch.[3] This effect can be used in Sonnet to indicate continuous progress without ever running out of notes!

For each iteration, starting at the root, we descend the tree, picking a left or right node at random and playing the next note in an ascending sequence. The same starting note is always used for the root. If the tree is reasonably well balanced, we should hear sequences of ascending notes with an abrupt transition each time we restart at the root. The longer the sequence, the deeper the particular route taken. Over time, it should not be too hard to tell whether all the sequences are roughly the same length, indicating a roughly balanced tree. If the tree contains a large number of nodes, then instead of playing a new note at each level, we can skip levels with no real loss of generality. The purpose of such skipping is to shorten the amount of time needed to play the tree.

10. CONCLUSION

I have introduced some mechanisms to facilitate the easy use of sound in an interactive monitoring and debugging session. The visual programming interface provides a simple way to allow executing code to trigger the required sounds. I have described some simple modulations that do not necessarily provide explicit mappings of interesting values to sounds, but rather allow the use of sounds to detect whether code is executing properly and to give hints as to where to look for trouble. A key point is that the user has a reasonably good idea of what he expects to hear once the program starts. Deviations from the expected results are clues that something is not working as expected.

This research is at a very early stage of development. Much work remains to be done. The experiments carried out give tantalizing evidence that the approach is useful but a fully built debugging environment will be needed to validate the approach. However, one subtle bug in another tool was discovered even with the primitive version of Sonnet. Would this bug have been discovered without Sonnet? The answer is probably yes but, on the other hand, the other tool was in use by many people for a long time without any hint of a problem.

REFERENCES

1. Busaclacchi, Perry J., and James R. Slagle. "A Visual Scheme-Based Distributed Programming Environment." In *Proceedings, 4th Conference on Hypercubes, Concurrent Computers and Applications*. Monterey, CA: Golden Gate Enterprises, 1989.
2. Cameron, E. J., B. Gopinath, P. Metzger, and T. Reingold. "InfoProbe a Utility for Animation of IC* Programs." In *Proceedings of the Hawaii International Conference on System Sciences*, vol. 2. Washington, DC: IEEE, 1989.
3. Deutsch, Diana. "Paradoxes of Musical Pitch." *Sci. Am.* (August 1992): 88–95.
4. Francioni, Joan M., Larry Albright, and Jay Alan Jackson. "Debugging Parallel Programs Using Sound." In *ACM Proceedings Parallel and Distributed Debugging*, 60–68. ACM, May 1991.
5. Francioni, Joan M., and Jay Alan Jackson. "Breaking the Silence: Auralization of Parallel Program Behavior." Technical Report TR 92-5-1, Computer Science Department, University of Southwestern Louisiana, Lafayette, LA, May 1992.

6. Hotchkiss, Robert S., and Cheryl Wampler. "The Auditorialization of Scientific Information." In *ACP Proceedings SUPERCOMPUTING '91*, November 1991.
7. Kramer, G., and S. Ellison. "Audification: The Use of Sound to Display Multivariate Data." In *Proceedings, International Computer Music Conference*, 214–221. Computer Music Association, 1991.
8. Myers, Brad A. "Displaying Data Structures for Interactive Debugging." CSL-80-7, Xerox Corporation, June 1980. Also an MIT Masters thesis.
9. Rabenhorst, D., D. Jameson, E. Farell, T. Linton, and J. Mandelman. "Complementary Visualisation and Sonification of Multi-Dimensional Data." Paper presented at the 1990 Symposium on Electronic Imaging Science and Technology, Santa Clara, CA, 1990.
10. Shepard, Roger N. "Circularity in Judgements of Relative Pitch." *J. Acous. Soc. Am.* **36(12)** (December 1964).
11. Sonnenwald, Diane H., B. Gopinath, Gary O. Haberman, William M. Keese III, and John S. Myers. "InfoSound: An Audio Aid to Program Comprehension." In *Proceedings of the 23rd Annual Hawaii International Conference on System Sciences*. Washington, DC: IEEE, 1990.

Tara M. Madhyastha and Daniel A. Reed
Department of Computer Science, University of Illinois, Urbana, Illinois 61801

A Framework for Sonification Design

One of the obstacles to widespread experimentation with sonification of data has been the lack of a standard model for sound generation, and a standard interface to control that model. This paper describes Porsonify, a tool kit that provides a uniform network interface to sound devices through table-driven sound servers. Sonifications can be constructed that encapsulate all device-specific functions in control files for each server. A user interface to configure sound devices and sonifications can be generated independent of the underlying hardware. This framework was easily integrated with Pablo, an environment designed to support the performance analysis of massively parallel computer systems, providing synchronized sound and graphics. Several sonifications of both multivariate data and time-varying performance data, created in this environment, are described. We conclude with a brief description of planned extensions, including integration with a virtual reality system.

1. INTRODUCTION

Incidental and ambient sounds supplement the information we derive from the visual dimension. Movies and television games rely on this; it would be difficult now to imagine watching a movie without a soundtrack, or television without sound. Video game designers also are aware that aural cues provide a sense of interaction. Just as the sounds created by mechanical parts in a pinball machine give concrete feedback to a player, electronically generated sounds at appropriate moments in video games reinforce virtual interactions. Frequently sound is used to present information about events one cannot see. For example, a fire alarm is a signal that people should leave the building, whether or not they smell smoke, feel heat, or see flames.

Given the proven ability of sound to convey information, it is interesting to consider it as a medium for presenting data. Sound can possibly highlight characteristics that cannot easily be seen, much in the same way that a movie soundtrack conveys information complementary to the imagery. The elements of sound (e.g., pitch, volume, duration, location, and timbre) can be used in the same way that visual elements (such as color, form, and line) are manipulated to present and analyze data in visual displays. *Sonification* refers to the use of sound to present data, the auditory equivalent of visualization.

Unfortunately, the two major technical obstacles to widespread experimentation, with sound as a mechanism to present data, have been the absence of a standard sound hardware platform and the corresponding lack of software to describe and manipulate the mapping of data to sound.

Porsonify is a sonification toolkit that addresses these issues. Porsonify describes a sound device as a set of commands that manipulate the state of the device with arguments controlled either by user configuration or by input data. In this way, the functions of a sound device are separated from the algorithm that generates a sonification. The device-specific details of a sound device are encapsulated in network servers, and the interface to local and remote sound devices is the same.

In this work, we provide an overview of Porsonify's design and the network interface to sound devices. We describe some examples of sonifications created with Porsonify and the Pablo Performance Analysis Environment. These examples illustrate the power of the simple building blocks that Porsonify provides; a wide range of complex effects can be achieved without writing any new sonification code.

2. PHILOSOPHY AND DESIGN OF PORSONIFY

Previous work in sonification suggests that a general-purpose mechanism to map parameters of data to characteristics of sounds on available sound hardware is essential for experimentation. Many of the psychoacoustic properties of different characteristics of sound are not well understood. Therefore, Porsonify was created to be easily portable to many sound hardware platforms, and to provide tools for creating new sonifications.

2.1 HARDWARE AND SOFTWARE CONSIDERATIONS

The variety of available sound hardware and the current absence of a universal control interface means that the choice of a sonification hardware platform has a dramatic effect on the kinds of sonifications that can be created.

All synthesis begins with the creation of waveform samples, which are made audible using a digital-to-analog converter. Software control can be used to generate these samples, providing the ultimate in flexibility. With synthesis software, input data can be used to control any aspect of sound; it can even be used to generate the waveforms themselves. However, most software synthesis techniques are too computationally complex to produce high-quality sounds in real time on common platforms. Thus, this method usually requires that some specification, or "score," be prepared in advance that describes the sound and its variables. This has the drawback that the data must be analyzed before a score can be generated and the sonification produced.

Generally, one wishes to accompany high-quality audio output with corresponding visualizations. Sonifications by themselves can be useful, but one consequence of their still-experimental nature is that they are frequently best heard in connection with corresponding graphics. This will become increasingly important in future virtual reality applications. Sound can be used to provide feedback in a virtual world in the same way it does in the real world, but this will require real-time sound synthesis.

Dedicated digital signal-processing hardware can make real-time software synthesis possible, but it is still relatively expensive. Hardware sound synthesis is a less costly compromise. Although embedding specific sound synthesis algorithms in hardware limits the ways that data can be used to control the sonification, specialization greatly reduces costs. Most commercial synthesizers support the MIDI (Musical Instrument Digital Interface) protocol, a communication standard that allows sound synthesizers to be interconnected and computer controlled. MIDI commands support such options as turning notes on and off and adjusting controls such as volume, sustain, and voice selection. Furthermore, even low-end professional synthesizers offer a wide range of features and controls that can be used for data sonification. For this reason, MIDI is emphasized in the Porsonify toolkit.

At this time, Porsonify supports two different MIDI synthesizers and the Sun SPARCstation audio device. However, this does not preclude integration of more sophisticated hardware. Although MIDI is practical now for experimentation, there is no way of knowing what hardware will emerge as a standard in the future. In Porsonify, an abstract model for representing a sound device has been chosen that allows a driver to be written for any available sound hardware.

2.2 PORSONIFY SOUND DEVICES

For the purpose of sonification, each sound device may be viewed as a set of "knobs," or controls, and "buttons," or commands, that direct the synthesis. A button indicates a command, and causes the sound device to change state (e.g., play a note or set the volume). A knob controls the arguments to a command. For example, knobs might specify the pitch and duration for the "Play a Note" command or the volume for the "Set Volume" command.

Almost any reasonably sophisticated sound device has a conceptual plethora of controls and commands. For example, one could create a command to play Beethoven's Ninth Symphony and a control to vary its measure by measure tempo, should this display prove useful in data analysis. In practice, however, the set of interesting controls is much smaller. Complex commands, such as playing musical patterns, can be built from primitives such as "note on" and "note off." For these primitives, we would like to control note pitch, duration, timbre, volume, and location. The characteristics of the sound device and synthesis technique determine the flexibility of control and the range of variation. For example, software synthesis permits actual data to be used to create the timbre. In contrast, hardware synthesizers typically provide a palette of fixed voices (e.g., piano, flute) and possibly some control over vibrato, sustain, and other selected timbre components.

The wide range of sound devices and their radical differences in function constrain the possible approaches to creating a sonification system. One can usually assume that every graphics device can draw a line. Unfortunately, no analogous assumption regarding sound device functionality is possible. The common denominator of sound devices is not a set of universally supported commands; the only guaranteed common attribute is that sound devices can make sounds. Although it would be possible to create a low-level sound command language that can express all the functionality of a set of sound devices, lack of sound device standardization, even using the

MIDI protocol, would force the revision of the language for every newly announced sound device. Letting the sound description language be the greatest common subset of the sound device functions suffers from another, more serious problem; in an area that demands experimentation, limiting the functions used to create sonifications is counterproductive. From this point of view, it is essential that underlying sound hardware functionality be accessible to the sonification designer while isolating device idiosyncrasies to the greatest extent possible.

The MIDI protocol does not describe what knobs a particular synthesizer supports; it describes how to control the knobs that are there. The Porsonify software is an extension of this metaphor. Rather than design a sound control language that describes a subset or superset of the capabilities of some collection of common, supported sound devices, we chose to define an interface that permits both the specification of sound device controls and commands and the communication of this information to the application. This permits construction of a sonification that is independent of the underlying sound device; the sonification need only understand the format of the sound commands and not their function.

2.3 NETWORK PROTOCOL

Network access to sound devices is important in an experimental environment. Typically, sound devices are less common than display devices; to maximize their utility, they should be as widely accessible as possible. Porsonify makes the interface to remote and local devices identical. In this way, one or more sound devices can be used together from any host workstation to create a sonification.

In Porsonify, access to sound devices is based on a server/client architecture, where client processes (applications) contact server processes to request services. In this case, *sound device servers* specific to each sound device encapsulate all the device-dependent code and accept *sonic messages* from client processes that manipulate the device. Within a server, the commands and controls the sound device supports are represented by a *sound device description*. Client applications interact with the sound device servers through a general-purpose network audio access daemon (server), or *naad*.

A client application can discover what devices exist on a given host by querying the *naad* on that host. In return, the client application receives a textual list of available sound devices (e.g., "Yamaha TG33 MIDI Tone Generator") and the names of the ports associated with the device-specific servers.[1] Figure 1 illustrates this communication.

---

[1] The port name is a network address used to open a connection to a server.

A Framework for Sonification Design

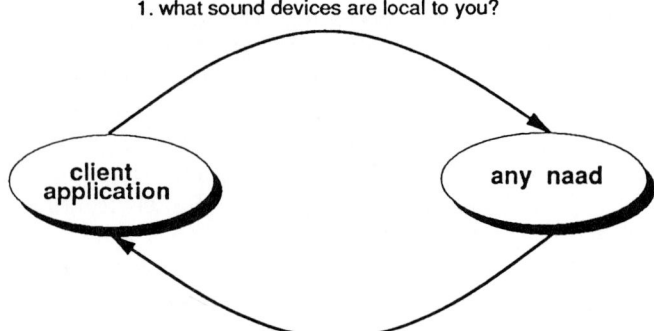

FIGURE 1 Discovering available sound devices.

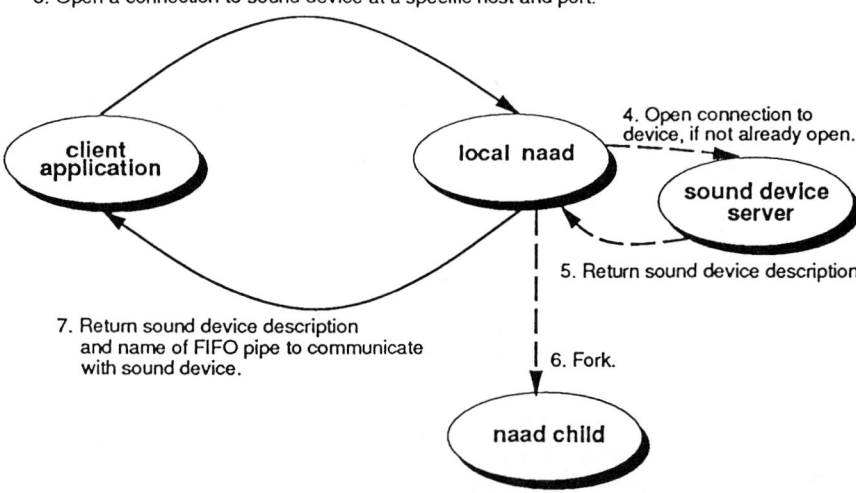

FIGURE 2 Opening a connection to a sound device.

Once the application is ready to open a connection to a specific device, it contacts the *naad* on the local host. It gives the name of the (possibly remote) host and the port associated with the device's server to the *naad* (step 3 in Figure 2). The *naad* checks whether any other application on the local host has a connection open to the sound device. If not, it must contact the sound device server (step 4). When the *naad* opens a connection to the server, it receives in return the sound device description, containing the commands and controls the sound device supports (step 5). Then the *naad* forks, so that its child can propagate sonic messages to the sound device while the *naad* continues to listen for commands from other processes.

Steps 4, 5, and 6 are unnecessary if the local host already has a connection open to this particular device; the *naad* will have cached a copy of the device description, and one of its children will already be connected to the sound device server. Finally, the parent *naad* returns the sound device description and the name of a UNIX named FIFO pipe (step 7). The client application will write commands to this pipe, and they will be read by the *naad* child and forwarded to the appropriate sound device server.

The sound device description that the client receives describes the sonic commands the sound device can execute and the valid arguments for each command. For example, "Play A Note" might be one such command, with arguments "Pitch" and "Duration." The device description would specify the range of valid pitches and durations. The client then uses this description to compose sonic messages that it sends to the sound device server via the *naad*.

Figure 3 illustrates how the client sends these messages to the *naad* child through the named FIFO pipe. The *naad* child forwards the messages to the sound device server to which it is connected.

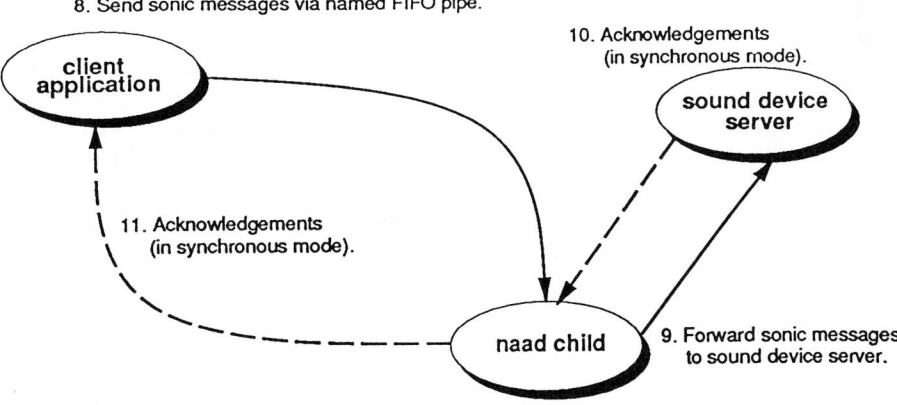

FIGURE 3 Sending messages to a sound device.

Client applications can interact with sound devices in two modes: synchronous and asynchronous. In synchronous mode, the client may send a batch of sonic commands and receive an acknowledgement to indicate that the commands have executed. In asynchronous mode, no acknowledgements are sent. Acknowledgements are vital when synchronizing audio with graphics, because they give the client a way to permit output sounds to complete before updating graphical displays. If the client is operating in synchronous mode, steps 10 and 11 occur and acknowledgements are passed back to the client.

2.4 SONIFICATION STRUCTURE

The second important component of Porsonify is support for writing sonifications. Algorithms can be used to generate sonic messages with the message parameters instantiated by input data. For example, a sonification can be as simple as "play a fixed note for each data point" or as complicated as "vary the tempo and volume of a specified melody according to the running average of the input data." However, it is desirable to express a reasonably large subset of possible sonifications without writing new code. Certain sonification patterns can be identified and categorized. The sonic messages passed between application and sound device server are specific to each device, but the effect of a sequence of sonic messages for one device can be approximated by another sequence of messages on a different device. These messages can be encapsulated with the sonification algorithm in control files. In this way, sonifications can be represented as a set of control files for a particular device.

Consider again the "knob" analogy drawn earlier for sound device functions. Through the Porsonify server interface, each sound device possesses a finite number of knobs, or controls. There are also a number of commands whose arguments are parameters that can be controlled with these knobs. For example, one control might be the volume knob. If a short note is played to represent each datum, the value of the datum might be used to control the volume of that note. This is the simplest type of sonification: a repeated sequence of commands with controllable parameters is executed for each data value. This sequence of commands can be as primitive as playing a single note, or more complicated (e.g., playing a sequence of chords or a motive, the pitches and rhythm determined by transformations on the input data).

Data might also be used to control which commands are executed. For example, certain extreme data values might trigger an alarm. This behavior can be combined with that described above; for example, not only might the data determine if an alarm is played, but it might control the sound of the alarm.

These behaviors (e.g., how data controls command parameters and/or command execution) are abstracted from the actual sound commands that are executed; in Porsonify, these behaviors are called *sonic widgets*. The following is a list of the Porsonify sonic widgets.

- *SimpleSonicWidget* executes a set of sound device commands for every input data value, using that data value to control one or more command arguments. This is the simplest sonic widget behavior, hence the name.
- *BooleanSonicWidget* tests the input data value and, if the test returns true, uses that data value to control the arguments of some set of commands for a sound device. The alarm behavior described above is realized using this sonic widget.
- *MultiSonicWidget* performs some calculation on the input data value and produces a value. This value selects which of several distinct sets of sound device commands is executed, using the original input data to control command arguments. For example, this widget would be used to implement a sonification that receives an error code and plays a corresponding alarm.
- *LoopingSonicWidget* performs some calculation on the input data value and uses the result to decide how many times to execute a set of sound device commands, again using the original input data to control command arguments.
- *InitializingSonicWidget* executes a special set of commands to initialize a sound device the first time a data value is received, and then acts like a *SimpleSonicWidget* thereafter.

We have spoken of tests and calculations on the input data values. These are necessary to realize the isolation of device-specific features in sonifications. Not only do different devices have different sets of commands, but they have different "knobs." It is likely that approximating a sonification on a different device requires not only substituting a new set of sonic commands, but manipulating the input data used to control its parameters or execution in a slightly different way. For example, one device might have volumes in the range 0–9, and another in the range 0–90. If the sonification on the first device used the value of the input data (in the range 0–9) to control the volume directly, the second sonification should scale the input data by a factor of 10 to create the same effect on the second sound device.

Such adjustments, tests, and calculations are called *transformation functions* in Porsonify. The current set of transformation functions includes Boolean tests and a variety of scaling and mapping functions. These transformation functions are used uniformly to perform the tests used in complex sonic widgets (e.g., *BooleanSonicWidget*, *MultiSonicWidget*) and to control sound command parameters in all sonic widgets. These functions are very simple; more sophisticated data manipulation and analysis may be done with tools intended for this purpose before the data is presented using Porsonify.

Lists of sound device commands, default arguments, and the transformation functions to control the data-driven arguments are encapsulated as *widget control files*. If a sonification is constructed using one or more sonic widgets, it can be ported to a different sound device simply by writing a widget control file to approximate the sounds on the new hardware.

For example, consider a simple sonification, called "ScaleSound," that plays a note for every input data value, mapping the value to note pitch. Because the same actions are taken for every input data value, this can be implemented using a *SimpleSonicWidget*. Figure 4 shows a widget control file for this sonification. A # indicates a comment that continues to the end of the line; comments can be delineated by /* and */. The first item in the widget control file is the name of the type of sonic widget the file describes. In this example, it is SimpleSonicWidget. The next two items are used by the *naad* to locate the sound device server for this widget control file. They are the name of the host on which the sound device server is executed, and the symbolic name of the network port on which it listens for applications attempting to make audio connections.

What follows is a list of commands for the particular sound device. A command and its arguments together form a *sonic message* (i.e., one such message might be "Set Volume 60," where 60 is a valid setting for

```
# Sonic Widget Type
SimpleSonicWidget

# hostname          portname
localhost           midiaudio

#                           MessageId      Arguments
/* Select Instrument */     3              75
/* Play Note        */      0              ?60
                                           ?1.0
                                           $0 ( /* Linear */  1
                                                /* A =    */  2.0
                                                /* B =    */  4.0
)
```

FIGURE 4 Sample "ScaleSound" widget control file with two variable arguments.

the volume). Arguments may be fixed, configurable, or variable. Fixed arguments appear as they are in the file. In Figure 4, the single argument for the "Select Instrument" command, after the message identifier 3, is 75. This means that the selected instrument is number 75. The sound device description that the sound server returns to the application process (step 7 in Figure 2) specifies the instrument to which the number 75 corresponds.

Configurable arguments are preceded in the file by a question mark (?). For example, the arguments to the "Play Note" command in Figure 4 are a pitch, duration, and velocity (note volume). The first two arguments, pitch and duration, are configurable. The default values are 60[2] and 1.0 (seconds). Obviously, it is possible to edit the values of these arguments in the widget control file itself; however, the distinction between fixed and configurable arguments is drawn so that a graphical user interface can be automatically constructed to edit configurable arguments using the sound device description and the widget control file.

Variable arguments are data-driven controls. They are specified by a dollar sign ($) and a vector index. The data input to a widget control file may be a single value (a scalar) or an array of indexed values (a vector). A vector index of 0 may denote either a scalar value, or the first element of a vector. In this way, several data values can control multiple facets of a sonification. In Figure 4, the information in parenthesis following the $0 is used to configure the transformation function associated with this argument. In this case, the transform is a linear transform that calculates $ax + b$ where $a = 2.0$ and $b = 4.0$. The values for a and b would normally be selected based on the expected range of the input data and the range of valid values for the argument to the sound command.

3. INTEGRATED SONIFICATION AND VISUALIZATION

Ideally, one would like to accompany sonifications with complementary or redundant visual displays to capitalize on both visual and aural pattern recognition abilities. The modularity of the sonic widget abstraction permits sound to be easily integrated with arbitrary applications, encouraging experimentation with sonification. The Porsonify software, coupled with the Pablo Performance Analysis Environment, is an excellent arena for experimentation with coordinated sonification and visualization of data.

---

[2] We follow the convention that 60 is C4 (middle C), and each integral value represents a half-step.

Pablo[9] is an extensible, portable performance analysis environment capable of supporting the analysis of data drawn from both shared and distributed memory parallel systems in the graphical analysis style of AVS[13] and Khoros.[15] It consists of a set of data transformation and visualization modules that can be interactively interconnected in a graphical workspace to form a directed, acyclic data analysis graph. These graphs can be configured, saved, and reloaded. At the highest analysis level, performance trace events are represented using a self-describing data format[1] that includes internal definitions of data types, sizes, and names. A trace file in this format consists of a set of record definitions (much like C structures) and a stream of record tag/data pairs. Each tag identifies the type of the record to follow.

Using the tags and record descriptions from the input trace data file, one can select record fields to be passed into the input ports of each module, and define new record types to be output to descendent modules. Although intended for analysis of performance data, the flexibility of the input data format and analysis graph configuration enable Pablo to accommodate any kind of data.

In typical Pablo operation, performance trace data is read from an input file and is processed by the nodes in the analysis graph. Because they have no children to which they can propagate processed data, the leaf nodes of the analysis graph usually display the data. Integration of the Porsonify sound toolkit with Pablo involved creation of a set of "sonic displays," selected as leaves of the data analysis graph in the same way as graphical displays (e.g., strip chart or matrix). There are two basic classes of sonic displays. The first class of displays are those that are composed of a single sonic widget. These operate in synchronous mode (i.e., they block until all sound commands have been executed) so that graphics display modules are synchronized with sound. A second class of auditory displays reads time-stamped sound commands that have been embedded in the trace data file. This is one advantage of the self-describing data format; a Pablo data analysis graph can generate a new trace file with time stamps and sound commands created from the original data, and this file can be later replayed. The time stamps can be dynamically scaled to "slow down" or "speed up" the display.

4. SONIFICATION EXPERIENCES

Once Porsonify was integrated with Pablo, it was easy to construct a variety of sonifications for different kinds of data. This section describes both some of the sound mappings that seemed most useful and general observations and conclusions about various styles of aural "displays." However, no formal experiments were conducted to determine the effectiveness of any of these methods of presentation. These sonifications simply illustrate the flexibility of the software and are suggestions to inspire further work in this area.

3.1 AUDIO CUES

One of the simplest to create, yet most sophisticated ways to use, sound is as an alarm, or cue, to signal some event. The meaning of the signal might be directly related to the data (e.g., a value has exceeded some threshold) or it might a completely different kind of information (e.g., the entire data set has been converted to a new format). The sound itself might have no inherent meaning, like a string tied around one's finger, or it might be a digitized voice message, more like a note attached to a door.

In Porsonify, the choice of sounds is completely arbitrary, within the capabilities of the available sound devices. The *BooleanSonicWidget* is typically used to implement audio cues; it represents the behavior "if some condition is true, execute some sound commands." More complicated symbolic cues, where multiple sounds are used to express different concepts, can be implemented using the *MultiSonicWidget*.

3.2 MULTIVARIATE DATA

Multivariate data with more than three dimensions is challenging to visualize, because more creative means than X-Y plotting must be used to represent the higher dimensions. Chernoff's faces,[5] for example, is a technique to aid data clustering by mapping variables of a data sample to facial characteristics (e.g., length of nose or size of eyes). Each face represents an individual sample.

Chernoff faces are analogous to Bly's approach to sonifying multivariate data, where the four variables of each data sample were mapped to pitch, volume, duration, and fundamental waveshape of a note.[2] The resulting notes enabled listeners to place each sample in one of three related sets. Sound mappings of this type share many of the same problems as Chernoff faces. Specifically, face parameters are not completely independent, and extreme values of some face parameters cause others to lose their effect.[6] In the same way, sound characteristics are interrelated; it can be difficult to distinguish the timbre of a very short or high-pitched note, or the pitch of a very short note. Some assignments to face parameters are better than others for illustrating set groupings; the same is true of sound. Not only do some sound parameters carry more weight than others (for example, pitch and rhythm are the most distinguishing characteristics of a melody and, thus, can be considered more significant than, say, volume), but the range of each sonic parameter also differs widely across sound devices and among other parameters on the same device. Certain mappings could imply loss of critical information.

In our experiments, the data we used to explore multivariate data sonification was taken from the *Places Rated Almanac*.[3] In this data set, each of 329 cities in the United States has a rating for climate and terrain, housing, health care and environment, crime, transportation, education, the arts, recreation, and economics, in addition to a latitude, longitude, and population. These twelve parameters provided ample opportunity to experiment with different mappings without attempting to solve any particular analysis problem.

3.2.1 DIRECT MAPPING.

Bly's direct mapping technique, where each variable of a data sample is mapped to a different characteristic of a note, is easy to replicate in Porsonify (although the Sun SPARCstation and MIDI hardware platforms support variation of sound characteristics different than those used in Bly's experiments).

Tr 47   This method is used to sonify the rated cities in five states (Illinois, California, Florida, New York, and New Jersey). Each city is represented by a single note. The recreation index is mapped to note pitch (one of 29 over four octaves in a C major scale). The population is scaled to generate a duration of up to five seconds for each note. The climate index determined the timbre; three instruments were chosen to represent the range of weather. Bad weather was a raspy digital sound, average weather was a flute, and above average weather was a pleasant mixture of bells and strings. The volume was controlled by the housing cost index. Longitude was mapped to stereo balance, since it is inherently locational; most people can envision maps of the United States where higher longitudes are to the right. In this way, each note represented four variables of each sample. (CD Track 47.)

In our experience, this can be a very effective technique for clustering and understanding higher dimensional data. Generally we chose to map each data dimension to characteristics of sound that corresponded to the perceived "importance" of the dimension. In the above example, the recreation index is mapped to pitch, a more prominent sound characteristic than volume, to emphasize it. However, even random mappings proved useful to gain a qualitative feel for a multidimensional data set.

One problem with this sort of presentation is that it overlooks potential interpretations of time. Specifically, in the Places Rated example, both the absolute and relative positions of notes in the sonification are quite arbitrary. Each note represents a city, and the notes are played in random order. Just as the position of a data point along the horizontal axis has meaning in a plot, the temporal placement of a note in a sonification can also be significant. Additional information could be provided in the above

---

[3] The Places Rated Data is used with the kind permission of Simon and Schuster, Inc.

sonification simply by sorting the notes by a fifth, "silent" variable, or by one of the four variables controlling the sonification. Furthermore, the duration of the entire sonification has no significance, except as the sum of the variables mapped to durations of notes. When listening to sequences of samples, the duration of the entire sequence is an important characteristic in itself. We can take advantage of this time dimension as described below.

3.2.2 SONIC SCATTER PLOTS. Notes can also be "plotted" in time, creating a multidimensional sonic scatter plot. One variable is selected as the time axis. The entire data set is sorted by that key. Next, the duration of the plot is chosen, and the key variable is scaled to that range of seconds. These scaled values represent logical time stamps for notes. The rest of the variables of the data sample are mapped to other characteristics of a note. In our example, population was selected as the time axis, and the arts index was mapped to pitch. Population in the Places Rated data set ranges from 62,820 to 8,274,961; if the sonification lasts for one minute, the population must be scaled to the range 0 to 60. These scaled values represent logical time stamps for notes; a note whose pitch corresponds to some other variable in the current sample is sustained until the time stamp of the next sample.

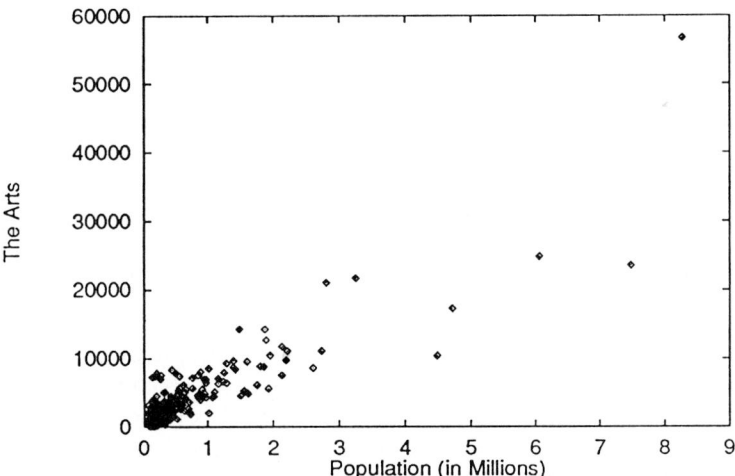

FIGURE 5 Population and the Arts (Places Rated Data).

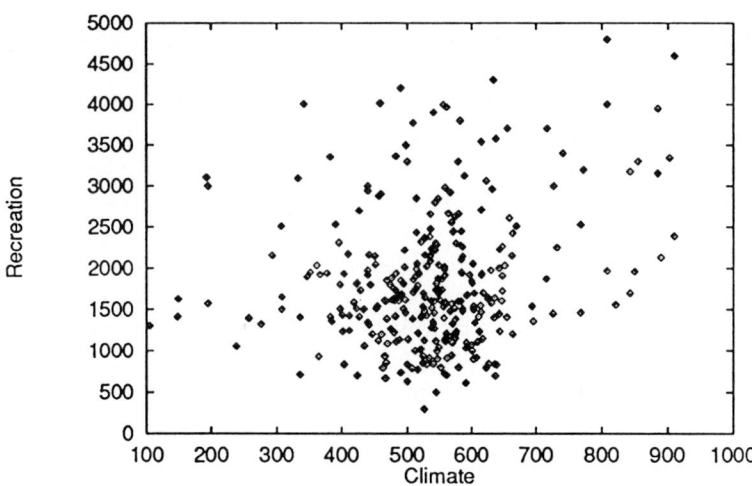

FIGURE 6   Climate and Recreation (Places Rated Data).

Tr 48   This kind of sonification is especially useful for detecting correlations between different parameters. For example, Figure 5 illustrates that the arts index is highly correlated with the population; the pitches played are generally increasing. It is also easy to intuit from the sonification the distribution of the key variable. In the sonic scatter plot of population and the arts, with poplulation as the sort key, one can hear that although the population range is large, most cities have relatively small populations; many notes are played within a very short time at the beginning of the sonification. It is also clear that the average arts index is low; most of the pitches heard are low, except for the ones corresponding to very highly populated cites (e.g., New York), at the very end of the sonification. (CD Track 48.)

Tr 49   The sonification generated by listening to two correlated variables can best be appreciated in contrast to the same sonification of two different variables (e.g., climate and recreation indices) which have little correlation, as shown in Figure 6. If the climate index is used as the time axis, it is obvious that most cities have average climates. This can be discerned from the clustering of many low pitches, indicating low recreation ratings, that occurs in the middle of the sonification. (CD Track 49.)

A sonic scatter plot can be extended to three or more dimensions in several ways. Perhaps the most direct approach is to use other variables to control the characteristics of each note. For example, stereo balance could be mapped to longitude, weather to timbre, and the recreation index to pitch, for a total of four dimensions. One consideration is that as the number of variables encoded in each note increases, the tempo of the sonification should decrease, to allow sufficient time for listeners to assimilate the extra information. On the other hand, the information encoded in each individual note might be secondary to overall patterns or trends in the data that can be percieved only at a very high presentation rate.

3.3 PERFORMANCE DATA

Using each note to represent a data point also is effective with other types of data. This technique was used to sonify an execution trace from a partial differential equations solver running on an Intel iPSC/860 hypercube with 64 nodes, a message-based parallel system. This trace file was captured using the PICL subroutine library[8] and was converted to a self-defining data format for use with Pablo. The trace file consists of records of different types; each record includes a time stamp and other information pertinent to the record type. Records document events such as message transmission and receipt between processors, message sizes, and processor idle time.

The PDE code is communication constrained; the processors are idle roughly 70% of the time, while they are waiting for messages they have sent to propagate to other nodes. To convey an impression of the frequent communication patterns, a note of fixed duration is played for each message "send" and each message "receive." The pitch of the note corresponds to the sending processor identifier (processors 0–63 were mapped into pitches 30–93). The location (left or right) in the stereo balance field indicates whether the note represents a send or a receive, respectively. The order of the notes corresponds to the ordering of sends and receives in the trace file, but there is no indication of how much time passed between events.

In the sonification of this trace file, message sends and receives are matched; although in the beginning there are bursts of sends (long sequences of notes in the left of the stereo balance field) and corresponding bursts of receives, communication reaches a stable state where a few sends are followed by a few receives. The pitches of these notes are scattered over the entire range; it is immediately evident that no single cluster of processors is responsible for a large portion of the communication.

This sonification can best be appreciated by contrasting it with the same sonification of another trace file from a program with a strikingly different communication pattern. For example, in a trace of a recursive matrix transpose, also executed on a 64-node hypercube, all processors first send and then receive. Therefore, all 64 pitches are heard in the left, and then all 64 are heard in the right.

Rather than playing a discrete note for each message send and receive, it is possible to start a note at the message send and terminate it at the corresponding receive. This approach is impractical for trace files with large

Tr 50

numbers of processors; too many notes must be sustained to detect any patterns. It provided interesting insight, however, into the behavior of a simplex code running on an 8-node Intel iPSC/2 hypercube.[4] This trace file was generated using the Crystal trace capture library.[10] As noted above, the pitch of a note corresponds to the processor identifier, however, nodes 0–7 were mapped to pitches in a C major scale to create more musically recognizable patterns. A note was started for each send and finished at the corresponding receive. All notes were balanced in the stereo field, and a synthesized instrument with a long, natural sustain (Bells and Strings) was selected to carry the notes without sounding strained. (CD Track 50.)

In each iteration of the simplex code, the processors cooperate to identify and distribute the global minimum. In this process, first all local minima are passed from the upper half of the hypercube to the lower half. Then the lower half passes the newly computed minima from its upper half to its lower half. The process repeats until one node (processor 0) has the global minimum, and it broadcasts that value to every other node. This computation creates a distinct pattern of notes that is repeated frequently. Broadcasts from processor zero can easily be detected as the same pitch repeated seven times in succession. The psychoacoustic effect of a broadcast is that the pitch corresponding to the broadcasting processor sounds extremely powerful until the broadcast message is received. It begins decaying as the individual notes are resolved.

Using this sonification, one can quickly detect message transmissions without corresponding receipt; they are audible as trailing, sustained notes. In fact, this sonification revealed an error that had not been detected in multiple years of visualization; it was erroneously recorded that a processor had sent a message to itself. A variation on the sonification helped to isolate the error, which was difficult to identify by pitch alone. A different voice was assigned to each processor. Among eight processors, it was possible to immediately detect which processor had an unmatched send by the timbre of the sustained note.

3.3.1 AURAL TIME LINES Trace data usually consists of time-stamped events. Considering this, it might seem counterintuitive that neither the duration of notes nor their starting times correspond directly to any timing information in the trace file. For example, in the matrix transpose trace, although the relative order of the message sends among the processors can be heard, the amount of time between each event could vary radically based on the number of intervening trace events. Time can be used to great advantage in conveying additional information. Rhythm is a natural mechanism for conveying a qualitative impression of the relative duration and timing of events, represented by the individual notes. Ideally,

[4]The simplex method is an algorithm used to solve linear optimization problems.[12]

time-stamp information could be used in conjunction with the sonifications described above to create a more realistic representation of communication patterns and processor activity.

Unfortunately, even if a logical mapping between some characteristic of the data is obvious, such as using event time stamps in a trace file to control when notes are played, it may not be practical to do this. Events might occur in very short, infrequent bursts over a long period of time, so that it would be impossible to hear what is happening during busy periods, and tedious to listen to the entire sonification. Linear scaling of time (e.g., letting each second of trace time be ten seconds of sonification time) alleviates the problem only if one can selectively listen to bursts of activity; otherwise, it can quickly become impractical to listen to the entire trace. Furthermore, if the sonification is to be played with a corresponding visualization, the time to update graphical displays and process data that do not contribute to a sonification can easily exceed the scaled trace file time. Nonlinear time dilation might be useful to highlight bursts of activity. In such cases, a graphical data-time clock could be used in conjunction with the sonification to illustrate time distortion. For these reasons, all aural time lines described here are in some sense an approximation to reality, distorted by processing time required by the visualizations.

To experiment with using data time stamps in sonifications, we obtained an execution trace of one code from the Stanford Parallel Applications for Shared Memory (SPLASH) benchmark suite[11] running under the experimental *Choices*[4] operating system. The benchmark was a many-body molecular dynamics simulation (WATER). *Choices* is an object-oriented operating system designed to support experimentation with resource management policies on shared and distributed memory parallel systems. To support this experimentation, *Choices* provides software tools to collect performance data from the operating system. The *Choices* context switching code was instrumented to record the transitions among *Choices* system processes and the WATER application processes. Each of the context switch records contains a time stamp, processor identifier, the duration of the previously running process, and the identifiers of the previously executing process and the process that is about to gain control of the processor.

These kind of time-stamped events lend themselves nicely to the construction of aural time-line sonifications. Every note represents an event whose duration equals the duration of the note. Other qualities of the note (e.g., timbre, stereo location, volume, or pitch) may be used to represent additional or even redundant information about the event. The time that notes are played in the sonification corresponds to the time they actually occur.

If pitch variation is used, this kind of sonification can draw upon people's natural ability to recognize melodies. It also uses the fact that sound is temporal to great advantage. At any time one hears only the events that are actually occurring. When a "normal" aural pattern has been established, abnormalities are immediately evident as breaks in the normal rhythm.

Several sonifications of the *Choices* context switch data, taken from a dual-processor Encore Multimax, were created using aural time lines, to characterize the behavior of processes under the *Choices* operating system. The purpose of these sonifications was to gain a qualitative sense of the length of time spent in each application process. A note with a pitch assigned to each process started when the process began to execute and stopped when that process was preempted by another.

We first played this sonification for each of the two processors individually. Since only one process can be running at a time, notes generated this way do not overlap; however, certain patterns were immediately evident. For example, the `IdleProcess`, an operating system process that runs when nothing else is ready to run on the processor, was assigned a low pitch. This could frequently be heard in the midst of of staccato cycles of higher pitches that represented other network daemons.

To listen to both processors together, stereo balance is used to distinguish events. All processes executing on the first processor are heard in the left of the stereo balance field, and all processes on the second are heard in the right. This sonification reveals an even more interesting pattern. When a process is preempted on one processor, it is usually quickly picked up by the other, effectively bouncing back and forth across the stereo field amidst the low tone of the `IdleProcess`.

As soon as the application process, marked by a comparatively high pitch, began to execute, aural patterns changed dramatically. Typically the application process would execute for substantially longer periods of time when nothing else was available to run; this was distinctly audible as sustained high and low notes, representing the application process and the `IdleProcess`, respectively. Furthermore, this implementation of the WATER code forked and executed in parallel on both processors. During this period, a high pitch could be heard in both the left and the right of the stereo field.

4. CONCLUSION

In practice, Porsonify has satisfied our goals for a sonification system. It provides a uniform network interface to sound devices through table-driven servers without hiding unique hardware functions. The device-dependent aspects of a sonification can be encapsulated in a single file. A graphical user interface to configure sound devices and sonifications can be generated independently of the underlying hardware. Porsonify was easily integrated with the Pablo Performance Analysis Environment, enabling experimentation with synchronized sonification and visualization.

However, the sonic widget set provided by Porsonify is but a first step in providing a framework for data-driven sonification. These sonic widgets do not offer control structures; they simply perform actions based on very simple data transformations. A general-purpose data analysis program (such as Pablo) reduces the data to the point where these sonic widgets can be applied. Instead, one might envision tightening the connection between data analysis and display, so that algorithms used to create sound can dynamically provide feedback to data analysis modules, changing the sonification as appropriate. For example, we are currently experimenting with an interactive graphical/sonification display module to assist with clustering of multivariate data, sonified with a direct mapping technique. When a cluster has been tentatively identified by audio and visual inspection, a new mapping can be selected to more clearly distinguish different clusters. In addition, a single control mechanism should exist for creating data-driven soundtracks and animations. For example, in a virtually created world, one could imagine weaving a path through a three-dimensional scatter plot, "brushing" against points that resound with different timbres and pitches, revealing additional variables in each sample.

In the future, we intend to integrate sound with virtual reality displays for data presentation. Aural cues in a three-dimensional audio space can be simulated with the aid of a Convolvotron,[14] which can position up to four audio sources in three-dimensional space in real time.

We intend to create high-level algorithms for reading a data set, for characterizing it, and, given some rules for a particular sound device, for generating appropriate data to sound mappings. This is a nontrivial endeavor; not only must one know the commands the device supports, but things like the ranges of different instruments and what kinds of voices sound good together. Once such rules are created, sonifications could be written based upon information analogous to color maps on visual displays. Such settings could be saved and downloaded to configure a sound device.

Clearly, much work remains to be done before sonifications are as recognized, accepted, and understood as their visual counterparts; however, the uniform interface to sound devices that Porsonify provides is a excellent platform for further development.

ACKNOWLEDGMENTS

This work was funded in part by National Science Foundation grants NSF CCR86-57696, NSF CCR87-06653, and NSF CDA87-22836 (Tapestry);

NASA ICLASS Contract No. NAG-1-613; DARPA Contract No. DABT63-91-K-0004; and by a grant from the Digital Equipment Corporation External Research Program.

REFERENCES

1. Aydt, Ruth A. "SDDF: The Pablo Self-Describing Data Format." Technical Report, Department of Computer Science, University of Illinois at Urbana-Champaign, 1992.
2. Bly, S. "Sound and Computer Information Presentation." Ph.D. Thesis, University of California, Davis, 1982.
3. Boyer, Richard, and David Savageau. *Places Rated Almanac*. New York: Rand McNally, 1985.
4. Campbell, Roy, Vincent Russo, and Gary Johnston. "The Design of a Multiprocessor Operating System." UIUCDCS-R-87-1388, Department of Computer Science, University of Illinois at Urbana-Champaign, 1987.
5. Chernoff, Herman. "The Use of Faces to Represent Points in k-Dimensional Space Graphically." *J. Amer. Stat. Assoc.* **68** (1973): 361–368.
6. Flury, Bernhard, and Hans Riedwyl. "Graphical Representation of Multivariate Data by Means of Asymmetrical Faces." *J. Amer. Stat. Assoc.* **76** (1981): 757–765.
7. Gaver, William W. "The SonicFinder: An Interface that Uses Auditory Icons." *Hum.-Comp. Inter.* **4** (1989): 67–94.
8. Geist, G. A. "A Portable Instrumented Communication Library, C Reference Manual." ORNL/TM-11130, Oak Ridge National Laboratory, 1990.
9. Reed, D. A., Ruth A. Aydt, Tara M. Madhyastha, Roger J. Noe, Keith A. Shields, and Bradley W. Schwartz. "The Pablo Performance Analysis Environment." Technical Report, Department of Computer Science, University of Illinois at Urbana-Champaign, 1992.
10. Reed, Daniel A., and David C. Rudolph. "Experiences with Hypercube Operating System Instrumentation." In *Intl. J. High Speed Comp.* (1989): 517–542.
11. Singh, J. P., W. D. Weber, and A. Gupta. "SPLASH: Stanford Parallel Applications for Shared-Memory." CSL-TR-91-469, Computer Systems Laboratory, DEECS, Stanford University, 1991.
12. Solow, D. *Linear Programming: An Introduction to Finite Improvement Algorithms*. New York: North-Holland, 1984.

13. Stardent Computer, Inc. *Application Visualization System, User's Guide*, 1989.
14. Wenzel, E. M., F. L. Wightman, and S. H. Foster. "A Virtual Display System for Conveying Three-Dimensional Acoustic Information." *Proc. Hum. Factors Soc.* **32** (1988): 86–90.
15. Williams, C., and Rasure, J. "A Visual Language for Image Processing." *IEEE Workshop on Visual Languages*. October, 1990.

Jay Alan Jackson and Joan M. Francioni
Computer Science Department, University of Southwestern Louisiana, Lafayette, LA 70504-1771; e-mail: jaj7298@ucs.usl.edu; 318/231-6768

Synchronization of Visual and Aural Parallel Program Performance Data

Understanding the behavior of a program that runs on a parallel computer poses a challenge to programmers due to the difficulties of analyzing multiple concurrent events. In order to test, debug, and tune the performance of a parallel program, it is necessary to study information such as interprocessor communication logs and processor utilization profiles. Visual tools have been developed to make this job easier, but often substantial effort is still required to interpret and comprehend multiple graphical and textual views. In this paper, we discuss the properties of parallel programs that are suitable for aural representation and present a number of examples of sound mappings that have been implemented. A prototype tool which provides synchronized visual and aural displays for depicting parallel program behavior is described, and the justification for, and the effectiveness of, this approach are discussed. In general, sound was found to be a natural medium in which to realize certain patterns and timing information related to the run time performance of a parallel program. By providing combined visual and aural cues, speed of recognition and retention of relevant details were observed to be enhanced over either method alone.

1. INTRODUCTION

To debug a computer program, it is necessary to understand the execution behavior of the program. In parallel programs, the program itself is made up of concurrently executing processes that interact with each other in various ways. Hence, to understand the execution behavior of the entire program, it is not only necessary to be able to follow the logic within each process but also to be able to follow the interactions among the individual processes. In general, this makes parallel program debugging much harder than sequential program debugging. Furthermore, parallel programs are intended to run faster than their sequential counterparts. Thus they must also be tuned for good performance.

During testing, debugging, and performance tuning of a parallel program, one seeks to gain insight into, and to develop an intuition about, the concurrent events of the program and their interrelations. As with sequential programs, you are not always sure exactly what you are looking for. Usually the programmer must consider the run-time events from many different perspectives in order to get a full appreciation of the program's behavior. The kind of behavior that is of interest includes such things as the relative amounts of time when processes are idle versus busy; information related to a communication between two processes; the density of communication activity over a certain period of time; the contents of a message sent from one process to another; the order in which incoming messages are processed; and so on. The main problem associated with trying to understand the behavior of a parallel program is that many things may be happening at the same time. The more concurrency there is, the harder it is to follow any single thread of logic or control.

The traditional techniques for representing parallel program run-time behavior include tables of statistics, profile information, call-graphs, and graphics of run-time events over time. All of these techniques have one thing in common: they are processed via our visual sense. Some recent studies, however, have begun to investigate the potential effectiveness of using sound to represent aspects of the run-time behavior of parallel programs.[1,3,2,6] The possibility that additional valuable insight into parallel program behavior can be obtained when our visual sense is combined with our aural sense is the motivation for these studies.

The representation of program data using sound is called an auralization of the program.[3] (The term sonification is also used in this respect.) In the same way that a visualization of program events is intended to help one gain an understanding of the program, an auralization is based on the actual execution data of the program. As such, the relative timing, duration, and patterns of sound events have a one-to-one correspondence with the relative timing, duration, and patterns of execution events.

Synchronization of Visual and Aural Parallel Program Performance Data   **293**

In this paper and accompanying audio CD, we describe and present a sample of parallel program auralizations. In addition, we describe the mechanism used in a prototype auralization tool that synchronizes both the sound and the graphic depictions of a program's run-time behavior. The background to this paper is presented in Section 2, followed by a description of the graphical tool used in the prototype auralization tool in Section 3. In Section 4 we describe how sound mappings of execution events are generated and synchronized with the graphic displays of Section 3. Example auralizations are described in Section 5, and concluding remarks are given in Section 6.

2. RELATED WORK

There have been few studies demonstrating the feasibility of mapping parallel program performance data to sound. One approach was studied by Sonnenwald et al. in their design of InfoSound.[9] The system allowed developers to create and store musical sequences and special sound effects, and then to associate those sounds to an application program's events. The system was tested on two programs—a telephone network service simulation and a parallel computation simulation—and in both cases was found to help users detect rapid, multiple event sequences that were difficult to visually detect using text and graphical interfaces. InfoSound was not developed as a general-purpose tool and, thus, the sound mappings had to be customized for each program. In addition, the project was discontinued after only these two tests. Nonetheless, it did provide a mapping of parallel computation events to sound, and it did demonstrate that comprehension of the run-time behavior of the program was increased.

The feasibility of mapping actual parallel program performance data to sound was first demonstrated by two separate reports at the same conference: Francioni et al.[1] and Reed et al.[8] Francioni et al. demonstrated the potential of sound to portray performance data via three separate sound mappings depicting communication events, processor utilization, and flow of control. Reed et al. presented an auditory display that was synchronized with a corresponding graphics display. In both cases, the sounds were directly mapped to run-time events of parallel programs, and the mappings were designed as general-purpose tools.

Evidence of how program auralizations were effective in portraying significant behavior information to the listener regarding the execution of a parallel program was given by Francioni and Jackson.[3] In that paper a set of useful characteristics for describing the run-time behavior of parallel programs was defined to include the relative timing of events, duration of events, patterns of events, phases of behavior over time, frequency over time of some behavior, balance of behavior over the entire system, and the specifics of an event. Since the goal of parallel program auralization is to represent as many of these program characteristics as possible in an effective way, it was shown that, for the most part, there is a one-to-one correspondence between the characteristics of parallel programs and the inherent properties of sound.

In general, an auralization of a particular parallel program's events is based on the actual program's run-time behavior as well as the specific way sounds are mapped to each event. For example, each process could be mapped to a distinct note of a single instrument, or to different instruments and the same note, or to different instruments and different notes, and so on. Thus, the level of effectiveness of an auralization is governed by the sound mappings used. As shown by Francioni and Jackson,[3] some sound mappings are more effective than others; some are not effective at all.

It was also shown by Francioni and Jackson[3] that the ability of sound to represent specific event information depends on the kind of specifics intended. When there are many events being represented simultaneously, the overall sound of the auralization will tend to dominate the sounds of individual events. Conversely, if the sounds relevant to a small number of processors are played at a slow tempo, detailed information can be deduced from a playback.

3. PARAGRAPH DEPICTIONS OF PROGRAM BEHAVIOR

It is possible to develop purely aural displays for the purpose of understanding parallel program behavior, based on the analogies that have just been described. However, one of the major difficulties with this approach is that, when one hears something in the playback that warrants further investigation, it is difficult to locate and select regions for closer inspection. That is, of course, why most tape players and CD players have built-in digital counters. Clearly then, some type of visual correlation is called for. Numerous visual displays have been developed that provide the means for presenting many of the characteristics mentioned in Section 3. These visual displays can be grouped into one of two categories: static or dynamic. In a static display, previous history is available but temporal information must be abstracted from the graph. Conversely, in a dynamic display, the temporal information is intrinsic in the animation whereas the history of past events is available only as that which can be remembered.

The ParaGraph visualization tool[5] is one example of a parallel program performance-monitoring tool that provides a multitude of both textual and graphical views of program execution data. All of the figures in this paper are ParaGraph displays. For the prototype auralization tool, ParaGraph was modified to include a sound component for two kinds of static displays: the Space-Time display and the Gantt display. These displays are described below.

Execution behavior of a parallel program serves as input to the ParaGraph tool in the form of a trace file following the PICL format.[4] PICL provides execution tracing information on events such as basic sends and receives, high-level communication operations, idle/busy time, and user-defined events. The trace file is collected during the run time of the parallel program and processed post mortem by ParaGraph. When running ParaGraph, a user opens some number of windows for different static and animated displays of the run-time behavior and overall statistics. Figure 1 is an example of a ParaGraph Space-Time display which depicts the timing and processor information of messages in a distributed-memory parallel program. In this diagram, time is on the horizontal axis and processors are along the vertical axis. The left endpoint of a diagonal line in the graph represents a send by that processor at that time; the right endpoint represents the receiving processor and time. The ParaGraph Gantt chart of Figure 4 depicts processor utilization with a dark bar representing when a processor is busy, and a white bar representing idle time. As in the previous figure, time is along the horizontal axis and processors are along the vertical axis.

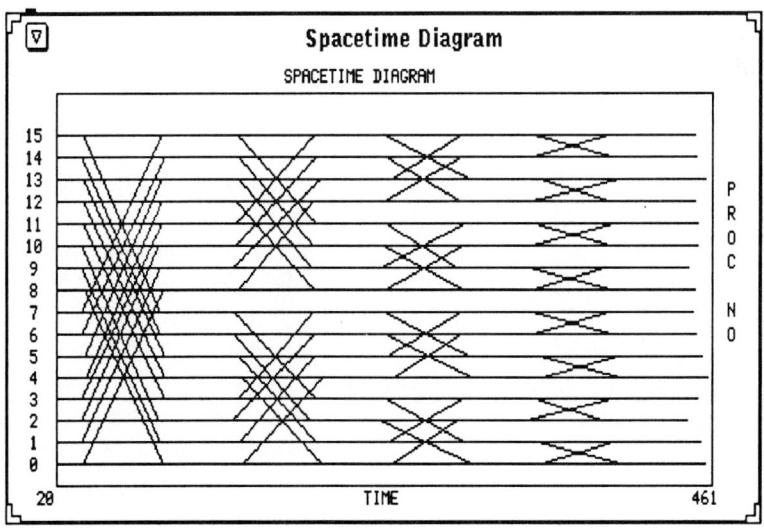

FIGURE 1 FFT Program.

4. PROTOTYPE AURALIZATION TOOL

The prototype auralization tool that was developed consists of two components: (1) several predefined event-to-sound mappings, where program-related events derived from a PICL trace file are mapped to MIDI (Musical Instrument Digital Interface) events[7]; and (2) the synchronization of the MIDI file sound generation with the ParaGraph graphical display. Details of each of these components are given in the following two sections.

4.1 EVENT-TO-SOUND MAPPINGS

The event-to-sound mappings are done by a set of C programs that are run post mortem on the trace data. In some cases, individual PICL-defined events are mapped directly to sound; in other cases, the sound mapping is derived from multiple PICL events. An example of a direct event-to-sound mapping would be as follows:

- PICL event: send, clock 0 523, node 0, to 1, type 4, length 4.
- MIDI event(s): playnote, time=523, note=C, channel=1, volume=90
  playnote, time=533, note=C, channel=1, volume=0.
  (The second MIDI event turns the previous note off after ten time units.)

No matter what kind of sound mapping is used, the event times in the resulting MIDI files correspond exactly to the times specified in the original trace file.

Five different event-to-sound mappings, described by Francioni and Jackson,[3] are summarized below. The recordings used in this paper are based on the first two mappings: Send-Receive and Idle-Busy.

SEND-RECEIVE. A basic sound mapping that captures the Space-Time type of behavior is to assign a particular timbre to each kind of event—sends and receives in this case—and play a note for each send and receive event that occurs. The particular note that is chosen can be a function of event attributes such as sending or receiving processor number, message length, and message type. Depending on the specific parameters used in the mapping, and depending on the speed chosen for the playback, different kinds of behavior can be depicted.

IDLE-BUSY. Corresponding to the Gantt chart, the ratio of idle-to-busy time can be depicted aurally by playing a certain sound for each processor whenever it is idle and playing no sound when the processor is busy (or vice versa). In either case, the sound should be sustained for the duration of the burst. This property dictates the kinds of voices that are suitable. An enhancement to this mapping is to also use the intensity dimension of sound (i.e., loudness) to reflect the length of an idle burst. In the resulting auralization, the beginnings and ends of the idle periods should be evident, as well as the relative number of processors idle at a time. With the enhancement, attention will also be drawn to the longer idle bursts due to their higher volume. Since it is also possible that the higher volume periods will drown out the lower volume ones, it is important to choose a voice that emphasizes the beginning of each idle period.

GROUP-SEND-RECEIVE. A variation of the basic send-receive mapping applicable to large numbers of processors is to use a grouping strategy for note assignment. The processors are combined into some number of groups, and individual notes are assigned to each group rather than each processor. Whenever a processor sends a message to another processor in its same group, the group's note is directed to the "intra-group" channel. Whenever a message transcends group boundaries, two notes, corresponding to both the sender and receiver, are directed to a different "inter-group" channel. The combined sound of the sender and receiver notes of an inter-group communication imparts information about which two groups are involved in the message. In addition, it can serve to cue the listener as to where the receiving node is located relative to the sender, given a visual display that is spatially oriented with the relative pitch of the group notes. By experimenting with different processor-to-group assignments, a user can study the behavior of a program in varying ways.

FLOW-OF-CONTROL. This sound mapping is intended to depict the flow of control within each processor. The mapping is straightforward: each processor is assigned a particular instrument which may or may not be unique, and a note is played for each monitored event. The intention is to emphasize the progression of a set of related events through the program. Thus, this mapping may be combined with a background sound depiction of overall behavior such as general send-receive communication, serving to "highlight" the intended events in the context of the whole.

METERS. It is often useful in performance monitoring to keep track of certain cumulative statistics over time, expressed as percentages. Percentages can aurally be represented simply by selecting notes in a range of pitches to represent corresponding values from 0 to 100 percent—low notes corresponding to low percentages, high notes to high percentages. This mapping is independent of the actual number of processors or events in the system and, hence, it is completely scalable. In addition, sound meters correspond well with visual displays where the sound represents the aggregate statistic and the graphics represent specific values in detail. Such an aural-visual combination serves to save screen space and also relieves the user of trying to simultaneously follow two or more graphical displays.

4.2 SYNCHRONIZATION OF VISUAL AND AURAL DISPLAYS

During presentation, the aural data is synchronized with either the ParaGraph Space-Time graph or the Gantt chart. Both of these displays are static displays and time is depicted along the horizontal axis. After the display has been drawn, the user defines a region of the display for aural presentation. During playback, a vertical bar, the height of the display, moves across the window from left to right, and the sounds associated with the current position of the bar are generated. Thus, the user knows what parts of the graph currently are being played.

Since the timing of aural events in a playback is exactly proportional to the timing of the actual trace events, all temporal relationships of the program's run-time behavior are preserved in the auralization. In addition, the ParaGraph tool is using the same trace file of events, so the auralizations are directly related to the graphics. The synchronization of the sound generation with the movement of the cursor bar over the graphics display is controlled by the times in the trace file and the hardware clock on the MIDI device controller. The ParaGraph program is modified such that a sound process is forked off at the time of playback to send the MIDI file to the MIDI device controller. The sound process sends the entire MIDI file to the controller as fast as possible. The controller has its own clock and a buffer for holding the MIDI events. According to the timestamps of the MIDI events, the controller then sends each event to the MIDI sound-generation device, e.g., a synthesizer, where upon the sound is generated immediately. At the same time as the MIDI device controller is sending MIDI events to the MIDI device, the ParaGraph parent process is busy moving the cursor bar over the graphics display. To control the speed of drawing the cursor bar, this process reads the clock from the MIDI device controller. Based on the clock value read, it moves the cursor bar up to the corresponding point on the time axis of the graph. Although this "catch-up" method of synchronization works fine in the prototype tool, it is not necessarily the best mechanism for a production tool. Sometimes, when the playback is very fast, the cursor bar will jump across the graph rather than move smoothly. For our experimentation purposes, however, the method is sufficient.

The tempo of the playback is at the user's control. Thus the user can listen to a fast version of a playback to get an overall impression of the program's behavior as well as slow the tempo down to hear details, similar to zooming in on a part of a visual display. The user also controls the channel-to-voice and channel-to-stereo field assignments. When the MIDI file is generated, each sound to be played is assigned to a specific channel. These channels can then be assigned to specific instruments and stereo fields based on the capabilities of the sound-generating device. So, for example, if all send events are assigned to channel 0, and all receive events are assigned to channel 1, then on playback, channel 0 can be assigned to the left stereo field and channel 1 to the right. In addition, the send events can be played by one instrument, while the receive events are played by another. Both channels can also be assigned to the same stereo field and/or instrument. This flexibility in channel assignment is important when experimenting to determine the most effective auralization for a program. Other controls at the user interface include changing the sound mapping, changing the volume, muting out specific channels, and setting the region of display for playback.

5. DESCRIPTION OF AUDIO SAMPLES

The audio portion of this paper includes playbacks of seven auralizations. The following is a description of the set up for each auralization and the matching ParaGraph figure. In Table 1 this information is summarized. Each auralization was generated from actual parallel program trace files. The programs were run on either an Intel iPSC/2 or an nCUBE parallel machine. The sounds were generated by a Yamaha TG77 tone generator.

In Figure 1 are represented the communications in a Fast-Fourier Transform program. Sound examples 1 and 2 are auralizations of this program using the send-receive mapping. All send events were mapped to one instrument (sampled sound), namely vibes, and all receives were mapped to another, namely marimba. Also, sends and receives were each directed to a separate stereo field. On each event, the note played was based on the processor that initiated the send or receive. As can be seen in the figure, this program has four distinct phases in which every processor sends a message and then every processor receives a message. In addition, the butterfly communication pattern is evident. The aural sound examples, on the other hand, provide for some different insights. In sound example 1 where the auralization is played at a relatively fast tempo, it sounds like all processors do a send at the same time, and they all do their corresponding receives together as well. In this way it can be heard that the duration of, and the time between, phases are relatively the same. When the auralization is slowed down for sound example 2, it is heard that the sends get more and more spread out at each subsequent phase, and that the order of the receives is an exact match. In other words, a processor that is later than some others to send off its message will be later in receiving its message of the same phase. This is not immediately obvious from looking at even a scaled down version of the graph.

Tr 53-55 In Figure 2 are indicated the communications of a matrix system solver run on 16 nodes. Sound examples 3, 4, and 5 are auralizations of this program using a variation of the send-receive mapping such that only send events are played. In the sound example 3, each processor is assigned the same note, and the voice used is one of indefinite pitch and short duration. The resulting sound is similar to a Geiger counter and is very useful in identifying the density of send events over the graph. From this auralization, it is evident that the first region of the graph has the most send communication activity going on. In the second auralization, sound example 4, each processor is assigned a different note and the voice is set as piano. Here, the flow of control can be heard to move through the lower to higher numbered processors. A potential problem with this type of mapping for larger numbers of processors is that, for a given instrument, it is unpleasant to hear certain notes played at the same (or similar) time. One solution to this is to use voices whose sounds are not associated with definite musical scales. As an example, the processors are each assigned a different note in sound example 5, but a percussive voice is used that depicts relative pitch information without creating any dissonance.

Tr 56,57 Figure 3 and sound examples 6 and 7 were generated by a Cholesky factorization parallel program using a cube topology for the communication. As in the previous example, the auralizations are based on the send-receive mapping where only send events are played. The voice used in both cases is a flute. In sound example 6, the tempo is relatively fast, and it is both aurally and graphically evident that one phase is repeated. In sound example 7, the tempo is slowed down considerably, and just the regions marked in the graph are played. This detailed investigation of the auralization helps to identify what really is the same in both regions and what is different.

Tr 58 The auralization of sound example 8, corresponding to Figure 4, is based on the idle-busy sound mapping and corresponds to a Cholesky factorization program using a ring topology. Specifically, the beginning of each idle burst is signified with a short, bell-like sound played at a note

Synchronization of Visual and Aural Parallel Program Performance Data

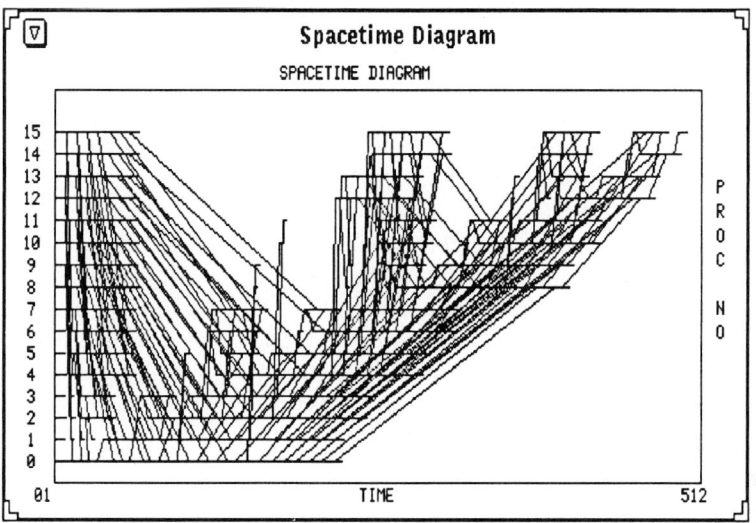

FIGURE 2   Matrix Computation.

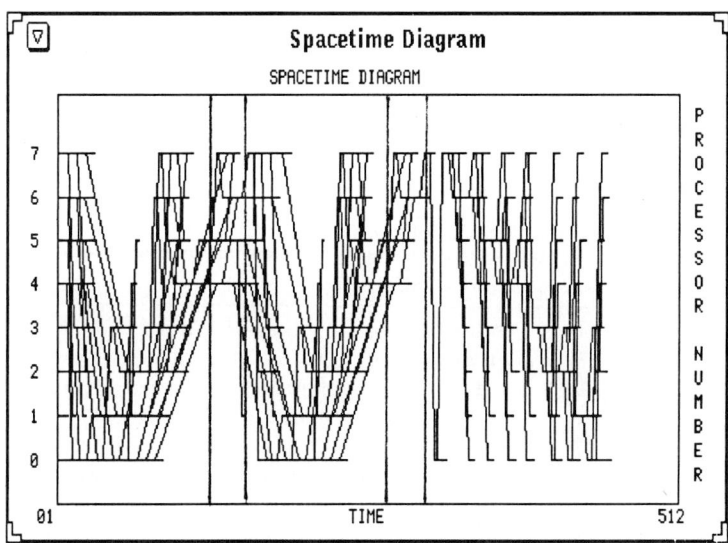

FIGURE 3   Cholesky—Cube.

corresponding to a particular processor. This is followed immediately by a sustained string sound, played at the same note and starting at a low volume. The loudness of the string sound then increases as the length of the corresponding idle burst increases, up to a predefined maximum loudness. In listening to this sound example, the beginnings and ends of the idle periods are evident, as well as the relative number of processors idle at a time. What is different from the bar chart is that the relative patterns among the processors at different points in time, versus the overall pattern of the complete display, can be detected more easily. For example, in Figure 4 it appears that all the processors are busy (dark) a large part of the time. But in reality, there are only two short intervals where all processors are busy—indicated by the tick marks. These intervals can be aurally identified as the only completely quiet times. Even though the duration of these quiet times is very short, the ear tends to detect them as the completion of a phase in the sound example.

Tr 59　　A second idle-busy example is given in Figure 5, sound example 9, based on a Sparse Matrix solver. As before, the sound example draws the listener's attention to certain characteristics of the utilization profile that are not as visually obvious. For instance, the fact that the initial idle bursts occur in pairs can be clearly heard. Also, timing differences at the start of certain idle bursts, that at first appear to be simultaneous when looking at the graph, can be easily detected upon hearing the sound example. In both these cases, it is possible to study the graphical display and observe the behavior discussed. However, the combination of the aural sound example and the graphical display facilitates the behavior being observed much sooner.

TABLE 1　Sound Example Configurations

Example	Figure	Mapping	Voice(s)	Tempo
1	1	send-receive	vibes/marimbas	fast
2	1	send-receive	vibes/marimbas	slow
3	2	sends only	Geiger	fast
4	2	sends only	piano	fast
5	2	sends only	percussive	fast
6	3	sends only	flute	fast
7	3	sends only	flute	slow
8	4	idle-busy	bell/string	medium
9	5	idle-busy	bell/string	medium

Synchronization of Visual and Aural Parallel Program Performance Data **303**

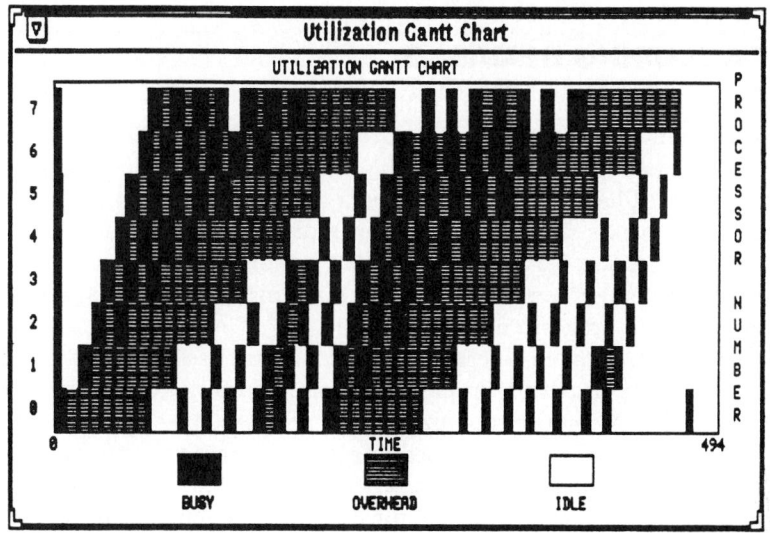

FIGURE 4   Cholesky—Ring.

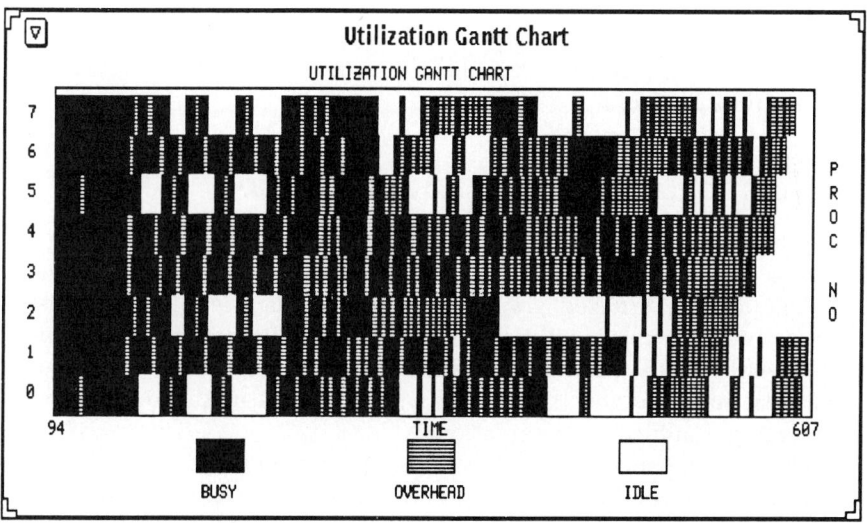

FIGURE 5   Sparse Matrix Solver.

6. CONCLUDING REMARKS

It is infeasible to directly observe the execution of a program. Thus, one studies the run-time behavior of programs by first mapping some set of execution events onto a set of observable events, and then studying the observable events. The observable events will always be different than the actual execution events since they represent a projection of the actual events onto a different space. Yet, it is known that important information regarding run-time behavior can be deduced by studying both visual and textual representations of program execution.

In this paper, we have described the mechanisms used to synchronize an auralization of a parallel program's run-time behavior with a corresponding visualization. As reported by Francioni and Jackson,[3] evidence suggests that the combination of the two media can more effectively portray significant behavior information than either one alone. This evidence is based on a survey taken at the Supercomputing '91 conference of a number of parallel programmers, as well as informal reactions by programmers over the past two years. In most cases, people were able to discern more quality information about a program's behavior when presented with an auralized graphical display than by studying the graphics alone. Furthermore, people did not feel that the auralizations interfered with their processing of the graphics. The general reaction to hearing auralizations that were not correlated with any graphics was that the sounds were interesting but were not very informative.

6.1 PRACTICALITY OF PROGRAM AURALIZATION

The practicality of using auralizations for understanding parallel program behavior depends on a number of factors. For one, users will get much more out of an auralized graphical display as they learn how to listen to these program representations. This does not mean all users will need professional musical training. Rather, it means users will have to get used to interpreting what they hear in the same way that most are now used to interpreting graphical representations.

The current sound capabilities of the average scientific workstation are not adequate to produce the kinds of auralizations used in this study. Instead, we used external sound equipment and a software device driver. The sound equipment necessary, however, is relatively inexpensive assuming the auralizations are defined as MIDI files. An adequate system, consisting of a serial-to-midi interface, a tone generator, and a set of stereo headphones, costs less than $1500. Although this type of equipment is unfamiliar to most computational scientists, it is affordable and standardized. Also, we have found the MIDI protocol to be more than adequate for our purposes to date. As workstations with built-in DSP (digital signal processing) chips capable of producing high-fidelity stereo sound suitable for use in multimedia applications become the standard, alternatives to MIDI-based sound generation will be feasible.

6.2 FUTURE WORK

This study has focused on investigating the appropriateness of using sound to depict the kind of information necessary to understand how a parallel program is working. Although the initial research has given promising results, the study is by no means complete. Further investigation into effective sound mappings of parallel program behavior information is necessary. Determining what parameters should be at the user's control for experimenting with auralizations versus which ones should be, and even can be, automatically generated based on trace data should also be studied.

The most effective auralizations will be a function of not only the sound mapping used but also the sound voicing used. Basically, sound voices can be categorized into one of three groups: musical instruments, natural sounds, and synthesized sounds. When trying to create aural analogies for certain program behavioral aspects, different kinds of sounds will be more appropriate than others. Research based both on the theory of sound, and experience with auralizations, is needed to identify families of sounds that are most effective for representing particular kinds of program behavior. Work by others investigating the auditory representation of scientific data should be applicable to this problem.

REFERENCES

1. Francioni, Joan M., Jay A. Jackson, and Larry Albright. "The Sounds of Parallel Programs." In *Proceedings of the Sixth Distributed Memory Computing Conference*, edited by Q. Stout and M. Wolfe, 570–577. Portland, OR: IEEE, 1991.
2. Francioni, Joan M., and Diane T. Rover. "Visual-Aural Representations of Performance for a Scalable Application Program." In *Proceedings of the Scalable High Performance Computing Conference*, 433–440. New York: IEEE, April 1992.
3. Francioni, Joan M., and Jay A. Jackson. "Breaking the Silence: Auralization of Parallel Program Behavior." *J. Parallel & Distrib. Comp.* (1993): 181–194.

4. Geist, G. A., M. T. Heath, B. W. Peyton, and P. H. Worley. "A User's Guide to PICL: A Portable Instrumented Communication Library." Technical Report ORNL/TM-11616, Oak Ridge National Laboratory, October 1990.
5. Heath, Michael T., and Jennifer Etheridge. "Visualizing Performance of Parallel Programs." Special issue on software for performance analysis. *IEEE Software* (September 1991): 29–39.
6. Madhyastha, Tara M. "A Portable System for Data Sonification." M.S. Thesis, Department of Computer Science, University of Illinois at Urbana-Champaign, 1992.
7. MIDI—Musical Instrument Digital Interface, Specification 1.0 (software), The International MIDI Association, North Hollywood, CA, 1983.
8. Reed, Daniel A., R. D. Olsen, R. A. Aydt, T. A. Madhyastha, T. Birkett, D. W. Jensen, B. A. A. Nazief, and B. K. Totty. "Scalable Performance Environments for Parallel Systems." In *Proceedings of the Sixth Distributed Memory Computing Conference*, edited by Q. Stout and M. Wolfe, 562–5569. Portland, OR: IEEE, 1991.
9. Sonnenwald, Diane H., B. Gopinaht, Gary O. Haberman, William M. Keese, III, and John S. Myers. "InfoSound: An Audio Aid to Program Comprehension." In *Proceedings of the Twenty-Third Hawaii International Conference on System Sciences*, Vol. II, 541-546. Hawaii: IEEE, 1991.

W. Tecumseh Fitch† and Gregory Kramer‡
†Department of Cognitive and Linguistic Sciences, Brown University, Providence, RI 02912.
‡ Clarity, Inc., Nelson Lane, Garrison, NY 10524 and Santa Fe Institute, 1399 Hyde Park Road, Santa Fe, NM 87501

Sonifying the Body Electric: Superiority of an Auditory over a Visual Display in a Complex, Multivariate System

Recent advances in the technology of computer sound generation allow sound to play a new role in human/machine interfaces. However, few studies have investigated the use of sound to display complex data in a practical setting. In this paper we introduce an auditory display for physiological data and compare it experimentally with a standard visual display. Subjects (college students) played the role of anesthesiologists, attempting to keep a computer-simulated "digital patient" alive and healthy through a series of operating room emergencies. Both the task and the stimuli were complex: subjects had to monitor eight continuously changing variables simultaneously, to identify problems (indicated by changes in one or three variables at once), and then to correct those problems. We found that subjects performed faster and more accurately when using the auditory display than when using the visual display. This difference was most pronounced with multivariate changes. We hypothesize that the auditory advantage may result from the inherent ability of the auditory system to process multiple auditory "objects" or "streams" simultaneously in parallel, in contrast to the visual system's propensity for processing multiple objects serially. If correct, this idea has important implications for the use of sound in computer interfaces.

1. INTRODUCTION

Audition and vision are highly developed, processing-intensive forms of perception that provide extremely rich channels for information to pass from the environment into our brains. However, sonic and visual information are quite different in nature. Vision is inherently spatial: a visual object's location in the environment is critical, and spatial coherence of its parts is typically what defines a visual object (especially for objects in motion). Its spatial nature makes the visual system well suited for analysis of a scene into its component parts. In audition, spatial location is important, but temporal coherence seems to play a more significant role in defining auditory objects or "streams."[3,7] We can listen to a chamber orchestra played through a single speaker and still easily delineate the separate instruments, illustrating the remarkable ability of the auditory system to extract multiple streams simultaneously from a complex acoustic morass without recourse to spatial cues. However, we can also hear and appreciate the composite sound of all the instruments at once, noting correlations and interactions between parts with exquisite sensitivity. Thus, the auditory system may also be well suited to the perception of complex patterns of interplay between multiple components.

Recent advances in technology have made computer-controlled sound a ubiquitous reality. Unfortunately, the current human-interface literature provides little guidance about how to best take advantage of these sonic capabilities. A few studies have compared auditory and visual displays for simple tasks and have demonstrated simple effects such as a reaction time advantage for sound over vision.[6,9] However, high-level advantages of audition, like the multiple-component integration advantage suggested above, are widely suspected to exist but have yet to be experimentally documented.

In this paper we introduce a new auditory display capable of representing eight variables simultaneously, which we used to display physiological data in a simulated operating room setting. Our task involved both complex stimuli (three of the eight variables changing at once) and a complex task (subjects had to "save the patient" by identifying problems and correcting them). Unlike previous studies which looked at tasks where sound is already in widespread use (radiation detection and sonar), we looked at a domain where sound is currently not widely used. In addition, this study explicitly compared performance with our auditory display to that with a standard visual display (a strip chart similar to those used in hospitals and laboratories around the world). We found that subjects answered faster and made fewer mistakes with the sonic display than with the standard visual display.

We shall briefly outline the task here, leaving the full details to Section 2. The subjects (college graduate and undergraduate students) were first given introductory training as anesthesiologists: they learned to simultaneously monitor eight physiological variables of a "digital patient," and to appropriately respond to various medical complications (such as anesthetic overdose or drop in body temperature). Because virtually everyone is already familiar with such terms as "heart rate," "body temperature," "overdose," and "blood pressure," we were able to minimize training time.

The familiarity of the variables also allowed us to make use of "self-labeling" auditory streams: heart rate was represented simply as the rate of occurrence of a repeated tone (which sounded like a heart beat), and breathing rate as the rate of amplitude modulation of band-passed noise (a sound that the subjects could immediately identify as breathing). Other, more abstract variables were then "piggy-backed" on top of these sounds: carbon dioxide level was represented as a change in timbre of the heart rate sound, while body temperature controlled the center frequency of the band-pass filter in the breathing sound (roughly speaking, the "pitch"). Our auditory display was thus a hybrid of realistic sounds (cf. Bill Gaver's "everyday sounds"[4,5]) and abstract sound. This novel approach combines the ease of use of familiar self-labeling sounds with the flexibility of abstract, nonrepresentational sounds.

In the visual display, labeling was accomplished by writing the name of the variable on the display (see Figure 1). Time originated at the left side of the display and moved toward the right, providing a visual history of the previous 15 seconds of activity. The rate of movement was the same for all variables, so simultaneous changes in multiple variables "lined up" and thus were very obvious.

We attempted to make the two displays equivalent in the sense that subjects could rapidly and easily learn to extract from them all the information required to perform the task. Obviously, the visual display had certain advantages over the auditory display: each visual stream was labeled (subjects had to remember the labels of the auditory streams), and changes in the visual display remained present as the strip chart moved across the screen, while auditory changes were transitory. On the other hand, subjects needed to look at and to focus on the visual display, but not the auditory display. We could have made the two displays more equivalent, for example, by removing the visual display labels or making the display more fleeting, or by making the sonic display audible only if the subject looked at the loudspeakers. However, we feel that attempts to make the displays exactly equivalent are doomed to failure since we have no way of knowing what "exactly equivalent" means across different sensory domains (see Section 4). Instead, we chose to use more natural displays which took full advantage of the possibilities offered by each domain, and ensured that the subjects were able to extract 100% of the required information from each display.

Using this computer-simulated "digital patient," we tested the speed and accuracy with which subjects responded to previously learned patterns of change. Subjects were each tested in the following three conditions: auditory display only (A condition), visual display only (V condition), and both auditory and visual displays simultaneously (AV condition). By comparing the performance in these different conditions, we hoped to gain insight into the types of information best carried by the two sensory channels.

2. METHODS

2.1 PHYSIOLOGICAL SYSTEM

Subjects monitored the following eight physiological variables in our "digital patient":

- body temperature (in °C),
- heart rate (in beats per minute or BPM),
- blood pressure (in mmHg),
- blood carbon dioxide level (in mmol/L),
- respiratory rate (in breaths per minute or BPM),
- atrio-ventricular dissociation (present or absent),
- fibrillation (present or absent), and
- pupillary reflex (present or absent).

The top five variables, of course, fluctuated within a continuous range of values; the last three were binary: either present or absent. Atrioventricular dissociation is a decorrelation between the two "pacemakers," which control cardiac contraction, where the heart nonetheless maintains a constant pulse (with a "syncopated" rhythm). Fibrillation, on the other hand, represents a random and incomplete cardiac contraction where both the atrial and ventricular signal become weak and no clear pulse is perceptible (random rhythm). The pupillary reflex is the contraction of the pupil in response to bright light; this reflex disappears in cases of brainstem depression, for instance, during an overdose of anesthetic. While most of the variables gave a constant audio and/or visual readout, subjects had to test the pupillary

TABLE 1 Possible Complications and Attendant Symptoms

Complication	Temp	HR	BP	CO_2	RR	AV	FIB	Reflex	Response
1. Temp Down	↓								Heat on
2. CO_2 Down				↓					↑ Ventilation
3. HR Up		↑							Digitalis on
4. O_2 Down		↑	↑			X			↑ Oxygen
5. Blood Loss	↓		↓				X		↑ Blood
6. Overdose			↓		↓			X	↓ Anesthetic

reflex in order to determine its state. This encouraged an active, exploratory involvement with the system on the part of subjects.

Table 1 lists each of the six possible medical problems, or "complications," that could occur.

There were two types of complications, simple and complex (or multivariate). Simple complications resulted in the change of a single physiological variable, while multivariate complications resulted in a change of three variables at once. Each multivariate complication consisted of a same-direction change in two continuous variables, along with a change in state of one of the three binary variables.

2.2 EQUIPMENT

The experiment was run using an Apple Macintosh IIcx computer with a 13″ color monitor and an Audiomedia 16-bit sound board (Digidesign Inc., Menlo Park, CA). Sound was produced in three ways. Eight-bit sampled sound was played back through the Macintosh sound system and speaker. Real-time synthesis occurred on the Motorola DSP 56001 chip on the Audiomedia board, and in an external Yamaha DX100 FM synthesizer. These signals were fed into the left and right channels of a Sony GX40ES stereo amplifier and played out through Camber .5 loudspeakers at a volume adjusted to the user's comfort. Subjects controlled the system via an external "control board" designed by the second author (a Lexicon MRC MIDI Remote Controller) that had four sliders and four buttons, each of which was clearly labeled with the parameter it controlled. The entire system was coordinated, and subject responses interpreted and recorded, by programs developed in the Max development environment (Opcode Systems, Menlo Park, CA). Synthesis on the DSP chip was controlled by Max-generated MIDI signals sent via the Macintosh MIDI Manager to a synthesis "patch" implemented in the Unison sound generation program.[1] The tutorial software was written using the Hypercard authoring system (Apple Computer, Cupertino, CA). All custom software was written by the first author. Subjects performed the task alone in a sound attenuated room.

2.3 SONIC MAPPING

In this study we used a novel system to represent the physiological variables in the sonic realm (examples of the auditory display can be heard on the CD that accompanies this volume). The basic auditory display consisted of two independent auditory streams: the heart signal and the breathing signal. These were "self-labeling" in the sense that they sounded very similar to the real thing: the heart signal was a low-pitched repetitive thudding sound synthesized by the Yamaha DX100 (with two different pitches representing the atrial and ventricular contractions). The breathing signal was synthesized by the DSP chip on the Audiomedia board; it was created by amplitude-modulating noise at the breathing rate, and the band-pass filtering the resulting signal. These two streams served as the "base stream" for the other "piggy-back" variables, which modulated aspects of the base stream sounds. Atrio-ventricular dissociation and fibrillation modulated the heart signal in a way that reflected reality: in atrio-ventricular dissociation the ventricular tone remained steady while the atrial tone varied randomly in its delay from ventricular; in fibrillation both tones were random. Thus, four of the eight variables were self-labeling.

The other four variables were abstract in the sense that their assignment to one stream or the other was arbitrary, and the correspondences between the physiological and sonic variables had to be learned and remembered. Body temperature was represented as the center frequency of the second-order band-pass filter on the breathing sound. Blood pressure controlled the pitch of the heart sound. Finally, the timbre or "brightness" of the heart sound reflected the CO_2 level and was controlled by changing the modulation index in a frequency-modulation patch on the DX100 FM synthesizer. Reflex was sonically encoded as a high tone from the DX100 synthesizer.

Our sonic mapping thus represents a hybrid between a totally abstract, more difficult-to-learn system, and a direct "representational" system (which is limited in flexibility). We believe that a system like this provides an optimal combination of ease of learning with flexibility and representational power. As our results show, the auditory display was readily learned and effectively used by subjects (who had no previous experience with auditory displays) in less than an hour.

TABLE 2 Summary of the Sonic Mapping

Physiological Var.	Mean	Mapped to	Changes to
Respiratory Rate	20 BPM	of AM on breathing noise	10 BPM
Body Temperature	37° C	Bandpass frequency on breathing	30° C
Heart Rate	70 BPM	Rate of heart sound (FM tone)	100 BPM
Systolic Blood Pressure	120 mmHg	Pitch of heart sound	80 or 160
CO_2 Level	25 mmol/L	Timbre of heart sound (mod. index)	18 mmol/L
AV Dissociation	Off	A pulse randomly delayed, V steady	On
Fibrillation	Off	Both A and V random	On
Reflex	On	High FM Tone Off	

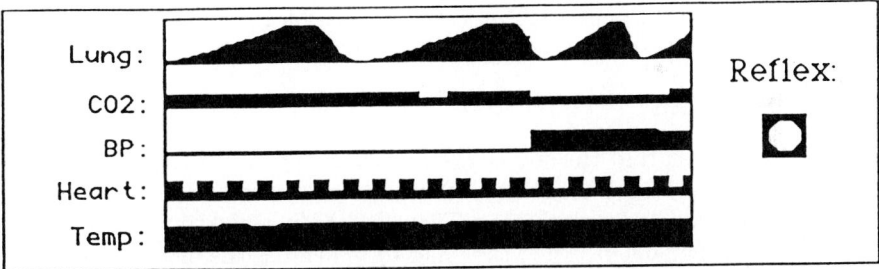

FIGURE 1 Illustration of the Visual Display. A precipitous drop in blood pressure is shown, along with a decrease in breathing rate. The display is shown here at about half size.

2.4 VISUAL MAPPING

The five continuous variables were displayed on a constantly moving strip chart on the Macintosh screen. Each variable occupied a single row of the display, and they were arranged to provide for minimal confusion between variables. Written labels were provided for each visual variable. Time originated at the left side of the display, which moved to the right. A total of 15 seconds of the immediate past was visible at any given time on the strip chart. The visual display is illustrated in Figure 1.

A readout of the pupillary reflex was provided as a separate display which vaguely resembled an eye. If subjects pressed the "reflex test" button and the pupillary reflex was intact, the simulated pupil contracted; otherwise, it did nothing. Atrioventricular dissociation and fibrillation were represented as corresponding changes in the heart signal.

2.5 SUBJECTS

Subjects were 13 Brown University undergraduates, graduates, and graduate students free of any medically diagnosed hearing disabilities, who aged from 20 to 37. Five of the subjects had musical experience (defined as at least four years of playing at least one instrument, including the voice, in the last ten years); four of the subjects had extensive experience in physiology which prepared them for much of the factual material learned in the training portion of the study. Subjects were paid for their participation.

2.6 PROCEDURE

The ultimate task that the subjects were asked to perform required several steps. A change occurred in the patient and was reflected as a change in the visual and/or auditory display. The subjects first had to identify which variable(s) had changed. Then they had to remember the medical problem or "complication" to which these changes corresponded. Finally, the subject had to remember the correct response to the problem and implement it using the "control board."

In preliminary trials we found that training subjects in this difficult task alone discouraged and frustrated them, so we developed a four-stage training/testing sequence to make the training more manageable. The experiment consisted of four stages arranged so that they grew gradually in difficulty and pressure, to slowly build up subjects' self-confidence. The sonic and visual displays were the same for each subject and each stage.

The first stage consisted of a self-paced Hypercard tutorial that methodically introduced the subject to the physiological terms and concepts required in the task. Using text and graphics, the tutorial first introduced each of the physiological variables and their significance (in layman's terms). Then, subjects were introduced to each of the possible complications and their attendant symptoms, along with the appropriate response. During this training phase, a window was available that allowed the subjects to "run" each complication using both the auditory and visual displays (subjects could control the presence or absence of either display). A pretest was given within the Hypercard stack to ensure that the complications and their responses were memorized. As in all tests in this experiment, subjects were required to answer all questions correctly before going on. Thus subjects were first afforded a self-paced, low-pressure tutorial on the terms and displays they would encounter in later stages. However, the tutorial was not available during these later stages, which had to be accomplished from memory. The amount of time spent in this first stage was automatically recorded.

In the next stage, we tested that subjects had mastered the use of the MIDI "control board." The subject was requested by the program to perform each possible corrective response, and stayed in this stage until she/he had performed all seven responses correctly in a row. Most subjects found this task very easy.

The third stage represented the first serious challenge. Subjects were tested on their ability to recognize each of the six possible complications, first in the AV condition (to allow them more easily to gain familiarity with the task), and then in the V and A conditions separately (order randomly selected). Thus the subjects had to recognize the problem, but did not have to correct it. In each condition, the subjects were required to identify all six complications correctly in a row before moving to the next condition. Subjects gave their answers by using the mouse to select from a menu of the six possible complications. Subjects were allowed unlimited time to make their selection. Feedback was given, informing the subjects of the correct answer. Most subjects found this task quite difficult and repeated each condition many times. This stage constituted the "real world" training for the final testing stage.

In the fourth and final stage, subjects were required to put all of their knowledge together. Again each complication was tested, but now subjects had to make the appropriate corrective response using the "control board," and to do so within 20 seconds (a countdown of "seconds remaining" was provided). For each trial the reaction time (measured to the nearest Macintosh 17-msec clock tick from the onset of the complication) and response were recorded to a text file for later analysis. In this stage, subjects were informed only of the correctness (by a green light and the spoken word "correct" from the Mac speaker) or incorrectness (by a red light and the word "wrong") of their response. Trials were self-paced (using the return key), so that subjects could rest between trials. Each subject was run in the AV, A-alone and V-alone conditions (order randomly determined). In each condition, all 6 complications were presented twice, so that there were 12 trials per condition and 36 trials in all.

In both the third and fourth stages, the structure of a trial was the same: a five-second baseline period with all variables set to normal was followed by a sampled bell sound which marked the beginning of the complication. All symptomatic changes corresponding to a given complication occurred simultaneously. In the third stage, subjects were required simply to identify the complication (from a menu of all possible complications), were allowed unlimited time to make their selection, and were told the correct response after each trial.

2.7 SUBJECT DEBRIEFING

After completion of the experiment, each subject was interviewed by the first author using a standardized questionnaire. Subjects were asked if they had had extensive prior experience with physiology, which rendered a significant amount of the training material familiar to them; if they answered "yes," they were considered "medically experienced." If subjects reported having had at least four years' experience playing at least one instrument (or voice) in the last ten years, they were dubbed "musically experienced." Subjects were asked if they had had any emotional responses, positive or negative, to the sounds used in the auditory display, and whether any of the variables (in either the A or V display) were obscured in any way. We asked which of the complications, if any, were difficult to remember or identify, and whether, in general, they found the task to be easy or difficult. Finally the subjects were invited to share any observations or criticisms they had about the experiment.

3. RESULTS

3.1 QUANTITATIVE ANALYSIS

For all analyses described below, response time (RT) data includes only correct responses; incorrect responses have been omitted. In discussing results from ANOVAs, comparisons between individual groups were determined with the Fisher PLSD test, with significance set at 95%. The basic effect we were interested in was the effect of condition (i.e., auditory vs. visual display) on reaction times and error rates. We include AV values in the results simply to give a basis of comparison for the two conditions of interest.

The distribution of response times was (as is normally the case) a normal distribution skewed towards shorter RTs (skewness = 1.31; kurtosis 1.59). Mean RT for all subjects was 6.33 seconds (SD = 3.53 s); this comparatively long mean response time suggests that our timing accuracy of ±17 msec was more than adequate.

Subjects responded significantly faster in the A condition than in V or AV conditions (single-factor repeated-measures ANOVA, $F(2, 24) = 14.71$, $p = .0001$). Additionally, subjects made significantly fewer errors in the A condition compared with the V condition (paired one-tailed t-test, $t(12) = 1.81, p < .05$). Neither testing order nor training order had an effect on this pattern (two-factor repeated-measures ANOVA, $F(1, 11) < 2.25, p > .15$).

Thus, in this comparison of visual vs. auditory display of the same variables, subjects answered significantly faster and more accurately when given auditory data alone than they did when given visual data alone. One possible reason for this difference is that (as many subjects pointed out during debriefing) visual data had to be serially accessed by fixating each trace on the strip chart sequentially, while changes in the auditory variables could be simultaneously (and thus more rapidly) perceived. This simultaneity of perception led many subjects to form an overall gestalt perception of a given event in the auditory domain. If this difference is partially responsible for the observed superiority of the auditory display over the visual display, we would expect that superiority to be more pronounced for complex changes (involving several variables) than for simple changes involving a single variable.

To test this hypothesis, we compared results for simple and complex complications. Not surprisingly, subjects made more errors and answered more slowly in the complex trials than they did in the simple trials: the mean RT for complex trials was 7.92 s vs. 6.0 s for simple trials, and there were 56 total errors in the complex condition vs. 29 in the simple. These differences were statistically significant (paired two-tailed t-tests, using mean RTs: $t(12) = 4.45, p < .001$; using total errors: $t(12) = 3.32$, $p < .01$).

More interestingly, the difference in subjects RTs between the A and V conditions was much more pronounced in the complex trials (Visual RT − Auditory RT = 1.56 s in simple trials and 4.09 s in complex trials). The A vs. V difference was significant at the 95% level in both cases, but achieved a much higher level of significance in the complex trials (simple trials: $F(2, 24) = 5.61, p = .01$; complex trials: $F(2, 24) = 23.27, p = .0001$). In terms of error rates, there was no significant effect of condition on error rate in the simple trials ($F(2, 24) = 0.35, p > .7$), while there was in the complex trials ($F(2, 24) = 4.85, p = .017$), as illustrated in Figure 2.

Somewhat surprisingly, the four individuals who had extensive prior experience with physiology did not perform significantly better than those with no experience ($F(1, 11) = 0.21, p > .65$). Of course, the sample of medically experienced subjects is too small for us to make much of this negatve result.

In contrast, the five musicians (defined as subjects with more than four years of continuous experience with one or more instruments in the last ten years) answered significantly faster than the eight nonmusicians (about 2.5 seconds faster on average; $F(1, 11) = 7.14$, $p = .022$). The difference between musicians and nonmusicians is illustrated in Figure 3. However, the basic pattern of more rapid RTs for auditory vs. visual conditions did not differ between the two groups; in each case, RTs in the A condition were about 2 seconds faster than in the V condition. Thus, musicians were faster in both display conditions, not just the auditory display.

Training times were not significantly correlated with error rates or RTs in this study; however, there was a negative correlation between stage 3 training time and error rate which approached significance ($p = .20$) and would perhaps have attained significance with a larger sample size.

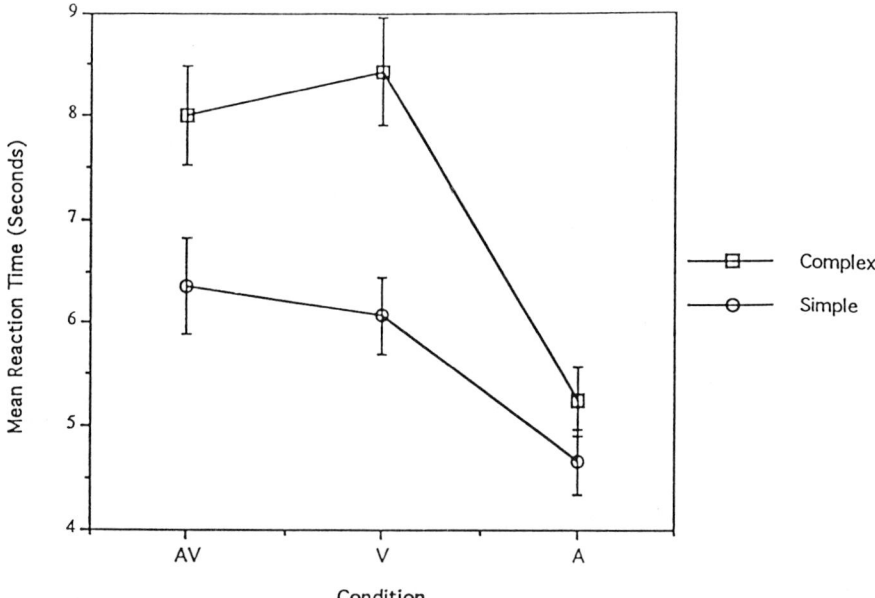

FIGURE 2 Comparison of subject performance on simple vs. complex complications. Error bars show standard error of the mean.

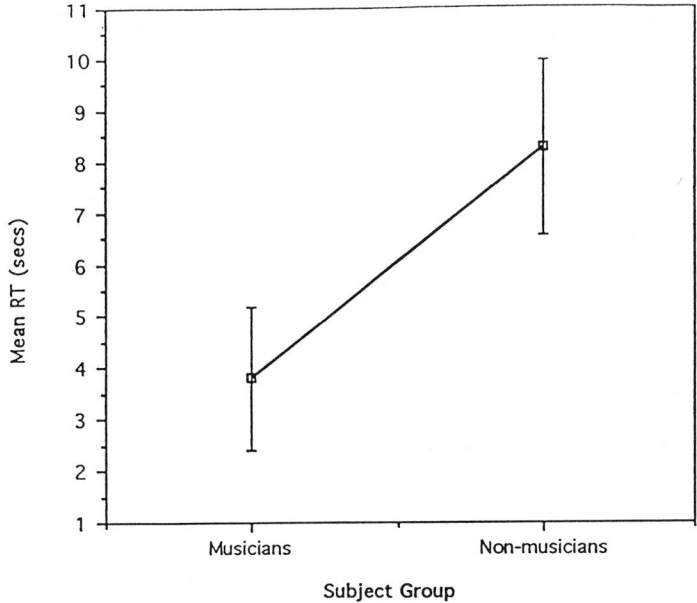

FIGURE 3 Comparison of Musicians vs. Non-Musicians. Error bars show standard error of the mean.

3.2 QUALITATIVE RESULTS

Most subjects described the task as difficult, but manageable after practice. Some subjects considered the task "very hard" because of the large amount of information to be remembered and processed. Thus, we apparently succeeded in creating a complex and demanding task (within the time confines of a psychological experiment). Subjects also tended to find the task interesting; several veteran experimental subjects volunteered that it was the most interesting experiment they had ever participated in.

Although many of the subjects expressed likes or dislikes of various sounds, there was no clear pattern: for example, some subjects liked the breathing sound and others found it obnoxious or "scary." Our subjects were not unanimous in their feelings about the particular sounds we used in the auditory display; we are thus unable to make any recommendations for future designers other than to provide flexibility and room for individual tastes.

Not surprisingly, most subjects cited the complex complications as being the most difficult. This subjective experience mirrored the quantitative results: complex complications yielded slower reaction times and higher error rates as well.

The most common comment made comparing the visual and auditory displays was that it was possible to develop an overall "gestalt" perception of a given complication with the auditory display; the covariance of a number of separate variables fused into a unitary perceptual entity. This led most subjects to report that their responses to complications in the auditory domain were more direct, while those in the visual display had to be sequentially and more analytically worked out. (A single subject reported the opposite impression: the verbal labels made the visual display more direct for him; in the auditory domain he had to puzzle out the correspondence between auditory and physiological variables. This was the only subject of the 13 to show a higher error rate and slower reaction times in the auditory condition.)

Although several subjects reported that the labels on the visual display helped them feel more confident, most subjects found the auditory display easier to use after practice. We expected the visual display to be easier to learn and use because of its familiarity, and were surprised when many subjects reported that they preferred the auditory display. Several subjects reported that they found the presence of both displays in the AV condition overwhelming, and they had to focus on one display or the other (usually the auditory).

In summary, the self-reports of subjects' experiences corroborate the quantitative results discussed earlier: in this complex dynamic system, the ability of the auditory system to simultaneously perceive a number of variables proved advantageous compared to the one-by-one perception necessitated by our visual display.

4. DISCUSSION

In this study, which is the only study we know of to directly compare a complex auditory display with a standard visual display, we found that subjects were able to respond more rapidly and more accurately to simulated operating room emergencies when using the auditory display than when using the visual display.

Our auditory display made use of some novel techniques. First, we used "self-labeling" streams, which correspond in many ways to the use of "everyday sound" that Gaver[4] has proposed. While visual objects can be easily labeled using writing, it is much more difficult to "label" a sonic stream. Self-labeled sounds correspond experientially to the variable that they encode and, thus, avoid this problem. Second, we developed the concept of a self-labeled base stream modified by "piggy-back" parameters. Using this concept, a single auditory stream can carry information about multiple variables. If the variables are logically related, this allows us to develop sonic displays where the auditory streams relate in a rich, hierarchical fashion to the underlying variables being "sonified." Thus our auditory display was a hybrid of realistic and abstract sounds; this approach combines the ease of use of everyday sound with the power and flexibility of nonrepresentational sound. Our results suggest that this is a fruitful marriage.

This study, like any study comparing displays across sensory modalities, must deal with the question of display quality: Do the differences found reflect *bona fide* differences between sensory modalities or simply differences in the quality of the two displays? It seems quite plausible that we could have improved our visual display: by adding color, for example, we could have made the visual readouts more distinctive. Alternatively, a completely different type of visual display could have been used, for example, one linking several variables to an animated object that updated its shape and color from moment to moment. Similarly, the auditory display could have been altered in various ways: spatial location in the stereo field could have been used, or the number of auditory base streams could have been changed. Unfortunately, we have no idea how to compare audition and vision directly in quantitative terms (does frequency in the sonic realm, which correlates with pitch, correspond to frequency in the visual realm, which corresponds to color? It seems unlikely). We can only compare informational equivalence (is all the information present and accessible in both displays?) and make all other aspects of the experiment invariant across displays.

We avoided the need to ensure exact display equivalence by focusing instead on ensuring informational equivalence: Each subject was tested before data was collected to ensure that he or she was capable of correctly identifying every complication with 100% accuracy using either the auditory or the visual display alone. In each display, the same variables were presented, and the same complications, requiring the same responses, were tested. Visual vs. auditory order was randomized in both the training and testing stages, so that learning effects would be minimized. Thus the differences seen in the last stage are reflective not of the raw information available in either display or of experimental artifacts, but reflect the speed and accuracy with which equivalent information could be accessed.

At the present embryonic stage of auditory display research, we think it is important to explore innovative ways of sonifying complex systems, and then rigorously compare subjects' performance in those systems with well-established visual display techniques. The present study shows that subjects performed better with our auditory display than with this *particular* visual display. Given that this particular visual display type is in wide use today, and has been for many years, we find this result to be significant.

However, sweeping statements about auditory vs. visual display in general will have to await a host of similar experiments with different displays, tasks, and subject populations (we are currently engaged in such studies). For now, this result is the only one that we know of showing superiority of *any* complex auditory display over a comparable visual display.

However, there is other data showing superiority of an auditory over a visual display in a much simpler task: the location of a radioactive emitter using a radiation monitor or "Geiger counter." In this study by Tzelgov et al.,[9] subjects were asked to locate a source of radiation hidden under a large table using auditory, visual, or "redundant" (AV) display modes. The auditory display made a simple beep for each radiation count the meter detected, while the visual display was a meter displaying the counts per second. Subjects were significantly faster, and made fewer errors, using the auditory display alone than with the visual display. Interestingly, in the redundant display mode, subjects seemed to rely on the visual display and thus actively worsened their own performance; Tzelgov et al.[9] suggested that a visual bias leads subjects, when given a choice, to selectively attend to visual rather than auditory information. The same explanation may account for our finding that AV performance was equivalent to the V condition. Apparently, subjects would have done better to close their eyes and use only sound, since performance in the A condition was significantly better.

Interestingly, several studies have demonstrated a bimodal advantage. For example, Bly[2] demonstrated that the addition of sound to a visual display can increase subject performance. Lewandowski and Kobus[6] designed a simple sonar task where a high or a low sound had to be detected and categorized either visually (using a time vs. frequency plot) or sonically (by direct listening) or with both. A speed/accuracy tradeoff was found: subjects answered faster but less accurately with sound alone than with vision. The bimodal (AV) condition resulted in performances as fast as audition and as accurate as vision! Further research will be necessary to determine why bimodal performance is good in some tasks and poor in others.

We shall now turn to an explanation for the findings reported here. We find it plausible that our results reflect fundamental differences in the information-processing style of the auditory and visual systems. In particular, we think that vision has evolved to process spatially defined objects in a sequential fashion, while audition is specialized to process multiple temporally defined objects or "streams" simultaneously. Our task required that the subject attend to many continuously changing variables at once and notice patterns of change in the overall system. As a result, it was well suited to auditory display. If we are correct in our belief in the strength of the auditory system for parallel-processing, auditory display represents a vastly underutilized resource in a world increasingly packed with information. But why should audition be better suited than vision for simultaneous processing of multiple data streams?

The primate visual system is highly adapted to fixate and to identify objects in the environment. Specializations like the fovea (an area in the retina of extremely dense and acute photoreceptors) and binocular vision, combined with the inability of light to curve around objects, encourage a serial mode of visual processing: we quickly and effectively scan the environment, serially locating and fixating objects via eye movements. In the course of evolution, the cortex seems to have "internalized" the phenomenon of eye movements, resulting in a much faster but fundamentally similar mechanism of visual attention. In the so-called Treisman task of visual attention, subjects must identify the presence or absence of a pre-designated target amidst a variable number of distractors. While performance for simple targets is independent of the number of distractors, complex targets involving a conjunction of two variables become more difficult the more distractors are involved: RTs (or error rates) increase linearly with the number of distractors. This suggests that complex visual tasks are performed using an attentional "searchlight" which scans each object sequentially.[8] The "pause time" seems to be about 20 msecs per distractor: much faster than eye movements but still fundamentally similar.

In contrast, the human auditory system lacks any strong directionality: our ears are fixed on our heads and open to sound coming from any direction (including around corners). Research into the mechanisms involved in delineating sonic objects is still in its infancy (see Bregman[3] for a comprehensive review), but it is already clear that location is by no means the most important cue. Instead, our ears are exquisitely sensitive to *temporal* coherence between separable portions of the auditory array, and such coherencies often define sonic entities. For example, a collection of simple sine waves may be heard as such (i.e., as several discrete "streams"); but, if we force temporal coherence by adding a small amount of vibrato or random jitter to each partial, they "fuse" into a unified percept of a single sound (see McAdams[7] for a detailed exploration of this "spectral fusion").

Similarly, while our eyelids and ability to move our eyes allows us to ignore visual information, extraneous or interfering auditory information cannot be *anatomically* filtered out, suggesting that neural/computational approaches to ignoring or habituating to uninteresting sounds will be better developed in the auditory system.

It seems that, given these constraints, millions of years of evolution have produced an auditory system well suited to absorbing and making sense of large volumes of information simultaneously ("in parallel"), and a visual system specialized for sequentially locating and identifying objects ("serially"). Although this dichotomy is a simplification (peripheral vision may function much more like the auditory system, and it is certainly possible to adopt a focused, serial approach to listening, for example, when isolating a single part from a symphony), it seems to have made evolutionary "sense" for the two main sources of long-distance sensory information we use to play complementary roles. An important task for cognitive scientists will be to explore and to delineate these roles, both because of their intrinsic interest and to provide guidance in the design of auditory and multimodal computer interfaces.

Audition has traditionally taken a back seat to vision in scientific or other types of data display (for example, there is no colloquial auditory equivalent of the term "visualization"). However, the considerations above and the results of this experiment suggest it is time to consider the idea that certain types of information might be better perceived through auditory than through visual channels. In particular, for systems where large numbers of variables are causally and temporally interconnected in ways that are not obvious, auditory displays may have a distinct advantage over traditional visual displays. As the data in our world grows more complex and more voluminous, we will do ourselves a disservice if we fail to avail ourselves of the powerful information-processing machine that constitutes our auditory system.

ACKNOWLEDGMENTS

This work was supported by an NSF Predoctoral Fellowship to T. Fitch and by a grant from Clarity, Inc. We sincerely thank Stephen Ellison for his assistance, both practical and conceptual, in developing the sonification system. We also thank Peter Eimas, Eric Nicolas, and Cynthia Romano for their comments on an earlier version of this paper.

REFERENCES

1. Bate, J. "A Multi-Processor DSP System for Real-Time Interactive Sound Processing and Synthesis." *Proceedings of the International Computer Music Conference.* Edited by B. Pennycook and B. Alphonce: Montreal, 1991.
2. Bly, S. "Sound and Computer Information Presentation." Ph.D. Thesis, University of California, Davis, 1982.

3. Bregman, A. S. *Auditory Scene Analysis: The Perceptual Organization of Sound.* Cambridge, MA: MIT Press, 1990.
4. Gaver, W. W. "Auditory Icons: Using Sound in Computer Interfaces." *Hum-Comp. Inter.* **2** (1986): 167–177.
5. Gaver, W. W., R. B. Smith, and T. O'Shea. "Effective Sounds in Complex Systems: The ARKola Simulation." In *CHI '91 Conference Proceedings*, edited by S. P. Robertson, G. M. Olson, and J. S. Olson, 85–90. Reading, MA: Addison-Wesley, 1991.
6. Lewandowski, L. J., and D. A. Kobus. "Bimodal Information Processing in Sonar Performance." *Human Perf.* **2(1)** (1989): 73–84.
7. McAdams, S. "Spectral Fusion, Spectral Parsing and the Formation of Auditory Images." Ph.D. Thesis, Stanford University, Stanford, CA, 1984.
8. Treisman, A., and G. Gelade. "A Feature Integration Theory of Attention." *Cog. Psych.* **12** (1980): 97–136.
9. Tzelgov, J., R. Srebro, A. Henik, and A. Kushelevsky. "Radiation Search and Detection by Ear and by Eye." *Human Factors* **29(1)** (1987): 87–95.

Kevin McCabe* and Akil Rangwalla**
*Sterling Software and **MCAT Institute

Auditory Display of Computational Fluid Dynamics Data

Auditory Display (AD) is the mapping of values in some data space onto parameters in acoustic space. This paper discusses some of the motivations and techniques for using auditory display in the analysis of data generated from computational fluid dynamics (CFD) simulations. Two simulations are used as case studies. In the first case, data from a simulation of the Penn State artificial heart pump is analyzed using a technique called parameter mapping. In the second case the tonal acoustics of rotor-stator interaction inside turbomachinery are directly simulated.

1. CFD BACKGROUND

Computational fluid dynamics (CFD) research at NASA Ames Research Center is typically done by solving Navier-Stokes and Euler equations for fluid flow on three-dimensional structured grid volumes.[1] These simulations produce data sets in the tens of megabytes range. Current data sets are undergoing a migration from three-dimensional to four-dimensional where the fourth dimension represents time.

1.1 STRUCTURED GRID FORMAT

A four-dimensional structured grid consists of a time series of three-dimensional arrays of spatial coordinate information. For a given time step, the grid dimension is given by grid$[IDIM][JDIM][KDIM]$, where $IDIM$, $JDIM$, and $KDIM$ are the maximum indices of the three array dimensions (referred to I, J, and K directions, respectively). The contents of the grid cell$[I, J, K]$ is a point described by Cartesian coordinates. There is an implicit connectivity between adjacent cells. Interior cells in the grid are connected to their six neighbors, while edge cells are connected to three to five neighbors depending on the I, J, K indices. Associated with each grid is a corresponding solution. The solution associates five physical quantities (density, momentum x, momentum y, momentum z, and internal energy) with each grid cell. Other scalar and vector fields (including pressure) are derived from this fundamental data. For this four-dimensional solution, a fixed I, J, K index defines a series of scalar and vector samples through time at a fixed point in space. It is these samples that are the target for using auditory display techniques.

1.2 VISUALIZATION TECHNIQUES

Traditionally, these data have been analyzed using scientific visualization techniques found in software packages such as Plot3D[3] and FAST.[2] These techniques include, but are not limited to, interactive three-dimensional viewing, two-dimensional plots, function-mapped cutting planes, surfaces of constant scalar value, volumetric images, particle traces and ribbons through vector fields, and topological icons.[4,5] While these techniques have been adequate in the visualization of steady state three-dimensional data sets, they do not always do so well with four-dimensional data. A function color-mapped surface that changes at the rate of 8000 Hz is of little or no value, for example.

The state of the art in terms of CFD data requires that new data exploration techniques be analyzed. It is in this context that auditory display is analyzed here.

2. MOTIVATIONS AND TECHNIQUES FOR USING AUDITORY DISPLAY

In general auditory display involves the mapping of values in some data space to values in acoustic space. These mappings range from zeroth order to nth order where n is the number of parameters in the audio space that are being controlled by the data. There are several motivations for using auditory display in the analysis of computational fluid dynamics simulations. These include but are not limited to:

- direct simulation,
- support for virtual reality,
- annotation of scenes and animations, and
- validation of other techniques.

These motivations are discussed in turn.

2.1 DIRECT SIMULATION

Of all the motivations for using auditory display, perhaps direct simulation or zeroth-order mapping, that is audification or is the most compelling. Here, pressure samples over time can be used directly to generate the audio waveform. This technique can be thought of as an audio microphone which is placed in the simulation. This "microphone" listens to the fluid flow as it develops over time. Output from this microphone can be compared to recordings from experiments or compared against what we expect to hear, and can yield valuable insight as to the quality and completeness of the flow solver and the modeling of the problem. In other words, if the output from this direct simulation does not sound like the experimental results, then we might ask if the problem is modeled correctly or completely, or perhaps if the flow solver is not generating a fine enough solution.

Direct simulation requires that the sampling rate is twice the highest frequency that one expects to hear (a result of sampling theory). For example, if one wanted to listen to what a jet turbine sounds like including frequencies up to 20 KHz (human hearing is in the range of 20 Hz to 20 KHz), the samples would have to be played back at 40,000 times a second (call this rate S). In order for the flow solver to provide a solution with this resolution, a delta time of $1/S$ would have to be used. Once the CFD solution is calculated, the pressure field should be derived from the fundamental solution variables and should be normalized $(-1.0, 1.0)$ over the entire data set. Now a "microphone(s)" can be placed somewhere in the pressure field in order to gather normalized pressure samples (placing a microphone means selecting I, J, K indices). For binaural output, two microphones should be placed apart from one another in a relationship that corresponds to the physical distance between our ears. Once the microphone is placed in the computational domain, one simply cycles through the normalized pressure field converting each of the computed pressures into a 16-bit signed number where a pressure of 1.0 corresponds to 32,768 (2^{15}) and a pressure of -1.0 corresponds to $-32,768$. The 16-bit values are written into contiguous locations in memory called a "wave table." Once all of the pressures have been written into the wave table, the scientist can listen to the solution by having a 16-bit digital to analog converter play the wave from memory at the frequency S. The DtoA converter will produce a line-level signal which could be amplified by conventional audio amplifiers and loud speakers.

2.2 VIRTUAL REALITY

Virtual reality, by definition, is the immersion of one's senses in the computer. Auditory display is an integral part of this process. The audio cues are essential in providing the feedback necessary to convince the user of the validity of the environment. Imagine the exploration of terrain data from Mars. The scientist in the virtual reality could have his/her interactions with the data be reflected off the surfaces being analyzed. These reverberations could supply valuable information as to the proximity of the surfaces. Additional scientists immersed in the reality could talk to one another through a virtual audio communication device. The integration of audio into virtual reality is growing, especially with the advent of computer hardware to support the computations necessary.

2.3 ANNOTATION OF SCENES/ANIMATIONS

Annotation of scenes and animations is another motivation for considering auditory display. Audio annotation in the context of this paper is defined to include narration and nth-order mappings of data into audio space. The value of a scientist's narration is clear. Encapsulating scientists' thoughts and analyses with a scene or animation serves to complete the visualization. The value of nth-order mappings as annotation is perhaps not as clear. Consider a mapping where a scalar value is used to control the pitch of the auditory display. One might think of this as an audio thermomete that could be placed in the scene/animation. This "thermometer" could be used to monitor the value of temperature. Similarly, key moments in time could be used to trigger some audio event. This is like a Geiger counter. This "Geiger counter" could be placed in an animation where it could monitor the frequency at which certain thresholds are encountered.

The nth-order mapping technique is sometimes referred to as parameter mapping, because the data is mapped into the control of different audio parameters.[1] Audio objects that use this technique will be referred to as audio indicators in this paper. Some of the audio parameters that can be controlled are:

- attack time
- sustain time
- release time
- decay time
- pitch
- amplitude
- timbre

MIDI (Music Instrument Digital Interface) is a hardware and software specification that was developed to control these parameters (and more) on digital music synthesizers. As it turns out, MIDI can be an adequate language, and is frequently used in the field of auditory display, particularly when using this parameter-mapping technique. To implement the audio "thermometer," for example, a MIDI note on command is sent to the synthesizer. The pitch of this note can be modulated in proportion to the data by sending a MIDI pitch bend command. If the data is normalized, we can label the normalized value at time t as NDt, and we can calculate the pitch at time t (PT) by:

$$PT = P0 + NDt \times PMAX$$

where $PMAX$ is the maximum pitch shift desired and $P0$ is the pitch at time 0. Once the MIDI events are generated, they can be sent out to a standard MIDI port or recorded to a standard MIDI file. Output sent to the MIDI port can be sent directly to a synthesizer. Output stored in a MIDI file must be read into MIDI sequencing software, which in turn sends the MIDI data to the synthesizer.

A certain degree of caution should be used in the selection of the sounds (voices) that are chosen on the synthesizer, because they can impart additional information not present in the original data. For example, the choice of a gunshot versus the sound of a harp invokes different perceptions. In one case more gunshots would seem to indicate that something bad was happening, but the same data mapped to the harp might indicate the something desirable was happening.

[1] Editors note: This differs from Scalletti's use of the term nth-order mapping (in this volume).

2.4 VALIDATION OF OTHER TECHNIQUES

There is a common misconception that visual images of data contain some kind of inherent truth. Of course this is not true and, as evidence, it is considered good practice in scientific visualization to contrast visual representations of data by using different visualization techniques and algorithms. For example, comparison of isosurfaces generated from a marching cubes algorithm with those generated from a tetrahedra-based algorithm show two different surfaces for the same data. It is in this context that auditory display should be considered as yet another technique that can be employed in the analysis if for no other reason than to validate (or call into question) the representations generated visually.

3. CASE STUDY (PENN STATE ARTIFICIAL HEART PUMP)

The parameter mapping (audio indicator) method was recently used in the study of the Penn State artificial heart pump at NASA Ames. The modeling of the pump[6] includes the pump chamber, inlet and outlet valves that open and close, and a pusher plate that moves up and down inside the pump chamber (Figure 1). From the solution generated, locations of blood cells, as well as pressure and vorticity at discrete points in the chamber as the pump goes through its cycle were derived. Critical points in the cycle include the moment that a valve opens and closes, and the moment that a blood cell encounters a level of vorticity that is considered unsafe.

The auditory display that was generated for this study included narration and three audio indicators. The first indicator was a continuous tone (thermometerlike) at a frequency of 440 Hz. This pitch was modulated by the normalized pressure on the pusher plate. The new pitch for each time step was calculated using the relationship given in Section 2.3. The second indicator was a tap on a wood block (Geiger counter-like) and was used to signal the moment when a blood cell encountered the threshold vorticity. The third indicator was the sound of a bass drum and it was used to identify the instant that a valve opened or closed.

A program was written on the Silicon Graphics workstation that read in the total pressure and vorticity at each time step, and output a corresponding standard MIDI file that could be read on a Macintosh. The program selected the indicators on a Korg M1 synthesizer by using the MIDI "program change" opcode. It then turned on Indicator 1 using the MIDI "note on" opcode. The program modulated the pitch of Indicator 1 by sending the MIDI "pitch bend" opcode with a pitch value that was

Auditory Display of Computational Fluid Dynamics Data

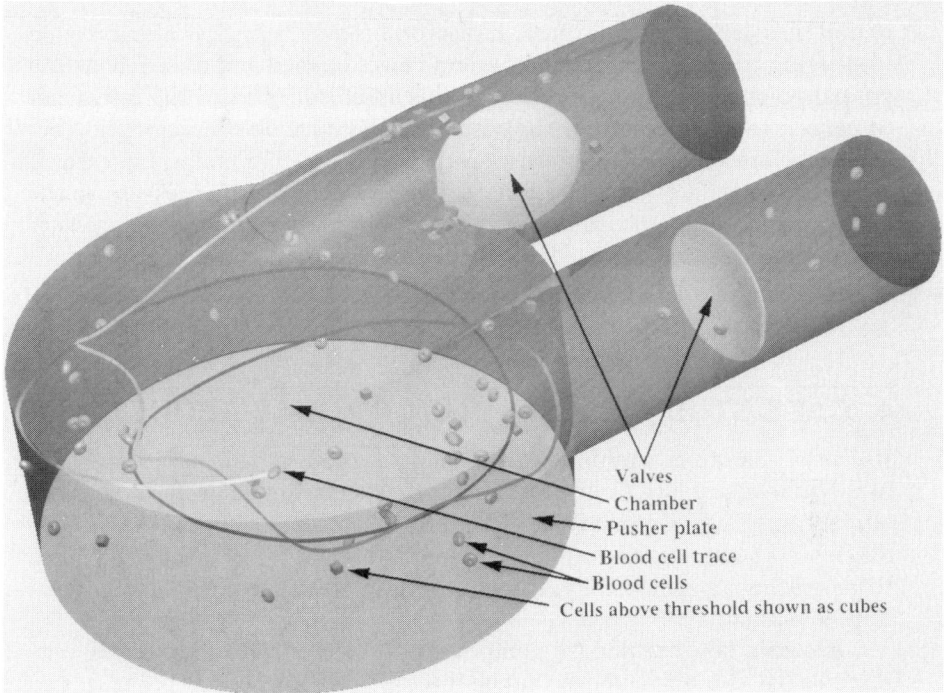

FIGURE 1   Penn State artificial heart pump.

calculated as described above. Indicator 2 was triggered using the MIDI "note on" opcode at every time step at which a particle encountered the threshold vorticity. Indicator 3 was triggered every time the inlet and outlet valves began to open or close. Once the standard MIDI file was written, it was transferred to the Macintosh and read into Vision MIDI[8] sequencing software. The sequence was recorded to videotape, and then edited onto a video which contained traditional visualization techniques that had been produced using the FAST software.

Though no formal study was done on the effectiveness of the auditory display, it was generally agreed by those involved in the project that this technique enhanced the understanding of the data in addition to the traditional visualization techniques that were used. In particular, Indicator 2 ("Geiger counter-like") seemed more effective at determining the frequency of blood cells crossing the threshold vorticity than was the visual display. The visual display changed the color of the blood cells from red to green as the threshold vorticity was encountered. It seemed harder to correlate the frequency of this change in color with the position of the pusher plate than it was using the audio cues. Indicator 3 (bass drum) was also useful in identifying the moment when the pump valves opened and closed. This is a critical point in the pump's cycle and this audio cue enabled the researcher to concentrate on events in the visual display and determine if they occurred before or after the valves opened and closed. Finally, Indicator 1 ("thermometer-like") was effective at indicating the overall pressure on the pusher plate, while a visual representation of the instantaneous pressure was simultaneously displayed on the plate itself.

4. CASE STUDY (ROTOR-STATOR TONAL ACOUSTICS)

Recently, the direct simulation technique discussed in this paper was used in the analysis of data from a tonal acoustic simulation done on rotor-stator interaction inside a jet turbine. Unlike the parameter mapping technique discussed above, this technique requires that some consideration be given to generating an acoustical solution (see Kramer's discussion of audification in this volume).

An axial flow turbine (or compressor) produces rotating pressure patterns called spinning modes that may propagate in a spiral path. In two dimensions, these spinning modes propagate at a nonzero angle to the axial direction. For any particular harmonic of blade-passing frequency, the interaction field can produce an infinite number of spinning modes. Each of these modes rotates at a different speed. Some of these modes propagate whereas others decay. The modes that are possible for some multiple of blade-passing frequency depends upon the number of stator and rotor airfoils. This study focuses on an examination of the modes present for a 3-stator/4-rotor case using a Navier-Stokes solution procedure.

In order to obtain pressure samples that vary over time, a flow solver that can produce time-dependent (unsteady) solutions must be used. Such unsteady two- and three-dimensional rotor-stator interaction codes have been developed. Rai[9] presented a two-dimensional calculation of rotor-stator interaction for an axial turbine. More recently, Rai[10] and Madavan et al. computed the fully three-dimensional flow fields for the same case. The hub, outer casing, and the rotor tip clearance were all included in the calculation. Rai[9,10] and Madavan et al.[7] solved the thin-layer Navier-Stokes equations in a time-accurate manner using an implicit, upwind-biased, third-order accurate method to compute the flow field. The ability of their codes to predict near field flow quantities such as the time-averaged pressure distributions on airfoil surfaces and the pressure amplitudes and phase on the surface of the airfoils was demonstrated. In addition, the two-dimensional codes were used to accurately predict the total pressure defects in wakes. More recently, these computer programs have also been used in the design process of turbomachines.[12] The present work is focused on investigating the ability of these codes to predict "correctly" the tonal acoustics in the flow field due to rotor-stator interactions.

The numerical study of tonal acoustics involves the study of the flow field in the upstream and downstream regions of the interacting rotor-stator airfoils. The unsteady, thin-layer, Navier-Stokes equations are solved using an upwind-biased finite-difference algorithm. The method is third-order accurate in space and second-order accurate in time. At each time step, several Newton iterations are performed, so that the fully implicit finite-difference equations are solved. Additional details regarding the scheme can be found in a paper by Rai and Madavan.[7]

Once an acoustic solution has been generated with the techniques described above, the data can be used to directly simulate the audio waveform. For this study, numerical pressure data were collected upstream and downstream of the rotor-stator pair. At every point in the grid, the pressure signal was sampled in time and stored (Figure 2). The flow is nearly periodic in time with a period corresponding to a blade-passing period. In this case, the numerical data contained 500 samples per blade-passing period. The time interval was a constant between samples. Data corresponding to a total of four blade passages (or 2000 samples in time) were collected. Ideally we would like to have data for as many blade passages as possible; however, four blade passages were chosen due to memory constraints. Since the pressure history was stored at every grid point (there were over 70,000 grid points), this resulted in pressure data of over 500 megabytes. Using four blade passages balanced the memory requirements with resolving any subharmonics that could be in the flow due to vortex shedding at the trailing edge of the rotor-stator airfoils. The samples were made audible by the technique described in Section 2.1 and upon playback were looped through continuously in order to provide a continuous tone. The looping technique was justified due to the periodic nature of the problem. Besides sampling in time, discretization in space is also important. The effect of grid coarseness on the numerical dissipation of propagating waves was studied by Rangwalla and Rai.[13] The higher harmonics have smaller wavelengths and, hence, sufficient grid discretization is required in order to accurately resolve these. It was found that at least ten grid points were required per wavelength in order to maintain the wave. This resulted in a faster numerical dissipation of the higher harmonics due to grid stretching as compared to the dissipation suffered by the lower harmonics.

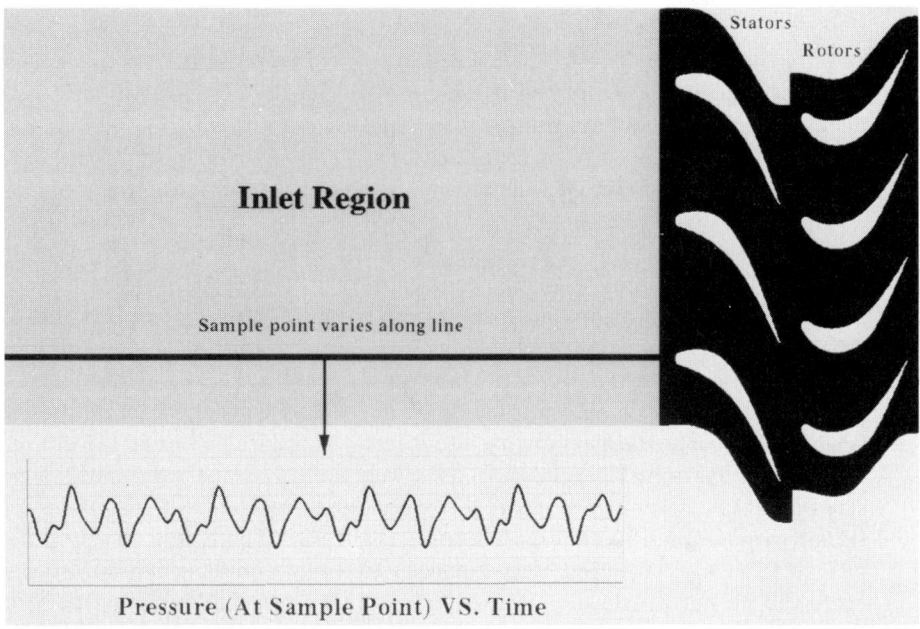

FIGURE 2 Rotor/Stator tonal acoustics.

Tonal acoustics for rotor-stator interactions have been analyzed before by Rangwalla and Rai.[11] The analysis consisted of using finite Fourier transforms in space and time in order to isolate the different modes. The propagation characteristics of these modes (such as decay rates and wavelength) were compared with that predicted by linear theory under the assumption that the underlying mean flow is uniform. The disadvantages of this approach are the following:

1. The assumptions made for the linear theory may not be valid.
2. The comparison between numerical results and theoretical predictions is made by x–y plots, where the amount of data that can be shown is limited. As mentioned before, rotor-stator interactions generate a plurality of spinning modes. An x–y plot can show the behavior of a few modes. The complete tonal acoustic field at a point in space needs to be represented by many x–y plots.

The auditory display technique used in this study has the potential for describing the nature of the acoustic field more completely than the x–y plots because it represents the actual signal and not some filtered portion of the signal. Also, it has the possibility of displaying certain intangibles such

Auditory Display of Computational Fluid Dynamics Data

as timbre. This technique verified some of the results that were already predicted by visual and numerical analysis of tonal acoustics.

For example, we know for this particular numerical case, the fundamental and first harmonics decay as we move farther away from the rotor-stator whereas the higher harmonics propagate (as long as they have not suffered from numerical dissipation due to grid coarseness). Additionally, we know that the amplitude of the sound increases as we near the rotor-stator and that there is additional harmonic content in the rotor wakes due to vortex shedding. We were able to verify all of these features by playing back samples from different locations in the rotor-stator geometry. The sound heard can be described as fundamental in nature, and not unlike the sound produced by an oboe. At first this seems to be in contrast with what we might expect, but when we consider that this is just one component of a jet turbine, with no combustion modeling, we begin to understand that this is the sound that we should expect to hear. The sound we heard is one component of the overall sound produced from a jet engine. Conversely, it seems as we add more components to the modeling of this problem, the nature of the sound will become more complex (less oboe-like and more jetlike) and this technique will become more informative and perhaps be necessary to understanding the data.

This technique has the potential of acquiring an overall understanding of the data quickly. When data of this size is analyzed, a general understanding of the data can indicate which specific questions to ask. This technique gave us the overall understanding quickly, perhaps quicker than examining several x–y plots would have.

5. SUMMARY

A variety of motivations and techniques for the use of auditory display were discussed. It seems clear that these techniques at the very least can serve to enhance traditional techniques of CFD visualization. There is also the possibility that auditory display could serve to validate CFD results or CFD visualizations.

In this work, auditory display of CFD data was used in two different ways for two different simulations. The first was for the simulations of the Penn State artificial heart pump where the animations were annotated by the auditory display. Annotation seems to be specially useful for those simulations consisting of many (discrete in time) simultaneous or nearly simultaneous events in time, where the scientist wishes to study the possibility of any correlation between the many events. (For the Penn State heart, some of the different events were the blood cells periodically crossing a threshold vorticity, the opening and closing of the pump valves and the overall pressure changes in time.) To rely only on visual animations may prove difficult since many distracting events may be occurring in the animation. Annotation can help the scientist filter out the unimportant events and zero in on the important ones.

The audio annotations of the heart simulations were implemented by writing a portion of the MIDI language on a Silicon Graphics Unix workstation. The library was used to generate a standard MIDI file which was transferred to a Macintosh to be played by conventional sequencing software. For completeness other components of the MIDI langauge should be implemented, allowing for other acoustic parameters besides pitch, amplitude, and duration to be controlled. Standard MIDI files were used because at the time, a MIDI interface was not available for the workstation. At some point, the library should be modified to output directly to a MIDI port, eliminating the need to transfer files.

The second way auditory display was used in this work was by directly playing back the pressure signals in time, that were obtained from the CFD solutions of rotor/stator interactions in a turbine. This method seems to be suitable for those data sets, where the object of study is continuously and smoothly varying in time, and where the time varying signal may contain several harmonics with different spatial behavior such as decaying characteristics. Relying only on a visual display may prove inadequate because of the complexity of the data and a visual plot may at best display the characteristics of just a filtered portion of the data. However, there are a few issues still to be resolved for this kind of display. Auditory display of a direct simulation may require large amounts of memory since a signal in time has to be sotred at every gird point in space. To overcome this, we have to cut short the time signal and loop it back during play back, store the signal for only a few preselected grid points and/or store just a sparse signal in time. Each of these options impacts the auditory display and will have to be properly assessed. For the rotor/stator case, we cut short the time period to the time it takes for the rotor to rotate through four blade passages and also we only stored the pressure data on a subset of the flow domain. Cutting short the signal and looping back does limit the subharmonic resolution and the quality of the sound produced. The other limitation of the case is the fact that the CFD simulation only models a few components of the whole machine and hence the actual auditory display produces the sound only due to the rotor/stator interactions and hence can sound artificial. In this study auditory display was just used to verify what we already knew about the tonal acoustic field for this particular case. It is hoped that with more exposure to this kind of display, we would be able to use it to enhance our understanding rather than just verify it.

Although untested, it seems that the binaural sampling described in 2.1 would further increase the effectiveness of these perceptions. With binaural sampling, our ears would integrate the waves from two adjacent sampling points into one comprehensive understanding of the entire volume (not easily done by comparing two x–y plots). With this technique, events of interest could be spatially located quickly.

REFERENCES

1. Bancroft, G. "Scientific Visualization in Computational Aerodynamics at NASA Ames Research Center." *IEEE Computer* (August 1989).
2. Bancroft, G., F. Merrih, T. Plessel, P. Kelaita, R. McCabe, and A. Globus. "FAST: A Multi-Processed Environment for Visualization of Computational Fluid Dynamics." Paper presented at Technology 2001 Conference, December, 1992.
3. Buning, P., et al. "Flow Visualization of CFD Using Graphics Workstations." Paper No. 87-1180, AIAA, 1987. In *Proc. 8th Computational Fluid Dynamics Conference*, held June 9–11, 1987.
4. Globus, A., et al. "A Tool for Visualizing the Topology of Three-Dimensional Vector Fields." Paper presented at Visualization '91 Conference, October, 1992.
5. Helman, J. and L. Hesselink. "Representation and Display of Vector Field Topology in Fluid Flow Data Sets." *IEEE Computer* (August 1989): 27–36.
6. Kiris, C., S. E. Rogers, D. Kwak, and I. Chang. "Computation of Incompressible Viscous Flows Through Artificial Heart Devices with Moving Boundaries." *Contemp. Math.* **141** (1993).
7. Madavan, N. K., M. M. Rai, and S. Gavali. "Grid Refinement Studies of Turbine Rotor-Stator Interaction." Paper No. 89-0325, AIAA, 1989.
8. MIDI 1.0 Detailed Specification, 1988. Available from the International MIDI Association 5316 W. 57th St., Los Angeles, CA 90056 USA.
9. Rai, M. M. "Navier-Stokes Simulations of Rotor-Stator Interaction Using Patched and Overlaid Grids." *AIAA J. Prop. & Power* **3** (1987): 387–396.
10. Rai, M. M. "Three-Dimensional Navier-Stokes Simulations of Turbine Rotor-Stator Interaction; Part 1 and 2." *AIAA J. Prop. & Power* **5** (1989): 305–319.

11. Rangwalla, A. A., and M. M. Rai. "A Kinematical/Numerical Analysis of Rotor-Stator Interaction Noise." Paper No. 90-0281, AIAA, 1990.
12. Rangwalla, A. A., N. K. Madavan, and P. D. Johnson. "Application of an Unsteady Navier-Stokes Solver to Transonic Turbine Design." *AIAA J. Prop. & Power* **8** (1992): 1079–1086.
13. Rangwalla, A. A., and M. M. Rai. "A Numerical Analysis of Tonal Acoustics in Rotor/Stator Interactions." *J. Fluids & Struc.* (1993): to appear.

Gottfried Mayer-Kress,[†] **Robin Bargar,**[‡] **and Insook Choi**[*]
[†]Center for Complex Systems Research, Department of Physics, 3025 Beckman Institute, 405 N. Mathews, Urbana, IL 61801; gmk@pegasos.ccsr.uiuc.edu (NeXT-Mail).
[‡]National Center for Supercomputing Applications, and School of Music, University of Illinois at Urbana-Champaign.
[*]Computer Music Project and Experimental Music Studios, Composition Division, School of Music, University of Illinois at Urbana-Champaign.

Musical Structures in Data from Chaotic Attractors

One of the most prominent aspect of data from natural phenomena is that of irregularity and complexity. Many universal aspects of such phenomena can be described in the context of chaotic dynamics, and chaotic attractors can serve as model generators of such data. Auditory representations used in conjunction with chaotic attractors can be designed to reveal the unique properties of nonlinear dynamical systems representing complex phenomena. The design of such an auditory representation can benefit from being informed by observations common to both chaotic and musical structure. Recurrence structures in chaotic systems, including intermittency and self-similarity, are compatible characteristics for drawing analogies to musical structures. In this paper we explore several designs for auditory representation of chaotic systems. These include both low-level methods (where the sequence of system states is mapped directly onto auditory parameters); and higher-level methods which map derived statistical quantities, such as the approximations of the probability distribution (measure) of an attractor into polyphonic auditory constructions. We focus on a few simple dynamical systems where we have a clear understanding of the structure of chaotic attractors, so that we can draw analogies between their representation using sound and the complex non-linguistic structures found in music. Using these analogies we can develop a new generation of auditory representation tools.

1. INTRODUCTION

1.1 AUDITORY REPRESENTATION AND MUSIC

Auditory representation of numerical data and music are often described as mutually exclusive.[20] The complexity associated with music is not considered useful for the purposes of scientific observation. The acceptance of auditory representation by the scientific community is said to depend upon the dissociation of data-driven synthesized sound from music. Low-level auditory representations lack sufficient frames of reference for clarifying complex events, creating the impression that extensive ear training is necessary to glean useful information from sonification. The scientific community thus far has largely rejected traditional sonification on these grounds.

We know from scientific visualization that the recognition of emergent patterns can be of heuristic value; patterns are recognized by observers with widely varying visual skills, and can stimulate further exploration which can then lead to scientific discovery that otherwise would have gone unnoticed. Properly designed color maps and display environments alleviate the scientist's need to develop powers of artistic interpretation. We suggest that higher-level auditory representations can provide a similar orientation. Data sonification produces abstract sound patterns, and music is created from complex nonlinguistic sound structures. It is relevant to investigate cases where the organizational principles of music might help reveal patterns imbedded in scientific data. This inquiry is enhanced in the study of chaotic attractors by the presence of chaotic structures that resemble musical structures.

1.2 WHAT MAKES AN ATTRACTOR CHAOTIC?

There has been some recent interest in exploring the connection between chaos and music.[15,12] Before we discuss these connections, let us first give an informal introduction into definitions of mathematical structures that we claim can be found both in chaotic systems and in music. When we refer to attractors, we want to describe two properties of the solution set of a dynamical system (see, e.g., Ruelle[19] for an introduction):

i. Boundedness: The system does not wander off but stays in a bounded domain. Counterexample: a random-walk, that is, at any point the system can progress in any arbitrary direction; eventually it will drift outside any bounded domain. Other examples are singularities, where systems blow up, usually through positive feedback.

ii. Nearby solutions will eventually approach the attracting set. If all starting values of a system will lead to solutions that approach the attracting set, then the attractor will be globally attracting. Note, however, that the domain of attractivity (basin of attraction) can be either very small or have regions where it becomes very narrow, i.e., small external perturbations will be able to kick the system out of the basin to a different attractor with potentially qualitatively different properties.

Thus the attractor gives the set of all possible solutions global constraints which determine their overall dynamical and structural properties. Not only is the probability distribution of the system determined, but also temporal correlations: which sequences of points on a trajectory appear with what probability.

The standard examples of attractors are:

i. point attractors, where the system is asymptotically static;

ii. periodic limit cycles, where the system asymptotically undergoes elf-sustained oscillations[1];

iii. quasi-periodic tori, basically superpositions of incommensurate periodic solutions; and

iv. chaotic attractors.

These attractors frequently have a fractal structure. Their most important property is that solutions on the attractor show sensitivity to initial conditions: If we start off two solutions very close together, within a very short time they will develop behaviors completely independent from each other. It is important to note that this time is "exponentially" short, but it is not zero. That means that we will be able to anticipate the behavior of the system for a finite time. A second important property is that we can fine-tune the rate at which these trajectories diverge. This rate is related to the loss of information about the state of the system at a future time. Thus chaotic systems cover the full range from (i) indiscernible from ordered systems all the way to (ii) indiscernible from completely stochastic, random systems.

The main parameters that quantify the predictability or the information production rate of a chaotic system are its fractal dimension and an averaged divergence rate called Lyapunov exponents. As a consequence of

[1] These can give rise to complicated "beating phenomena."

chaotic behavior we also observe a decay of the autocorrelation function. Here we can again observe everything from minimal decay to an oscillating function that looks completely regular to the other extreme of a decay to zero after only one step. We want to stress the fact that this intrinsic instability is a property of the individual solutions within the chaotic attractors. The attractors themselves can be highly ordered and very robust and stable.

2. LOW-LEVEL REPRESENTATIONS
2.1 CONTINUOUS FUNCTIONS

THE WEIERSTRASS FUNCTION. Fractal functions have been used in computer music to some extent. The Weierstrass function is a standard example of a fractal function[13] and it is defined as a sum over sine functions with arbitrary high frequencies, corresponding to arbitrary small scales:

$$W(t) = \sum_{n=0}^{\infty} t^{Hn} A_n \sin(\frac{2\pi t}{t^n} + \phi_n) \qquad (1)$$

where H, A_n, ϕ_n are parameters that determine the exact shape of the function. Over a wide range of parameters, the function has irregular structures on many scales, a property that is also present in music. In Figure 1, we show the graph of the Weierstrass function for the parameters: $r = 0.618$; $H = 0.8$; $A_n = \begin{cases} 1 & \text{for } 0 \leq n \leq 9 \\ 0 & \text{else} \end{cases}$; and $\phi_n = \begin{cases} .5 & \text{for } 0 \leq n \leq 9 \\ 0 & \text{else} \end{cases}$.

In this and other applications[2] the Weierstrass function is used as a continuous function of the time parameter t. In that respect it generates a continuous stream of numbers, similar to the case of solutions of ordinary differential equations (ODEs; see below). Chaotic values from systems of continuous change, such as the Weierstrass function and ODEs can be sonified in two ways. The function can be directly converted into an audible waveform; conversely, the function output can be mapped to a synthesis output parameter such as pitch. In both cases the conversion to sound requires discretization of the continuous function. The first method obtains discretization by mapping the output of the original function to the digital audio sampling rate. The second method maps a minimum change in function output to a minimum change in a synthesized sound parameter, such as pitch; this method allows for flexible changes in discretization.

[2] The Weierstrass function is used, for example, in the fractal melody generator *Ensemble* by Michael McNabb.

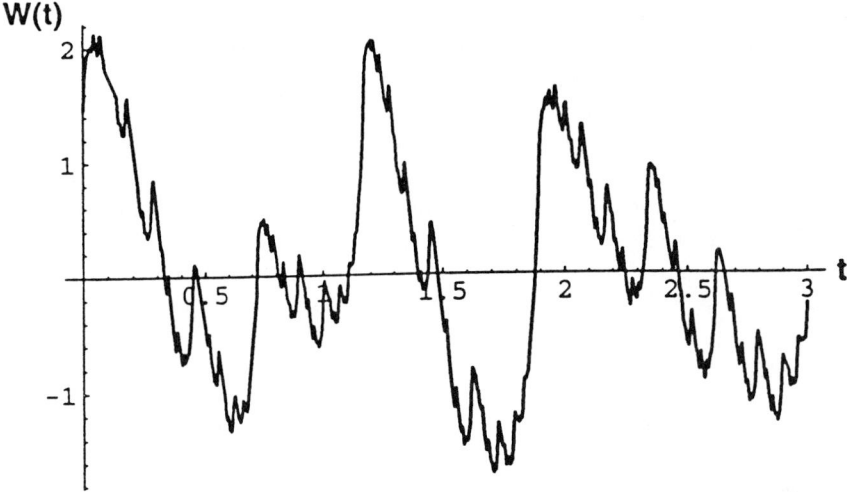

FIGURE 1 Fractal Weierstrass function $H(t)$.

For the Weierstrass function the choice of step size for pitch quantization is analogous to choosing a distance from which to view a fractal curve. The smallest step size is that at which pitch change sounds like a continuous glissando; continuous change fixes us at a minimum distance from the fractal curve, beneath which there is no resolution of detail. Larger increments will produce discrete pitch changes, as if the distance from the fractal curve has increased, providing lower-resolution approximations of continuous change. In order to perceive a discrete change in sinusoidal tones the increment of change for the average listener needs to be greater than 3% for frequencies around 100 Hz and 0.5% around 2000 Hz; this minimum increment is referred to as the *Just Noticeable Difference (JND)*.[18] The choice of increment significantly affects the perceived pattern of the sequence.

SOUND SYNTHESIS. In the above method the limit of minimum resolution is imposed by the parameter pitch. Pitch is a perceived phenomena, an output of sound synthesis algorithms, not an input parameter. Pitch results from synthesis input parameters such as frequency and pattern of oscillation; pitch perception depends upon multiple iterations of a pattern stored as a numerical table called a waveform. In most synthesizers the waveform acts as a display device; the wave-table values produce a predetermined timbre, and are not related to the data being represented; the waveform is not altered during the iteration process. The minimum temporal resolution for data display is the rate at which the wave table is iterated. Scientific data is used to modulate the iteration rate of the fixed wave table, changing the pitch of the fixed timbre. A less common method for representing continuous change is to establish a discretization that is beneath the perceptual threshold of resolution of detail. Software synthesis techniques allow attractor values to be stored as a continuous wave table, and auditioned at a specified sampling rate, which often ranges from 20,000 to 48,000 samples per second. This technique discards periodic repetition of an invariant wave table and the perception of steady-state pitch which periodic wave tables produce. Displaying numbers at the sample rate radically alters the display scale; individual values cannot be discerned; differences in patterns are heard as a combination of timbre, pitch, and loudness changes.

Sounds described as oscillating air pressure may, like dynamical systems, show periodic, quasi-periodic, or aperiodic, noisy behavior. Traditionally these states have been represented and synthesized by direct Fourier transforms modulated by control signals such as periodic waveforms and noise generators. Fourier synthesis does not permit descriptions of transient events yet transience is centrifugal to hearing and to natural sound and characterizes articulation and liveness; it is necessary for complexity, which Risset[16] identifies as "the fundamental challenge of sound synthesis." Both Risset and Chowning use piecewise linear transformation of control signals in frequency modulation to generate complex dynamical timbres from static sources.[3,16]

Patterns from chaotic systems differ significantly from the Fourier series patterns commonly used for sound synthesis. The Fourier transform generates harmonic spectra by compositing sine waves, thus the changes of value from one sample to the next are regular, and the resulting sound is pitched and steady-state. In data from chaotic attractors, abrupt changes between values tend to create "buzzy" or "noisy" timbres when converted into a waveform. This noise is coupled with a pitch component from the periodic aspects of the pattern. While the majority of examples in this paper utilize wave-table synthesis, the application of chaotic patterns to sound synthesis can benefit composers who desire nonlinear specifications for sound (see Truax,[21] Chafe,[2] and Rodet[17]), and scientists seeking complex representations of complex dynamics.

Tr 63,64

THE CHUA OSCILLATOR. The Chua oscillator is an ordinary differential equation (ODE)[3] which possesses chaotic attractors for certain ranges of

[3] They are given by

$$\dot{x}_1 = \alpha^e (x_2 - h(x_1, p^e)) \quad (2)$$
$$\dot{x}_2 = x_1 - x_2 + x_3 \quad (3)$$
$$\dot{x}_3 = -16x_2 \quad (4)$$

parameters (see Chua[4] and Peter[11]). One way this ODE system may produce sounds is direct waveform synthesis: successive values are stored in a table which is then converted into sound.[4] While gathering samples a Chua parameter value may be systematically shifted; the resulting sounds will reflect the relative periodicity of the attractor for those parameters. At 20,000 or more samples per second, each value receives a brief audition. Examples (A) reproduce Chua attractors for several parameter values. Careful attention will reveal that the "noisy" portions of the function offer periodicity as well as irregularity; the sound contains noise plus pitched elements that sound filtered. This reflects the periodic skeleton which probabilistically influences chaotic attractors (see Section 2.2). Rodet observed that transient onset velocity and richness of quasi-periodic states may be controlled in the Chua circuit.[17] The Chua circuit produces over 30 strange attractors uniquely determined by 7 parameters.[5,6] Currently, our group work in progress concerns the wide range of attractors produced by the Chua circuit and their potential for generating differentiated timbres in sound synthesis.

2.2 DISCRETE FUNCTIONS

ITERATED MAPS: THE LOGISTIC MAP. Different algorithms have been used **Tr 65** to produce sequences of sound events. Random number generators have been used as well as more interesting maps.[5] One of the simplest dynamical systems that generates interesting, nontrivial behavior and chaotic attractors is the logistic map:

$$x_{n+1} = f_r(x_n) = rx_n(1 - x_n). \tag{5}$$

The sequence $(x_0, x_1, x_2, \ldots)$ is contained in the unit interval $0 \leq x_n \leq 1$ if the parameter r is chosen in $0 \leq r \leq 4$. In Figure 2 we see the graph of $f_r(x_n)$ for $r = 3.9$. We also indicate graphically how a point x_0 is mapped onto x_1, reinserted as a new argument of the map (on the horizontal axis) mapped onto x_2, etc. Figure 3 illustrates how small differences in x-values

---

where $\alpha^e = 9.5$. $h = 2/7x_1 + 0.5(m_0^e - 2/7)(|x_1 + 1| - |x_1 - 1|)$ is a piecewise linear function, where $m_0 = -1/7$.[4] For this parameter setting the dynamics of the system is of spiral-type chaos.

[4] The technique of making samples audible by direct playback has been referred to as 0th-order sonification (see Scaletti, in this volume), or audification (see Kramer, this volume); however, criteria for distinguishing higher orders are unclear and we want to be cautious about ranking representations based upon scientific metaphor rather than acoustically differentiated features.

[5] See, e.g., C. Chafe's "ZorroDemo" for the NeXT which is available via anonymous ftp from ccrma-ftp.stanford.edu in /pub.

can be amplified by a chaotic map: We start with a small interval I_1, map it onto I_2, and observe that its length has more than doubled. I_2 is mapped onto I_3 with a further spreading. This behavior can be made audible by assigning, for example different voices to different initial points. In this type of sonification, intervals between pitches are presented simultaneously, representing the distance between two points, as well as sequentially representing the paths of both points. Together the auditory streams depict divergence and convergence. Note that even in the case of chaotic systems exponential stretching only takes place in the average. There are regions in the map (close to the critical point $x_c = 0.5$, the maximum of the map with zero slope) where nearby points get even closer. This behavior is responsible for the generation of periodic windows in a chaotic parameter range as explained in Section 2.2.

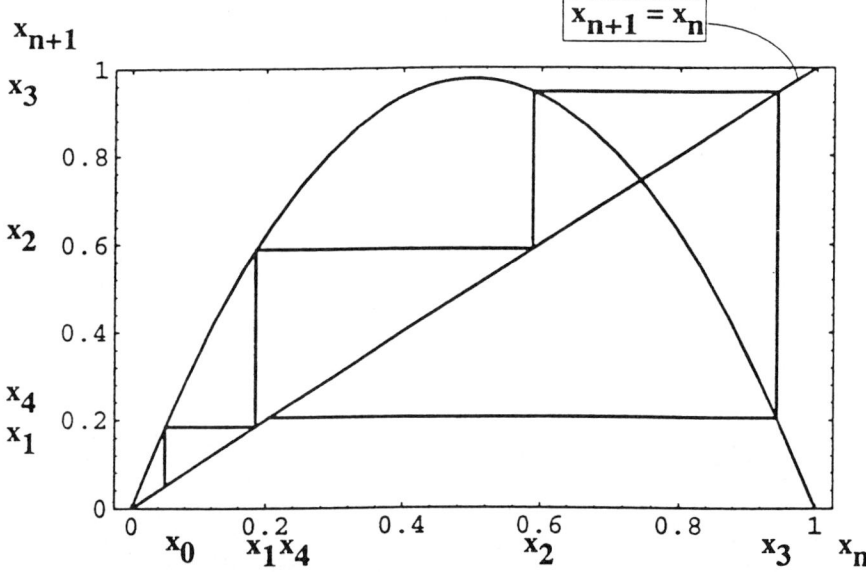

FIGURE 2 Logistic map and iterates of point x_0.

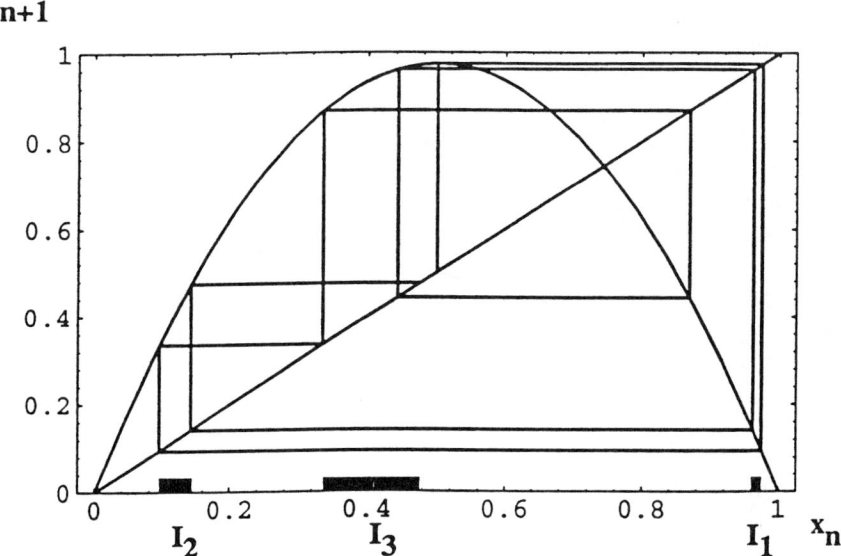

FIGURE 3 Sensitive dependence to initial conditions: Points contained in a small interval I_1 spread exponentially fast.

PERIODICITY AND RECURRENCE. Iterated maps generate discrete values, suggesting an intuitive association to pitch-oriented note sequences. Intermittent recurrence of quasi-periodic patterns is the best way to experience emerging dynamical behavior. Auditory displays are dynamical, representing sequential structures; sound may complement graphical displays, which can only provide statistical representations of values accumulated during time intervals. Accumulated numbers or points on a graph do not represent sequential order, making it difficult to determine the relevant aspects of their emerging order. Sound can preserve temporal information which is lost in graphical attractor reconstructions or statistical representations.

We have chosen the simple example of the logistic map because of its universal properties.[8] Its chaotic dynamics can be organized around the distribution of (infinitely many unstable) periodic orbits. For any chaotic attractor of the logistic map (indexed by its r-value), an arbitrary small parameter perturbation will lead to stabilization of one of the infinitely many periodic attractors. If the period is very high, we will not be able to discriminate that period from a chaotic attractor. For small periods, however, the statistical properties of the corresponding attractor will be very strongly influenced by that nearby low attractor of small periods. In this case the period may be experienced in a sequential display, audibly perceived as a repeating pattern or as an intermittent recurrence of a recognizable pattern.

Tr 5,6 The sound examples (E) reveal the presence of periodicity as an organizing principle. Intermittency is time-dependent and cannot be experienced statistically, without dynamical display. Changes in r-value are reflected by changes in the degree of periodicity that can be heard in intermittent patterns. Hierarchy may be introduced by combining low-level representations.

THE BIFURCATION DIAGRAM OF THE LOGISTIC MAP. When we compute the iterates of a map from a starting value x_0 we will observe in the case of the logistic map that the iterates x_n eventually are approaching a subset A_r of the unit interval. We will call the bounded set A_r the *attractor* of the logistic map for the parameter value r. The attractors of the logistic map can be either regular (fixed or periodic) or chaotic: For small values of r, almost all initial points will eventually approach a fixed point or a periodic point set. Increasing the value of r we observe a *period-doubling sequence*; i.e., the system cycles through 2,4,8,16, etc. points, and the spreading between the points becoming smaller and smaller. In a simple mapping of x_n onto musical pitch, we would only recognize up-down dynamics with fine modulations in the actual pitch. This periodic behavior becomes chaotic when r is larger than $r_c = 3.56\ldots$. The type of chaos that we observe in this Feigenbaum scenario is called *chaotic bands*, or *band attractors* since the system still follows the basic up-down alternating pattern, but the modulation is now chaotic and irregular. The fine modulations during these cycles may be emphasized by expanding the sonification to include simultaneous pitches, which represent former as well as current points.

ORGANIZATION OF CHAOS IN THE LOGISTIC BIFURCATION DIAGRAM. We observe an intricate structure in the organization of the periodic windows and chaotic attractors visible in the bifurcation diagram. There is a direct way of understanding this organization through graphical means. This structure can prove important for a systematic exploration of the chaotic patterns that can be used as a preprocessing for a subsequent structuring according to musical principles. The basic observation is that the orbit through the critical point x_c is most dominant: because of the vanishing slope of the function, there exists strong attraction of nearby orbits to that of the critical point. One can view this phenomenon as a competition between the order-generating property of the critical point and the chaos-generating properties of the regions of the graph with a slope of large magnitude: If the critical point "wins," we observe a periodic

Musical Structures in Data from Chaotic Attractors

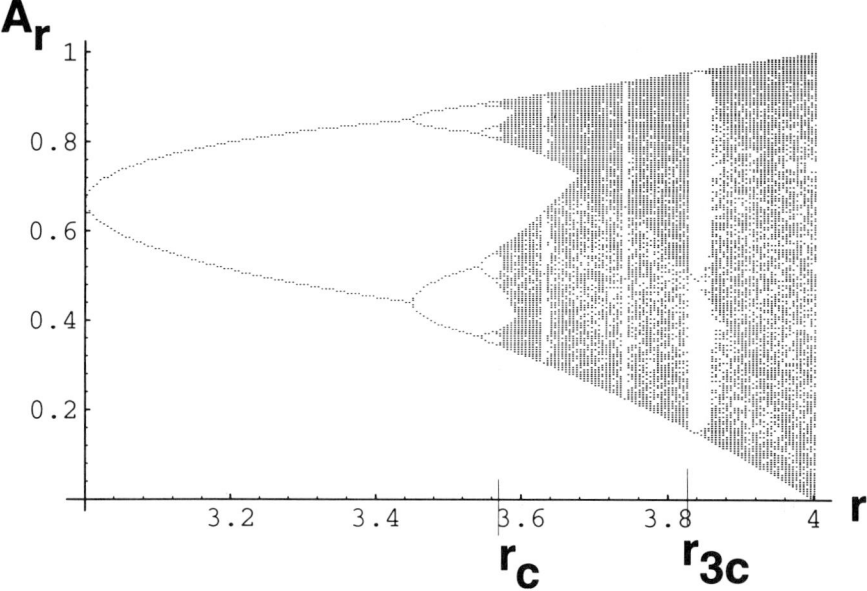

FIGURE 4 Bifurcation diagram of the logistic map. Note the critical parameters r_c for the onset of Feigenbaum chaos and r_{3c} for the onset of period 3 intermittency.

orbit; if the chaotic regions dominate, we observe chaotic attractors. Now if enough iterates of the critical points come close to the critical point, then its influence is large and we can expect ordered behavior. This is especially the case if the critical point is periodic itself, then it is maximally stable; we call those orbits *superstable periodic orbits*. We know that all periodic windows are organized around these superstable periodic orbits. This leads to a simple criterion for finding parameters for periodic orbits of arbitrary period $p > 1$: Find $r \in [0, 4]$ such that:

$$x_{n+p} = f_r^p(x_n) = x_n \qquad (6)$$

where: $f_r^p(x_n) = f_r(f_r^{p-1}(x_n))$ and $f_r^0(x_n) = x_n$ denotes the pth iterate of x_n under f_r. In Figure 5 we see the numerically generated histogram of the probability density function of the logistic map for (a) $r = 4$ where the system is homeomorphic to a Bernoulli-shift, the stochastic process that models a coin flip, and (b) for $r = 3.9$, where the system is still fully chaotic (1-band attractor) but the probability distribution is much more complex. We can see how, for $r = 3.9$, the two singularities that domin-

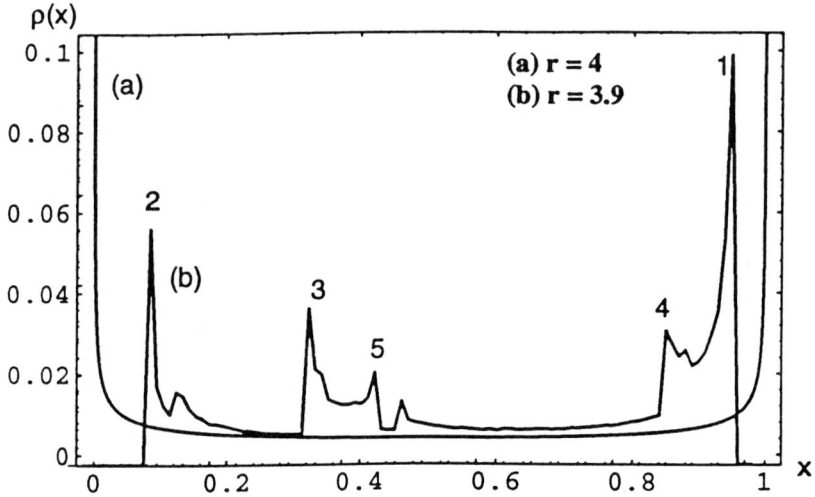

FIGURE 5 Histogram of the probability density function of the logistic lap for (a) $r = 4$ and (b) $r = 3.9$.

ated all patterns in the $r = 4$ case unfold into a sequence of singularities, the strength of which is represented by the height of the corresponding peak. The musical interpretation would be a structure in the occurrence of sequences of variable length: with the highest probability we have the occurrence of the peak at location "1." With a smaller probability this will be followed by a value at location "2," etc. That means with a decreasing probability we will observe recurrent patterns of finite length followed by chaotic variations.[6] We can identify the strongest peaks for each parameter by high density lines in the bifurcation diagram (Figure 4). They can be explained by examining the critical orbit, i.e., the orbit with initial condition $x(0) = 0.5$ for which $f'(x0) = 0$.

From Figure 6 we can predict at which parameters which sequences will occur with what relative frequency: We know that for the case (say, at parameter $r_{n,c}$) that some iterate of the critical point returns (i.e., for some $n > 1$, $f^n(x_c) = x_c$), then we observe a super-stable periodic orbit of period n. In that case the probability distribution is concentrated at the periodic orbit (Dirac δ-measure). In the neighborhood of $r_{n,c}$ we find parameters

---

[6] This striking auditory phenomenon for the most audible case of period 3 intermittency, triggered several years ago in one of us (GMK) the association with jazz music: the periodic 3 cycle structure sounded like a theme and the chaotic bursts were like improvisations which would return back to the theme at irregular time intervals.[14]

$r_{n,pd}$ and $r_{n,sn}$ for which the system will undergo a period-doubling bifurcation (at $r_{n,pd}$) and a saddle node bifurcation (at $r_{n,sn}$). In the first case $r_{n,pd}$, each singular peak will split up into two peaks (i.e., we hear a subharmonic modulation of the structure). In the saddle-node case $r_{n,sn}$ the transition is more dramatic: the system will repeat the period n cycle for a duration T_n which decreases in the average like $T_n \sim 1/\sqrt{|r - r_{n,sn}|}$. The pattern that we hear will be a repetition of the cycle for some time, then a chaotic burst with an unpredictable duration (the average of which, however, can be estimated), then a recapture into the cycle, etc., at irregular intervals.

In Figure 7 we show such a solution for the case of period $n = 3$ and a parameter $r_a = 3.8283$ (a) and $r_b = 3.8284$ (b). The saddle-node bifurcation for period $n = 3$ occurs at $r_{3,sn} \cong 3.831854$. From Figure 6 we can also extract the location and (!) structure of higher periodic orbits (Figure 8). Note that, for example, period 7 occurs at several parameters but with different sequence representations. The bifurcation diagram provides efficient statistics for predicting the location of periodic and chaotic attractors. These statistics mask the dynamical aspects of attractors and foreground their accumulation.

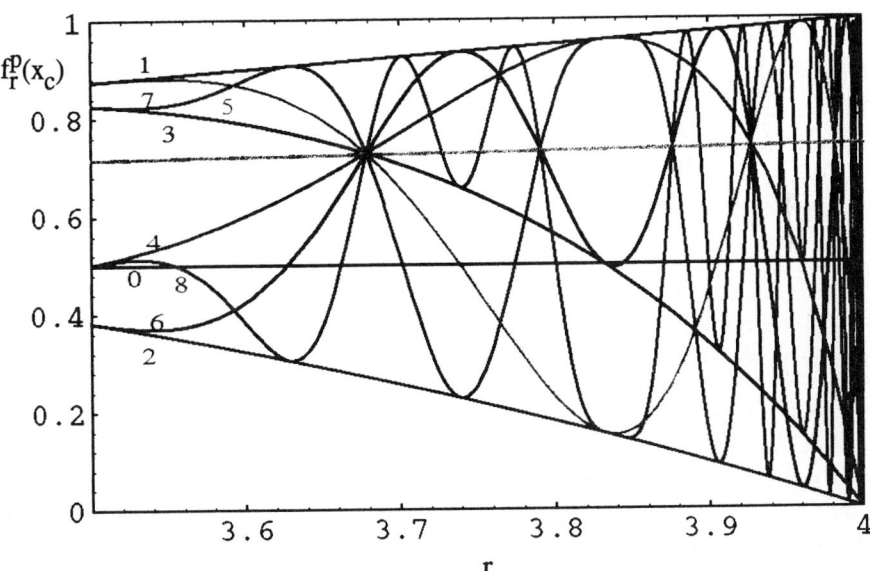

FIGURE 6 Orbit through the critical point for the logistic map for the first eight iterates. The darkness of the lines increases with the order of the iterate.

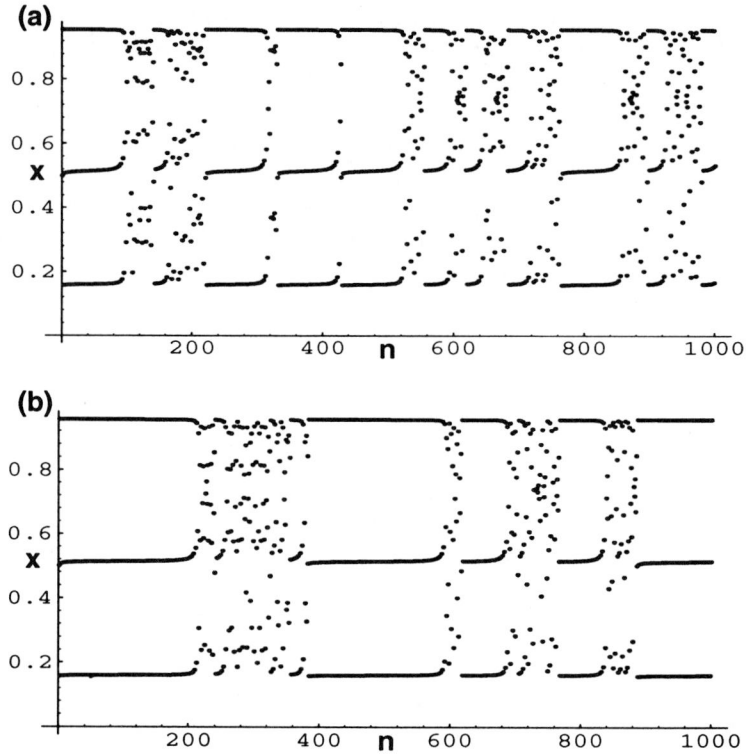

FIGURE 7 Intermittency close to the period 3 window: (a) $r_a = 3.8283$ and (b) $r_b = 3.8284$.

Tr 74 To render the entire diagram in sound requires the invention of a dynamical characteristic that statistics do not provide. One dynamic property that can be added is the act of reading the diagram from left to right, that is, sonifying the attractors while continuously increasing a bifurcation parameter value. Example (G) demonstrates a rendering of an (analog) electronic realization of the Chua oscillator, taking attractor values as sound samples and shifting as the appropriate bifurcation parameter the value of a resistor in the circuit. The dynamics of the resulting attractors are audited at a sampling rate of 44,100 Hz (the technique discussed in Section 2.2) (see Figure 9). The resulting sound gives a broad overview of the proportions of relative chaotic and periodic behavior across the scope of r-values specified. It does not provide details of the dynamical behavior of individual attractors stabilized at fixed r-values.

Musical Structures in Data from Chaotic Attractors

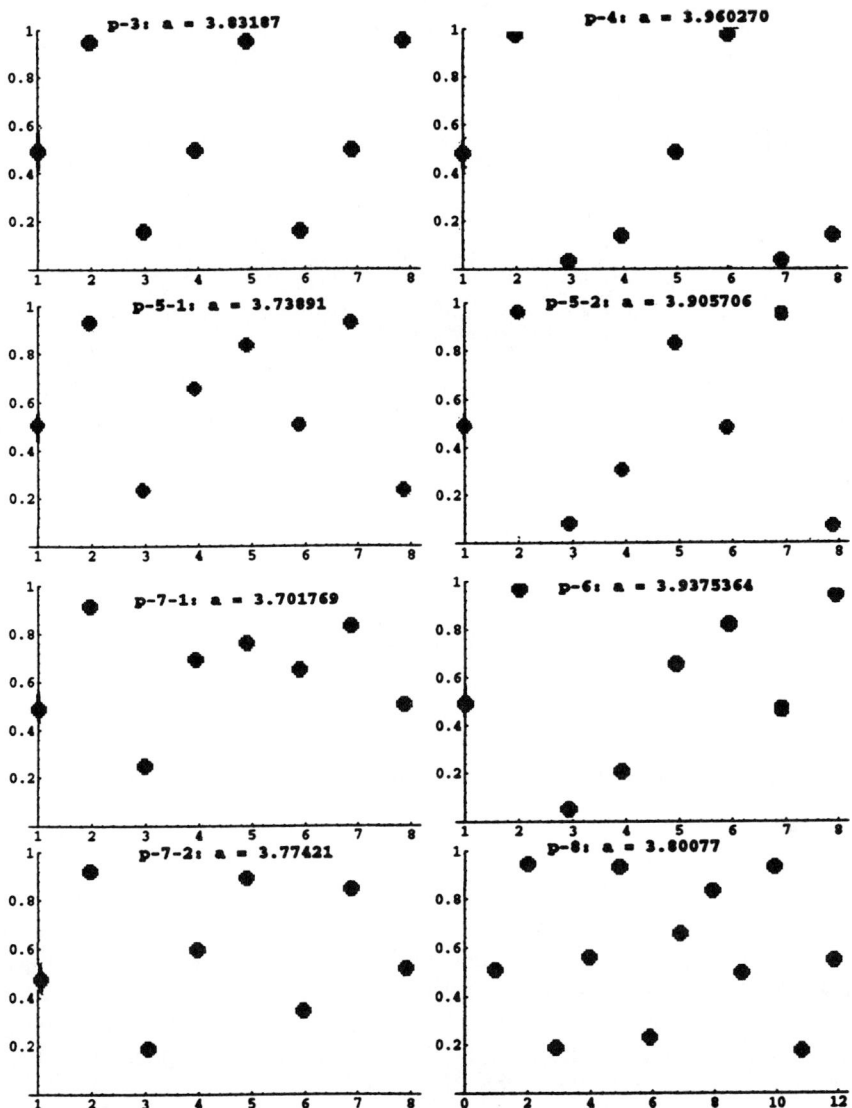

FIGURE 8 Several periodic orbits of the logistic map at different periods. Note that several periodic orbits occur for several parameters and different ordering structure.

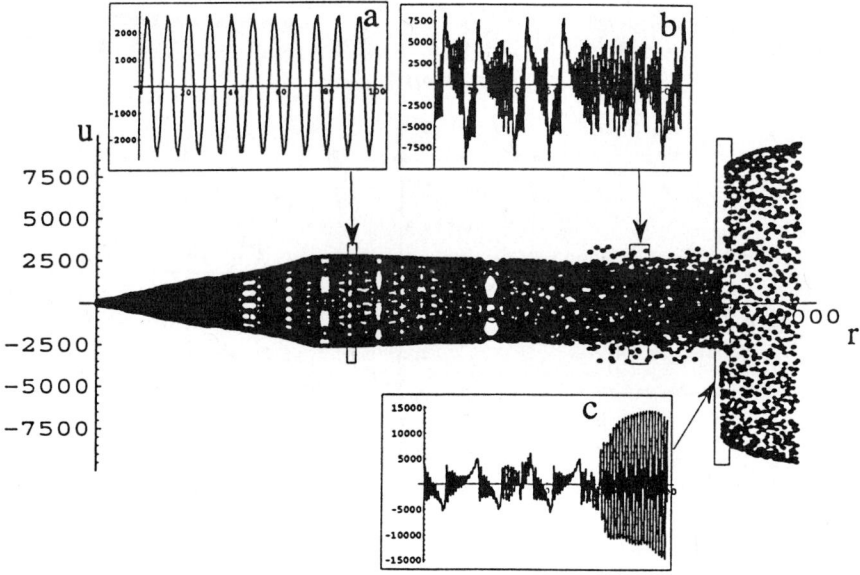

FIGURE 9   Bifurcation diagram of the Chua oscillator.

3. HIGHER-LEVEL REPRESENTATIONS

Perceived order and disorder, recurrence, and complexity are common features observed in both chaos and music. These features can be perceived in music because the music has been intentionally designed to reveal them. This is to say that it is not an accident for a listener to perceive a certain combination of order, disorder, and recurrence in music. If these are perceived in auditory display of scientific data, then it will also be as a result of an intentional design. It has been said that sonification techniques should not be confused with compositional activity. The consideration here, designing a system to make sounds that reveal complex structures, is a common goal of music composition and auditory display. In this light, it is not fruitful to imagine that, for a system to make sound, even the most direct forms of auditory display can exist without consideration of composition activity and design principles.

3.1 RECURRENCE STRUCTURES IN MUSIC

INTERMITTENCY. Sequential order is a significant characteristic of complex events. Low-level auditory representations (direct conversion of data to sound samples) capture local ordering; these tend to be too evanescent to depict varying degrees of sequential significance over broader time spans. To construct a way for listeners to notice relations between local change and changes over longer durations, we have developed methods for combining statistical and sequential representations. (The use of statistics in music composition dates from 1958.[10,22][7]) These techniques are intended to use the chaotic properties of intermittency and self-similarity in ways that allow them to signal significant change over large as well as local time spans, not unlike the role these structures perform in music.

The link between chaos and music lies on the issue of unpredictability. Predictability describes expectation and a method of observation. When an observer's expectation is not met with consistency, the observer loses the power of prediction; retroactively this loss may contribute to evaluating existing formulations about the method of observation, which represents an idea about the data. The heuristic value is the offer of an opportunity to orient differently such that the observer learns another way to observe. In chaos theory, the observer's loss of expectation is correlated to the exponential increase of information in the system. Entropy, in these systems, is a measurable parameter which controls the degree of unpredictability. In music, unpredictability is also brought about by specific structures. In both cases the location of the unpredictable event cannot be determined until it is identified by an observer. Representation tools which improve the ability to detect predictability may describe a dynamical system in ways which probabilistic tools based on linear correlations and spectral properties cannot or only inadequately describe.

In music, chaos and order emerge from the events the composer composes. Chaotic conditions may be associated with those conditions that create regions of predictability in an intended communication. One way to observe chaos in music is by examining music production. Notation and performance constitute the material component of music (including music produced by machine). Intermittent patterns in music notation, and in sounds produced during performance, have a range of predictability resembling chaotic dynamics. When the music is performed the listener brings much more complex interactions into the piece. A performance produces its unique recurrence with each listener; a score produces its unique recurrence with each performance. The attractor of all performances around a score

---

[7] There are earlier examples of random selections applied to help decide sequences of musical events. These are not considerable as examples of composing by designing statistical specifications of probabilities.

and all descriptions around a performance will be bounded according to the initial conditions provided by that score or performance. This suggests a similarity to *small-scale* chaos, where most of the chaotic fluctuations are in a small range around a primary structure, and these fluctuations are organized.

Composers create correlations between events which are intermittent. Listeners anticipate repetition and are occasionally successful. The listener, in addition to the composer, determines when a recurrent event is a repeat. In the neighborhood of repetition we find the dynamic regions of intermittency and nonrepetition. Chaotic characteristics are perceived in these regions due to expectation.

The experience of unpredictability contributes to a listener's understanding of structure and order in music. This suggests that the experience of unpredictability constructed in an auditory representation might also contribute to the understanding of the structure of a chaotic system. In section 3.2 we will discuss such representations.

SELF-SIMILARITY. A listener distinguishes one musical piece from others. Fine ears distinguish the late musical compositions by Mozart from Beethoven's early pieces. Musicians use the term style when they are able to identify, out of a pool of musical compositions, a certain mode of articulation or acoustic configuration consistently present in a certain era or in certain composer's works. What makes these figures observable in a listener's mind? One way to address this question is in terms of self-similarity. Self-similarity is characteristic of music design and music perception. Self-similarity appears in music (i) in terms of material such as melodic or rhythmic grouping, and (ii) in terms of structure such as pitch-related plans. Harmonic progression, for example, occurs at the phrase level and also as a formal design. The compositional task is to articulate a way to make music refer in both terms (material and structure) to itself so that it can reside in a listener's memory in its nontrivial way.

Looking to music to provide examples of techniques useful for auditory display, self-similarity may be considered helpful for allowing a listener to hear relational aspects of complexity. Self-similarity may be present as a display device regardless of the extent to which it is present in the chaotic patterns represented. Musical events can be understood as occurring in numerous simultaneous layers, some brief and some lengthy.

Self-similarity occurs between macroscopic patterns and the microscopic patterns that comprise them. The simplest examples are created by a technique called melodic sequence, where a short sequence of notes with a particular pitch contour is several times repeated to create a longer sequence of short sequences. Each repetition of the basic sequence is displaced to a new pitch level; the contour of the macro-sequence is created

FIGURE 10 Several nested layers.

by a pattern of displacements that replicates the contour of the basic sequence. Replications may be nested several layers deep (examples (a) and (b) of Figure 10).

Self-similarity seems to play a role in the theory and practice of tonality, the organization of music according to pitch. Tonality describes the ordering of pitches into subgroups which can be made to sound more or less stable. One or another of these groups can be made prevalent across a broad time span; accented periodic iteration, assisted by the listener's memory, is the strongest determiner of prevalence. This cumulative prevalence is called harmony. Nonadjacent iteration is stronger than adjacent iteration in determining harmony at a structural level; the time span between iterations encourages the listener to remember pitches and predict their return. During long-term perception of the structure from this interplay between receiving harmony and a listener's memory and prediction, structural coherence emerges. This assists the perception of extended events: by iteration an event is perceived as sound ongoing, though other events intervene. In Figure 11, extracted from the Beethoven string quartet excerpt, we show an ascending line that is both localized and broad.

Iterations over long periods, given strong accents, create the broadest regions of prevalence of a harmony (Figure 12, stemmed notes with unfilled heads). The rate of change from one prevalent group to another is harmonic rhythm. Regularity of harmonic rhythm can create expectations of harmonic change even at large time intervals. Pitches and pitch groups may intervene where they do not belong to the prevailing harmony. These secondary groups may act as subregions that locally contradict the harmony but do not displace it (as they have a shorter period and a shorter history of iteration). A subregion may itself contain subgroups; in Figure 12 the filled note heads, stemmed and unstemmed, indicate two layers of subgroups; notes under arched lines indicate the most local events. Self-similarity occurs when the same vertical or horizontal pattern appears in a broad region and in a nested subregion. The structural level of a large-scale musical form called a "sonata" often can be reduced into 3 to 5 harmonic regions.

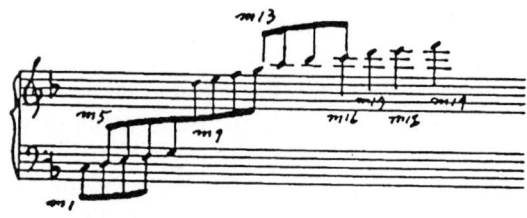

FIGURE 11 Beethoven string quartet excerpt.

FIGURE 12 Prevalence of a harmony.

Nested events are called dissonances when their degree of contradiction is audible, creating the perception of instability. The process of eliminating an audible contradiction by leading the pitches to members of a more prevalent set is called resolution of the dissonance. When one subregion seems to lead to another, it is called harmonic progression. Each of the horizontal sequences in Figure 12 indicates a progression, at a brief or a large time interval.

Pitch structures that establish layered time regions are a functional aspect of all common practice period music.[8] They have resulted from research in the form of many compositions during a long period of development of techniques for articulating patterns using sound. In the same manner the division of the octave into 12 equal steps might be considered useful for pitch differentiation in an auditory display; patterns articulating multiple time scales learned from musical structures may be adopted in implementing the analogous hierarchy in time organization for sonification. Examples of music composition considerations applied in auditory display are discussed by Bargar (in this volume).

[8] The common practice period of "classical" music extends roughly from 1700–1900.

Self-similarity in music is composed in response to an understanding of human perception, with the intent of establishing auditory structures that resist the decay of memory and withstand being obscured by the continuing iteration of sound. Self-similar auditory structures allow observers to make predictions, relevant to the dynamical behavior of an observed system, and to remember what has happened, relevant to the statistical characteristics of that system.

3.2 STATISTICAL REPRESENTATIONS

SYMBOLIC DYNAMICS. If we are not interested in observing the precise state of the system, we can partition the interval into a finite number of bins and label each of these bins by a symbol. For some systems one can show that there exists a "generating partition" which allows a complete equivalence between the symbolic dynamics (obtained by recording the sequence of bins visited by the system and assigning to each a string of symbols or a letter), and the original orbit (see, for example, Hao[9]). One can show that for the logistic map the partition induced by the critical point x_c is such a generating partition. If we refine the partition through preimages x_i^p of x_c (i.e., there is an integer $k > 0$ such that $f^k(x_i^p) = x_c$), we can describe each of those states with a new symbol (if we want a partition, say, for a 12-tone system) and observe the dynamics of n-cylinders, i.e., finite sequences of n symbols. For typical chaotic systems we will observe that not all possible sequences of length n will be observed. Their number, however, will grow exponentially with the length n of the sequence. This asymptotic (for $n \to \infty$) growth rate will determine the complexity in the sense of "dynamical entropy" (often known as Kolmogorov-Sinai (K-S) entropy, see Hao[9] for details) of the observed (infinite) sequence.[9]

Tr 75,76

AUDITORY REPRESENTATION OF n-CYLINDERS. Histograms and n-cylinders offer similar structures for sonification: a fixed set of bins each representing a unique range of chaotic values, and each assigned a weight according to its statistical prevalence in the system. In the standard use of n-cylinders, points on an attractor eventually converge to their characteristic, invariant distribution. Each chaotic attractor of the logistic map will have a unique distribution and, thus, a unique sound. For sonification we have assigned each bin a unique pitch and use that bin's weighting to determine loudness and timbre with which a bin is heard. All bins are presented

---

[9] Empirical investigations of pieces of classical music seem to indicate that the measured K-S entropy is close or equal to zero.[7] In that case, the musical patterns should become completely predictable for pieces that are of infinite length.

simultaneously. To represent dynamic behavior, the bins are displayed repeatedly as the attractor converges. One method measures the norm[10] of the differences $\| \vec{p}_1 - \vec{p}_2 \|_\alpha$ of two histograms $\vec{p}_1, \vec{p}_2$ recorded at different times t_1, t_2, with, say, $t_2 = t_1 + \Delta t$, and Δt a few seconds. This norm of the difference in the histograms measures the degree of change from the previous state in an averaged manner; when the change $\| \vec{p}_2 - \vec{p}_1 \|_\alpha$ has exceeded a certain predetermined threshold the new histogram is displayed (example H1). A second method plays only the bins that contain new information, i.e., the histogram $\vec{q}$ defined by $q_j = p_{1,j}$ if $p_{2,j} - p_{1,j} \neq 0$, when the threshold has been exceeded. A third technique is to sonify the quantity of the changes from the previous state, rather than rendering the quantities of the state itself, i.e., the histogram defined by: $\vec{p}_2 - \vec{p}_1$.[11] To avoid the accumulation of a steady-state sound as the histogram converges, a further technique is to only examine a sliding window of the most recent p points on the attractor (example H4) where we choose p between 20 and 500. These techniques emphasize the sequential ordering of events.

TWO COUPLED LOGISTIC MAPS. The concept of histograms can be generalized to arbitrary dimensions. As an example we want to consider the effect of coupling two logistic maps with bifurcation parameters r_1, r_2 and coupling parameter c:

$$\begin{pmatrix} x_{n+1} \\ y_{n+1} \end{pmatrix} := \begin{pmatrix} r_1 x_n (1 - x_n) + c(y_n - 0.5) \\ r_2 y_n (1 - y_n) + c(x_n - 0.5) \end{pmatrix}. \tag{7}$$

A typical attractor of this mapping is displayed in Figure 13. In part (a) we have plotted a few hundred iterates of the initial condition $(x_0, y_0) = (0.4, 0.51)$. We have pointed out two structures within the attractor that can be observed: The "o" symbols indicate that nearby in parameter space (for r_1, r_2 unchanged but c perturbed by $\Delta c = 0.00326$, to be precise.) we have a stable period 7 attractor. Graphically, this structure is completely hidden in the more complete attractor display (b). Although the graphical representation gives us a fairly good picture about the stationary, statistical properties, such as its distribution, the dynamical features are basically lost. They become directly evident in an auditory display, here symbolically in Bidlack's score representation.[1] We can clearly see how the system

---

[10] If we represent a histogram of m bins as a vector $\vec{p} = (p_1, \ldots, p_m)$, we can choose a norm $\| \cdot \|_\alpha$ as: $\| \vec{p} \|_\alpha := \sum_{j=1}^{m} | p_j |^\alpha$. The value of α determines how much large components contribute: for $\alpha = 1$ each component contributes with the same weight, for $\alpha \to \infty$ only the largest component contributes (max-norm).
[11] Note that $p_{2,j} \geq p_{1,j}$ for all $j, 1, \ldots, m$.

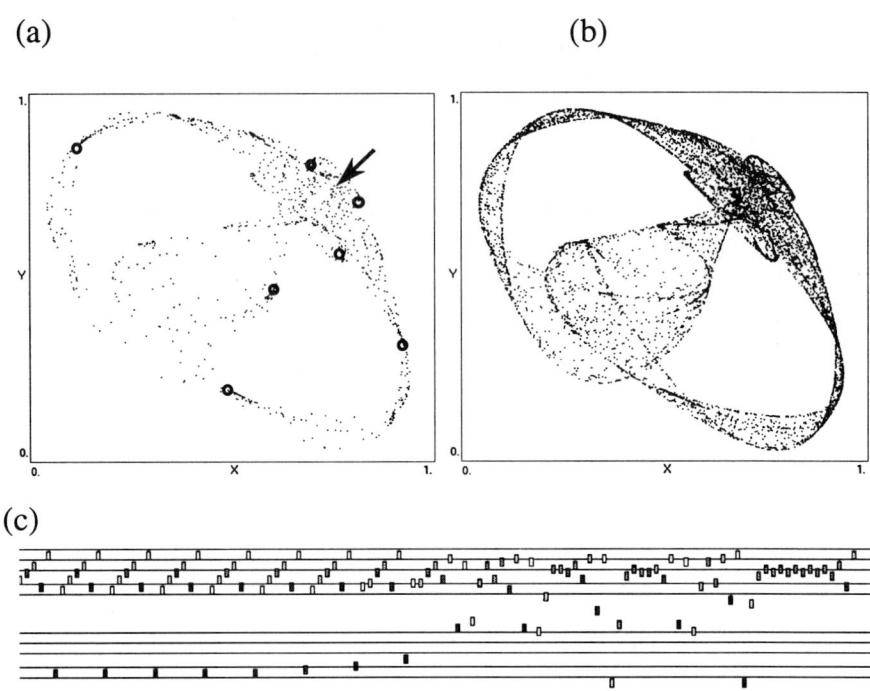

FIGURE 13 Representation of a chaotic attractor of the coupled logistic map of Eq. (7) ($r_1 = 3.0$, $r_2 = 3.1$, $c = 0.456$). (a) Distribution of a few hundred points on the attractor. The "o" symbols indicate the approximate location of a period 7 cycle which will become stable for $c = 0.45926$. The arrow points to the approximate location of an unstable fixed point. (b) Same attractor as in (a) after accumulation of more points. (c) Score representation of the x-components of the map. The plots were produced with a modification of R. Bidlack's Chaos program.

stays close to the period cycle for nine repetitions but then escapes and gets trapped close to the unstable fixed points (located close to the center line, top system in this display) to which it returns twice.

HISTOGRAMS OF THE COUPLED LOGISTIC MAPS. For the two-dimensional system of Eq. (7), the points on the attractor are distributed in a region of the plane. The histogram would then describe the density of those points in a given area. We can see that there are no points distributed outside the well-defined structure of the attractor. This imposes strong constraints on the observation of pairs (x_n, y_n); e.g., for a given value of x the possible values for y are constrained and vice versa. Similar to the one-dimensional case, we face the problem of perceiving this correlation in a individual acoustic rendering of each point. We therefore search for representations of approximated histograms in the plane. To this end we partition both x and y directions and thus obtain a covering of the attractor by rectangular bins, the value assigned to them would just correspond to the fraction of points in this bin. The audio representation that we have chosen assigns to each bin with the same x-index a fundamental frequency or pitch. Relative to that frequency we assign to each bin with a given y-index an intensity value corresponding to its spectral contribution. For example, in Figure 13(b), we would expect a polyphonic sound that has a timbre distribution that has more higher harmonic contributions at low fundamental frequencies than at high frequencies. At intermediate fundamental frequencies we expect contributions from intermediate and high harmonics. The same consideration with respect to update of the histogram representation after passing a threshold can be carried over to this two-dimensional case without any alterations (see example I).

4. CONCLUSIONS

From a standpoint of composition, the specification of degrees of predictability within a deterministic system provides rich resources for creating complex meaning. Sonification inquires, in what ways might these resources help characterize their system of origin. Composers take into consideration the limits of a listener's ability to predict, and create rearticulations of events at broader time spans. Chaotic attractors are not composed; their musiclike structures are not designed to represent themselves hierarchically. Auditory display may be designed to couple musiclike structures with compositionally informed representations. The design challenge is to display the rate of change of the system such that local detail does not prohibit the perception of larger dynamical structures. To do this, statistical representations are coupled with dynamical display characteristics. Our implementations utilized symbolic dynamics and histograms, which update according to the rate of change of the system being observed.

ACKNOWLEDGMENTS

We would like to thank Greg Kramer for his work with the organization of ICAD'92 and Alfred Hubler and Nick Weber for making their Chua circuit available to us. One of us (GMK) would like to thank T. Brown for stimulating this interdisciplinary project, C. Chafe for useful discussions, R. Bidlack for his help with his Chaos application, and Kathie Alblinger for her help with the manuscript.

REFERENCES

1. Bidlack, R. "Chaotic Systems as Simple (but Complex) Compositional Algorithms." *Comp. Music J.* **16(3)** (1992): 33–47.
2. Chafe, C. Presentation at the University of Illinois at Urbana-Champaign, Urbana, Illinois, Fall 1992.
3. Chowning, J. "The Synthesis of the Singing Voice." In *Sound Generation in Winds, Strings, Computers*, edited by J. Sundberg. Stockholm: Royal Swedish Academy of Music, 1980.
4. Chua, L. O., and G.-N. Lin. "Intermittency in a Piecewise-Linear Circuit." *IEEE Trans. Circ. & Syst.* **38** (1991): 510–520.
5. Chua, L. O. "A Simple ODE With More than 20 Strange Attractors." Memorandum No. UCB/ERL M92/141, University of California at Berkeley, Berkeley, CA, August 1992.
6. Chua, L. O. "Global Unfolding of Chua's Circuits." Memorandum No. UCB/ERL M93/7, University of California at Berkeley, Berkeley, CA, January 1993.
7. Ebeling, W., and G. Nicolis. *Entropy and Frequency of Words in Strings from a Dynamical Perspective*. **2** (1992): 635.
8. Feigenbaum, M. J. "Universal Behavior in Nonlinear Systems." *Physica* **7D** (1983): 16–39.
9. Hao, B. *Elementary Symbolic Dynamics and Chaos in Dissipative Systems*. Singapore and Teaneck, NJ: World Scientific, 1989.
10. Hiller, L., and L. M. Isaacson. *Experimental Music*. New York: McGraw-Hill, 1959.
11. Kennedy, M. P. "Experimental Chaos via Chua's Circuit." Memorandum No. UCB/ERL M91/95, College of Engineering, University of California at Berkeley, Berkeley, CA, 1991.
12. Kolleritsch, O., ed. *Musikalische Gestaltung im Spannungsfeld von Chaos und Ordnung*. Wien: Universal Edition, 1991.
13. Mandelbrot, B. B. *The Fractal Geometry of Nature*. New York: W. H. Freeman, c. 1983.
14. Mayer-Kress, G. "John Coltrane and the Swing—Chaotic Dynamics as Composition Tool in Computer Music." Lecture given at the Center for Music Experiment, University of California at San Diego, San Diego, CA, Summer 1987.

15. Nodaira, I. "To Establish Logic From Chaos—Music Composition with Itineraire." *Revue Musicale* **N421** (1991): 255–262.
16. Risset, J.-C. "Timbre Analysis by Synthesis." In *Representations of Musical Signals*, edited by G. De Poli, A. Piccialli, and C. Roads. Cambridge, MA: MIT Press, 1991.
17. Rodet, X. "Nonlinear Oscillator Models of Musical Instrument Excitation." In *Proceedings of the 1992 International Computer Music Conference*, 412–413. San Francisco: International Computer Music Association, 1992.
18. Roederer, J. C. *Introduction to the Physics and Psychophysics of Music*, 2nd ed. New York: Springer-Verlag, 1979.
19. Ruelle, D. *Chance and Chaos*. Princeton, NJ: Princeton University Press, 1991.
20. Scaletti, C., and A. B. Craig. "Using Sound to Extract Meaning from Complex Data." In *Proceedings of the SPIE Conference "Extracting Meaning from Complex Data,"* edited by E. Farrell, Vol. 1459. Bellingham, WA: SPIE, 1991.
21. Truax, B. "Chaotic Non-linear Systems and Digital Synthesis: An Exploratory Study." In *Proceedings of the 1990 International Computer Music Conference*, 100–103. San Francisco: International Computer Music Association, 1990.
22. Xenakis, I. *Formalized Music*. Bloomington, IN: Indiana University Press, 1971.

Chris Hayward
Southern Methodist University, Department of Geology, Dallas TX

Listening to the Earth Sing

Techniques for auditory monitoring and analysis of seismic data are described. Unlike many other kinds of data, seismograms may be successfully audified with a minimum of processing. The technique works so well because both sound in air and seismic waves in rock follow the same basic physics, that described by the elastic wave equation. Both exploration seismology, which examines with only the upper few miles of the earth, and planetary seismology, which examines the larger structures including the earth core, may make use of auditory display. Previous published work is limited to two papers now nearly 30 years old, which examine the utility of audio display to the discrimination problem of earthquakes and nuclear explosions. The applications are much broader though, including training, quality control, free oscillation display, data discovery, large data set display, event recognition, education, model matching, signal detection, and onset timing. Problems in audifying seismograms arise when the subsonic wide dynamic range signals must be rescaled to the audio without introducing distracting artifacts. Simple processing techniques including interpolation, time compression, automatic gain control, frequency doubling, audio annotation and markers, looping, and stereo are used to create seven example audio data sets. These seven examples illustrate the use of audio in presenting synthetic seismograms, shallow reflection data, quality control during field recording, noise analysis for earthquake observatories, earthquake analysis for events from various distances, nuclear explosions, and stereo display of seismic array data. The use of audio for seismic quality control, analysis, and interpretation will develop only when audio displays become integrated into the daily tools of seismologists.

1. INTRODUCTION

Seismograms like the ones shown in Figure 1 are graphs of small vibrations of points on the earth plotted against time. They are one of the essential tools of geophysicists studying the interior of the earth. An audio rendering of these can provide unique insight into seismic modeling results, seismic source and wave propagation characteristics, field-recording quality control, and training and education. This rich and unique opportunity to improve the quality of modern seismic interpretation is almost unexplored and yet is inexpensive, simple, and elegant.

In this volume are many examples in which data traditionally graphed or plotted is presented as sound. Stock market prices may be scaled and converted to musical notes or hospital patient status may be converted to analogs of everyday sounds, what Kramer (in this volume) and others have defined as sonification.

However, few people have been successful in directly treating data points as sound waveform samples and simply playing them, a process that Kramer defines as audification. One of the reasons audification fails for arbitrary data such as stock market figures or daily temperatures is that even a slow playback rate of 8,000 samples/second requires many data points to make a sound of any duration. The resultant short sound does not reveal valuable information to the listener. Even for long sequences, the sounds do not resemble natural or familiar noises because arbitrary data points seldom obey the physics of natural sounds transmitted through air.

Seismic data, however, is an almost perfect case for audification. Seismic data sets are large—seldom less than a few thousand samples and as much as several billion samples. A simple audification will play from several seconds to more than an hour. It will sound like a recording of natural environmental sounds because sounds transmitted through air (acoustic waves) have similar physics to seismic vibrations transmitted through the earth (elastic waves). The physics is similar enough that mathematical models that describe sound transmission through gas are successfully used for seismic modeling.[6] This suggests that doing the opposite, audifying the seismic data, should produce natural sounds such as those heard in the environment.

The advantage of a direct physically consistent auditory display is that as an analog to natural acoustics, it can take advantage of the vast human experience in interpreting noises. For example, a sharp explosion followed by decaying echoes includes information that is interpreted as the size and shape of the echo chamber. A set of echoes followed by the explosion is recognized as something physically ridiculous or artificial.

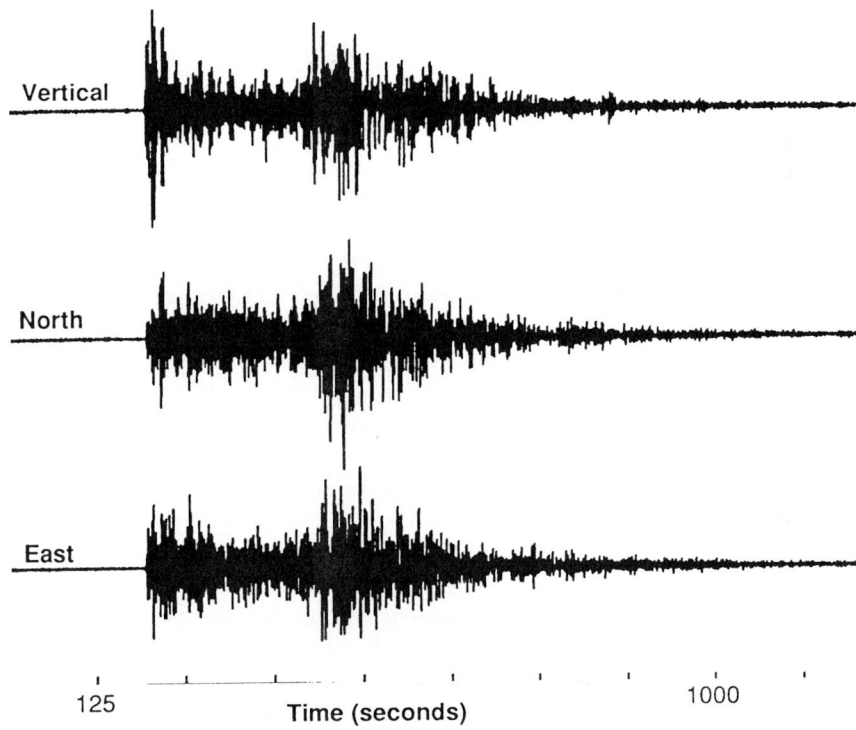

FIGURE 1 Three Component Seismogram of a Nuclear Explosion at NTS Recorded at Lajitas, TX. Each trace records ground velocity in one direction. Data from three GS13 seismometers was recorded at 40 samples/second using a 24-bit digital recorder. The initial impulse represents the arrival of the compression wave. The energy about 4 minutes later is from the shear wave arrivals.

Audified seismograms have been produced as a curiosity by many seismologists. However, no one has demonstrated a common application where auditory presentations are superior to graphical displays or suggested the areas in seismology where such success is likely.

2. BACKGROUND
2.1 COMPARISON OF SOUND WITH SEISMIC WAVES

Most seismic interpretation assumes that the recorded seismograms are elastic waves[1] and, therefore, follow the wave equation[3][2]:

$$\frac{\partial^2 y}{\partial t^2} = c^2,$$

where c is constant and $\nabla^2 \equiv \partial^2/\partial x_i^2$. In the equation, c is the velocity that the disturbance travels through the material. This same equation describes the transmission of sound through air. For sound in air c is about 1,100 feet/second. In rock, there are several different kinds of waves that propagate out from the source at different velocities. The fastest compression-rarefractions, or P-waves, travel at 5,000 feet/second in rocks near the surface, and over 40,000 feet/second deep in the earth's mantle. Slower shear wave distortions or S-waves travel at about half the P-wave speed. For sensors and sources near the earth's surface, there are also Rayleigh and Love surface waves that travel at lower speeds.

There are four major differences between the recordings of sound in air and of seismic waves in the earth. Except for unusual circumstances (such as thunder), sounds we experience travel through a simple material, air, with only one velocity. Seismic waves travel at different velocities in each rock type. Sounds have only one type of wave (compression-rarefraction), while seismic waves have several traveling at different velocities. Surface waves that travel along layer boundaries and at the earth's surface are a dominant feature in seismic waves, but are rare in sound. Seismic waves are scaled differently, with dominant frequencies of 40 Hz, much lower than audiable sound.

[1] This is not true in special cases very near the source where the explosion, impact, or earthquake permanently deform or fracture the rock.
[2] For a more general discussion to the elastic equations of motion as used in seismology, see Aki and Richards.[1]

2.2 THE GOALS OF SEISMOLOGY

Geophysicists interpret seismograms to infer the characteristics of the rocks along the energy's travel path or to learn the characteristics of the energy source. Vibrations from seismic sources travel through the earth reflecting and refracting from different layers of rock and, finally, shaking a sensitive displacement, velocity, or acceleration gauge (seismometer or geophone) that records vertical, north-south, east-west, or all three motions.

Studies in the field of seismology can usually be classified by the scale of the phenomena or of the object being studied. Most studies can be classified as either exploration seismology or as planetary seismology.

EXPLORATION SEISMOLOGY. *Exploration seismology* uses a controlled source such as a hammer or small explosive charge, and records reflections and refractions from the rock layers a few feet to several miles below the surface (see Table 1 and Figure 2). Interpretation of these recordings may be used for mineral exploration, characterization of shallow geologic structure for civil engineering studies and geologic hazard estimates, as a way to construct geologic maps of the subsurface or to understand shallow crustal and near-surface processes.

For typical studies, explosions or hammer blows are repeated and recorded on tens to thousands of geophones at the surface to build up a seismic cross section much like a sonar depth section. After each recorded shot, the geophones and energy source are moved forward a short distance.

During interpretation it is assumed that every explosion or hammer blow is identical and every geophone is recording the undistorted ground motion underneath it. Any differences are interpreted as the result of the geologic structure underneath the geophone and explosion. In practice, geophones are sometimes not firmly planted in the ground, which results in a distorted signal. Sometimes hammer blows bounce or are weak. If these problems are unrecognized, they may be mistakenly interpreted as having geologic causes.

PLANETARY SEISMOLOGY. *Planetary seismology* uses recordings of volcanic eruptions, large explosions, and earthquakes (see Table 2). Recordings are interpreted to understand the gross structure of the planet such as the nature of the core, mantel, or crust; the deep structure of continents; and the location and nature of seismic sources (earthquakes, volcanoes, or large explosions).

Planetary seismologists depend on records from seismic observatories located in quiet areas around the globe. Each observatory makes precisely timed high-quality recordings of ground motions that may be associated with recordings from distant stations. These may then be used to develop the location and time of the event.

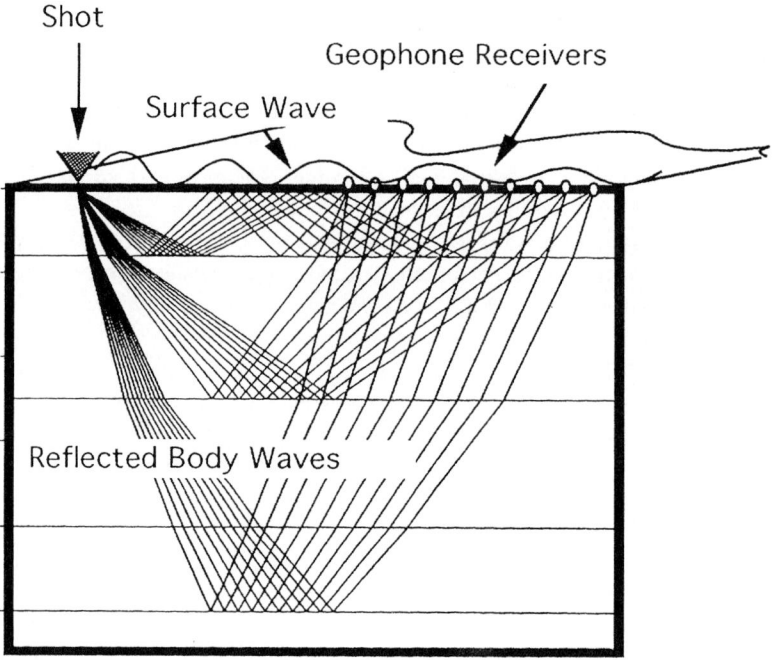

FIGURE 2 Seismic Exploration Data Collection. Seismic energy radiates from an explosion or hammer blow and is then reflected and refracted in the underlying rock. In this example the reflected energy is recorded by ten geophones (sensitive velocity gauges) on the ground surface. In addition to the body wave energy traveling through the rock, surface wave energy travels along the ground (much like ocean waves) and is recorded.

Each observatory records a large number of earthquakes each day. Seismic analysts perform a winnowing of the events, throwing out events that are uninteresting and saving others for more intense scrutiny.[3] It is assumed that critical events will be recognized. Since analysts always work against time deadlines (tomorrow's data is going to arrive in 24 hours), in practice whether a particular event gets interpreted depends on how much other activity there is and whether the event looks interesting during the first few minutes (or seconds) of study.

[3] Most stations have some degree of automated analysis, sometimes to the point of locating all events. However, at some point an analyst will be required to review the data.

TABLE 1 Characteristics of Exploration Seismic Signals.

Sample Rates	250 to 4000 Hz
Bandwidths	Usually less than 3 octaves 10–100 Hz common 20–500 Hz high resolution 20–2000 Hz special studies
Dataset Size	12,000 to several billion samples .2 seconds to 5 seconds/record Usually < 4000 samples/record Multiple shots per reflection point
Interpretation	Every peak is significant and waveform details (width or phase) are often also interpreted
Dynamic Range of Recording	8 bits to 24 bits are commonly used

TABLE 2 Characteristics of Planetary Seismic Signals.

Sample Rates	1 to 120 Hz Multiple sample rate streams are common
Bandwidth	1-hour period to 40 Hz (17 octaves)
Data Set Size	Continuous recordings 3–100 Megabytes/day Triggered recordings < 1 Megabytes/day
Interpretation	Precise timing and identification of phase (P, S, etc.) is critical Much interpretation is based on the envelope and spectra of the event
Dynamic Range of Recording	24-bit recording common Most signals easily represented by 16 bits

Seismograms also reveal characteristics about the energy source (typically earthquakes or explosions, but also footsteps, hammer blows, or vibrating machinery). One such application, which is heavily funded, is nuclear test-ban treaty verification. Explosions have particular signatures on seismograms that analysts may be trained to recognize. Even small nuclear explosions may be detected at distant seismic stations (most test nuclear explosions in Nevada are easily detectable in Texas). Being able to recognize the difference between a recording of an earthquake and that of an explosion is the *explosion discrimination problem* of seismology.

2.3 SIGNAL CHARACTERISTICS

Signals recorded for exploration seismology are short and have a limited dynamic range and bandwidth (see Table 1) compared to most sounds. They are more like a recording of a hand clap or gunshot made in a canyon and later analyzed for echoes. The source impulse is short and the interesting portion of the recording is the echo. Equipment limitations set the dynamic range of the recorded signal. Field equipment trades-off fidelity and dynamic range against channel capacity. Geophysicists adjust the recording levels and filters to capture a particular signal of interest and sometimes distort other noise and signals.

Signals recorded for planetary seismology are continuous with wide dynamic range and bandwidth compared to most sounds (see Table 2). The recording problem is like that of attempting to record all sounds that happen in a room, from cannon shots to pin drops, without knowing when such sounds may occur. When they arrive from great distances, earthquake signals become drawn out for many minutes (like distant thunder). Field equipment is designed for broad bandwidth and extreme dynamic ranges.

3. PREVIOUS WORK

Seismic data audification is not new. Many geophysicists with the opportunity and equipment have listened to seismograms. Early recordings made on FM tape recorders were often played at high speed to locate recorded earthquakes. From discussions with experienced seismologists of the 1960s, this seems to have been quite common, although it is not well documented. Time-compressed earthquake recordings appeared on an early Cook Laboratory sound-effects record. Other time-compressed recordings were used to shift seismic recordings up to a frequency band so that speech sonograph equipment could be used to do spectral analysis. Recently, a Mexico City earthquake has been audified by Joe Dellinger at the Hawaii Institute of Geophysics and offered to the Internet.

Even though seismic audification is well known as a curiosity, it has never been pursued broadly as a routine alternative to graphic presentation of all forms of seismic data. Extensive literature searches revealed only two published papers on seismic audification. Both investigated the use of audification to discriminate between earthquakes and explosions. The initial pilot study was by Speeth[8] and a more complete study was conducted by Frantti and Leverault.[5]

Speeth observes that listeners can recognize speakers' voices during telephone conversations, despite wide variances in telephone circuit characteristics, tone, and quality. He hypothesized that if it were possible to discriminate between earthquakes and nuclear explosions from recordings close to the source, discrimination at longer distances and along travel paths through varying geology would be like recognizing two different speakers through differing telephone lines. Speeth's results using a data set of only five explosions and eight earthquakes demonstrated that nine observers with 30 to 60 training events achieved 90% correct recognition. This seemed promising. Unfortunately, all the earthquakes were recorded at one seismic station and all the explosions at another. Also all the earthquakes were farther away than the explosions. This strongly suggests the discrimination results could be artifacts of the data set selection.

Frantti repeated the work in 1965 with 21 subjects and 1500 earthquake-explosion questions. Subjects correctly identified the difference between earthquakes and explosions about 65% of the time. There was a wide variance in individual performance. Two observers averaged about 70%, while one continued to improve with training, never showing the plateau that others had. It seems some people may be able to use their experience and background to develop their understanding of new audio signals. Although a 65% accuracy is poor compared to the performance of trained analysts working with paper seismograms, because Frantti's subjects had little training in geophysics, this may not be a fair comparison.

Several interesting leads and conclusions resulted from the research. The best discrimination was found to be at a time compression of 128, which yielded a P-to-S[4] separation of about 1/4 second on the playback. Listeners found that the events had to be repeated four times in order to understand the sound. During some experiments, when recordings of vertical motion were played back on one channel and recordings of horizontal on

---

[4] P and S are two phases which travel at slightly different velocities. Their separation in time is a measure of the distance to the explosion or earthquake. Their amplitude ratio can be a measure of how much the event is like an explosion or earthquake.

the other, the discrimination seemed to be better.[5] Discrimination was as good for recordings played forward as for those played backward. Playing earthquakes backward sounds unnatural (something like a piano recording played backward) but observers did no worse than with a more familiar forward recording, suggesting that previous knowledge about natural sounds were not used in discrimination decisions.

Frantti suggested the results could have been improved had listeners been limited to records from a single recording station and been able to compare the unknown record against a catalog of known events. A trained analyst reading paper records in effect does this, but the comparison is with remembered events.

Audifying seismic data in 1965 was cumbersome. The tape playback system had to be built. Once the tapes were mastered, listeners had little control over data playback. In contrast, seismic analysts have a great deal of control over how long and at what scales a particular event is displayed before making a conclusion. Quite frequently they also have a suite of reference displays.

Neither study used trained seismologists as subjects nor compared the success of untrained individuals to untrained individuals doing more classical analysis. Training a seismic analyst is a long process, taking up to a year at a single station. Analysts are generally best when they deal with a few stations that they know well rather than a large collection of stations. The current understanding of auditory seismogram display is so immature that it is difficult to design meaningful statistical tests using human subjects. Instead, it may be more advantageous to extend the understanding of auditory analysis by developing and documenting the successful use of such tools by trained analysts in an operational setting.

Frantti only published the one report. There has been very little published work since. At this point the field looks unexplored.

4. APPLICATIONS OF AUDITORY DISPLAY

For some common seismological tasks, audified seismic data should have distinctive advantages over the traditional plotting of wiggly lines on paper based on the nature of the task itself. Some of these tasks are mundane parts of seismology, but a demonstration that audified seismograms are a superior display will push research into more esoteric subjects.

[5] One classical discriminate for earthquakes and explosions is the ratio of the amplitude of the first arrival (which is on the vertical only) to the later shear arrivals, which also are on the horizontals.

4.1 TRAINING

Some people may learn to identify particular seismic events more quickly by listening to the signal rather than by seeing a wiggly trace. Others may learn better by using a combined graphic and auditory illustration. Just as we learn the alphabet with the ABC song, the audio presentation may no longer be necessary once the concepts and images are firmly in place. Since it takes up to a year to train an analyst, an improved educational tool represents a significant savings of money.

4.2 QUALITY CONTROL

For exploratory seismic field recording, audified data could efficiently deliver information when the operator is visually occupied. During field data recording an observer will normally be busy directing crew activity and scheduling shots as well as doing on-the-fly trouble shooting and maintenance. Under such conditions, careful field quality control is sometimes difficult. Particularly insidious are problems that occur slowly or intermittently. Scanning visual displays takes time and concentration away from the primary goal—collecting data. It is useful under such conditions to be able to quality control the data in the background. An audio presentation can alert the operator to problems that may then be addressed with other presentations.

For planetary seismology, quality control includes choosing good sites for seismic observatories and monitoring the background noise levels at existing observatories. This is currently done by comparing spectrograms at noisy and quiet sites. Auditory displays, with their natural ability to summarize the spectral information as a "coloring," provide a quick easy method for monitoring noise levels at new and existing stations. They are particularly well suited for monitoring noise generated by resonant structures, such as telephone lines, buildings, or pipelines.

4.3 FREE OSCILLATION DATA

One interesting application of audio data is for free oscillation data. The earth rings when struck, like any solid object such as a bell. The characteristics of the oscillations depend on where and how the earthquake occurred, the material properties of the surrounding rock, and the location of the seismic recorder. It is not yet clear what could be illustrated in an audio display but, in view of the analogs to bell-like objects, the possibilities are intriguing.

4.4 DATA DISCOVERY

Data discovery includes the processes of examining seismic waveforms for information of interest and forming the initial hypothesis about the causes of the disturbances. Such hypothesis will be tested and explored further with additional specialized processing and displays. This initial look forms the jumping-off point for imaginative new observations and ideas. By enriching the current displays with audio in a simple direct manner with few processing biases, there is less possibility that interesting information will be overlooked.

4.5 LARGE DATA SET DISPLAY

Large data set display is a method of interacting with a large data set to extract the information of interest. Seismic data sets are large, often more than a billion samples for a single experiment. They must be reduced and summarized or excerpted for interpretation. This reduction and selection, besides being time consuming, means that significant features in the data may be missed. Auditory displays summarize the information in a different manner than common plots, allowing a second chance to recognize relationships in the seismic data.

4.6 EVENT RECOGNITION

Event recognition and classification were the problem classically attacked in the 1960s by Speeth and Frantti. There are several motivations to this attack. People seem to recognize events and voices from limited data. It is also an important seismological problem with good funding. Since classifying an event requires a mental comparison with remembered models, providing a larger set of symbols based on both visual and aural cues should result in better classifications. Being able to simultaneously present both auditory and graphic displays may also decrease the time required to make a correct classification.

A long dispersed surface wave train buried in noise is one of the more difficult events to identify on a time-series display. The energy is so spread out over time that the identification is often difficult without using spectrograms or phase-matched filters (if it is even known that the event is present). Such sequences are easily heard and recognized on the audio seismogram.

During seismic storms, a single small area will, in a short time, produce a large number of similar small earthquakes and aftershocks mixed with the normal background noise. Recognizing that an event comes from a seismic storm instead of another location is important during test periods.

One of the most successful audio application may be for identifying noise and equipment problems. Equipment and noise sources sound distinctly different from seismic sources. They tend to be periodic and persistent. When trying to locate these sources, several parameters may need to be adjusted. Being able to listen to the output allows the changes to be made quickly, without having to study plots and frequency diagrams.

4.7 EDUCATION

Education is potentially a very rewarding application. The explanation of how seismic energy travels through the earth can be enhanced with examples and explanations that draw on everyday listening experience, such as echoes or the sounds that objects make when hit. The intuitive grasp of physical acoustics that we accumulate in our life can be carried over to certain aspects of seismology.

4.8 MODEL MATCHING

Much of seismology is devoted to determining reasonable earth structures. Once an initial hypothesis is formed, numerical models are developed to test and further improve the hypothesis. Various methods of "fitting" data to models are available, but it is often more appropriate to build a guided model based on interactive input. The model is iterated and displayed versus the real data while the user makes certain adjustments to model parameters. Model and real data may be summarized with a goodness of fit, but may also be display overlaid. It is natural in such a case to also display the audification in side-by-side comparisons.

4.9 SIGNAL DETECTION

Signal detection is the identification of segments of continuous recordings that may have seismic earthquake signals on them. These segments are intensively studied to identify the onset time used for event location.
    The first problem related to signal detection is the recognition of small quiet signals buried in various types of noise. The converse problem is recognizing when noise segments contain no signals. Automatic seismic signal-detection algorithms can exceed the performance of a trained seismic analyst when background noise conditions are stable. Under varying noise conditions, analysts often do better. People are experienced in picking out specific sounds in a noise background, so it is natural to expect that the signal detection problem is well posed for audio presentations. Since the

automatic detector performance exceeds that of human analysts, it suggests the traditional analysts' display could be improved.

Related to the detection of signals is the recognition of multiple signals once a segment has been detected. In some cases several signals, each from a different source, may overlie each other. In other cases a signal may be complex, created by an aftershock sequence (all from the same source). Signals from different sources usually have strong differences in spectral character, so recognizing an overlying signal should be easy on the auditory display which emphasizes spectral differences.

4.10 ONSET TIMING

Once a segment is identified as containing a signal, the event's onset time (the time of the first arrival of energy) must be determined. For impulsive signals, it is quite easy, but the signal emerges slowly out of a background noise for signals whose source is emergent, or for sources at particular distances. Determining the correct onset tends to be a matter of experience and training related to projecting the rise of the signal back into the noise sequence. A combined audio and visual presentation may help, since small changes in the spectral character of the noise may indicate the first arrivals of the signal. Once an analyst is trained, a combined visual and audio presentation might decrease the review time.

5. PROBLEMS IN AUDIFYING SEISMOGRAMS

Ideally, processing techniques should preserve the physics of the input data while making all interesting features audible. This is required if the final display is to take advantage of our native ability and experience in interpreting natural sound. To be audible the sound must fit within human frequency and amplitude limits for comfortable listening (which is usually wider than the commonly available reproduction equipment).

It is obvious from its characteristics (see Table 1) that exploration seismic data requires processing to make events audible. These characteristics present the following problems when audifying raw seismic data:

- **The information is nearly subsonic.**
  Some recorded data includes audio frequencies, but the dominant energy for most reflections is centered near 30 Hz where it is difficult to distinguish small changes in pitch and timbre.
- **The bandwidth is narrow compared to natural sounds.**
  Seismologists go to great lengths to record broadband signals. Good

definition over three octaves is considered excellent. This is far short of the ten-octave range of natural sounds.
- **Both the direct wave and reflected signals are short compared to natural speech and most environmental sounds.**
  Most reflection recording is less than 3 seconds with much of the interest in the first second or two.
- **Classical seismic interpretation (where each waveform peak is interpreted) is nearly impossible in a straight audification.**
  In reflection seismic interpretation, coherent peaks on adjacent traces are interpreted as reflections of a rock layer. The times of these peaks and the interval to the next set of peaks are carefully measured and mapped. This measurement of each wiggle may be more amenable to sonification or visual display.
- **The dynamic range of some signals is extremely large (>100 dB) over a relatively short period.**
  This problem has to be handled by graphical displays too. One method used for plots is to clip waveforms when they exceed a particular amplitude.

Earthquake seismograms (see Table 2) present their own set of problems:

- **The signals are nearly all far infrasonic.**
  Typical teleseisms (distant signals) have their dominant energy below one 1 Hz and only the closest earthquakes and blasts have energy above 10 Hz.
- **Earthquake signals span more than 17 octaves.**
  This presents problems not only for auditory displays, but for almost any other form of raw data display. Researchers have found it difficult to record or deal with displays that include more than a few octaves of signal. Seismographs capable of recording broadband signals have recently become available, but interpretation and display techniques still rely on presenting different filtered bands.
- **The dataset size can be huge.**
  Because stations record continuously for years, the dataset size grows. If instrumentation has not changed, selected waveforms from many years of recording may be appropriate to a particular interpretation problem. Indexing, retrieving, and displaying these selected segments is a problem in itself.
- **The dynamic range of interesting earthquake signals is large.**
  Today, 24-bit recording systems with a dynamic range > 140 dB are common. Fortunately, in many cases 16 bits will adequately represent the waveform.

6. PROCESSING TECHNIQUES

A number of technologies are available to make seismic audifications. Some of them specifically address the problems discussed in Section 5 above. Other techniques are designed to make the audification easier to use.

6.1 PREPROCESSING

Preprocessing consists of adapting the raw seismic field data into a form that can be output to an audio interface. Most data requires a simple amplitude rescaling and DC removal to adjust the range of the recorded data to the range of the output D-to-A converters. Usually high-quality interpolation is also needed to produce an acceptable sample rate without distorting the input waveform.

Tr 80

6.2 TIME-COMPRESSION

Some shallow exploration reflection data has enough high frequency to be audible if played at recorded speed (Track 80, 0:02.7-0:29.9). Most earthquake records are not directly audible since the significant energy is all below 10 Hz. To be heard the seismograms must be processed to move the dominant energy into the audible bands. This process must preserve any of the relations to acoustic physics. For example, a low pianolike sound must remain pianolike when shifted in frequency. It must preserve the characteristic attack and decay of impulsive sounds as well as harmonic relationships. Two tones separated by an octave must maintain an octave separation shifted up in frequency.

Time-compression is playing data faster than it was recorded. It is a simple direct method of making low frequencies audible and it preserves the physics (the result still follows the wave equation). A data stream recorded at 40 samples/second and played at 8000 samples/second has a 200-times compression. This allows 3 hours of recording to be played back in less than a minute, comparable to the amount of time to quickly study 3 hours of plotted data.

6.3 AUTOMATIC GAIN CONTROL (AGC)

Some type of amplitude scaling has to be done to bring the signals in the range of the D-to-A. Many seismic signals have a dynamic range in excess of comfortable listening levels, so signals may be processed through an AGC prior to playback. AGC increases the volume during quiet periods and decreases it for large signals analogous to automatic level controls on recording equipment. Several different forms of gain control were tried. The most successful method was to divide each sample by the average absolute value of a small window of surrounding samples:

$$(S'_k) = s_k \sum_{i=k-n}^{k+n} |s_i|,$$

where n is half the window size.

Window lengths are typically 200 samples or about 1/40 second for most traces. This gain adjustment is rapid compared to most audio recordings. It changes the nature of the recorded data so that differences in spectral properties of signals are more easily discriminated.

It was also necessary to preprocess traces with AGC before doing frequency doubling. Otherwise the dominant signal swamped all other subtle arrivals.

6.4 FREQUENCY DOUBLING

While time-compressing data does move events from the subaudible to the audible, it brings with it another set of problems. Signals from some earthquakes may only last a minute on the original recording. Speeding up the recording by 200 times will result in a sound that only lasts a fraction of a second, so short that it is difficult to study. It may be recognizable as a single sound but it cannot be taken apart into its components.

Some early experimenters avoided the problem by using the minimum compression to make the seismogram audible. The resultant sounds were dominated by very low frequencies. Others shifted the frequencies into the audio range by multiplying the signal by a carrier to produce an FM signal. Unfortunately, this is an additive shift in frequencies. Thus, if 20 Hz is shifted to 440 Hz, then 40 Hz will be shifted to 480 Hz. This does not preserve the harmonic relationships. It also produces a section that is narrower bandwidth than the original (in terms of octaves) and a sound that is quickly fatiguing.

Close study requires a way to slow the playback without changing the pitch and the harmonic relationships. The operation of frequency doubling shifts any pure tone up one octave. The audio track may then be played at half speed to get the original pitch in a signal with twice the duration. The technique works well for pure sine waves, but complex waveforms have some unexpected results. Fortunately most seismic waveforms are simple enough that their character is preserved.

The development of this method may be illustrated with elementary trigonometry. Assume a pure cosine and sine signal $x = \cos(t)$, $y = \sin(t)$. Using double-angle relations

$$\cos(2t) = \cos^2 t - \sin^2 t,$$
$$\sin(2t) = 2\sin(t)\cos(t),$$

the doubled signals could be produced as

$$x_{\text{doubled}} = x^2 - y^2,$$
$$y_{\text{doubled}} = 2xy.$$

For an arbitrary signal s, we may generate the complex analytic signal,[9] S, whose real component is the original seismogram, and whose imaginary component is the Hilbert transform of s. If s were a pure sine wave, $H(s)$ would be a pure cosine wave. The frequency-doubled analytic signal S is then $S = s^2 - H(s)^2 + i2sH(s)$, where $H(s)$ is the Hilbert transform of s. In a like manner the envelope of a seismic signal is computed as

$$e(s) = \sqrt{s^2 + H(s)^2}.$$

Experience shows that for impulsive signals, the process of frequency doubling makes the signals more bursty, since it, in effect, squares the envelope of the signal. In order to reduce this effect on the frequency-doubled signal, the new envelope is divided out and replaced with the old.

$$S_{\text{corrected}} = \frac{S}{e(S)} e(s).$$

This preserves the amplitude decay of the original signal.

For pure sine and cosine waves these expressions are exact. For other signals, the results are not always as expected. For example, speech signals, doubled and played back at half speed which one might expect to sound simply like a slow talker, sound instead like a badly clipped tape played back at half speed. For the simpler signals in seismology the technique is adequate. Frequency-doubled and half-speed displays may be easily related to the normal displays up to an expansion of about four times. Beyond this, the signals all decay into something that sounds like a large metal sheet being shaken.

One problem with frequency doubling is that it accentuates the dominant energy and suppresses more subtle arrivals. Informal experiments indicated that it improves the display if digital automatic gain control is applied prior to frequency doubling.

6.5 ANNOTATION AND MARKERS

Just as a good graphic needs annotation such as labels and tick marks, so does a good auditory display. In bimodal presentation the graphic and audio must have some common markers to indicate the relation of *time* on the audio track to *position* on the graphic. The obvious solution, the animated cursor on a workstation display, is difficult to mentally synchronize with the correct corresponding sounds. There appears to be a high correlation when random signals are displayed on the video screen, an effect well known to animators and movie arrangers.

Therefore, a system of aural tick marks and annotation was overlaid. This turned out to be more difficult than expected. When the annotation has a markedly different character than the data, it separates perceptually from the data stream and is no longer synchronized. It becomes difficult to tell if an event occurs before or after a particular aural cursor.

Undamped bell tones were used in the initial experiments. They were replaced with drumlike sounds generated by passing an impulse through a narrow bandwidth, first-order Butterworth filter. Families of marker tones were produced by varying the bandwidth while keeping the low-cut frequency constant.

For sections rendered in stereo with multiple markers, each set of markers was panned to a slightly different location in the stereo field. This helped separate the markers from each other and from the data.

6.6 LOOPING

Each selection may be repeatedly played in a short loop. Three or more iterations may be required to understand the sound. Although the individual tracks are about a minute long, it is the selections of only a few seconds within the track that should be looped for study. For some recordings, using an equalizer to boost or cut different frequency ranges while the loop is playing will help extract particular sounds.[6]

6.7 INTERACTIVITY

Initial investigations of seismic audification used a custom interactive program with the ability to snip out pieces of a seismogram for processing and auditory display. Unfortunately, too much interactive flexibility is a problem when working with many sample data sets. It was unusual that an optimum setting for one seismogram could be used on the next. Instead it

---

[6] The reader might try programming their CD player to create this looping with the examples provided on the accompanying CD. It was avoided on the CD to better utilize the available playing time

was more valuable to play back a wide variety of seismograms with a single processing setup and pick out the obvious similarities and differences. This kept the processing relatively constant through training, providing the listener with a stable auditory environment for cross-sample comparisons.

6.8 STEREO PLACEMENT

Simple stereo processing, wherein a separate signal is played into each ear, was marginally successful. There seems to be more interpretable information in the sounds. In my own experience, repeated listening continues to uncover additional signals and relationships longer on stereo signals than on monaural signals.

Some signals were played back in stereo to convey the direction of a seismic source. The result was only partially successful. While the actual input signals were similar to naturally occurring acoustic phenomena, the selection of two signals for left and right ears was quite arbitrary and not analogous to the stereo effect heard in the environment. More advanced processing or a better choice of inputs may be required for useful stereo.

For other signals, a second audio channel was used to provide an additional annotation. Signals were panned from left to right to represent the seismograms recorded from different sensors on the ground. In this case, stereo position was created by simple amplitude differences. The result is partially successful, but better results might be had with additional stereo cues.[2]

When listening to the examples, there will be a difference in the interpretability through headphones and speakers for the stereo selections. Listening to each channel separately several times will help separate the sounds.

7. THE EXAMPLES

Seven audio examples on CD tracks 78–84 illustrate the problems and rewards of the techniques of seismic data audification.

7.1 MODEL DATA AS A REFERENCE

Model traces are synthetic seismograms created from an arbitrary earth model (see Figure 3) of layers of rock with varying properties (velocity, density, and attenuation). This model assumes a single explosion buried just below the surface. Thirty-nine vertical geophones evenly spaced along a line from the source position record the seismograms resulting from reflections and refractions of the source explosion.

Models are an appealing starting example because all parameters may be completely controlled, to the extent the assumptions in the modeling process themselves are correct. A model developed from an interpretation of field data may be used to validate the interpretation. If the synthetic seismograms from the model match the field data, then the interpretation could be correct. If not, then model parameters may be adjusted until the resulting synthetic seismograms match the field data. In other cases, results from two different models may be compared to determine if two modeling techniques yield similar answers, or if two different models create recognizable differences in the seismogram. Models also form reasonable

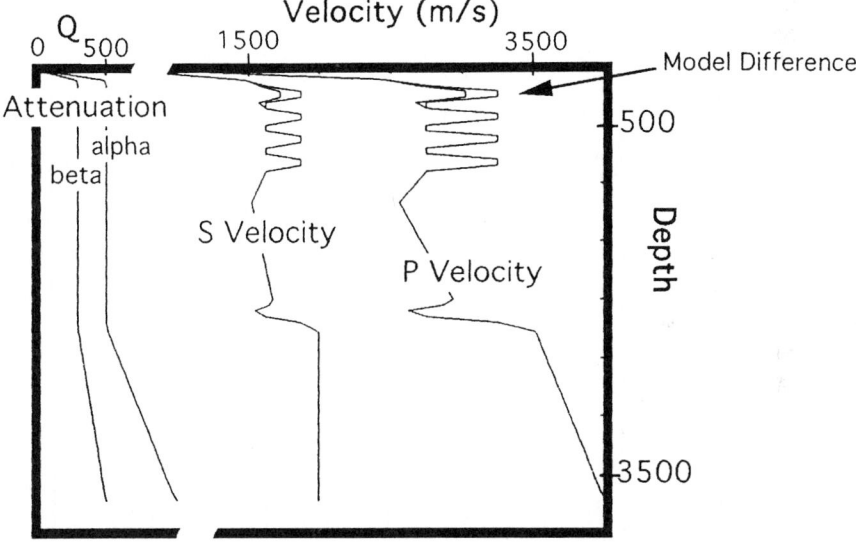

FIGURE 3 Synthetic Seismic Model Description. A complex velocity model was used to describe a possible layer cake geology. Attenuation models (Q) are shown on the far left. The geologic section was assumed to be a loose, dry, low-velocity material at the surface, underlain by four thin interbedded fast and slow layers, a low-velocity zone, and several zones with increasing velocity. The model was not representative of any particular geology, but was selected to represent a section with realistic complexity.

benchmarks for audification. The audification process may be tuned with perfect model data before attempting it with field data.

The first selection (Track 78, 0:01.6-0:07.0) contains synthetic data traces played back at 48 times compression to produce recordings from eight separate geophones every second. The dominant feature is the energy fade away as seismic traces at longer distances are played. Notice that when seismograms of varying distances are plotted at the same scale (see Figure 4), traces near the source are clipped while those at long distances are not visible. Because the audification has a wider dynamic range, it

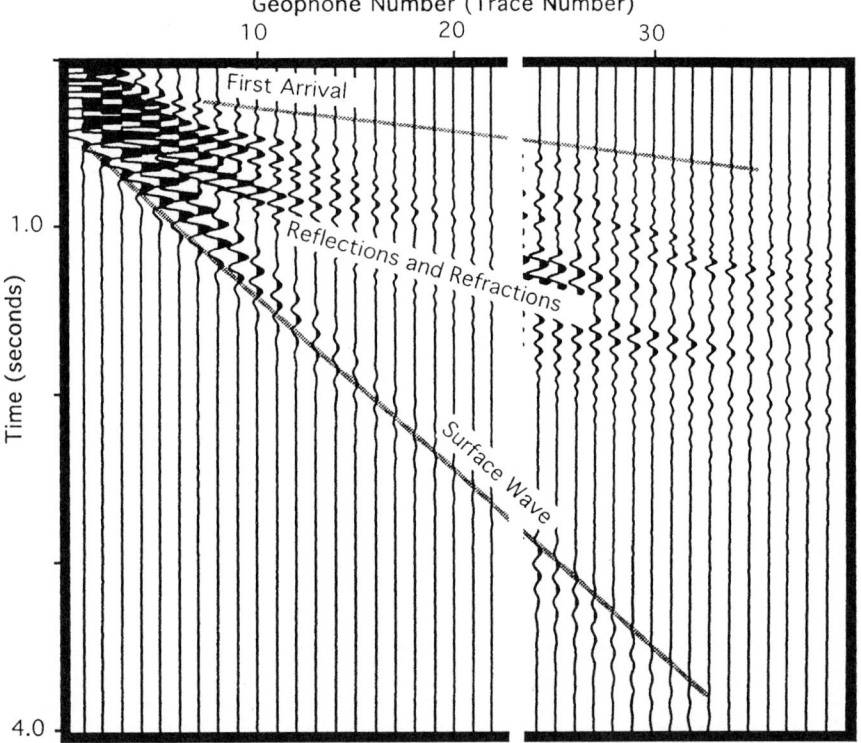

FIGURE 4 Synthetic Seismic Model Section. This synthetic seismic section is created from the velocity model in Figure 3. Each vertical trace is the velocity time series recorded from a single geophone on the ground. Deflections to the right are upward motion. Traces beyond 22 have been plotted with a larger gain to make the deflections apparent.

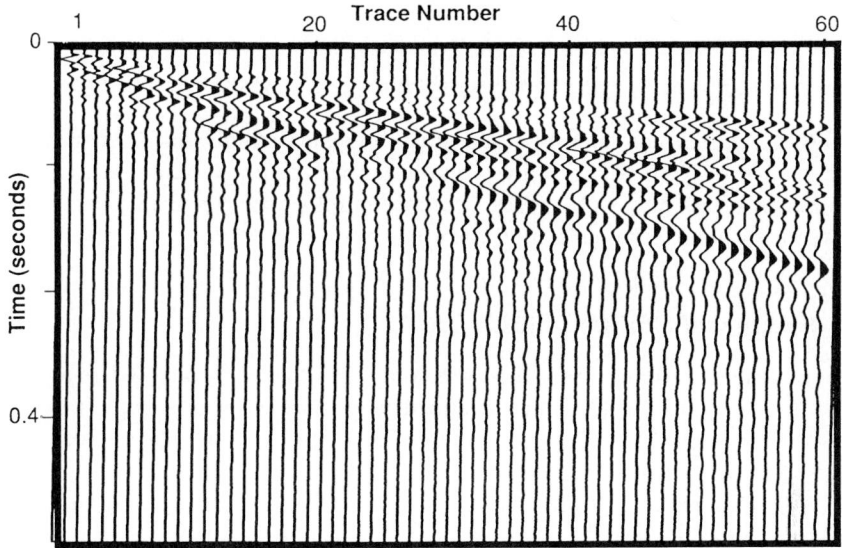

FIGURE 5 Shotgun Seismic Section. This shallow seismic section records the response from a shotgun blast into the earth near trace 1. The recording was done in three groups of 20 traces each. All traces were recorded with the same gain and are plotted with a common scale factor.

presents the closest traces without clipping, while leaving the distant traces audible. Typically, seismograms are normalized to the energy in each trace to remove the effect of energy decay with distance (sometimes called a spherical spreading correction). This may also be done in the auditory domain (Track 78, 0:08.1–0:13.8) so that time and distance changes in the recorded seismic signal produce clear timbral changes in the audification.

The seismograms are automatic-gain-controlled and frequency-doubled to examine the traces in more detail (Track 78, 0:14.9–0:18.6), then played back at half speed (Track 78, 0:19.4–0:25.8) for a time-compression of 24 times. The process may be repeated (Trace 78, 0:28.0–0:34.5) to arrive at a compression of 12 times.

Comparing the seismograms resulting from two different models is usually done by overlaying plots of the two in different colors. A naive method for audio is to play each model into one ear (Track 78, 0:36.1–0:41.8); this is a dichotic presentation. It does not take full advantage of our built-in stereo processing and is more analogous to displaying slightly differing plots on each eye. Differences in the two signals cause an apparent stereo shift from left to right. This is most pronounced at the distant traces.

Imbedded in the signals is a low-frequency surface wave[7] that travels more slowly (see Figure 4) than the other body waves. A second set of drumlike tones has been added to point out this later arrival (Track 78, 0:45.0–1:05.8). Each annotation tone immediately precedes the surface wave and each trace is also marked at the start. The annotation tone is positioned in a separate stereo location by panning it towards one channel away from the other tones and data. This example emphasizes the utility of audio markers to point out particular features.

Tr 79

7.2 SHALLOW SEISMIC FIELD DATA

These same techniques are valid for field data (Track 79) collected as a part of an investigation of the shallow subsurface. A number of seismic recordings were collected along a line of increasing distances from the source, just like the model traces (see Figure 5).

Field recording equipment can only record a few channels simultaneously (20 channels for Track 79, 0:3.5–0:11.5) and it is, therefore, necessary to repeat the explosions after the geophones are moved to build up the complete section. The assumption is that the source explosions are identical and that the geophones are identical so that each trace should sound similar to adjacent traces and the section (Track 79, 0:3.5–0:11.5) should be similar to the model (Track 78). This first example of a shotgun fired into the ground[4] is scaled like the first model section. The differences every 20 traces are a result of slight differences in the three shots fired to create the 60 traces. There are also small differences from trace to trace indicating minor differences in the way individual geophones are attached to the ground.

The second section (Track 79, 0:12.3–0:22.1; see Figure 6) is a composite section formed from six hammer blows recorded into twelve geophones on each blow. Every sixth trace comes from the same hammer blow. Thus traces 1, 7, 13, and so on are from hammer blow 1, traces 2, 8, 14, and so on are from hammer blow 2, and traces 3, 9, 15, and so on are from hammer blow 3. Sharp changes in the sound from trace to trace or regular patterns are artifacts of the field procedure and indicate that the prior assumptions are not completely valid. Extreme variations and prominent patterns indicate inadequate field quality control. Rapid repeating (looping) of each trace emphasizes sharp changes from trace to trace

---

[7] Surface waves are confined to the uppermost layers of earth and concentrate energy near the upper boundary, much like familiar waves in the ocean. Body waves are those waves that travel through the material, much like familiar sound waves.

Listening to the Earth Sing

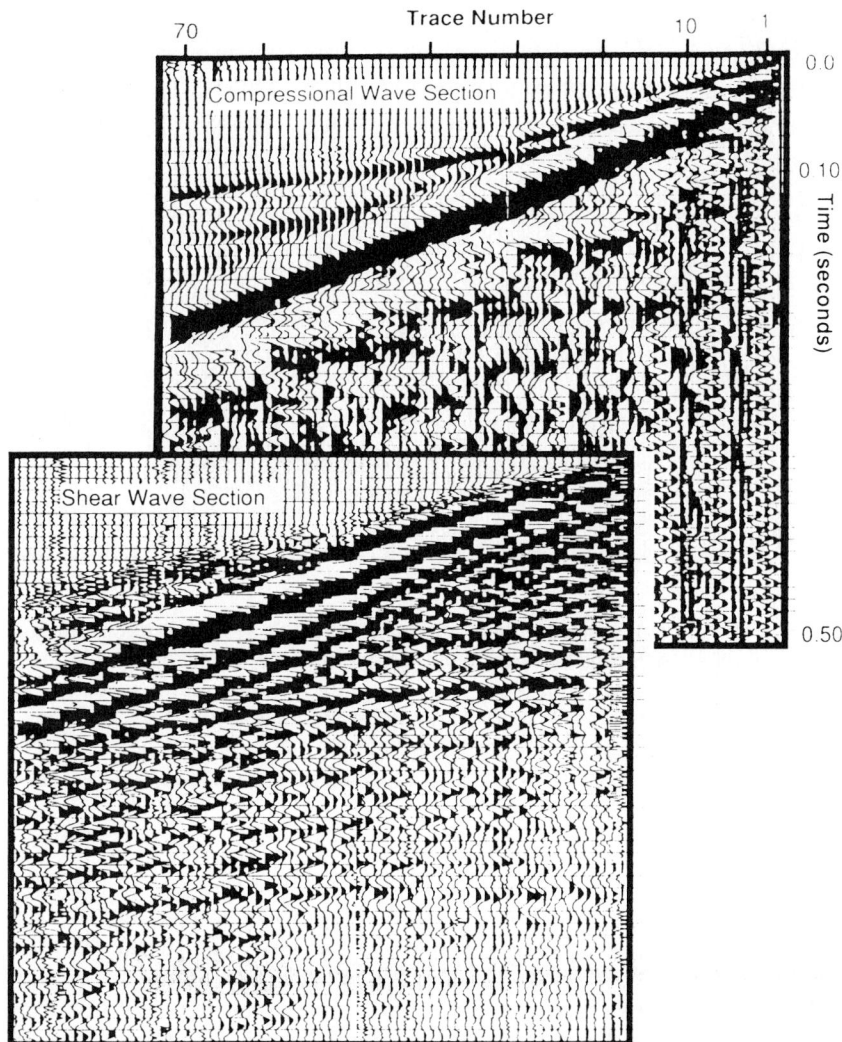

FIGURE 6   P and S section from Caldwell Ranch. At Caldwell Ranch 72 channels were recorded in 6 groups of 12. Traces are numbered from right to left. Data was acquired with 10 Hz geophones for the compressional section and 50 Hz geophones for the shear section. Channel gains were optimized for reflected energy expected at 0.20 seconds. Earlier arrivals were clipped. This section was processed with AGC before being plotted. On the compressional section several of the first few traces are ringy.

394 Chris Hayward

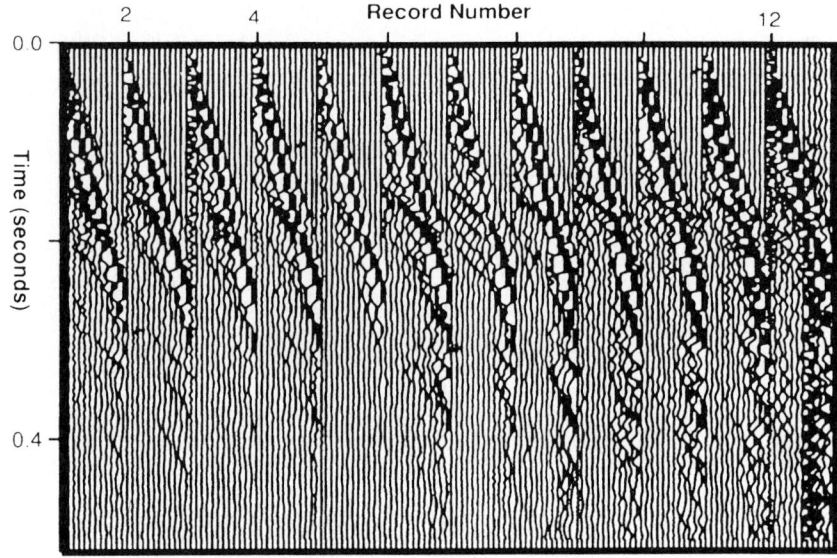

FIGURE 7 Twelve Consecutive Shot Records. Each set of 12 traces was acquired with a different hammer blow. Channel gains were adjusted in the field to try to keep the recorded energy constant. With perfect acquisition and flat-lying bedrock, each record should be identical to the adjacent ones. With this particular equipment, the field operator can view only one record (12 traces) at a time.

(Track 79, 0:22.7–0:37.7). Applying AGC to the section (Track 79, 0:38.8–0:48.6—left channel) better emphasizes some of the weaker signals. In this case, a second set of shear wave seismograms is used to measure depth to the water table. Interpreting these joint experiments requires recognizing arrivals common to the two sections and arrivals that appear to be similar but show slightly different velocities and arrival times on the two sections. On plots these two may be overlaid in color. For audio, each is played into one ear (Track 79, 0:38.8–0:48.6).

7.3 SHALLOW SEISMIC REFLECTIONS

Common depth point recording (CDP) is a standard exploration method that records a hammer blow into a line of evenly spaced geophones (in these examples, 12 geophones). The hammer source point and the geophones are shifted a measured distance in-line and a second recording made. This is repeated hundreds of times until the line of geophones may be moved several miles from its initial position. Again sudden changes are indications of field problems. It is likely the seismic observer was unable to find time to do careful quality control. In this case the audio channel forms another important method to warn of problems.

A set of four sample records illustrate some field problems that may be encountered (Track 80, 0:02.7–0:29.9). This high-frequency seismic section is unusual in that much of it is audible with no time compression or special processing. These records illustrate significant quality control problems not detected on the visual field display during the field collection phase (field crew of seven people, newly trained operator, sledge hammer source, four to six shots per minute). **Tr 80**

A set of 12 consecutive records (Track 80, 0:32.0–0:56.8; see Figure 7), played at 8 times compression, illustrates the ease of quality control. Traces are played in stereo with a simple amplitude pan relative to their ground position. Ideally the listener would hear the sound originate at the corresponding geophone position. The audio evidence of field problems could even be radioed to each crew member, allowing each individual to monitor his contribution to recording quality. For small (two- to four-person) operations it may even allow elimination of one crew member.

7.4 NOISE AT EARTHQUAKE SEISMIC STATIONS

For seismic observatories recording earthquakes, a key quality control issue is finding quiet locations. Seismically noisy locations are generally poor candidates for recording small distant earthquakes. Typically a noise survey is made at several candidate sites by recording background noise for days or weeks. These recordings (see Figure 8) are then analyzed to produce frequency spectra (see Figure 9) of the background noise.

An audio selection of background noise from four different stations (Track 81, 0:01.8–0:18.0) demonstrates that each station has a distinctive character. Because this data is dominated by 1/20 to 5 Hz frequencies, it requires much higher time-compression than exploration data—200 times in this case. The distinctive sound suggests that site noise conditions may be recognized by listening for a few minutes to a compressed recording. **Tr 81**

Following the four noise examples are two extended selections (Track 81, 0:19.4–1:05.4) from two of the sites represented in the short recording. These two stations are immediately recognizable from the prior 5-second samples. The stations have distinctly different and recognizable noise spectra, although this is not immediately apparent on the time-series (see Figures 10 and 11). Some individuals may learn to recognize characteristic sounds quicker than plots of characteristic spectra. **Tr 81**

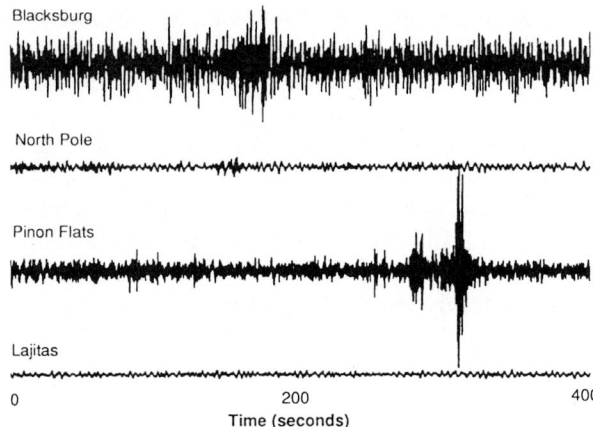

FIGURE 8 Background Noise at Four Seismic Observatories. All seismograms were plotted with a common scale factor. Each seismogram is of the vertical GS13 recorded at 40 samples/second into a 24-bit digitizer. Segments were selected from random times and may not be representative of long-term noise conditions.

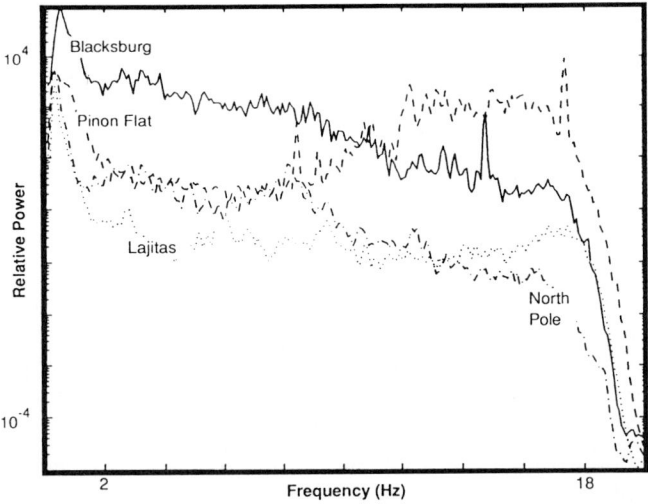

FIGURE 9 Background Noise Spectra at Four Seismic Observatories. Four hundred seconds of data from each of four stations was used to calculate the average noise spectra. Strong peaks at Blacksburg and Piñon Flats are indications of resonant man-made structures.

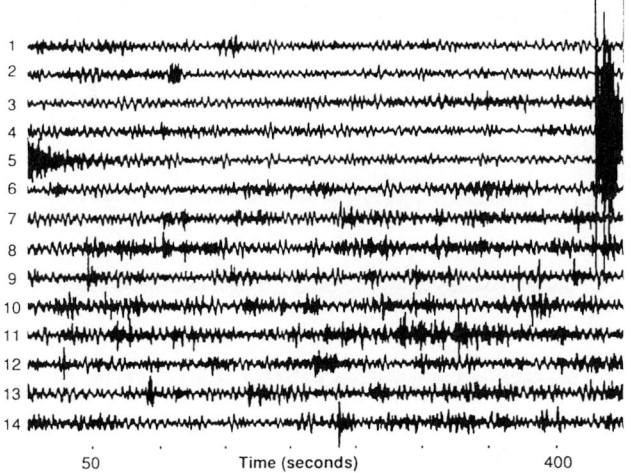

FIGURE 10 Seismogram from North Pole, Alaska Near Fairbanks. Each horizontal segment is 450 seconds of data and is contiguous with the segment below. A small earthquake is recorded on segments 4 and 5. A microseismic storm or other source of noise occurs from segments 8 through 14.

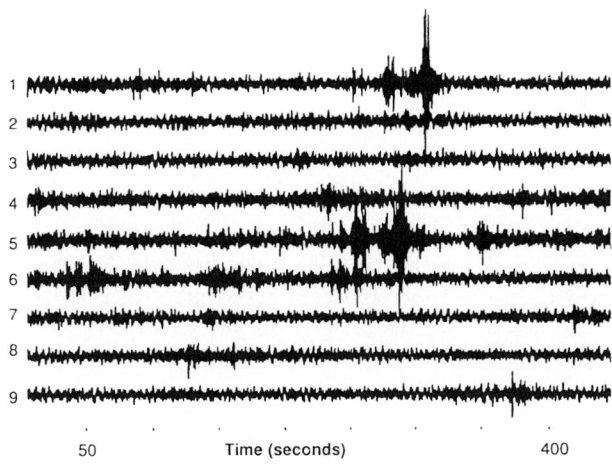

FIGURE 11 Seismogram from Piñon Flats, California. Some of the bursts on the seismogram are small earthquakes although the seismogram is also contaminated with noise.

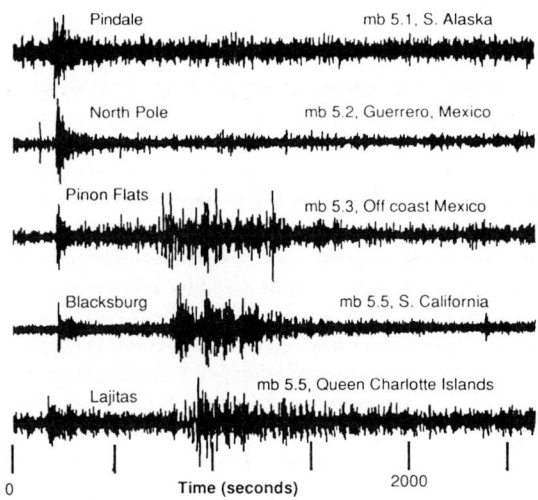

FIGURE 12 Seismograms from Earthquakes 27 degrees away from the seismic observatory. The initial impulse is the compressional wave arrival. The second package of energy (seen best on the Blacksburg station) is the surface wave arrivals. Each seismogram was normalized with a separate gain factor. The separation of the compressional wave arrival and surface wave arrival is nearly constant for a given distance.

Tr 82

7.5 SELECTED EARTHQUAKES

Recordings of selected earthquakes, all about the same distance (27 degrees) from the recording station, illustrate the characteristic features of audio earthquake seismograms. Five different earthquakes recorded at five different stations (Track 82, 0:03.7–0:20.5; see Figure 12) have been time-compressed 800 times. On plots the primary control of the appearance of a particular recording is the distance from the event. At most distances an audio seismogram of an earthquake has three parts: two sharp sounds (reminiscent of a heavy door with a latch closing) followed much later by a drawn out, low-frequency surface wave that usually shifts from higher frequencies.

A selection of earthquakes recorded at from 12 to 101 degrees (Track 82, 0:23.3–0:52.8; see Figure 13) illustrates the progression as the station is farther and farther away. At long distances part of the energy is subaudible, even at 800 times compression. Doubling the compression to 1600 times (Track 82, 0:54.0–1:05.7) demonstrates that all three parts are present even at far distances. These low-frequency surface waves, while audible, may not always be easily visible on a plot without special filtering. Because relative distances may be recognized on either audio or graphic seismograms, the display choice may be optimized for the operator preference.

Tr 83

7.6 NUCLEAR EXPLOSIONS

Two past nuclear explosions at White Sands, New Mexico, recorded in Lajitas, Texas, are played back at 200 times compression (Track 83, 0:01.4–0:13.5; see Figure 1). These relatively close events (13 degrees) do not have a strong surface wave which is typical of explosions. They also reverberate compared to previous recordings. The reverberant structure is an indication of an unusual geology along the travel path from the test site to Lajitas.

Stereo recordings (Track 83, 0:15.0–0:21.7) were created by playing vertical ground velocity into the center channel, north-south velocity into the left channel, and east-west velocity into the right channel. Shear waves and surface waves have strong horizontal components, and they should demonstrate a stereo preference. The apparent "roominess" of the sound was unexpected and is yet unexplained. Sounds within the explosion seem to shift from channel to channel in a complex way; the patterns are more

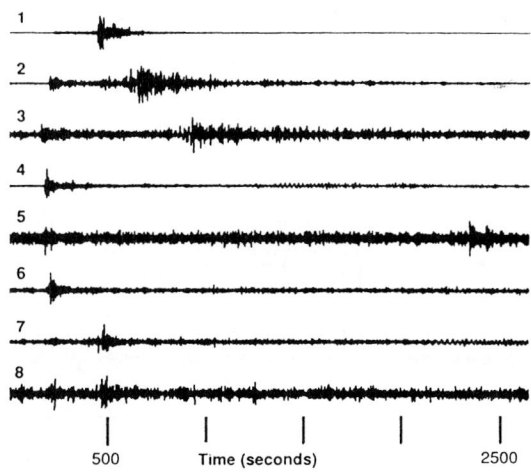

FIGURE 13 Eight Earthquakes from 12 to 101 degrees away. Eight earthquakes illustrate the progressive separation of the initial and later arrivals with increasing distance. The low-frequency dispersed surface wave is easily visible on seismograms 4 and 7 (for a listing of the events and exact distances, see the CD annotations).

apparent when the explosions are frequency doubled and slowed down (Track 83, 0:21.7–0:25.7).

7.7 ARRAY DATA

A particularly rich application for stereo presentation is recorded array data. Seismic arrays usually include 10–30 seismometers spaced in a regular pattern, such as concentric rings. Recordings of such arrays may be beamformed during processing to enhance signals in particular directions or to estimate the direction to the origin of a particularly interesting earthquake. If an audio seismogram from each element of an array were played back simultaneously, a situation similar to a noisy cocktail party or choral choir would result. The goal is to be able to pick out individual voices to find those elements that are unusually quiet or noisy. Another possibility is to beamform the array to enhance particular directions and then position those signals correctly in the stereo field. Ideally, the output of a head motion sensor attached to the display user could be used to control the beamform direction. Turn your head left and signals to the west would be enhanced.

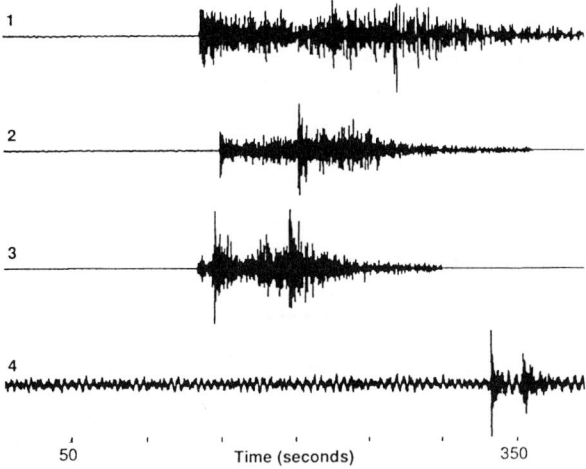

FIGURE 14 Center Element for a Stereo Experiment. The four displayed seismograms are for one of the vertical channels used in forming the stereo audification. The events are listed in the CD annotation. Each event is plotted with an independent scale factor.

Listening to the Earth Sing

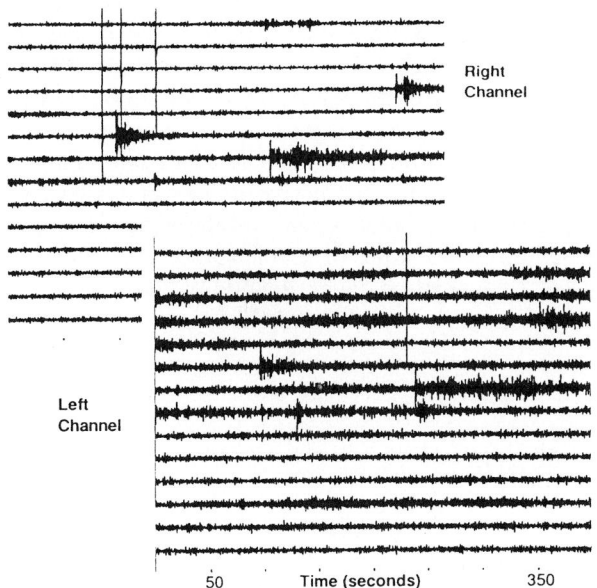

FIGURE 15 Two Channels from the 100-Km Array Audification. The right channel contains a number of data glitches (three are obvious on the plot). The left channel contains one. Seismic events on the two channels arrive at slightly different times. The event on segment 4 of the right channel has no corresponding signal on the left channel. This was either a very local signal, or a signal that is obscured by high-noise conditions on the left channel.

For example, on audio track 84, stereo imaging was employed as follows: Three elements, spaced in a triangle 2 kilometers on a side, were selected. One was played in the center channel, one on the left channel, and one on the right channel (Track 84, 0:01.5–0:30.1; see Figure 14). The stereo separation is prominent for all these events although it is difficult to associate a preferred direction for a particular earthquake. A single earthquake will include sounds with different delays between the channels because different wave types travel at different velocities. This is not something that is common to acoustic events.

The first three events are all dominated by low frequencies. In acoustic signals, low frequencies are difficult to locate and, therefore, we have less success in accurately locating these signals. The final earthquake is a high-frequency signal. The stereo audio clearly suggests two different directions although both sounds originate from the same signal.

The final example (Track 84, 0:33.3–1:09.9; see Figure 15) is an audification of a three-element array with each element separated by 100 kilometers. The energy arrives with significant delays between stations. The separation is great enough that a waveform recorded at one station may look significantly different from those at the other two. The immediately obvious characteristics are the distinctive noise patterns related to each channel. Channel noise conditions appear unrelated to each other. Glitches and pops occur at unrelated times on the left and right channel, but never the center channel. The earthquake at the end of the recording appears to occur almost simultaneously on all channels.

8. CONCLUSIONS

These examples of audification suggest its value for doing diagnostic quality control in field reflection seismology and in earthquake seismology. A method for moving the frequencies of seismic waves from subaudible to audible has been demonstrated. Repeated listening to a short waveform helps to better understand the signal and to increase the effective change in character from sound to sound. Simple processing yields the most natural interpretable sounds. Stereo presentations seem to include more extractable information and have potential, although the simplified method used here is inadequate for most signals.

Annotation is particularly important. Good annotation allows the listener to match events on a graphic display with a correlated audio. In multiple annotations, tones must be separated from each other by slight frequency differences or positions in stereo. Annotation tones must be similar to the signal or the listener will perceive two unrelated streams and it will be difficult to relate the timing between the two. Good annotation avoids the problem of finding terminology to describe sounds to a reader. Examples may be played with audible markers just prior to the sound of interest in a manner analogous to adding arrows to a graph.

The wiggle plot results in a poor representation of frequency information, but the auditory display gives a good idea of the overall characteristics and evolution of the spectra. Just as it is possible to process and plot the seismogram to extract spectral information and plot the spectrogram, the seismogram data could be used to control a sound generator (i.e., sonified) to present each wiggle. However, this approach does not fully utilize the natural advantage of each style of presentation. Obvious events (those with high signal-to-noise ratios) are clear in both the auditory and graphic displays. More subtle events may be initially recognized in only one domain, but may then be processed to enhance them in the other domain.

The natural advantages of each seem to encourage the use of graphics to display time domain variations and the use of audio for frequency domain information.

9. FUTURE EXPERIMENTS

A number of experiments in several subject areas seem worthwhile. A natural extension of Speeth's earlier work would be to bury known seismic signals in background noise and then to compare audio and visual identification. Of particular interest, however, is the real-world performance of analysts using their standard visual analysis tools with and without audio.

A second rich area for exploration is the use of stereo auditory displays. Single stations could include three-component motion detectors so that processing information from these detectors would determine direction. Even more exciting would be the use of stereo ,pmotpromg with array data, when head position could be used to beamform the output of the array. Alternatively, the raw output of the array could be processed to set up the station locations in spatial locations. The use of stereo might allow identification of compression and shear arrivals. Shear energy will predominate the horizontal channels that could be placed in a distinct location in audio space.

Combined audio and visual displays could be useful in the routine analysis of seismic data. If an analyst is given audio tools, the question is "under what conditions are they useful and when are they distracting?" If the tools are to some extent "adjustable," what sorts of adjustments are most useful to the analyst?

The use of audio suggests new possibilities for seismic attribute displays in exploration data, including direct hydrocarbon indicator displays of various types, audio attribute displays, and amplitude vs. offset displays. Of particular interest are displays for modeling and quick look, when the audio tag may indicate whether the visual display is useful.

The audio seismogram is still in its infancy. The tools to produce it are currently cumbersome and not part of the daily tool set available to seismic analysts. The real proof of the utility of the tool will come only when auditory display is a natural part of seismic analysis. This will require making audio tools accessible to working seismologists in a natural well-designed interface.

ACKNOWLEDGMENTS

The author gratefully acknowledges the unending patience and encouragement of Gregory Kramer who would not let this work die. Without his continued efforts and substantive input this paper would still be an uncompleted collection of notes and outlines. Thank you, Greg. The initial material development was a direct result of the First International Conference on Auditory Display held at the Santa Fe Institute. The comments and suggestions from participants helped define the problems and method of attack.

REFERENCES

1. Aki, K., and P. G. Richards. *Quantitative Seismology: Theory and Methods*. W. H. Freeman, 1980.
2. Begault, Durand R., and Elizabeth Wenzel. "Techniques and Applications for Binaural Sound Manipulation in Human-Machine Interfaces." *Intl. J. Aviation Psych.* **2(1)** (1992): 1–22.
3. Bullen, K. E., and Bruce A. Bolt. *An Introduction to the Theory of Seismology*, 4th ed. Cambridge: Press Syndicate of the University of Cambridge, 1985.
4. Clark, R. D. "Gunning for Data." *The Leading Edge* **3(11)** (1985): 20.
5. Frantti, G. E., and L. A. Leverault. "Auditory Discrimination of Seismic Signals from Earthquakes and Explosions." *Bull. Seismol. Soc. Am.* **55(1)** (1965): 1–26.
6. Hilterman, Fred J. "Three-Dimensional Seismic Modeling." *Geophysics* **35(6)** (1970): 1020–1037.
7. SMU Staff. "Development of an Intelligent Seismic Facility and Preparation for Participation in the Conference on Disarmament Group of Scientific Experts Technical Tests." Semi-Annual Technical Report, Southern Methodist University, Defense Advanced Research Projects Agency Contract #MDA 972-88-K-0001, 1991.
8. Speeth, Sheridan Dauster. "Seismometer Sounds." *J. Acous. Soc. Am.* **33(7)** (1961): 909–916.
9. Taner, M. T., F. Koehler, and R. Sheriff. "Complex Seismic Trace Analysis." *Geophysics* **44(6)** (1979): 1041–1063.

Sara Bly
Xerox PARC, 3333 Coyote Hill Road, Palo Alto, CA 94304;
bly@parc.xerox.com

Multivariate Data Mappings

An on-going issue in data exploration is how best to represent the data to support finding its structure and patterns. Visualization techniques are traditionally useful but audio representations are being explored as well. This paper describes an exercise in which three different aural mappings were presented for the same six-dimensional data. Informal observations of the use of these mappings indicated the importance of identifying structure in the data and of integrating data exploration techniques.

1. INTRODUCTION

An on-going issue in data exploration is how best to represent the data to support finding its structure and patterns. Graphical displays are a standard means of presenting data, and there is considerable literature about the area of graphical data representation (e.g., Tufte,[7] Friedman et al.,[3] Everitt,[2] and Tukey[8]). Whatever the technique, a first concern is to understand as much as possible about the data itself. For example, plotting logarithmic data on an x,y graph is usually not as enlightening as plotting it logarithmically. With multivariate data, it is important to know not only the relationships between various variables but also which variables are predominant in carrying the information about that data.

Audio representation of multivariate data (sonification) raises additional problems. In particular, one must have knowledge of acoustics as well as of the data itself. Acoustical parameters are not independent, nor are they equally prominent to the listener. Given the need to understand as much as possible about the relationships among both data and acoustical parameters, it is a significant challenge to determine how to map data to an audio representation. At this time, there are no known methods for determining a best mapping.

One technique for studying data mapping is to create data with known characteristics (constructed data) and then apply various mappings to that data. As an exercise in provoking ideas and discussion, two data sets were generated for exploring data-to-acoustic mappings. Before the International Conference on Auditory Display (ICAD '92), participants were given the data and invited to bring representations to the workshop. During the workshop, three mappings were presented to attendees. This paper provides an explanation of the data, the results of informally listening to that data, and some points of discussion raised by these data mappings.

2. MULTIVARIATE DATA SETS

Two types of data were selected for the exercises: static data in the realm of discriminant analysis and time-varying data in the spirit of pattern detection. The data sets were constructed algorithmically so that definite structure existed within each data set. Six dimensions were chosen as sufficient to be multivariate and difficult to present graphically. For both types of data, the structure or patterns were completely determined by the six dimensions.

2.1 DATA SCENARIO A: FINDING THE GOLD

The first exercise was representative of static multidimensional data typically encountered in discriminant analysis problems. There are two distinct sets of the data. The data are six-dimensional, each data 6-tuple is a static sample, and the order of the samples is not significant. The problem is to find a way to identify unknown samples as belonging to Set 1 or to Set 2. The following scenario was used to motivate the exploration of the data:

"Can you find the gold?" It is hypothesized that six different aspects of the land in which gold may be found are determinative of whether or not gold is there. The first 20 data variables (each 6-d) are from sites known to have gold; the second 20 data variables are from sites known not to have gold. For each of the remaining 10 data variables, decide for each whether or not it is from a site with gold.

The data was taken from a previous study.[1] A set of 100 samples was generated from a normal random deviate generator and then separated into the two distinct sets. Only samples in which all six variables had positive values between 0.0 and 3.5 were included. A sample, $s = (x1, x2, x3, x4, x5, x6)$, belonged to Set 2 if and only if

$$x2^2 + x3^2 + x4^2 + x5^2 + x6^2 \leq 1.5^2$$
OR
$$x1^2 + x3^2 + x4^2 + x5^2 + x6^2 \leq 1.5^2$$
OR
$$x1^2 + x2^2 + x4^2 + x5^2 + x6^2 \leq 1.5^2.$$

At least five of the six variables in each sample of Set 2 had a value less than 1.5 and at most one of the variables $x1$, $x2$, or $x3$ could have a value greater than 1.5. Note that the two sets are completely distinct only in six-space; any representation in fewer dimensions will overlap Set 1 and Set 2. Attachment A contains 20 known samples from Set 1 and 20 known samples from Set 2 used as training sets. Attachment B lists three sets each with ten unknown samples. These three sets were used as the test samples in the exercise.

2.2 DATA SCENARIO B: IS IT RAINING?

The second data set was representative of time-varying multidimensional data typically encountered in pattern recognition problems. The data are six-dimensional, each data 6-tuple is one sample in the sequence, and the order of the samples is fixed. There are two sets of data (additional test data sets were generated but not used). In both sets, there are known areas identified as patterns within the data. The problem is to find a way to identify the pattern areas in the data. The following scenario was used to motivate the exploration of the data:

"Should I roll up the car windows?" It is hypothesized that six different measurements in the air affect the likelihood of a thunderstorm occurring. The six measurements are recorded daily during three different periods, each for 100 days. In the first two periods of 100 days, we know when thunderstorms occurred (see data descriptions below). In the third set of 100 days, on which days do you advocate rolling up your car windows?

A data set of 2,000 samples was generated by a random walk in 6-space from 0.0 to 10.0 in each dimension. The pattern area is defined as the interior six-space where $0.0 \leq x1 \leq 9.0$, $0.0 \leq x2 \leq 2.5$, $2.0 \leq x3 \leq 4.5$, $1.0 \leq x4 \leq 10.0$, $4.5 \leq x5 \leq 7.0$, and $7.0 \leq x6 \leq 10.0$. The random walk is designed so that movement has some slight predictability. A variable only moves 0.3 in any direction 80% of the time. However, 20% of the time, a variable may move up to 3.0 in 5 time steps. Two sets of 100 samples each with the storm areas identified were available to participants.

3. MULTIVARIATE DATA MAPPINGS

Three different data-to-sound parameter mappings were applied to the data in "Scenario A: Finding the Gold." (Although participants had tried mappings for "Scenario B: Is It Raining?," no one offered to submit a mapping for the workshop session. In general, it was extremely difficult to find a mapping that identified patterns given the information.) The intent was that participants explore various mappings of data to sound without using standard data analysis techniques. In any actual exploratory data work, an analyst would expect to use various math analysis techniques. In the case of these constructed data sets, such functions will immediately detect the patterns in the data.

The three different mappings were presented to approximately 25 participants in ICAD '92. For each mapping, the known sets were played for training. A set of ten test samples was then presented. These ten test samples differed for each of the three mappings. Attachment B contains the three test sets.

The results support previous work demonstrating that nonspeech audio can effectively represent multivariate data. The intent of this exercise was to demonstrate mapping techniques and to raise questions for discussion. In that sense, the constructed database was useful. In the broader sense, however, this exercise should be considered in the context of ongoing work in sonification. In general, techniques are designed for interactive exploration of data and for applying as much knowledge as possible to the data itself. This exercise prevented both of those crucial aspects of sonification. Thus it is probably not surprising to note that the more time that was spent on the data exploration, the more positive the results in terms of being able to identify the samples correctly.

TABLE 1 The results of the test for Mapping 1 were essentially random.

# correct	10	9	8	7	6	5	4	3	2	1	0
# participants	0	0	0	3	3	7	3	5	?–3	0	0

3.1 MAPPING 1

The first mapping was done using the granular synthesis algorithm[4,5] on a Kyma system.[6] Although sounds are typically designed over a couple weeks to represent a given database, these were generated in only a few hours. The resulting data had each of the six dimensions mapped arbitrarily to six of the nine possible algorithm parameters. The parameters used were (1) the number of grains in the cloud, (2) the initial center frequency of the cloud, (3) the maximum deviation of the grain frequency from the center frequency, (4) the change in the center frequency over the duration of the cloud, (5) the change in frequency deviation over the duration of the cloud, and (6) the inter-grain distance. In Table 1 is shown the distribution of correct identification of samples.

3.2 MAPPING 2

The second mapping was done using a Yamaha TG33 MIDI tone generator driven by the Porsonify toolkit (see Madyastha and Reed in this volume). Several hours were spent trying different mappings of data dimensions to sound parameters. The final mapping was selected as the one that best seemed to differentiate (to the listener) the two sets of known samples. The sound parameters selected were balance, timbre, sustain, pitch, duration, and volume. Note that in the presentation session, the two audio speakers were relatively close together. Thus, balance was not as useful as it was for the initial single listener using headphones.

In Table 2 is shown the distribution of correct identification of samples. The results show that the audio representation helped listeners discriminate between the two sets of data. Seventy-five percent of the respondents correctly identified more than half of the test samples. This result is analogous to earlier work[1] and suggests that straightforward mappings of data to audio are immediately effective in offering clues for data exploration.

TABLE 2 The results of the test for Mapping 2 were definitely better than chance.

# correct	10	9	8	7	6	5	4	3	2	1	0
# participants	0	1	6	7	4	4	2	0	0	0	0

TABLE 3 The results for Mapping 3 indicated clear discrimination between data sets.

# correct	10	9	8	7	6	5	4	3	2	1	0
# participants	3	7	7	5	?-1	1	0	0	0	0	0

3.3 MAPPING 3

The third mapping was done using the Kyma environment,[6] and the data mapping was explored over the course of a few days. In addition to trying different combinations of mapping data dimensions to sound parameters, various mathematical functions were applied to the data and the results mapped to sound parameters. The mapping finally chosen was one which mapped the sum of the squares of the six dimensions to pitch. For each data sample, a standard reference pitch was played followed by a pitch based on the sum of the squares. A sum of 2.5 was used as the dividing point between sets and as the reference pitch data value.

In Table 3 is shown the distribution of correct identification of samples. This mapping was extremely effective in allowing listeners to discriminate between the two sets of data. In fact, since the algorithm used to separate the two sets depended on summing the squares of the variables, using the sum of the squares was an excellent use of data characteristics. As a result, only one acoustical parameter was then necessary and almost any mapping scheme would have worked.

This mapping illustrates several interesting points about sonification:

- First, identifying any structure in the data is a crucial first step for data representation. Clearly this mapping derived considerable benefit from the attempts to discover some structure in the sample values.
- Second, this scheme does not allow completely correct identification of the test samples. The sum of the squares, $x1^2 + x2^2 + x3^2 + x4^2 + x5^2 + x6^2 = 2.5$, does not separate the two sets. In general, the sum of the squares for Set 2 will be small relative to the sum for Set 1. However,

Multivariate Data Mappings

noting again the algorithm in Section 2.1, it is possible for one of the values, $x1$ or $x2$ or $x3$, to be arbitrarily large in Set 2, thus yielding a sum of squares as large as those in Set 1. Note that Sample 4 in the Set 2 known samples (Attachment A) has a value of 6.9 for the sum of the squares, thus suggesting it belongs in Set 1. As it happens, the ten test samples were fairly easily distinguished by the sum of squares method. Sample 8 of the test for Mapping 3 (Attachment B) has a value of 2.518 for the sum of the squares, making it right on the border of the distinction.

- Third, reducing the data to a single variable (the sum of the squares) greatly simplifies the representation. This point was made explicit to workshop attendees when the value of the sum was mapped to "no" (for Set 1) or "yes" (for Set 2).
- Finally, this experience highlights the difficulty of knowing when a pattern has been found. Since only a few samples did not fit into the sum of squares scheme, it was a question whether or not those few samples were outliers or "mistakes" in the data. With actual data, one might have trouble knowing when a sufficiently good structure has been found.

4. DISCUSSION

The two major issues arising out of the data-mapping exercise concerned the process of data exploration and the use of constructed data. First, it is clear that the question of mapping data parameters to acoustical parameters depends on more than arbitrarily mapping one to the other. Serious consideration must be given as to which factors will make the process of data exploration, especially sonification, most effective. Second, given the importance of finding structure in the data, the use of constructed data with built-in structure may be problematic. However, a well-defined structure can be a useful tool in studying data mapping. When and how to use constructed data depends on the questions under study.

The mapping exercise presented in this chapter demonstrates both that sonification can be used for data representation and that knowledge of the structure of the data is crucial in providing an effective mapping. Three points about the data exploration process are worth raising in particular. One, for most multivariate data exploration problems, there is little known structure in the data. Therefore, one should take advantage of all techniques for analyzing the data (for example, principal component analysis). Any time that the data can be reduced to fewer dimensions, the value of the representational techniques is increased. Furthermore, dominant features can be mapped more appropriately to acoustical parameters. In general, sonification should be integrated with other data exploration methods and not viewed as an isolated technique. Two, the sound parameters themselves are not independent of one another. A good understanding of psychoacoustics is critical in designing environments for sonification. Just as relationships among data parameters are important to understand when constructing a mapping, so are the relationships among acoustical parameters. Three, an interactive environment is critical for the exploration process. If the structure were known, the problem would be solved. Given that it is not, the scientists/analysts should have every opportunity to examine the data in a wide variety of displays. Such an environment can integrate a variety of representations of the data (graphical and aural) as well as a variety of techniques (mathematical analyses, for example).

The use of constructed data, that is, data that has been generated algorithmically, can be useful but misleading. Because most actual data analysis problems are concerned with data in which there is no known structure (or perhaps even no existing structure), it is often not possible to study acoustical mappings systematically using that data. Constructed data can be useful for determining whether or not the mappings to acoustical parameters have the desired effect and for testing whether or not a particular mapping does, in fact, provide a useful representation of the data. Constructed data can be misleading in that the structure provided by the algorithm generation often makes it easy to analyze the data itself and thus render the mapping exercise trivial. The data for Scenario 1 is perhaps most useful for exploring different sonification techniques to distinguish between data samples. Thus, the use of six straightforward acoustical parameters in Mapping 2 versus the granular synthesis algorithm in Mapping 1 provides an interesting comparison for which constructed data is helpful.

ACKNOWLEDGEMENTS

Tara Madhyastha (University of Illinois), Carla Scaletti (Symbolic Sound Corporation), and Stuart Smith (University of Massachusetts at Lowell) made the data-mapping exercise possible by creating the three mappings and bringing them to ICAD '92 for presentation. They are very much appreciated for their willingness to participate and to offer their work for testing. All the workshop attendees were great participants in the exercise and took it in the spirit of exploration.

ATTACHMENT A

TABLE 4 Twenty samples known to be from Set 1.

1.	1.394	0.775	0.080	0.446	0.541	1.631
2.	0.807	1.411	0.515	0.851	0.681	0.749
3.	1.355	0.099	0.235	2.054	0.656	2.127
4.	0.498	2.482	0.370	0.743	1.580	0.415
5.	1.197	0.411	3.198	0.370	1.241	0.373
6.	1.955	0.433	1.390	0.067	1.314	0.858
7.	0.330	0.572	0.815	0.574	0.778	1.606
8.	2.019	0.317	0.737	1.144	0.193	0.705
9.	0.996	0.426	0.613	1.612	0.157	0.711
10.	0.775	0.359	0.589	0.874	0.281	1.618
11.	1.335	0.280	1.014	0.407	2.072	0.138
12.	0.251	2.937	0.902	0.905	1.296	0.118
13.	1.182	0.186	0.503	1.302	0.807	1.188
14.	0.782	0.134	1.671	2.251	2.005	0.678
15.	1.206	1.025	0.267	0.584	1.056	0.116
16.	0.674	0.867	0.567	1.283	1.531	1.046
17.	0.207	0.244	0.969	0.523	1.060	1.342
18.	0.907	1.190	0.834	0.281	1.129	0.029
19.	0.453	0.733	1.138	0.840	1.078	0.324
20.	0.086	0.699	0.167	1.241	0.980	0.222

TABLE 5 Twenty samples known to be from Set 2.

1.	0.838	0.812	0.178	0.064	0.417	0.897
2.	0.178	1.062	0.085	0.716	0.088	0.871
3.	0.671	0.237	0.388	0.689	0.899	0.600
4.	2.388	0.698	0.355	0.006	0.691	0.437
5.	0.551	0.377	0.007	0.112	0.051	0.246
6.	0.242	1.417	0.275	0.405	0.590	0.810
7.	0.404	1.758	0.631	0.247	0.060	0.767
8.	0.443	1.472	0.297	1.058	0.046	0.455
9.	0.188	1.112	0.245	0.628	0.782	0.547
10.	0.007	0.286	1.223	1.339	0.300	0.475
11.	0.640	0.393	0.642	0.941	0.678	0.123
12.	0.737	0.427	0.796	0.723	0.194	0.149
13.	0.565	0.571	0.005	0.340	0.507	0.168
14.	0.976	0.342	0.098	0.356	0.532	0.461
15.	1.165	0.510	0.212	0.980	0.736	0.574
16.	0.821	0.291	1.185	0.465	0.061	0.392
17.	2.195	0.776	1.027	0.518	0.517	0.135
18.	0.438	1.427	0.183	0.050	0.370	0.563
19.	0.243	0.717	0.531	0.597	0.376	1.170
20.	0.042	0.199	0.745	0.364	0.399	1.086

ATTACHMENT B

TABLE 6 Test for Mapping 1.

1.	0.333	0.692	0.176	0.138	0.354	0.058	(Set 2)
2.	0.554	1.232	0.074	0.198	0.358	0.065	(Set 2)
3.	1.028	0.576	0.070	0.077	2.401	0.162	(Set 1)
4.	0.217	0.885	0.259	1.087	0.739	1.005	(Set 1)
5.	0.507	0.862	0.198	0.309	0.578	1.107	(Set 2)
6.	0.570	0.866	0.301	0.434	0.526	0.674	(Set 2)
7.	0.554	1.232	0.074	0.198	0.358	0.065	(Set 2)
8.	0.530	0.042	0.852	0.250	0.053	0.185	(Set 2)
9.	0.497	1.590	1.365	1.175	1.278	0.101	(Set 1)
10.	0.167	1.261	0.031	0.658	0.306	1.289	(Set 2)

TABLE 7 Test for Mapping 2.

1.	0.554	1.232	0.074	0.198	0.358	0.065	(Set 2)
2.	0.292	0.699	0.076	0.153	0.303	0.064	(Set 2)
3.	0.130	0.096	1.686	0.108	1.020	1.198	(Set 1)
4.	0.462	1.383	0.047	1.059	0.344	0.798	(Set 2)
5.	0.886	0.366	0.570	1.571	2.040	1.357	(Set 1)
6.	1.224	1.481	1.835	0.318	0.510	1.500	(Set 1)
7.	1.170	0.812	0.500	0.119	0.979	0.334	(Set 2)
8.	2.239	0.036	0.790	0.663	0.025	0.357	(Set 2)
9.	0.117	0.524	0.950	1.131	0.022	1.142	(Set 1)
10.	0.530	0.042	0.852	0.250	0.053	0.185	(Set 2)

TABLE 8 Test for Mapping 3.

1.	0.130	0.096	1.686	0.108	1.020	1.198	(Set 1)
2.	0.853	1.876	0.985	0.792	0.558	0.488	(Set 1)
3.	0.596	0.413	0.240	1.077	1.439	0.760	(Set 1)
4.	2.046	0.333	0.235	0.689	0.840	1.816	(Set 1)
5.	1.700	0.577	0.229	1.114	0.432	0.978	(Set 1)
6.	0.292	0.699	0.076	0.153	0.303	0.064	(Set 2)
7.	0.217	0.885	0.259	1.087	0.739	1.005	(Set 1)
8.	0.286	0.585	0.120	1.361	0.398	0.263	(Set 2)
9.	0.904	0.669	0.487	0.783	1.665	0.031	(Set 1)
10.	0.530	0.042	0.852	0.250	0.053	0.185	(Set 2)

REFERENCES

1. Bly, Sara A. "Sound and Computer Information Presentation." Ph.D. Thesis, University of California, Davis, 1982. Also Technical Report UCRL-53282, Lawrence Livermore National Laboratory, 1982.
2. Everitt, B. *Graphical Techniques for Multivariate Data*. Amsterdam: North-Holland, 1978.
3. Friedman, J. H., J. A. McDonald, and W. Stuetzle. "An Introduction to Real-Time Graphical Techniques for Analyzing Multivariate Data." In *Proceedings of the Third Annual Conference and Exposition of the National Computer Graphics Association, Inc.*, Vol. 1, 421, 1982.
4. Roads, Curtis. "Granular Synthesis of Sound." In *Foundations of Computer Music*, edited by C. Roads and J. Strawn, 144–159. Cambridge, MA: MIT Press, 1985.
5. Roads, Curtis. "Asynchronous Granular Synthesis." In *Representations of Musical Signals*, edited by G. diPoli, A. Picialli, and C. Roads, 143–186. Cambridge, MA: MIT Press, 1991.
6. Scaletti, C. "The Kyma/Platypus Computer Music Workstation." In *The Well-Tempered Object: Musical Applications of Object-Oriented Software Technology*, edited by S. Pope. Cambridge: MIT Press, 1991.
7. Tufte, E. R. *The Visual Display of Quantitative Information*. Cheshire, CT: Graphics Press, 1983.
8. Tukey, J. *Exploratory Data Analysis*. Menlo Park, CA: Addison-Wesley, 1977.

William W. Gaver
Rank Xerox Cambridge EuroPARC, 61 Regent Street, Cambridge CB2 1AB, UK;
gaver@europarc.xerox.com

Using and Creating Auditory Icons

Auditory icons are everyday sounds that convey information about events in the computer or in remote environments by analogy with everyday sound-producing events. Several examples of interfaces that use auditory icons demonstrate that they can add valuable functionality to computer interfaces, particularly when they are parameterized to convey dimensional information. But because they are based on a new approach to sound and hearing that emphasizes perceptual and acoustic attributes of auditory event perception, they are difficult to create and manipulate if standard synthesis and sampling techniques are used. In order to support their creation, new synthesis algorithms are introduced which are controlled along dimensions of events rather than those of the sounds themselves. Several algorithms, developed from research on auditory event perception, are described in enough detail here to permit their implementation. They produce a variety of impact, bouncing, breaking, scraping, and machine sounds. By controlling them with attributes of relevant computer events, a wide range of parameterized auditory icons may be created.

1. INTRODUCTION

Over the last several years, I have been developing a strategy for creating auditory icons, everyday sounds mapped to computer events by analogy with everyday sound-producing events (e.g., see Gaver[7,13,14]). Auditory icons are like sound effects for computers: Objects make sounds as they are selected, dragged, bumped against one another, opened, activated, and thrown away. But they are not designed merely to provide entertainment; rather, they convey information about events in computer systems, allowing us to listen to computers as we do to the everyday world.

For instance, selecting a file in a direct manipulation system might make the sound of an object being tapped. The type of file can be indicated by the material of the object, and the size of the file by the size of the struck object. Auditory icons of this sort are similar to visual icons in that both rely on an analogy between the everyday world and the model world of the computer. Because they exploit the power of such organizing metaphors, auditory icons may be learned as easily as visual icons are. In addition, because listening and looking provide complementary kinds of information, auditory icons can be created that will both complement and supplement visual icons.

In this chapter, I review work on auditory icons with the aim of showing why they are useful and how they can be designed and implemented. First, the idea of auditory icons is explained in terms of its foundation on a new approach to sound and hearing, and some of the advantages of this approach to conveying information are outlined. Then several systems that have used auditory icons are described to illustrate their functionality. Finally, methods for creating auditory icons are described, and a number of synthesis algorithms are introduced that make their specification and implementation relatively straightforward.

2. EVERYDAY LISTENING AND AUDITORY ICONS

Auditory icons are an application of a new approach to sound and hearing that stresses the experience of hearing events in the world, rather than sounds per se (e.g., see Gaver[7,13,14]). For instance, if one drops a crystal vase while carrying it in a darkened room, one is less likely to attend to the pattern of pitched impulses it produces than one is to try to ascertain what it landed on and whether it broke. If one hears a sound while walking down a deserted alley at night, one is less likely to be concerned with its timbre or duration, and more concerned with determining what its source is and whether that source is potentially threatening. In each case, the experience of listening to the attributes of the sound—its pitch, loudness, duration, or timbre—is an example of musical listening. The experience of listening to determine the source itself—whether it involves a hard or soft surface, a bouncing or breaking object, a threatening or harmless source—is an example of everyday listening.

Although normal audition is often concerned with everyday listening, very little psychological research has concerned what we hear of events in the world and how we hear them (though see Jim Ballas' work, as represented in this volume, as well as that by Gibson,[15] Heine and Guski,[17] Jenkins,[20] Warren and Verbrugge,[28] and Vanderveer[27]). There seems to be two reasons for this: First, research on psychoacoustics has grown out of a concern with music, and has focused on musical sounds and attributes. Second, studies of audition have been constrained by sensation-based theories of perception and the supposed primitives of sound they suggest. The result is a strategy preoccupied with elemental sensations, while questions concerning auditory event perception—if recognised at all—are left to higher-level cognitive accounts. But because students of cognition are often more interested in general questions of mechanisms and representations than in the content of cognition, focused examinations of everyday listening often fall between the cracks separating psychoacoustics and cognition.

2.1 THE ECOLOGICAL APPROACH TO PERCEPTION

My approach to everyday listening is suggested by Gibson's[16] ecological perspective on perception. The ecological approach emphasizes that perception should be understood in terms of the fit between the organism and a structured environment. A key concept for this approach is that of direct perception, the notion that we obtain information about events in the world via possibly complex patterns of energy—in the case of vision, which Gibson was primarily concerned with, we see complex, meaningful things in the world because light specifies them.

From this perspective, our perceptual experience may be explained in terms of these patterns, not in terms of representational structures in the mind which are subject to the effects of memory and inference.

This is a controversial claim, and it is not the job of this chapter to pursue it. But it provides a perspective that allows everyday listening to be explored directly. Instead of starting from the traditionally assumed primitive dimensions of sound and corresponding dimensions of sensation, we may assume that listeners can hear information about complex event attributes directly, and try to describe the acoustic patterns that underlie these complex perceptions.

2.2 A NEW FRAMEWORK FOR SOUND

What this implies is the development of a new framework for describing sound and hearing, one that complements more traditional approaches (see Gaver[13,14]). The vast variety of sounds we hear in the world may be characterized in terms of their sources, their attributes in terms of source attributes. Instead of talking about pulses and buzzes, pitches and timbres, we can talk about impacts and machine sounds, size and material. Instead of relating sensations to simple acoustic attributes, we can relate source perception to more complex acoustic patterns. Most importantly (at least for the purposes of this chapter), we can build auditory cues from this framework. Instead of mapping information to sounds, then, when thinking about auditory icons it is more appropriate to think about mapping information to events.

This is the strategy behind auditory icons, then: To map computer events and at tributes to the events and attributes that normally make sounds. Returning to the example from the beginning of this chapter, selecting a file is an event in the computer; it is mapped to the everyday, sound-producing event of tapping an object. The size and type of the file are attributes of the computer event; they are mapped to attributes of the sound-producing event (in this case the size and material of the virtual object that has been tapped). In general, the result is to relate interface sounds to their referents in the same way that natural sounds are related to their sources and, thus, to allow people to use their existing everyday listening skills in listening to computers.

2.3 PARAMETERIZED AUDITORY ICONS

Auditory icons not only reflect categories of events and objects as visual icons do, but can be parameterized to reflect their relevant dimensions as well. That is, if a file is large, it sounds large. If it is dragged over a new surface, we hear that new surface. And if an ongoing process starts running more quickly, it sounds quicker.

Thus auditory icons convey multidimensional information by mapping the dimensions of the information to be displayed to dimensions of everyday events. In this way, any one sound can convey a great deal of information. Moreover, "families" of auditory icons can be created by exploiting the organization inherent in everyday events. For instance, if the material of a sound-producing event is used to represent the type of object, all auditory icons concerning that type of object would use sounds made by that kind of material. So text files might always sound wooden, whether they are selected, moved, copied, or deleted. In this way, a rich system of auditory icons may be created that relies on relatively few underlying metaphors.

2.4 AUDITORY ICONS AND ENGAGEMENT

When the same analogy underlies both auditory and visual icons, the increased redundancy of the interface can help users to learn and remember the system. In addition, making the model world of the computer consistent in its visual and auditory aspects increases users' feelings of direct engagement[18] or mimesis[22] with that world. The concepts of direct engagement and mimesis refer to the feeling of working in the world of the task, not the computer. By making the model world of the computer more real, one makes the existence of an interface to that world less noticeable. Providing auditory information that is consistent with visual feedback is one way of making the model world more vivid. In addition, using auditory icons may allow more consistent model worlds to be developed, because some computer events may map more readily to sound-producing events than to visual ones.

3. SYSTEMS THAT USE AUDITORY ICONS

The benefits of auditory icons can be more fully illustrated by examples of systems that have used them. A number of systems have been created that illustrate the potential for auditory icons to convey useful information about computer events. In particular, these systems suggest that sound is well suited for providing information:

- about previous and possible interactions,
- about ongoing processes and modes,
- useful for navigation, and
- to support collaboration.

In the following sections, a variety of systems that have used auditory icons are briefly described. Their order is both chronological and, not surprisingly, also reveals the increasing functionality that auditory icons can provide.

3.1 THE SONICFINDER

The SonicFinder[8] is the first interface to incorporate auditory icons. Developed for Apple Computer Inc., it is an extension to the Finder, the application used to organize, manipulate, create, and delete files on the Macintosh. The Finder is automatically run when the machine is booted and, thus, is probably the program most frequently encountered by Macintosh users. Because of this, and because the SonicFinder was easily portable (requiring no special hardware, it could be distributed on a single 800K floppy disk), many people have encountered the SonicFinder and have provided useful feedback about its utility.

Creating the SonicFinder required extending the Finder code at appropriate points to play sampled sounds modified according to attributes of the relevant events. Thus a variety of actions make sound in the SonicFinder: selecting, dragging, and copying files; opening and closing folders; selecting, scrolling, and resizing windows; and dropping files into and emptying the trash can. Most of these sounds are parameterized, although the ability to modify sounds is limited. So, for instance, sounds that involve objects such as files or folders not only indicate basic events such as selection or copying, but also the object's types and sizes via the material and size of the virtual sound-producing objects. In addition, the SonicFinder incorporates an early example of an auditory process monitor in the form of a pouring sound that accompanied copying and that indicates, via changes of pitch, the percentage of copying that had been completed (see Cohen, in this volume).

Because the Macintosh is a single-processing machine with a fairly simple interface, the sounds used in the SonicFinder basically provide feedback and information about possible interactions (as well as more general information about file size and type, dragging location, and the like). Nonetheless, it provides a valuable example of the potential of auditory icons, showing that sounds such as these can be incorporated in an intuitive and informative way.

Apple has never released the SonicFinder (although it did appear on a developer's CD-ROM). There are several reasons for this. Most fundamentally, the perceived benefit to disk-space ratio was not high enough: because it uses sampled sounds, the SonicFinder could never be reduced below about 100K in size—prohibitively large to release in the days before high-density floppy disks and CD-ROMs. Although many people found the auditory cues useful, others found them irritating or thought of them as merely entertaining. This provides a valuable example of the real-world challenges that designers of auditory interfaces must face. Nonetheless, the interface has spread around the world in what Buxton has termed the "research underground," and has hopefully helped to demonstrate the potential and appeal of auditory interfaces in general.

3.2 SOUNDSHARK

Although the SonicFinder is useful in incorporating auditory icons into a well-known and often-used interface, its simplicity may lead people to underestimate the functions that auditory icons may serve. For this reason, Gaver and Smith[11] demonstrated auditory icons used in a large-scale, multiprocessing, collaborative system called SharedARK, and dubbed the resulting auditory interface SoundShark.

SharedARK is a collaborative version of ARK, the Alternate Reality Kit. Developed by Smith,[26] ARK is designed as a virtual physics laboratory for distance education. The "world" appears on the screen as a flat surface on which a number of 2.5-dimensional objects may be found. These objects may be picked up, carried, and even thrown using a mouse-controlled "hand." They may be linked to one another, and messages may be passed to them using "buttons." Using this system, a number of simple physical experiments may be performed. In addition, SharedARK allows the same world to be seen by a number of different people on their own computer screens (and is usually used in conjunction with audio and video links that allow them to see and talk to one another). They may see each other's hands, manipulate objects together, and thus collaborate within this virtual world.

SharedARK is a multiprocessing system, with the potential for several "machines" or self-sustaining processes to run simultaneously. In addition, it provides a very large world to users, in that the space for interaction is many times larger than the screen (depending on available memory, it may cover literally acres of virtual space). Users move around this space by moving their hand near the edge of the window, causing it to scroll over adjacent territory. To help with navigation, a "radar view" is presented which shows a much-reduced representation of the world and objects within it.

This interface was extended by adding auditory icons to indicate user interactions, ongoing processes, and modes; to help with navigation; and to provide information about other users. Sounds were used to provide feedback as they were in the SonicFinder: Many user actions were accompanied by auditory icons which were parameterized to indicate attributes such as the size of relevant objects. In addition, ongoing processes made sounds that indicated their nature and continuing activity even if they were not visible on the screen. Modes of the system, such as the activation of "motion," which allows objects to move if they have a velocity, were indicated by low-volume, smooth background sounds. Collaborators could hear each other even if they couldn't see each other, which seemed to aid in coordination. Finally, distance between a given user's hand and the source of the sound was indicated by the sounds' amplitude and by low-pass filtering, aiding with navigation. The apparent success of this manipulation led us to develop "auditory landmarks," objects whose sole function was to play a repetitive sound that could aid orientation.

SoundShark was implemented using an external sampler that was triggered and controlled via MIDI. This allowed a number of features that would have been more difficult to achieve had the sound playback been handled by the workstation. Multiple sounds could be played simultaneously, and manipulations such as varied attack times and low-pass filtering could be used. Nonetheless, the use of external hardware meant that the system was not as portable as the SonicFinder.

3.3 ARKOLA

Our experiences with SoundShark suggested that auditory icons could provide useful information about user-initiated events, processes, and modes, and about location within a complex environment. To test this, we developed a special application within SoundShark which we used as a basis for observing people's use of the system. This application, developed in collaboration with Tim O'Shea, was a model of a soft-drink plant called the ARKola bottling factory.[12] It consisted of an assembly line of nine machines which cooked, bottled, and capped cola, provided supplies, and kept track of financing. The plant was deliberately designed to be too large to fit on the computer screen, so participants could only see about half the machines at any given time. In addition, we designed the plant to be fairly difficult to run, with the rates of the machines requiring fine tuning and with machines occasionally "breaking down," necessitating the use of a "repair" button.

Each of the machines made sounds to indicate their function. For instance, the "nut dispenser" made wooden impact sounds each time a nut was delivered to the cooker, the "heater" made a whooshing flamelike sound, the "bottler" clanged, and the "capper" clanked. In addition, the rate of each machine was indicated by the rate of repetition of the sounds it made, and problems with the machines were indicated by a variety of alerting sounds such as breaking glass, overflowing liquid, and so forth.

With as many as 12 sounds playing simultaneously, designing the sounds so that all could be heard and identified was a serious challenge. In general, we used temporally complex sounds to maximize discriminability, and designed the sounds to be semantically related to the events they represented. Two strategies were found to be useful in avoiding masking. First, sounds were spread fairly evenly in frequency, so that some were high pitched and others lower. Second, we avoided playing sounds continuously and instead played repetitive streams of sounds, thus maximizing the chance for other sounds to be heard in the gaps between repetitions.

Six pairs of participants were asked to run the plant with the aim of making as much "money" as they could during an hour-long session. Each pair ran the plant for two hours, one with and one without auditory feedback (with the order, of course, being counterbalanced). We observed their performance from a "control room" via video links as they ran the plant, and videotaped their activities for later analysis.

Our observations indicated that sounds were effective in two broad areas. First, they seemed to help people keep track of the many ongoing processes. The sounds allowed people to track the activity, rate, and functioning of normally running machines. Without sound, people often overlooked machines that were broken or that were not receiving enough supplies; with sound these problems were indicated either by the machine's sound ceasing (which was often ineffective) or by the various alert sounds. Perhaps most interesting, the auditory icons allowed people to hear the plant as an integrated complex process. The sounds merged together to produce an auditory texture, much as the many sounds that make up the sound of an automobile do. Participants seemed to be sensitive to the overall texture of the factory sound, referring to "the factory" more often than they did without sound.

The second set of observations related to the role of sound in collaboration. In both the sound and no-sound conditions, participants tended to divide responsibility for the plant so that each could keep one area on the screen at all times. Without sound, this meant that each had to rely on their partner's reports to tell what was happening in the invisible part. With sound, each could hear directly the status of the remote half of the plant. This seemed to lead to greater collaboration between partners, with each pointing out problems to the other, discussing problems, and so forth. The ability to provide foreground information visually and background information using sound seemed to allow people to concentrate on their own tasks, while coordinating with their partners about theirs.

Sound also seemed to add to the tangibility of the plant and increased participants' engagement with the task. This became most evident when one of a pair of participants who had completed an hour with sound and were working an hour without remarked "we could always make the noises ourselves...." In sum, the ARKola study indicated that auditory icons could be useful in helping people collaborate on a difficult task involving a large-scale complex system, and that the addition of sounds increased their enjoyment as well.

3.4 EAR: ENVIRONMENTAL AUDIO REMINDERS

Where SoundShark and ARKola explored the use of auditory icons to support collaboration in software systems, another system, called EAR (for Environmental Audio Reminders), demonstrates that auditory icons are also helpful for supporting collaboration in the office environment itself.[10] This system plays a variety of nonspeech audio cues to offices and common areas inside EuroPARC to keep us informed about a variety of events around the building. It is one element of ongoing research at EuroPARC on environmental interfaces, which are aimed at merging the power of the computational and everyday environments. EAR works in conjunction with the RAVE audio-video network,[12,2] which connects all the offices at EuroPARC with audio and video equipment using a computer-controlled switch, and Khronika,[23] an event server that uses a database of events in conjunction with software daemons to inform us of a wide range of planned and spontaneous, electronic and professional events. EAR, then, consists of sounds triggered by Khronika when relevant events occur, which are routed using the RAVE system from a central server (in our case, a Sparcstation) to any office in the building.

A wide variety of sounds are used to remind us about a range of events. For instance, when new e-mail arrives, the sound of a stack of papers falling on the floor is heard. When somebody connects to my video camera, the sound of an opening door is heard just before the connection is made, and the sound of a closing door just after the connection is broken. Ten minutes before a meeting, the sound of murmuring voices slowly increasing in number and volume is played to my office, then the sound of a gavel. And finally, when one of my colleagues decides to call it a day, they often play the "pub call" to my office, the sound of laughing, chatting voices in the background with the sound of a pint glass being filled with real ale in the foreground.

Many of the sounds we use in EAR may seem frivolous because they are cartoonlike stereotypes of naturally occurring sounds. But it is precisely because they are stereotyped sounds that they are effective. More "serious" sounds—such as electronic beeps or sequences of tones—would be likely to be less easily remembered than these. In addition, we have taken some care in shaping the sounds to be unobtrusive. For instance, many of the sounds are very short; those that are longer have a relatively slow attack so that they enter the auditory ambience of the office subtly. Most of the sounds have relatively little high-frequency energy, and we try to avoid extremely noisy or abrupt sounds. So though the sounds we use are stereotypes, they are designed to fit into the existing office ambience rather than intruding upon it.

In sum, the auditory cues used in the EAR system can be unobtrusive, informative , and valuable. They serve to indicate events in the same way that they might be heard in everyday life, with the added advantage that the events cued are chosen by users. They allow us to hear distant events, or events that don't naturally produce informative noises, helping to blur the distinction between the electronic and physical environments. By informing us about ongoing events in the building they help to ease the transition between working alone and working together.

3.5 SUMMARY

These systems demonstrate the wide range of functions that auditory icons can perform. They can provide information about user actions, about possibilities for new actions, and about nonvisible attributes of objects in the system. They can provide background information about processes and modes in more complex systems. Continuous or repetitive sounds that are varied according to distance may serve as auditory landmarks, supporting navigation in complex systems. Finally, auditory icons can work with graphic displays, supporting a smooth flow between individual and cooperative work.

These examples also demonstrate the range of systems that may benefit from auditory icons, from traditional desktop graphical interfaces to more complex virtual realities and process simulations, to systems that introduce computational power into the everyday environment itself. By building these systems, using them ourselves, and observing others use them, we have gained a great deal of valuable information about their utility, their problems, and issues for their design.

Finally, these systems illustrate the broad range of sounds that may be used as auditory icons, from the simple impact and scraping sounds that are designed for graphical user interfaces such as the SonicFinder, to the more complex and continuous process sounds used in SoundShark and ARKola, to the very complex and stereotypical sounds used in EAR. They also illustrate some of the issues that must be tackled when designing sounds for interfaces, particularly the tension inherent in designing sounds that are simultaneously identifiable and subtle, memorable but not annoying. The rest of this chapter focuses on the creation of such auditory icons, describing a variety of synthesis algorithms that may make their design more tractable.

4. CREATING AUDITORY ICONS

As the examples above illustrate, auditory icons are everyday sounds designed to convey information about events in the computer, or remote events in the office environment, by analogy with everyday sound-producing events. This strategy relies on a new framework for understanding sound, one that focuses on the unique perceptual and acoustic dimensions characterizing everyday listening. This makes the approach quite different from most of the others described in this book: instead of parameterizing sounds along the traditional dimensions used to describe sound and music, such as pitch, loudness, and timbre, auditory icons must be conceived and manipulated along dimensions of virtual sources, such as material, size, and force.

Unfortunately, it is difficult to parameterize auditory icons because it is difficult to control a virtual source of a sound along relevant dimensions. Standard synthesis techniques have been developed for creating music, affording changes of a sound's pitch, loudness, and duration, and thus lend themselves for creating the variety of musical messages described elsewhere in these proceedings.[1] But they do not make it easy to change a sound from indicating a large wooden object, for instance, to one specifying a small metal one. It is easy to create a wide variety of beeps and hums using standard synthesis techniques, but difficult to create and manipulate sounds along dimensions that specify events in the world.

4.1 USING SAMPLED SOUNDS

Because of the limitations of standard synthesis techniques, the auditory interfaces described above have relied on digital sampling in their implementations. Desired sounds were captured either from actual events or from a variety of commercially available sound-effect CD's by recording them on a computer or sampler. Most of the sounds were then edited in a variety of ways, for instance by shortening or lengthening them, filtering them, or changing their amplitude envelopes. Finally, they are played back under the control of the interface, with their attributes manipulated in order to create parametrized icons.

Sampling enables the use of much more complex and realistic sounds than can be created by readily available synthesis algorithms. However, there are several drawbacks of sampling that limit its utility as a technique for creating and using auditory icons:

- It is difficult to capture an actual event that sounds like what is desired, because sounds are invariably coloured by the technologies used to record them. In addition, it is often difficult to eliminate background noise or to find the precise sound one desires, whether recording live events or using sound effect recordings.
- Shaping recorded sounds along dimensions relevant for auditory icons is difficult because available software is designed for making music. Most editing packages allow simple pitch and loudness transformations, filtering and envelope changes. But they do not make it easy to change the sound of an idling engine, for instance, to that of the same engine

[1] Musical messages are auditory cues that rely on the dimensions of sound usually used in creating music, such as pitch, loudness, or duration. A variety of names have been suggested for this strategy; Kramer, in this volume, refers to such sounds as abstract.

working hard. Although it is possible to change a given sound to specify a range of sources, this relies largely on craft and experience: one has to work around the available tools rather than with them.

- Real-time modification of sounds on playback is even more limited. Typically one can manipulate the pitch, loudness, and duration of sounds using programs available on workstations, and perhaps change the filtering and amplitude if using a sampler, but again these are not the dimensions most relevant for creating auditory icons.
- The amount of memory needed for complex auditory interfaces is often prohibitive (on the order of 10K bytes per second of sound). Although memory is becoming less expensive, it is well to remember that one of the main reasons the SonicFinder was never released was because of the memory it required. These limitations constrain the possibilities for designing auditory icons, and make their creation difficult and time-consuming.

For these reasons, sampling is limited as a technique for creating auditory icons. Instead of treating interfaces that rely on sampling, then, the focus of the following sections is on a new type of synthesis algorithm developed as a result of basic research on auditory event perception. Several examples of these algorithms are described in sufficient detail to allow readers to implement and explore them. These algorithms, based on my ongoing research,[13,14] allow sounds to be specified in terms of their sources rather than their acoustic attributes. They promise to overcome both the limitations of traditional synthesis algorithms and of sampling by allowing parameterized auditory icons to be specified along dimensions of virtual source events.

4.2 ANALYSIS AND SYNTHESIS OF EVENTS

Creating algorithms that allow synthesis of virtual events implies an understanding of the acoustic information for event attributes—how sounds indicate the material or size of an object, for instance. Event attributes often have very complex effects on sounds, effects that must be described as functions of frequencies and amplitudes over time that specify the partials, or frequency components, that make up a sound. If these functions are understood, source attributes can be specified directly, instead of by using separate controls over partial frequencies, amplitudes, and durations. But how can we determine what these functions are?

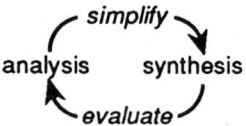

FIGURE 1 Traditional analysis and synthesis.

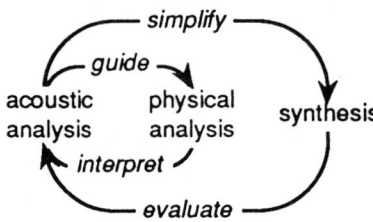

FIGURE 2 Analyzing and synthesizing events requires physical as well as acoustic analyses.

One approach to this problem is suggested by the analysis and synthesis methods[25] used by computer musicians to capture the relevant properties of traditional instrument sounds (Figure 1). This approach involves recording sounds that vary along dimensions of interest and analyzing their acoustic structure using Fourier analysis or similar techniques. Hypotheses about acoustic information suggested by the analysis can be tested by synthesizing sounds based on simplified versions of the data. For instance, if one supposes that the temporal features of a sound indicates the event that caused it, but that its frequency makeup is irrelevant, one might use the amplitude contour from the original sound to modify a noise burst. The hypothesis can then be assessed simply by listening to the result.

In practice, however, it is often difficult to identify the acoustic information for events in the mass of data produced by acoustic analyses. Thus it is useful to supplement them with analyses of the mechanical physics of the event itself (see Figure 2). Studying the physics of sound-producing events is useful both in suggesting perceptible source attributes and in indicating the acoustic information for these attributes. Acoustic analyses help both in checking the adequacy of physical models and in evaluating particular parameters. Finally, the resulting models can provide the basis for synthesis algorithms that allow sounds to be specified in terms of sources attributes.

Computer musicians, too, have increasingly turned to physical modeling to guide synthesis (see, e.g., Borin et al.,[1] Cremer,[4] McIntyre et al.,[24] and Jaffe and Smith[26]), and their work offers a great deal of insight into analytical, modeling, and synthesis techniques. However, relatively little of the content of their models is directly applicable to the creation of auditory icons. The domain of sounds they have studied—primarily those produced by musical instruments—is relatively limited. The aim of their models is not to explicate acoustic information for events, but instead to recreate the sounds of traditional musical instruments for musical purposes. Finally, acoustic instruments are often mechanically very complex. Not only does this imply that physically modeling them may be very difficult, but that the physical dimensions necessary for musically accurate modeling may not correspond to those that are phenomenally perceptible. For these reasons, it is appropriate to turn to a consideration of simpler, everyday sound-producing events that might be useful for creating auditory icons.

In the following sections, I discuss several case studies of events that have been studied in this manner, and the synthesis algorithms that have resulted are described. I choose the algorithms for their utility in creating auditory icons, and describe them in order of their complexity. The starting point of this discussion is the sounds made by mechanical impacts, which involve a simple interaction of objects. Next I explain the production of more complex bouncing, breaking, and spilling sounds specified by the temporal patterning of a series of impacts. A third algorithm allows the same virtual object to be hit and then scraped by distinguishing objects from the interactions that cause them to make sound. Finally, I describe an algorithm for producing machinelike sounds, which suggests that high-level attributes of complex events may be synthesized directly.

5. IMPACT SOUNDS

Many of the sounds we hear in the everyday world involve one solid impacting against another. Tapping on an object, placing it against another, letting it fall—all involve impact sounds. In the interface, impact sounds are useful in the design of a variety of auditory icons that indicate events such as selecting a file, moving it over another, or attaching one object to another.

Several studies have explored the perceptible attributes of impact events and the acoustic information about them. In this section, these studies are reviewed briefly, then the information they provide is used to create a synthesis algorithm that allows impact sounds to be specified along dimensions of the virtual source.

5.1 MALLET HARDNESS, MATERIAL, AND SIZE

Freed and Martins[6] studied people's perception of the hardness of mallets used to strike objects. They recorded the sounds made by hitting cooking pans with mallets of various hardnesses, asked people to judge hardness from the sounds, and used a model of the peripheral auditory system to analyze the acoustic correlates of their judgements. They found that the ratio of high- to low-frequency energy in the sounds and its change over time served as the most powerful predictors of subjects' hardness judgements. To a good approximation, then, mallet hardness is conveyed by the relative presence of high- and low-frequency energy.

The acoustic information available for the length and material of struck wood and metal bars and people's abilities to perceive these attributes were studied by Gaver.[7] The sounds made by wood and metal bars of several different lengths were recorded and analyzed, and a model of the physics of impacts that combined analytical solutions to the wave equation for transverse vibrations in a bar (e.g., see Lamb[21]) with empirical measurements of damping and resonance amplitudes was developed. This model was used both to aid interpretation of the acoustic analyses and to synthesize new experimental stimuli.

The material of the bars made several effects on the sounds they made. Perhaps most important, materials have characteristic frequency-dependent damping functions: the sounds made by vibrating wood decay quickly, with low-frequency partials lasting longer than high ones, while the sounds made by vibrating metal decay slowly, with high-frequency partials lasting longer than low ones. In addition, metal sounds have partials with well-defined frequency peaks, while wooden sound partials are smeared over frequency space. These results accord with Wildes and Richards'[29] physical analyses of the audible effects of the internal friction characterizing different materials, which show that internal friction determines both the damping and definition of frequency peaks.

Changing the length of a bar, on the other hand, simply changes the frequencies of the sound it produces when struck, so that short bars make high-pitched sounds and long bars make low ones. However, the effects of length may interact with the effects of material. For instance, frequencies change monotonically with length, but the frequency of the partial with the highest amplitude depends on material and thus may change nonlinearly with length.[7] These nonlinearities—and the perceptual confusion they cause—may be avoided by simplifying the model so that partial amplitudes do not depend on material.

5.2 A SYNTHESIS ALGORITHM FOR IMPACT SOUNDS

These results may be captured in a synthesis model that uses frequency and amplitude functions to constrain a formula for describing logarithmically decaying sounds. This formula describes a complex wave created by adding together a number of sine waves with independent initial amplitudes and logarithmic decay rates:

$$G(t) = \sum_n \Phi_n e^{-\delta_n t} \cos \omega_n t \qquad (1)$$

where $G(t)$ describes the waveform over time, Φ_n is the initial amplitude, δ_n the damping constant, and ω_n the frequency of partial n.

This formula has two properties that make it a useful foundation for synthesizing auditory icons. First, its components map well to event attributes. Second, it can be made computationally efficient using trigonometric identities.

5.3 MAPPING SYNTHESIS PARAMETERS TO SOURCE ATTRIBUTES

By constraining the values used in this formula, useful parameters can be defined which correspond well to the attributes of impact sounds discussed above. The formula involves three basic components: the initial amplitudes of the partials Φ_n, their damping $e^{-\delta_n t}$, and their frequencies $\cos \omega_n t$. These can be set separately for each partial. However, these three components also correspond to information for mallet hardness and impact force, material, and size and shape respectively (see Table 1). Thus it is more useful to define patterns of behaviour over the partials for each component.

For example, the partial frequencies ω_n can be constrained to patterns typical of various object configurations. The sounds made by struck or plucked strings, for example, are harmonic, so that $\omega_n = n\omega_1$. The sounds made by solid plates, in contrast, are inharmonic and can be approximated by random frequency shifts made to a harmonic pattern. The sounds made by solid bars can be approximated by the formula $\omega_n = (2n+1)^2/9$. Finally, the sounds made by rectangular resonators are given by the formula $\omega_n = c/2\sqrt{(p^2/l^2 + q^2/w^2 + r^2/h^2)}$, where c is the velocity of sound, l, w, and h are the length, width and height of the box respectively, and p, q, and r are indexed from 0.[29] An algorithm based on Eq. (1), then, can be constrained so that one of these patterns is used to control the partial frequencies ω_n. In addition, ω_1 can be specified such that $\omega_1 \propto 1/\text{size}$ to reflect the size of the object (this affects all the other partial frequencies).

The initial amplitude of the partials, Φ_n, can be controlled by a single parameter corresponding to mallet hardness. Recalling that Freed and Martin's[6] results identified the ratio of high- to low-frequency energy as a predictor of perceived mallet hardness, we might maintain a linear relationship among the partial's initial amplitudes, and use the slope from

TABLE 1 Mapping Parameters to Events

Term	Effect	Event Attribute
Φ_n	initial amplitudes	mallet hardness; force or proximity
$e^{-\delta_n t}$	damping	material
$\cos \omega_n t$	partial frequencies	size; configuration

Φ_1 to control perceived hardness. Thus $\Phi_n = \Phi_1 + h(\omega_n - \omega_1)$, where h is the slope—note that h may be negative, so that higher partials have less amplitude than low ones. Φ_1 (and thus all the amplitudes) may also be changed to indicate impact force or proximity.

Finally, the damping constants for each partial (δ_n) can be controlled by a parameter corresponding to material. A useful heuristic is to set $\delta_n = \omega_n \delta_0$, so that high harmonics die out relatively quickly for highly damped materials and last longer for less damped materials (e.g., metal, which has low damping, tends to ring; wood, which is highly damped, tends to thunk). This strategy is suggested both by Wildes and Richards,[29] and by my own research.[7]

In sum, Eq. (1) can be controlled by parameters that make effects corresponding to attributes of impact events. Controlling overall frequency corresponds to the object's size, while the pattern of partial frequencies corresponds to its configuration. The overall initial amplitude corresponds to the force or proximity of the impact, while the pattern of partial amplitudes corresponds to mallet hardness. Finally, the degree of damping corresponds well to the virtual object's material. By controlling these five parameters, then, a wide range of sounds can be created which vary over several useful dimensions.

5.4 AN EFFICIENT ALGORITHM FOR SYNTHESIS

Equation (1) is useful in allowing parameters to be defined in terms of source events. It is also attractive because it can be implemented in a computationally efficient way.

An efficient implementation of this formula relies on Euler's relationship $e^{i\omega t} = \cos \omega t + i \sin \omega t$ to rewrite Eq. (1) as:

$$\begin{aligned} S_n &= \mathrm{Re}[(a_n + ib_n)(p + iq)] \\ &= \mathrm{Re}[(a_n p - b_n q) + i(b_n p + a_n q)] \end{aligned} \quad (2)$$

where S_n is the nth sample, a_0 is the initial amplitude, $b_0 = 0$, $i = \sqrt{-1}$, $p = e^{-\delta t} \cos \omega t$, and $q = e^{-\delta t} \sin \omega t$. (A full derivation is available upon request.)

Samples can thus be generated by calculating p and q, setting a and b to the initial amplitude and 0, and applying Eq. (2). The output sample is the real part of the result, and a and b are updated to the real and imaginary parts respectively (see pseudocode in Figure 3). Computationally expensive sines and cosines need only be calculated once, and only four multiplications, one addition, and one subtraction are needed for each partial for a given sample. The efficiency of this implementation allows fairly complex impact sounds to be generated in real time on many computers.

6. BREAKING, BOUNCING, AND SPILLING

The impact algorithm can serve as a fundamental element in algorithms used to synthesize more complex sounds. For instance, an early example of analysis and synthesis of sound-producing events is Warren and Verbrugge's[28] study of breaking and bouncing sounds. In this study, they used acoustic analyses and a qualitative physical analysis to examine the auditory patterns that characterize these events, and verified their results by testing subjects on synthetic sounds.

Consider the mechanics of a bottle bouncing on a surface (see Figure 4(a)). Each time the bottle hits the surface, it makes an impact sound that depends on its shape, size, and material (as discussed above). Energy is dissipated with each bounce so that, in general, the time between bounces

```
p = cos(freq * 1/samplerate) * power(e, -1 * damping.rate * 1/samplerate);
q = sin(freq * 1/samplerate) * power(e, -1 * damping.rate * 1/samplerate);
a = initial.amplitude;
b = 0;

repeat for duration.in.secs / samplerate:
        anew = a * p - b * q;
        bnew = b * p + a * q;
        a = anew;
        b = bnew;
        output = anew;
end repeat;
```

FIGURE 3 Pseudocode for efficient generation of a logarithmically decaying cosine wave (Eq.(2)).

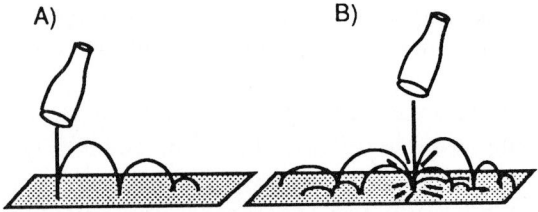

FIGURE 4 Bouncing (a) and breaking (b) sounds are characterized by the temporal patterning of a series of impacts (after Warren and Verbrugge[28]).

and the force of each impact becomes less. Thus bouncing sounds should be characterized by a repetitive series of impact sounds with decreasing period and amplitude.

When a bottle breaks, on the other hand, it separates into several pieces of various sizes and shapes (Figure 4(b)). Thus a breaking sound should be characterized by an initial impact sound followed by several different, overlapping bouncing sounds, each with its own frequency makeup and period.

Acoustic analyses of bouncing and breaking sounds confirm this informal physical analysis. In addition, Warren and Verbrugge[28] found that people were able to distinguish tokens of bouncing and breaking sounds that were constructed by using these rules to splice tapes of impact sounds together.

6.1 SYNTHESIZED BREAKING, BOUNCING, AND SPILLING

To create bouncing sounds, then, we need simply imbed the impact algorithm in an other that calls it at logarithmically decaying intervals. To create breaking sounds, the bouncing algorithm is imbedded in another algorithm that calls it with parameters specifying sources of different sizes at times corresponding to several logarithmically decaying time series.

Several new event parameters become relevant for these algorithms: The initial height of the virtual object is indicated by the time between the first and second bounce, its elasticity by the percentage difference of delays between bounces, and the severity of breaking by the number of pieces produced. In addition, the asymmetry of the perceived object can be varied by adding randomness to the overall temporal pattern.

It becomes clear upon listening to sounds synthesized using this algorithm that, although Warren and Verbrugge[28] claimed that information for breaking and bouncing depends only on temporal patterning, the perceived event depends on the virtual materials involved as well. For instance, if impacts specifying wooden objects are produced in a temporal pattern typical of breaking, we are liable to hear spilling rather than breaking. Similarly, if each of several virtual objects has different material properties, we again hear several spilling objects rather than breaking.

In sum, the impact algorithm described above can be used not only to generate the sounds made by mallets of different hardnesses striking virtual objects of a wide variety of shapes, sizes, and materials, but can also serve as the basis for more complex bouncing, breaking, and spilling sounds. As such, it serves as a research tool that allows the space of such sounds to be explored. Moreover, it provides an efficient method for generating families of related auditory icons. For instance, parameterized impact, bouncing, breaking, and spilling sounds might be used to differentiate and provide details about the results of actions involving icons, windows, containers, and so forth.

7. FROM IMPACTS TO SCRAPING

The sounds made by impacts and patterns of impacts are generally useful for creating auditory icons, but it is desirable to have access to sounds made by a wider range of events. In particular, it would be useful to generate the sounds made by the same object being interacted with in different ways. Using such algorithms, auditory interface designers might map a particular file to a particular object, and then hit, bounce, or scrape it depending on the relevant computer interaction.

In order to create such algorithms, it is necessary to separate the specification of a virtual object from that of the interaction that causes it to produce sound. This turns out to be possible because objects tend to vibrate only at certain invariant resonant frequencies. For example, the spectrogram in Figure 5 shows the sound made by a piece of glass being hit and then scraped across a rough surface. Note that despite the different temporal patterns of the sounds, the resonant modes of each are the same. These modes specify the object, then, while interactions determine the temporal pattern and amount of energy introduced to each.

Because the effects of interactions and objects are distinct, each can be modeled separately. The resonant modes of a virtual object may be modeled as a bank of filters that allow energy to pass at particular frequencies. Interactions, then, can be specified by the pattern of energy passed through the filter bank.

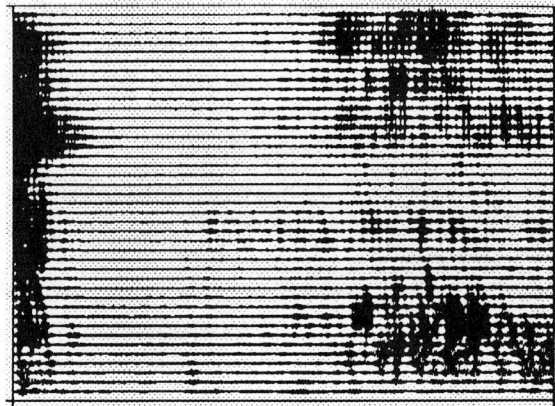

FIGURE 5 Spectrogram of a piece of glass being hit and then scraped: The resonant frequencies remain invariant over different interactions.

7.1 MODELING OBJECTS AS FILTER BANKS

A simple formula for a bank of one-pole filters is:

$$y_n = \sum_m \Phi_m(c1_m x_n + c2_m y_{n-1} - c3_m y_{n-2}) \qquad (3)$$

where Φ_m is an amplitude scalar for partial m, y_n is the nth output, x_n is the nth input, and:

$$c1_m = (1 - c3_m)\left[\frac{1 - c2_m^2}{4c3_m}\right]^{.5},$$

$$c2_m = \frac{4c3_m \cos 2\pi f_m}{c3_m + 1},$$

$$c3_m = e^{-2\pi b_m};$$

where f_m is the peak frequency and b_m the peak bandwidth of partial m.

The parameters used to control the impact algorithm can also be used to control this sort of filter bank. However, manipulating filter bandwidths to control damping actually provides more information for material than do simple manipulations of sine wave damping.

Bandwidth b is proportional to damping: the narrower the resonance peak of the filter, the longer the resonant response to excitation. This correlation between damping and the smearing of partials in frequency space corresponds well to the characteristics of sounds made by materials such as wood or metal.[7,29]

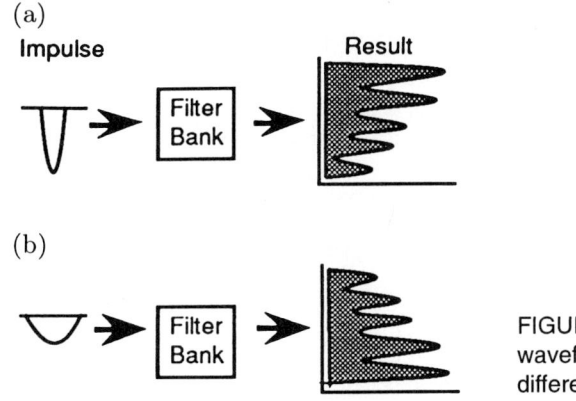

FIGURE 6 Sample impulse waveforms characterizing different impacts (see text).

7.2 SIMULATING INTERACTIONS WITH INPUT WAVEFORMS

A virtual object can be defined by the characteristics of the filter bank described above. The waveform passed through the filter bank, then, models the interaction that causes the object to sound. In this section, two sorts of input waveforms that have been explored are described. The first models impact forces, the second scraping.

When objects are struck, the input forces are characterized by short impulses such as those shown in Figure 6. The energy of such impulses is spread out over many frequencies: the pulse width reflects low-frequency energy, while its angularity reflects high-frequency components. This corresponds to Freed and Martin's[6] characterization of mallet hardness. Hard mallets introduce force suddenly to an object, deforming it quickly, and thus introduce a relatively high proportion of high-frequency energy to the resonant object. Soft mallets, in contrast, deform as they hit the object, introducing energy relatively slowly, and thus the corresponding impulses are characterized by a high proportion of low-frequency energy. Shaping the impulses used to excite a filter bank, then, is a physically realistic way to control perceived mallet hardness.

When an object is scraped, force is applied more continuously. Scraping has been relatively unexplored in terms of its physical or perceptual attributes. However, an informal physical analysis suggests that the pattern of force on an object generated as it is scraped across a surface can be approximated by band-limited noise, where the center frequency of the noise corresponds to dragging speed, and the bandwidth to the roughness of the texture (see Figure 7). Although these parameters are only approximate, being less well motivated physically or psychologically than those used to model impacts, experience shows that a wide variety of realistic scraping noises can be produced using these heuristics.

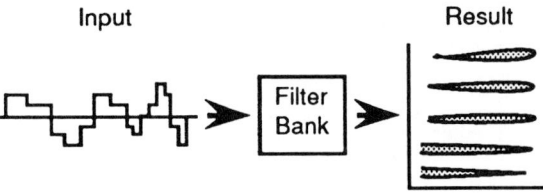

FIGURE 7 A sample force waveform characterizing a scrape with increasing speed.

In sum, the filter-based algorithm described in this section is based on a physically plausible model of sound-producing events. By separating the parts of the model that specify the object from those specifying the interaction, a wide range of virtual sound-producing events can be simulated. The model can create any of the impact sounds that the algorithm described in the last section can. In addition, it can also be used to create a variety of scraping sounds (and, potentially, any other sound involving solid objects).

The ability to generate the sounds of the same object being caused to sound by different interactions offered by this algorithm has great potential for the creation of auditory interfaces. It allows the design of parameterized auditory icons in which the same interface object (e.g., a file) might make sounds indicating a variety of events (e.g., selecting, dragging, opening).

8. MACHINE SOUNDS

Just as complex interactions such as scraping can be modeled by a few summarizing parameters, so might still more complex events be captured succinctly by high-level descriptions. For instance, another class of sounds useful for auditory icons are those made by small machines. Sampled machine sounds were used effectively in the ARKola simulation,[11] indicating ongoing processes that were not visible on the screen. More generally, they might be used to indicate background processes such as printing or compiling in more traditional multiprocessing systems.

A detailed account of the mechanical physics of machinery seems prohibitively difficult. But just as the scraping waveforms described above model the overall parameters of a complex force rather than each of the contributing details, so an approximate model of machines might capture some of the high-level characteristics of the sounds they produce. In particular, three aspects of machine sounds seem relevant for modeling: First, the overall size of the machine is likely to be reflected in the frequencies

Using and Creating Auditory Icons

of sounds it produces; second, most machines involve a number of rotating parts that can be expected to produce repetitive contributions to the overall sound; and third, the work done by the machine can be expected to affect the complexity of the sound.

8.1 FM SYNTHESIS OF MACHINE SOUNDS

I have explained an efficient algorithm for creating a variety of machinelike sounds that capture these properties. The basic strategy is to synthesize a sound using complex tones that vary in a repetitive way, indicating cyclical motion. The rate at which the virtual machine is working, then, can be indicated by repetition speed, the size of the virtual machine by the base frequency, and the amount of work by the bandwidth of the sounds (see Figure 8).

This class of sound may be synthesized efficiently using Frequency Modulation (FM) synthesis.[3] FM synthesis involves modulating the frequency of a carrier wave with the output of a modulating wave. This produces a complex tone with a number of frequency components spaced equally around the carrier wave and separated from one another by the modulating frequency. The number of components (and thus the bandwidth of the sound) depends on the amplitude of the modulating wave (see Figure 9). Thus machine sounds can be created simply by associating the carrier frequency with the size of the virtual machine, setting the maximum amplitude of the modulator to the amount of work done by the virtual machine, and modulating the amplitude of the modulator according to the speed of the virtual machine (see pseudocode in Figure 10).

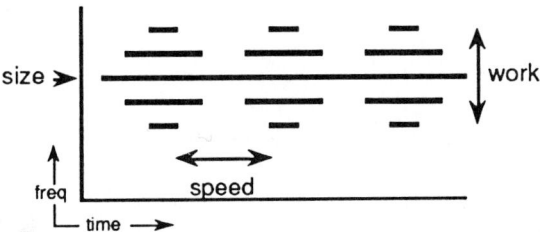

FIGURE 8 Machine sounds can be characterized by a complex wave that varies repetitively over time.

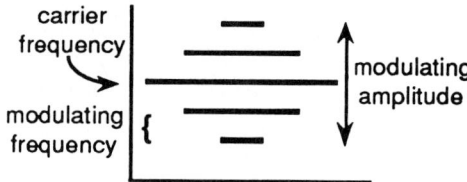

FIGURE 9 FM synthesis, which allows simple control over complex sounds, is well suited for generating machinelike sounds.

```
mod.wave.amp = work * sin(speed * time/samprate);
mod.wave.sample = mod.wave.amp * sin(mod.freq * time/samprate);

output = amp * sin((size + mod.wave.sample) * time/samprate);
```

FIGURE 10 Pseudocode for generating machinelike sounds characterized by speed, size, and work.

The resulting sounds are pitched humming noises that pulse at the speed of the virtual machine. When "work" is low, the throbbing is subtle; when it is high, it becomes quite pronounced. Moreover, the quality of the sounds can be varied by changing the ratio of modulating to carrier frequency: when the two are an integral multiple of one another, the resulting sound is harmonic; when they are not, the sound is inharmonic or noisy. Using this algorithm, then, a wide variety of machinelike sounds can be produced for use as indicators of ongoing processes in multiprocessing systems.

9. CONCLUSIONS

The algorithms described here allow the synthesis of a variety of everyday sounds specified in terms of attributes and dimensions of the events that cause them. Because they are based on a combination of acoustic and physical analyses, and use relatively sophisticated synthesis techniques, they capture a great degree of the richness and complexity of their naturally produced counterparts. Because they are specialized for the classes of event they are to simulate, they are efficient and can generate sounds in real time on many computers. Finally, because they have been designed with potential applications in mind, the events they simulate are those useful for auditory icons.

These algorithms vary in their physical accuracy. Some are based on quantitative physical analyses, while others are based on more qualitative, informal descriptions of events. Moreover, even the quantitative analyses are only approximate. For instance, the physics of a struck bar of wood is much more complex than implied by the simple account given here. Insofar as these algorithms are approximate, the sounds they produce will differ from those made by real events.

Nonetheless, these algorithms do produce quite realistic sounds. Listeners comment that they have the impression of hearing an actual event rather than a synthesized sound. Insofar as the sounds do differ from those made by real events, they may be considered as "cartoon sounds," sounds which capture the relevant features of their sources just as visual caricatures (or graphical computer icons) capture those of theirs. For the purposes of simulating sound-producing events, then, these algorithms are adequate. For the purpose of creating auditory icons, they show great potential, combining flexibility, intuitive controls, efficiency, and relevance.

I have specified these algorithms in sufficient detail here that readers may implement and explore them, in the hope that they will spur further research on parameterized auditory icons. These algorithms open many possibilities for the design of rich auditory interfaces. Impact and scraping sounds can be used to increase the tangibility of graphical objects in direct manipulation interfaces. Bouncing, breaking, and spilling sounds can be used to indicate events in virtual reality systems. Machine sounds might allow us to hear a remote printer as our job reaches the queue, and characteristics of the sound might tell us how fast the job is printing or how much time it will take. In sum, using these algorithms we can design interfaces that we can listen to the way we do to the everyday world.

ACKNOWLEDGMENTS

Many thanks to Dave Woodhouse and Roy Patterson for help with the physical and acoustical analyses presented here, and to Don Norman, Anne Schlottmann, Bill Buxton, William Mace, Allan Maclean, Wendy Mackay, Tom Moran, and Michael Turvey for valuable discussions about this research.

REFERENCES

1. Borin, G., G. De Poli, and A. Sarti. "Algorithms and Structures for Synthesis Using Physical Models." *Comp. Music J.* **16(4)** (1993): 30–42.
2. Buxton, W., and T. Moran. "EuroPARC's Integrated Interactive Intermedia Facility (iiif): Early Experiences." In *Proceedings of the IFIP WG8.4 Conference on Multi-User Interfaces and Applications*. Held September 1990 in Herakleion, Crete, 1990.
3. Chowning, J. "The Synthesis of Complex Audio Spectra by Means of Frequency Modulation." *J. Audio Eng. Soc.* **21** (1973): 526–534.
4. Cremer, L. *The Physics of the Violin*. Translated by John S. Allen. Cambridge, MA: MIT Press, 1984.
5. Draper, S., K. Waite, and P. Gray. "Alternative Bases for Comprehensibility and Competition for Expression in an Icon Generation Tool." In *Proceedings of Interact'90*. Held in Cambridge, UK, August 27–31, 1990. Amsterdam: North Holland, 1991.
6. Freed, D. J., and W. L. Martens. "Deriving Psychophysical Relations for Timbre." In *Proceedings of the International Computer Music Conference*. Held in The Hague, The Netherlands, October 20–24, 1986. San Francisco, CA: International Computer Music Association, 1986.
7. Gaver, W. W. "Everyday Listening and Auditory Icons." Ph.D. Dissertation, University of California at San Diego, San Diego, CA, 1988.
8. Gaver, W. W. "The SonicFinder: An Interface that Uses Auditory Icons." *Hum.-Comp. Inter.* **4(1)** (1989): 67–94.
9. Gaver, W. W., and R. B. Smith "Auditory Icons in Large-Scale Collaborative Environments." In *Proceedings of Human-Computer Interaction—Interact'90*, 735–740. Held in Cambridge, UK, August 27–31, 1990. Amsterdam: North Holland, 1990.
10. Gaver, W. W. "Sound Support for Collaboration." In *Proceedings of the Second European Conference on Computer-Supported Collaborative Work*. Held in Amsterdam, September 24–27, 1991. Dordrecht: Kluwer, 1991.
11. Gaver, W. W., R. B. Smith, and T. O'Shea. "Effective Sounds in Complex Systems: The ARKola Simulation." In *Proceedings of CHI 1991*. Held in New Orleans, April 28–May 2, 1991. New York: ACM, 1991.
12. Gaver, W. W., T. Moran, A. MacLean, L. Lvstrand, P. Dourish, K. Carter, and W. Buxton. "Realizing a Video Environment: EuroPARC's RAVE System." In *Proceedings of CHI'92*. Held in Monterey, CA, May 3–7, 1992. New York: ACM, 1992.

13. Gaver, W. W. "Synthesizing Auditory Icons." *Proceedings of INTER-CHI'93*. Held in Amsterdam, April 24–29, 1993. New York: ACM, 1993.
14. Gaver, W. W. "What in the World Do We Hear? An Ecological Approach to Auditory Source Perception." *Ecol. Psych.* **(5)1** (1993).
15. Gibson, J. J. *The Senses Considered as Perceptual Systems*. Boston: Houghton Mifflin, 1966.
16. Gibson, J. J. *The Ecological Approach to Visual Perception*. Boston: Houghton Mifflin, 1979.
17. Heine, W-D., and R. Guski. "Listening, the Perception of Auditory Events? An Essay Review of *Listening: An Introduction to the Perception of Auditory Events by Stephen Handel*." *Ecol. Psych.* **3(3)** (1991): 263–275.
18. Hutchins, E. L., J. D. Hollan, and D. A. Norman. "Direct Manipulation Interfaces." In *User-Centered System Design: New Perspectives on Human-Computer Interaction*, edited by D. A. Norman and S. W. Draper, 87–124. Hillsdale, NJ: Lawrence Erlbaum, 1986.
19. Jaffe, D., and J. Smith. "Extensions of the Karplus-Strong Plucked String Algorithm." *Comp. Music J.* **7(2)** (1983): 56–69.
20. Jenkins, J. J. "Acoustic Information for Objects, Places, and Events." In *Persistence and Change: Proceedings of the First International Conference on Event Perception*, edited by W. H. Warren and R. E. Shaw, 115–138. Hillsdale, NJ: Lawrence Erlbaum, 1985.
21. Lamb, H. *The Dynamical Theory of Sound*, 2nd ed. New York: Dover, 1960.
22. Laurel, B. "Interface as Mimesis." In *User-Centered System Design: New Perspectives on Human-Computer Interface*, edited by D. A. Norman and S. W. Draper, 67–85. Hillsdale, NJ: Lawrence Erlbaum, 1986.
23. Lövstrand, L. "Being Selectively Aware with the Khronika System." In *Proceedings of ECSCW'91*. Held in Amsterdam, The Netherlands, September 25–27, 1991. London: Kluwer, 1991.
24. McIntyre, M. E., R. T. Schumacher, and J. Woodhouse. "On the Oscillations of Instruments." *JASA* **74** (1983): S52.
25. Risset, J. C., and D. L. Wessel. "Exploration of Timbre by Analysis and Synthesis." In *The Psychology of Music*, edited by D. Deutsch. New York: Academic Press, 1982.
26. Smith, R. B. "A Prototype Futuristic Technology for Distance Education." In *Proceedings of the NATO Advanced Workshop on New Directions in Educational Technology*. Held in Cranfield, UK, November 10–13, 1988.
27. Vanderveer, N. J. *Ecological Acoustics: Human Perception of Environmental Sounds*. Dissertation Abstracts International 40/09B, 4543. University Microfilms No. 8004002, 1979.

28. Warren, W. H., and R. R. Verbrugge. "Auditory Perception of Breaking and Bouncing Events: A Case Study in Ecological Acoustics." *J. Exp. Psych.* **10** (1984): 704–712.
29. Wildes, R., and W. Richards. (1988). "Recovering Material Properties from Sound." In *Natural Computation*, edited by W. Richards. Cambridge, MA: MIT Press.

Meera M. Blattner,†* Albert L. Papp III,† and Ephraim P. Glinert‡
†Department of Applied Science, University of California, Davis, and Lawrence Livermore National Laboratory, Livermore, CA 94551
‡Dept. of Computer Science, Rensselaer Polytechnic Institute, Troy, NY 12180
*Department of Biomathematics, M.D. Anderson Cancer Research Center, University of Texas Medical Center, Houston

Sonic Enhancement of Two-Dimensional Graphics Displays

By studying the specific example of a visually cluttered map, we discover general principles that lead to a taxonomy of characteristics for the successful utilization of nonspeech audio to enhance the human-computer interface. Our approach is to divide information into families, each of which is then separately represented in the audio subspace by a set of related earcons. Animations are used to introduce these earcons to the user, so as to link each earcon in his/her mind with a visual representation. From then on, the earcon suggests the corresponding visual representation to the user, including any real-world sound that may be associated with it, even though the earcon itself is not a real-world sound.

1. INTRODUCTION

Animated displays such as those used for scientific visualization typically show many features that the user needs to understand concurrently. Graphics displays that are not animated (such as maps, blueprints, x-rays and satellite pictures) present a different kind of challenge, in that they often contain "hidden" information that for a variety of reasons is not visible at any given moment. In this paper, we explore ways to use sound in the interface to enhance graphics displays, in particular those that fall into the latter category. By studying the specific example of a visually cluttered map, we hope to discover general principles for the successful incorporation of nonspeech audio in the interface. To this end, we examine in Sections 2 and 3 two problems that cut across a broad spectrum of applications:

1. *Structuring sounds as a language.* Visual nontextual means of communicating information, such as icons and diagrams, have been popular since the development in the 1970s of graphical user interfaces. Even so, the structure of visual languages is not well understood.[31] We do know that iconic information must be limited in complexity if we are to avoid cumbersome pictographic languages. Diagrams, and pictures, can be very complex and often communicate ideas that are difficult or impossible to express in words. But graphic design, and icons in particular, must be visually simple and quickly understood if they are to be effective.[30,34] The sounds we examine in this paper are closer to visual symbols or icons than to text, diagrams, or pictures. We limit their complexity to make them easy to understand. Our approach is that these auditory messages may be structured as a language and best understood through their syntax, semantics, and lexical properties.
2. *Parallel processing of auditory information.* We hear a multitude of different types of sounds all the time. This is true in nature more than in our homes or offices, where the sounds of man-made objects predominate. Humans are sensitive to sounds such as insects, rain, wind, birds, footsteps, rustling of leaves, voices, etc. Our auditory systems evolved over millions of years. Sound, together with our other sensory input, provided us with an edge that allowed us to survive in hostile environments. Awareness and comprehension of the auditory world around us occur in parallel. One of the advantages of vision is that we can process visual information in parallel. Reading while listening to spoken language seems to require more focused attention. Dreyfus and Dreyfus[13] point out that, as we become expert in our tasks, fewer conscious decisions need to be made. These tasks are moved to the background of our mental processes, and we can do more things at once. Anderson[1] writes that:

"Rather than thinking of attention as single-minded ... it probably is more accurate to think of it as not having the capacity to perform two *demanding* tasks simultaneously."

Can we utilize our ability to use parallel processing in understanding auditory information more effectively?

In Section 4 we present and discuss our implementation of sound-enhanced maps. We conclude, in Section 5, with a taxonomy of characteristics that are important for the effective sonic enhancement of two-dimensional graphics displays.

2. STRUCTURING SOUND

Different approaches have been employed when using sound for auditory data display, on the one hand, and for messages or audio cues, on the other. This is because auditory data display has been concerned with the mapping of data points (in an n-dimensional space) to audio output, while the use of sound for messages or cues has been more concerned with the syntax and semantics of the audio output. An example of auditory data display is the work done by Mansur, Blattner, and Joy,[24] in which points on an x-y graph were translated into sonic equivalents and in which pitch is plotted along the x-axis and time along the y-axis, with a nonlinear correction factor.

Auditory display techniques are used, for the most part, to enable the listeners to picture, in their minds, real-world objects or data. There is a lot of leeway for the interpretation of how this may be successfully accomplished; human factors studies are required to understand what may actually be heard. In auditory data display, syntax is determined by the way the data objects present themselves, rather than on the basis of some internal, and often abstract, structure.

Auditory messages or signals were used by people before the discovery of electricity. Bells, bugles, trumpets, and drums sent information to the countryside or announced the arrival of an important person or messenger.[4] Auditory cues used to reduce visual workload were studied by Brown, Newsome and Glinert.[9] The purpose of these cues was to identify the location of groupings of letters on a screen; hence, these cues may be considered "auditory pointers" to items on the screen.

Audio messages have been studied by Gaver,[16] by Blattner,[6] and by Blattner, Greenberg, and Kamegai.[5] Gaver used "real-world" sounds, called *auditory icons*, to convey messages. His examples include objects colliding, breaking, and so on. The fact that Gaver used "real-world" sounds does not imply anything about the meanings of these sounds in his system, which could be an abstraction (e.g., "your computer is going down").

Blattner, Sumikawa and Greenberg,[6] on the other hand, used *earcons*, which are tones or sequences of tones, as a basis for building messages. Earcons and auditory icons are synonymous terms in the sense that they both denote the auditory form of an icon or visual symbol. The distinction lies in the approach to their construction. Whereas Gaver uses the word "icon" in its original meaning (that is, highly representational images), Blattner, Sumikawa, and Greenberg based earcons on similarities between their auditory messages and abstract visual symbols.

The difference between these approaches is the difference between lexical, syntactic, and semantic levels of a language. Syntax is the formal composition of elements in a language. Semantics is the meaning given to the structures in the language. The lexical level refers to the properties of the base elements. The lexical level of auditory messages is concerned with the attributes of sounds: frequency, pitch, duration, loudness, etc.

Most of the work done with the auditory display of data is concerned with the attributes or parameters of sound. This effort has been concentrated on the lexical level. Because earcons are not a direct translation of data into audio, they are able to use a formal syntax in the composition of messages—messages in regard to real-world sounds. So auditory icons are studied primarily on the semantic level. The syntax of auditory icons does not ordinarily arise, because such icons are sampled sounds composed of very complex waveforms and not easily parameterized (see Gaver, in this volume). Hence, the important contribution of the work on auditory icons to the treatment of messages as a language is our deeper knowledge of the "meaning" we give to sounds.

2.1 THE LEXICAL PROPERTIES OF SOUND

The lexical level of language is its alphabet or basic symbol set. Sound attributes play that role in nonspeech audio. Yeung[35] used nine sound attributes (parameters and variables) for the identification of sonic objects: pitch, volume, timbre, duration, changes in frequency, changes in amplitude, changes in duration (heard as rhythm), direction in space, and damping. In her dissertation on sound and computer information processing, Bly used the parameters of pitch, volume, duration, wave shape, attack, and harmonics to relay information about various data sets. Blattner, Sumikawa, and Greenberg[6] used rhythm, pitch, timbre, register, and dynamics. Kramer and Ellison[20] used a "parameter nesting" technique for amplitude and frequency, along with other auditory variables, to create as many as four or five different dimensions. Smith, Grinstein, and Pickett[28] used multiple notes with the characteristics of pitch, peak loudness (volume), and timbre (attack rate, decay rate, and depth of frequency modulation) to make their data audible. Lunney and Morrison[22] used parameters including pitch, duration (rhythm), timbre (attack, decay, and wave form), and loudness (volume). In their experiments, Rabenhorst et al.[26] used detuning (the sounding of two notes that are close together in pitch), stereo balance (relative volume of the stereo channels), direction, and attenuation.

Most of these auditory data display projects used variance in the same physical sound properties as attributes. The characteristics are frequency (the major contributor to perceived pitch), amplitude (directly related to perceived volume), and rhythm or duration of notes in a sequence. More than half of the experiments used timbre characteristics such as wave form, attack rate, decay rate, and other enveloping techniques as parameters that alter the sound in a perceivable manner. The sound source was used as an attribute in some of the experiments. Wenzel[32] separates sound sources into the three spatial dimensions by modeling the pinnea of the ear. Spatial effects such as blurring, thickening, peaking, distancing, muffling, thinning, distortion, and self-animation are created by Ludwig, Pincever, and Cohen[21] to give the illusion of three dimensions.

With the large number of possible parameters for sound, sonic objects can differ greatly. However, the parameters or attributes of sound are less generally agreed upon than the eight retinal dimensions of Bertin[2]: color, size, shape, value, texture, orientation, and relationship in the plane. If we confine our attention to the most basic sonic parameters, we can use as few as five (e.g., rhythm, pitch, timbre, dynamics, and direction). However, those five may differ somewhat depending upon our interpretation of sonic information (i.e., pitch or frequency), and they are not independent.

The experimental results described above show that human perception of auditory phenomena can distinguish between two objects for which 5–8 distinctive attributes are quite different. Use of a greater number of attributes can create greater separation of sounds.

2.2 EARCONS: A SYNTAX FOR AUDITORY MESSAGES

In the musical world, a short sequence of tones is called a *motive*. In the construction of earcons, a motive is used as a building block for larger groupings. The advantage of these constructions is that the musical parameters of rhythm, pitch, timbre, dynamics (loudness), and register can be easily manipulated. The motives can be combined, transformed, or inherited to form more complex structures. The motives and their compounded forms are called *earcons*. However, earcons can be any auditory message, such as real-world sounds. In the application used in this paper, auditory maps, the messages are information about data rather than a direct translation of data into sound. One advantage of structuring audio messages in this way is that they can be used with any basic sonic unit, such as tones or sampled sounds.

A motive may be an earcon, or it may be part of a compounded earcon. Let A and B be earcons that represent different messages. A and B can be *combined* by juxtaposing A and B to form a third earcon AB. For example, if A is an earcon for "file" and B the earcon for "deleted," then AB is the earcon for "file deleted." Earcon A may be *transformed* into earcon B by a modification in the construction of A (for example, by changing the pitch in one of its notes). If A were an earcon for "the computer is up," the transformed earcon might instead inform the listener that "the computer is down."

A family of earcons may have an *inherited* structure. Although many different types of hierarchical structures can be created, the one proposed in Blattner, Sumikawa, and Greenberg[6] follows. A *family motive A* is an unpitched rhythm of not more than five notes. To use A to define a family of messages, the family motive is first elaborated with pitch $A + p \rightarrow B$. The result B is then preceded by the family motive A, to form a new earcon AB. Hence, the earcon has two distinct components, an unpitched motive followed by a pitched motive with the same rhythm. A third earcon, ABC, can be constructed by combining AB with a third motive C whose pitch and rhythm are identical to those of AB but which possesses an easily recognizable timbre $A + p + t = B + t \rightarrow C$. If, for example, A were to indicate a printer problem, then AB could indicate the printer is out of paper. It is desirable to structure the meaning of inherited earcons hierarchically through the use of refinements that correspond to hierarchical musical structure, in a manner analogous to that in which a refinement process is used in semantic networks and object-oriented programming. The exact nature of the refinements is not specified in the construction of earcons.

Various methods of structuring sounds are used in musical composition. Music theory is too complex to be discussed here. Suffice it to say that many of the basic principles used in the construction of earcons are similar to those used in the composition of music: transformations, hierarchical structures, combining sounds, and polyphony.

An earcon is usually associated with an object. By an object we mean any identifiable structure, whether it is an abstraction such as an error message or a representation of something in the physical world. To display two or more earcons simultaneously, we have to identify their temporal locations with respect to each other. Two primary methods are considered below: overlaying one earcon on top of another, and the sequencing of earcons. Some sort of merging or melding into new sound could also be envisaged (e.g., the pitch of two notes could be combined into a third pitch), but we have not attempted this. The display of multiple earcons is discussed in Section 3.

Data translated directly into sound require less explanation or motivation than abstractions such as earcons. Although auditory icons that make real-world sounds usually can be recognized quickly, several experiments have shown that earcons are preferred over many other types of sonification. Jones and Furner[18] compared earcons, auditory icons, and synthesized speech; their results showed that subjects preferred the sounds of earcons but were better able to associate auditory icons to commands. Brewster Wright, and Edwards (in this volume) found earcons to be an effective form of auditory communication. They recommended six basic changes in earcon form to make them more easily recognizable by users. These changes were: (1) use synthesized musical timbres; (2) pitch changes are most effective when used with rhythm changes; (3) changes in register should be several octaves; (4) rhythm changes must be as different as possible; (5) intensity levels must be kept close; and (6) successive earcons should have a gap between them.

Earcons are necessarily short, because they must be learned and understood quickly. Earcons were designed to take advantage of chunking mechanisms and hierarchical structures that favor retention in human memory. Furthermore, they use recognition rather than recall.[1] The tests run by Brewster, Wright, and Edwards had no training period associated with them (the earcons were heard only once before the test); in spite of this, the subjects could use them effectively.

Will sounds as abstract as earcons be accepted by the majority of users? The advantages are clear. Earcons are easily constructed on almost any type of workstation or personal computer. The sounds do not have to correspond to the objects they represent, so objects that either make no sound or an unpleasant sound still can be represented by earcons without further explanation. In the next section, we discuss some of the techniques we are using to reduce learning and recognition times for earcons, so as to make them useful for the majority of computer users.

2.3 THE SEMANTICS OF SOUND

A great many, indeed most, sounds that could be used to form a basic unit (motive) for earcons have not been examined for this purpose yet. Some sounds are naturally occurring such as bumping, scraping, and breaking, as described by Gaver.[16] This basic set can be enlarged to include both the sounds of nature (birds singing and wind blowing) and human sounds

---

[1] An excellent description of human memory can be found in Anderson.[1]

(hands clapping and coughing). Gaver divides auditory icons into the symbolic, the metaphorical, and the nomic. Elsewhere in this volume, Ballas discusses the role of representation in sound as linguistic analogies: exclamation, deixis, simile, metaphor, and onomatopoeia. We will not be concerned with these divisions, because any association with an identifiable and familiar sound is sufficient for our purposes.

Also in this volume, Cohen discusses a system called "ShareMon" that he implemented based upon Gaver's ideas of auditory icons, in which he created a variety of sounds used to notify users of background activity (e.g., entry into a network). The auditory icons were sounds such as those of knocking, a key turning, footsteps, and a cough. One of the major drawbacks of Cohen's implementation, as determined by his tests on users, is the realism of his sampled sounds. A cough often made users turn to see who had entered the room. Even though the auditory icons were relatively easy to identify, the user still had to identify the relation of the sounds to the messages they were delivering.

At this point we return to study (visual) icons, to see if we can learn anything from them. Icons must represent abstractions, too. Although a variety of techniques are employed, most of the visual symbols transmit an ambiguous message that must be resolved by inserting words or letters within the image. Familant and Detweiler[14] describe some of the incongruities and difficulties that system designers have experienced with the construction of icons. For example, the FINDER applications of the Macintosh operating system include an icon that looks like a personal computer (one of Apple's low-end Macintosh products). Nevertheless, icons are tremendously useful; even a somewhat inappropriate icon triggers the memory of an object or event more quickly than a purely abstract figure such as a square or circle. Because we are used to *seeing* physical objects, many icons are pictures of objects.

Sound is produced when objects interact (e.g., bump into one another). That may be one of the reasons we have difficulty associating objects with sound. We usually *infer* the presence of the object from the sound. Of course, telephones ring and we don't see anything move; in fact, however, the sound is the result of movement. We know that sound and vision are complementary informational media.[16] A well-known recall technique is the so-called *method of loci*, in which a list of unrelated items is associated with a location such as a room or a path.[1] The association of sounds with visualized objects strengthens our ability to recall those sounds, particularly if the objects are associated with locations.

2.4 USING SOUND AND IMAGES IN CONCERT TO CONVEY MEANING

We are experimenting with introducing earcons to the user as an animation with sound. Sound is the result of movement and the interactions of objects, so a static icon may not serve well as a visual representation of a sound. The animation describes the object/event that the motive introduces. For example, suppose our visual and auditory symbol was to be a bouncing ball. The motive that forms the basis of the earcon is the sound of the bouncing ball.

Often, physical events that are quite different give rise to similar sounds. One might say these are naturally occurring auditory homonyms. For example, a hammer hitting a table may have the same sound as a car door slamming. Hollywood takes advantage of this observation on a regular basis. But such similarities of sounds interfere with the listener's ability to identify the intended meaning of an auditory icon. A picture or animation of the event would help clarify the intended association between the sound and the object, so that an auditory icon that represented the disposal of paper would never be confused with the sounds of an emergency such as fire.

Tr 89

Another advantage of pairing an animation with an auditory icon is that the auditory display need only resemble the object or event but need not sound exactly like it. For example, a bouncing ball can be represented by a tom-tom drum. In the construction of earcons, we are using an animation or a picture with the sound to introduce the user to the sound. Following the suggestions of Brewster, Wright, and Edwards (in this volume) to create richer timbres, we used sampled sounds from musical instruments. These sounds were then manipulated using the parameters of rhythm, pitch, dynamics, and register.

How can we get earcons that bear a resemblance to real-world objects, such as a bouncing ball, from this? Recall that in the ShareMon system mentioned above, the realism of the sounds used in the auditory icons led to confusion on the part of the users. Therefore, instead we prefer tones that resemble the events in question but that would never be confused with the real-world sounds. An added bonus is that our sounds are easier to manipulate. We chose to use the timbres of musical instruments for our earcons because users seem to prefer them. However, more testing is required to confirm this observation. The point here is that the sounds do not have to be very realistic if the user is presented with a method of determining what the sounds represent.

In our implementation of auditory maps, we had three families of earcons associated with buildings: access privileges, computer equipment, and the administrative unit in charge of the building. This choice of information immediately provided a challenge, because all three types are abstractions and neither icons nor earcons have a natural representation associated with them. To represent access privileges (security level), we chose a knocking sound represented by a tom-tom drum. Computer equipment is represented by electrical noises with a wave sound of four notes, and administrative unit is a rustling tree (our administrative units are hierarchical!) with three increasing notes. It would be difficult to associate these sounds with images, unless each were initially presented to the user with the appropriate visual image.

The animated icons with earcons are implemented as *metawidgets*[3,17]: autonomous objects that can display themselves in a variety of ways which include (combinations of) graphics and sound. The metawidget, or the meta-icon in this case, chooses its particular representation depending upon the other events on the screen and the state of the system. When the map first opens, a series of meta-icons with both their auditory and visual representations is displayed for the area the map is showing. Later on only the auditory display is retained, as new areas of the map are examined and the corresponding meta-icons are displayed both visually and sonically for them.

3. SOUND SEPARATION OF COMPLEX OBJECTS

What are complex objects? We will define a complex object as one with many attributes. How quickly can a complex object be recognized, as compared to a simple object or one with fewer attributes? Many readers will be familiar with Chernoff's faces, a unique method for representing multivariate data. Humans are particularly sensitive to expressions on human faces and can assimilate the whole face as a single "chunk" of information. Chernoff[11] encoded multivariate data into human faces and found that the data could easily be classified; in this case, the faces had expressions determined by eyes, nose, mouth, and head shape.

Buxton, Gaver, and Bly[10] point out that most real-world sounds are complex and difficult to analyze. Examples are hybrid sounds (such as filling a glass), temporal patterning (such as footsteps), and combinations of sounds (such as engine noises). In spite of this complexity, human beings are well suited to identifying such sounds.[16] This makes a good case for introducing natural sounds as part of our sonic experiences in the interface.

Many musical instruments have complex timbral effects as well. The shakuhachi (Japanese flute), for example, produces a rough sound with many complex overtones that is quite unlike the pure, clear sound of the western flute. The sounds of musical instruments synthesized by computer

Sonic Enhancement of Two-Dimensional Graphics Displays

have not been able to simulate the complex overtones of real musical instruments. Bregman[8] also points out that in music, a chimeric[2] effect is often desired even though the auditory system tries to avoid and separate these sounds.

One of the most interesting and complex features of auditory information is its perceived continuity (that is, stream formation). Individual tones may no longer be heard as a sequence of tones, but rather become a pattern known as a stream. Deutsch[12] states that:

> "Music presents us with a complex, rapidly changing acoustic spectrum, often resulting from the superposition of sounds from many different sources. The primary task that our auditory system has to perform is to interpret this spectrum in terms of the behavior of external objects. This is analogous to the task performed by the visual system when it interprets the mosaic of light patterns impinging on the retina in terms of objects producing them."

Bregman[8] has a thorough and readable description of the perception of streams. Deutsch discusses grouping mechanisms in music and musical illusions. Williams[33] has devised a computational model to study the effects of streaming. Bregman, Williams (in this volume), and Deutsch discuss the Gestalt grouping principles of proximity, similarity, good continuation habit or familiarity, belonging, common fate, and closure that contribute to the streaming effect.

Under what circumstances is it desirable to hear one complex object versus two or more simpler objects? Visual and auditory grouping mechanisms will determine how we hear or see our scenes. If, on the other hand, we structure our own views, is it more desirable to use one complex object or combinations of simpler ones? Streams will play an important role in the techniques used below. The creation of multiple sound objects is the creation of streams of sonic events. Bregman[8] divides auditory scene analysis into two types: primitive and schema-based.

> "Primitive segregation employs neither past learning nor voluntary attention. It is present in infants and, therefore, probably innate. ...Schema-based [auditory scene analysis is] developed for particular classes of sounds. [The schemas] supplement the general knowledge that is packaged in the innate heuristics by using specific learned knowledge."

A great deal of our perception of streams depends upon schema-based listening. This includes our ability to understand speech. We need to acquire

---

[2] The chimera is a mythical animal formed as a composite of other animals.

a deeper understanding of how we develop schemas for sound organization before we can develop a theory for the sonification of complex objects.

Natural sounds are created by the motion of real-world objects. Their separation depends upon the physical properties of the materials that constitute the objects and their interactions with each other. Three-dimensional separation of sound is an important factor in our daily lives. Blattner, Greenberg, and Kamegai[5] theorized about sonification of turbulence in terms of musical texture (the number and relationship between the melodies in a musical fragment), and the roles of natural sounds vs. related synthetic sounds. To create musical texture, they used such techniques as differentiating register, amount of activity, and timbre. Sounds are best separated into streams when their application indicates separate objects are being represented. For example, in the study of turbulence, low tones imitate a slowly moving laminar fluid while vortices are distinguished by a different timbre and registral space.

Earcons must be differentiated if they are to be heard concurrently. The simplest method of doing this is to separate earcons at the time they are created to be distinct from all other earcons that may be heard at the same time. For this reason, the area over which earcons are heard must be limited. The safest use of concurrent sound probably is to confine its use largely as background sound of a very simple nature. However, our implementation assumes that up to four earcons could be heard at the same time. Informal tests show that listeners can identify up to three concurrent earcons without great difficulty. The greater the difference in the attributes of earcons, the greater their separation. Timbre, rhythm, and pitch must be clearly distinguished for separation. Families of earcons should be put into different registers (octaves). Timbre seems to be the most important factor in earcon identification as reported by Brewster, Wright, and Edwards (in this volume) and families should be represented by different instruments (if instruments are being used for timbre).

4. IMPLEMENTATION OF AUDITORY MAPS

To test our theories, we chose to combine auditory display techniques with two-dimensional maps. Maps are primarily used for orientation, navigation within, and analysis of, geographic terrain. However, a broad range of additional information may be of interest in some cases: average annual rainfall, soil composition, location of mineral deposits and other natural resources, location of rail lines, location of historical sights, various economic factors, elevation, etc.

Because they need not be static, computerized maps can take advantage of many more methods for displaying information than can traditional paper maps. Enlarged windows can appear at a point of interest; numerical data can pop up on demand and disappear when no longer needed. Animation and pseudo-color can be used to track or call attention to specific information. Nevertheless, because the addition of visual data requires that space be allocated, a saturation point will eventually be reached beyond which interference with text and graphics already on display cancels out any possible benefit. In such cases (and others), it may be advantageous to present some of the data in a sonic representation. Auditory maps were employed by Kramer[19] to enhance Magellan's view of Venus, through use of sound to convey information such as the emissivity (i.e., radiation) and gravity of the area being viewed. The auditory output did not disrupt the view of the underlying landscape.

With the help of an experimental system implemented on a Silicon Graphics INDIGO workstation, we examined a variety of ways for transforming traditionally visual cartographic data into sonic representations. The data used comprise a map of Lawrence Livermore National Laboratory. This particular map was selected because it was in the form of a machine-readable database of buildings. The floor plans are visible, as are geographical data such as roads, parking lots, etc. The many pieces of information associated with each building include sewer lines, water lines, power lines, number of computers and people housed within, the department or administrative unit in charge, construction type, level of security clearance required to work there, job titles of those in the building, etc.

As the INDIGO's cursor is dragged over the image, relevant information is presented to the user. Alternatively, the mouse may remain still and information is requested by the user clicking on an appropriate button. This immediately leads to a problem: sonic information must be presented in a "short form" when the mouse is in motion but can be presented in a "long form" when it is stationary. The short form can't encode sufficient information to distinguish between items within a family, whereas the long form can easily do so.

The functionality required involved retrieval not only of the location of particular items, but also of area information. We also needed a way to provide summary data to users. It would be too slow and inefficient to scan an entire scene with a mouse! So we had to develop methods for scanning areas and presenting multiple data. Areas can be indicated by bounding circles or rectangles, or a free-form curve. An effective method of implementing an area selector is to use techniques similar to those commonly found in drawing packages; the user first clicks on an icon representing area, and then moves or drags some shape with the cursor. We did not implement the area selector this way because of time constraints. Summary data is

460 M. M. Blattner, A. L. Papp III, and E. P. Glinert

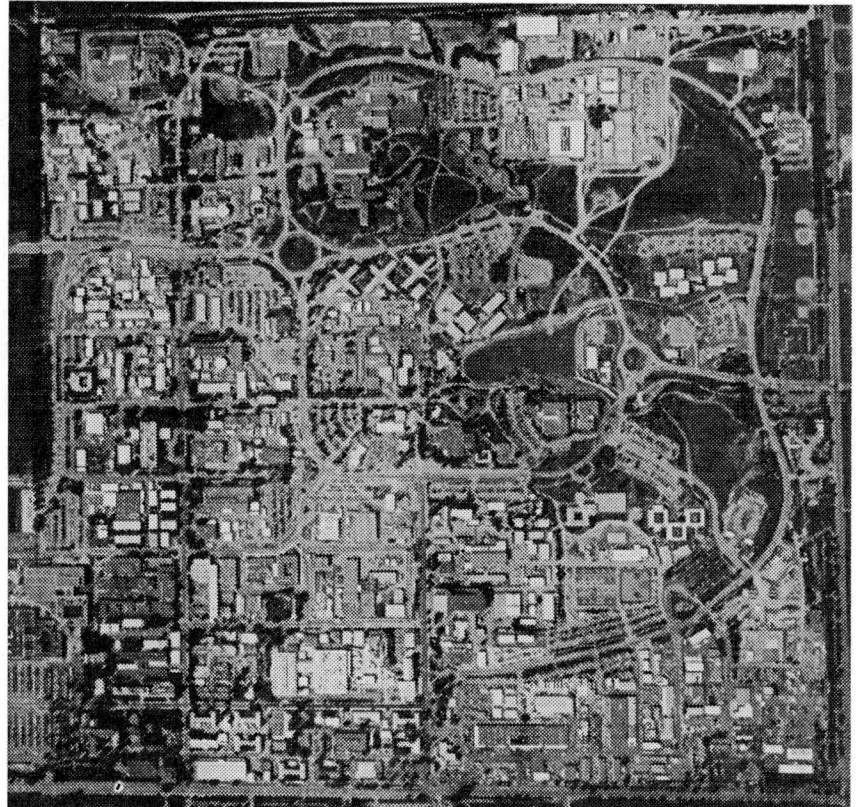

FIGURE 1 The initial view of the auditory map: an aerial view of Lawrence Livermore National Laboratory.

used to indicate that there were many items of a certain type in an area. We chose a simple method to handle summary data, by a linear mapping of earcon volume (loudness) to the magnitude of the numeric value.

5. TAXONOMY: AUDITORY CHARACTERISTICS FOR DISPLAYS

Our experience with the implementation discussed in the previous section allows us to identify the characteristics relevant to the effective use of non-speech audio in two-dimensional graphics displays.

- **Unit of Sound**: sampled vs. synthesized timbre.

    Sampled sounds are digital recordings of sounds, while synthesized sounds are created on the computer through mathematical functions and thus have easily modifiable sound attributes of the type discussed at the beginning of Section 2. Iconic sounds are for the most part digitized and edited recordings.

    Our implementation: The implementation of auditory maps presented here uses sampled waveforms to create timbre. The waveforms are samples of acoustic instruments and the intention is that they can be identified and discriminated readily by the listener. These waveforms are manipulated the same way that synthesized earcons are. The simple earcons vary in length from one to four notes; compounded earcons may be longer.

- **Sound Attributes**: timbre, rhythm, pitch, register, and dynamics.

    These categories were discussed in Section 2.1. For a more complete discussion of these attributes see Blattner, Sumikawa, and Greenberg.[6]

    Our implementation: We used timbres of various musical instruments to create earcons to help differentiate the sounds, as suggested by Brewster, Wright, and Edwards (in this volume). The sampled waveforms are altered by varying the frequency multiplier and the amplitude factor, to create different pitches and volumes. The frequencies range from 100 Hz to 2,000 Hz. None of the earcons varied in loudness within themselves. Instead, we interpreted summary data using dynamics; that is, sets of data with more like items are louder than sets with smaller amounts of like data.

- **Structural Composition**: techniques for building complex objects from simpler ones.

    The three methods described in Section 2.2 are "combined," "transformed," and "inherited."

    Our implementation: Simple sounds as well as complex ones built by combination, transformation, and inheritance from motives and other earcons were used in auditory maps. There is a simple earcon of a tom-tom drum pounding, which represents a building restriction. The pounding is transformed in pitch and frequency to indicate different access restriction levels of buildings. Higher restrictions are represented by faster, higher pitched knocking. The computer earcon is a four-note earcon using a flute timbre which is transformed over the x-axis, y-axis, and both axes. A simple three-note earcon in one pitch using a saxophone timbre indicates an administrative building. The second level motives inherit all the properties of the first earcon except that they have different pitch changes. It is important to note that each family of earcons shares the same timbre. Since timbre is one of the most easily recognized attributes of sound, one can immediately identify the family of an earcon just by recognizing what instrument is used in playing that earcon. Combination of these different earcons can be used to build new, more complex, earcons. For instance, a combination of three earcons can indicate a physics buildings with a clearance level of "confidential" that houses Sun computers.

- **Semantic Interpretations**: objects mapped to auditory and visual icons.

Messages (objects) may be mapped to sequences of notes with no real-world association, just as they may be mapped to a visual symbol (abstraction). However, iconic representations can also be selected that bear some resemblance to a real-world association, in order to help the user remember the sounds for the object, as described in Section 2.3 above. Since the visual and auditory modalities reinforce each other, using them together during the training period would greatly strengthen the user's ability to remember auditory objects. In this phase, there is no substitute for a good imagination and a sense of humor.

Our implementation: We used three families of objects for information about buildings for our earcons: access privileges, administrative unit, and computers. Only computers have an obvious visual icon. We chose knocking for access privileges, and a schematic tree diagram (with the root on top) for administrative units. To associate earcons with the visual symbol often isn't easy. Knocking was the simplest and computers the hardest. Knocking became a uniform thumping with a tom-tom drum, the administrative unit walked up a tree with three notes on a saxophone, and computers were represented by a whirring noise sounded out by four notes on a flute. Our animations were sequences of pictures with a new picture for each note.

- **Temporal Combinations**: sequential, concurrent, and combined.

Sequential combinations are sounds that are heard one after another. Concurrent sounds are those that are either played simultaneously or that partially overlap in time. Combined sounds are those whose attributes are combined into a single new sonic item. An approach that is used to combine data is to consider every mapping from a data item to an n-dimensional coordinate system, where the coordinates are sonic attributes.

Our implementation: The user may choose to listen concurrently or sequentially. However, if more than four earcons need to be sounded in concurrent mode, then the first four will play concurrently after which, as each earcon ends, another will begin. We have not implemented combined data at this time.

- **Speed Selection**: auditory display while moving or while stationary.

  Audio information can be displayed in two modes: moving and stationary. When the user moves the cursor over a region, there is not sufficient time to play its earcon. A series of short, truncated sounds inform the user that there are items of interest in that location. Stationary mode is indicated when the mouse is clicked and the long form of the data under the cursor can be displayed. Selection while moving can be turned off if the user wishes.

  Our implementation: Each earcon family is represented in its moving mode by a very short earcon of the same timbre as is used in the family's stationary earcons. It is this short earcon which is played during selection while moving. Currently, selection by moving must be initiated with a mouse click and the selection of long or short modes of earcons must be explicitly chosen. However, our plan is to have this switch made when the user stops moving the cursor.

- **Area Selection**: auditory display of a datum or of an area.

  One problem with examining a map under a cursor is that the point designated may be so small that it is difficult to find objects. Instead, an area selector can be used. The shape of the area selector can be some geometric shape, free-form, or constrained to conform to the shapes of rooms and buildings. The area selector can be dragged over the map to find the sources of auditory information. For reasons mentioned in Section 4, designing earcons to play concurrently over a large area may be too difficult to implement. So the size of an area should be limited.

FIGURE 2 A zoomed in view of some of the buildings on the site with an area selector positioned over one of the buildings.

Our implementation: The current version of our system allows the user to choose the area selector from a menu; the size of the selector is determined by arrow keys on the keyboard. A planned new version of the system will have a button on the area selector that will enlarge the size when dragged. Currently, only objects that are entirely within the area selector (which may be of any size) are heard.

- **Type of Data**: discrete values and functions over areas.

The data being observed can have a variety of properties. Two-dimensional data can be sparse, in that only certain locations have any value at all, or they can be a function over a broad area. An example of a discrete datum is a historical marker. An example of a continuous function is the amount of rainfall; state or country borders provide an example of a step function.

Sonic Enhancement of Two-Dimensional Graphics Displays

Our implementation: The data we used are step functions, in that a single value was associated with an entire building. Areas outside of a building had no auditory information associated with them.

- **Numeric Properties**: single data value, multiple data values, and summary data.

The introduction of an area selector raises the issue of multiple data items to be sonified. Two solutions are to display the items as multiple items, either in sequence or overlaid, or to create sonic output that represents summary data.

Our implementation: We implemented all the possibilities, with the loudness of the sound being controlled with the size of a data set.

- **Auditory Dimensions**: one, two, three, and four.

In this paper we are concerned only with two-dimensional graphics displays. However, the auditory display can be in four dimensions, if time is considered a dimension. One-dimensional sound may be obtained in a variety of ways, as described above. Two-dimensional sound is either represented in two spatial dimensions or in one spatial dimension varying with time. Three-dimensional sound is represented as three spatial dimensions or two spatial and one in time. Four dimensions can be represented as three dimensions in space and one in time.

Our implementation: We did not implement any spatial dimensions.

- **Integration with Graphics**: coordination of image and sound.

Sound can be used alone or in conjunction with graphical effects on the underlying display. Objects can be thought of as having both visual and sonic aspects. The visual can be iconic, diagrammatic, textual, or some kind of alternate representation of the data.

Our implementation: We did not add any graphics to the auditory maps besides the area selector, which is a circle that is used to specify the area from which information is gathered. An extension to our implementation might pinpoint the locations of computers within a building when the earcons are heard; as the map program is accessed, the visual animations associated with the various earcons are displayed.

- **Information Selector**: menus, voice, command line, direct manipulation, etc.

  Information selection refers not to what is actually on display, but rather to the palette of interaction styles with which the user configures the combined auditory/graphics display.

  Our implementation: Items are selected from a hierarchical menu. When an item is selected, the waveform of the associated earcon and other properties that may be of interest are displayed. The "play" button is clicked to audition an earcon while the menu is displayed.

- **Editing Tools**.

  These tools assist the user in constructing and editing new earcons. There are five parameters (rhythm, pitch, timbre, register, and dynamics). The user should be able to construct these from selections in a menu or by a simple visual language. Other editing problems include attaching sounds to locations and separating sounds so that they can be played concurrently without interfering with each other. There are some editing tools available but, due to their cost and other implementation problems, very few are in use.

  Our implementation: We are implementing a graphical editing tool for earcons, in which a color-coded bar helps us to visually differentiate the attributes of motives. Timbre, rhythm, pitch, register, and dynamics are all represented as segments on the bar. Each attribute is associated with a different color; similar attributes have similar colors.

6. DISCUSSION

Several novel concepts have been proposed and/or explored in this chapter. Using two-dimensional maps as our testbed, we divided information into families, each of which was then separately represented by a set of related earcons that vary in controlled ways (as described above in the section on structural composition). All earcons with hidden objects lying under the area selector could be heard concurrently (note that the problem of concurrent sound has hardly been touched, due to the complexity of the theory that will undoubtedly be required to deal with it). Animations were used to introduce earcons to the user. This technique links each earco in the user's mind with a visual representation. From then on, the earcon will suggest that visual representation to the user, including any real-world sound that may be associated with it, even though the earcon itself is not a real-world sound.

Human factors testing remains to be carried out on the effectiveness of auditory maps. However, the authors have gained some experience in using this system and creating different earcons to supply information. One of the objectives of this project was to determine if any of the compounding techniques for earcons (combining, transforming, or inheriting) was more effective than the others. We concluded that any one of the three techniques can be used effectively to structure earcons, but a poor choice of sonic attributes will make an earcon difficult to recognize. For example, our first experiment with transformed earcons was to reflect the pattern across the x-axis; that is, if the notes in the motive are considered points on a graph, the reflection will give rise to another pattern. This was one of the techniques used by twelve tone composers.[27] These reflected patterns are generally difficult to recognize. The pounding or knocking sound made by the administrative earcons with transformed pitch and frequency was very effective and could be recognized easily.

Although the taxonomy in Section 5 was derived from a particular application relating to maps, we believe that it applies in general. The function of the audio information is to identify the hidden data to the user. One of the questions that we haven't addressed is the relation of earcons to visual representations of the information on the display. Should visual marks indicate the location of items as well as the audio items? If so, should they be fixed on the scene, or can they fade away after the cursor has moved over them? Another issue is the proper role for voice in a display. These are but two examples of the many open questions that remain to be addressed in this field.

ACKNOWLEDGMENTS

This research was supported, in part, by the United States Department of Energy through Lawrence Livermore National Laboratory under contract W-7405-Eng-48. The third author was also supported, in part, by the National Science Foundation under contracts CDA-8805910, CDA-9214887, and CDA-9214892.

REFERENCES

1. Anderson, John R. *Cognitive Psychology and Its Implications*, 2nd ed. New York: W. H. Freeman, 1985.
2. Bertin, J. *Semiology of Graphics*. Madison, WI: University of Wisconsin Press, 1983.
3. Blattner, M. M., E. P. Glinert, J. A. Jorge, and F. Ormsby. "Metawidgets in the Multimodal Interface." *Proceedings of COMPSAC'92*, 115–120. Conference held in Chicago, IL, September, 1992. Los Alamitos, CA: IEEE, 1992.
4. Blattner, M. M., and R. M. Greenberg. "Communicating and Learning Through Nonspeech Audio." *Multimedia in Education*, edited by A. D. N. Edwards and S. Holland. NATO ASI. Series F, Vol. 76, 133–144. Berlin: Springer-Verlag, 1992.
5. Blattner, M. M., R. M. Greenberg, and M. Kamegai. "Listening to Turbulence: An Example of Scientific Audiolization." In *Multimedia Interface Design*, edited by M. M. Blattner and R. Dannenberg, 87–104. New York: ACM Press, 1992.
6. Blattner, M. M., D. A. Sumikawa, and R. M. Greenberg. "Earcons and Icons: Their Structure and Common Design Principles." *Hum.-Comp. Inter.* **4** (1989): 11–44.
7. Bly, S. "Sound and Computer Information Presentation." Ph.D. Thesis, University of California, Davis. Published as Technical Report UCRL-53282, Lawrence Livermore National Laboratory, Livermore, CA, 1982.
8. Bregman, A. S. *Auditory Scene Analysis*. Cambridge, MA: MIT Press, 1990.
9. Brown, M., S. Newsome, and E. P. Glinert. "An Experiment into the Use of Auditory Cues to Reduce Visual Workload." In *Proceedings of the CHI'89 Conference on Human Factors in Computer Systems*, edited by K. Bice and C. Lewis, 339–346. New York: ACM Press, 1989.
10. Buxton, B., B. Gaver, and S. Bly. "The Use of Nonspeech Audio at the Interface." Tutorial presented at CHI'91. In *Proceedings of the CHI'91 Conference*, S. P. Robertson, G. M. Olson, and J. S. Olson. New York: ACM Press, 1991.
11. Chernoff, H. "The Use of Faces to Represent Points in k-Dimensional Space Graphically." *J. Am. Stat. Assoc.* **68(342)** (1973): 361.
12. Deutsch, D. "Grouping Mechanisms in Music." In *The Psychology of Music*, edited by D. Deutsch, 99–134. Orlando, FL: Academic Press, 1982.
13. Dreyfus, H. L., and S. E. Dreyfus. *Mind over Machine*. New York: The Free Press, 1986.

14. Familant, M. E., and M. C. Detweiler. "Iconic Reference: Evolving Perspectives and an Organizing Framework." *Intl. J. Man-Mach. Stud.* (1993): to appear.
15. Frysinger, S. P. "Applied Research in Auditory Data Representation." In *Extracting Meaning from Complex Data: Processing, Display and Interaction*, edited by E. J. Farrell. Proceedings of SPIE **1259** (1990): 130–139.
16. Gaver, W. W. "Auditory Icons: Using Sound in Computer Interfaces." *Hum.-Comp. Inter.* **2(2)** (1986): 167–177.
17. Glinert, E. P., and M. M. Blattner. "Programming the Multimodal Interface." To appear in *Proceedings of the 1st ACM International Conference on Multimedia (MULTIMEDIA '93)*, Anaheim. 1993. New York: ACM Press, 1993.
18. Jones, S. D., and S. M. Furner. "The Construction of Audio Icons and Information Cues for Human-Computer Dialogues." In *Contemporary Ergonomics*, edited by T. Megaw. Proceedings of the Ergonomics Society's 1989 Annual Conference. 1989. London: Taylor & Francis, 1989.
19. Kramer, G. Personal communication, 1992.
20. Kramer, G., and S. Ellison. "Audification: The Use of Sound to Display Multivariate Data." In *Proceedings of the International Computer Music Conference*, Montreal, Canada. 1991. San Francisco, CA: ICMA, 1991.
21. Ludwig, L., N. Pincever, and M. Cohen. "Extending the Notion of a Window System to Audio." *Computer* **23** (1990): 66–72.
22. Lunney, D., and R. C. Morrison. "Auditory Presentation of Experimental Data." In *Extracting Meaning from Complex Data: Processing, Display and Interaction*, edited by E. J. Farrell. Proceedings of SPIE **1259** (1990): 140–146.
23. Mansur, D. L. "Graphs in Sound: A Numerical Data Analysis Method for the Blind." Master's Thesis, University of California, Davis, CA, 1984. Published as Technical Report UCRL-53548, Lawrence Livermore National Laboratory, Livermore, CA, 1984.
24. Mansur, D. L., M. M. Blattner, and K. I. Joy. "Soundgraphs: A Numerical Data Analysis Method for the Blind." In *Proceedings of the 18th Hawaii International Conference on System Sciences*, 163–174. 1985. Los Alamitos, CA: IEEE, 1985.
25. Mezrich, J. J., S. Frysinger, and R. Slivjanovski. "Dynamic Representation of Multivariate Time Series Data." *J. Am. Stat. Assoc.* **79** (1984): 34–40.
26. Rabenhorst, D. A., E. J. Farrell, D. H. Jameson, T. D. Linton, and J. A. Mandelman. "Complementary Visualization and Sonification of Multidimensional Data." In *Extracting Meaning from Complex Data:*

Processing, Display and Interaction, edited by E. J. Farrell. *Proceedings of SPIE* **1259** (1991): 147–153.
27. Schoenberg, A. *Style and Idea.* London: Williams and Norgate, 1951.
28. Smith, S., G. Grinstein, and R. Pickett. "Global Geometric, Sound, and Color Controls for Iconographic Displays." In *Extracting Meaning from Complex Data: Processing, Display, Interaction II*, edited by E. J. Farrell. *Proceedings of SPIE* **1459** (1991): 192–206.
29. Sumikawa, D. A. "Guidelines for the Integration of Audio Cues into Computer Interfaces." Master's Thesis, University of California, Davis, CA, 1985. Published as Technical Report UCRL-53656, Lawrence Livermore National Laboratory, Livermore, CA, 1985.
30. Tanimoto, S. L., and E. P. Glinert. "Designing Iconic Programming Systems: Representation and Learnability." In *Proceedings of the 2nd IEEE Computer Society Workshop on Visual Languages*, June 25–27, Dallas, 54–60. Los Alamitos, CA: IEEE, 1986.
31. Ware, C. "Visual Information Display." Unpublished panel position paper prepared for INTERACT'90, Cambridge (England), August 27–31, 1990. In *Proceeding of the 3rd IFIP Conference on Human-Computer Interaction*, edited by D. Diaper, G. Cockton, D. Gilmore, and B. Shackel. Amsterdam: North Holland, 1990.
32. Wenzel, E. M. "Three-Dimensional Virtual Acoustic Displays." In *Multimedia Interface Design*, edited by M. M. Blattner and R. Dannenberg, 257–288. New York: ACM Press, 1991.
33. Williams, S. M. "STREAMER: A Prototype Tool for Computational Modeling of Auditory Grouping Effects." Research Report No. CS-89-31, Department of Computer Science, University of Sheffield, 1989.
34. Wood, W. T., and S. K. Wood. "Icons in Everyday Life." In *Social, Ergonomic and Stress Aspects of Work with Computers*, edited by G. Salvendy, Vol. 1, 97–104. Proceedings of the 2nd International Conference on Human-Computer Interaction, held in Honolulu, August 10–14, 1987. Amsterdam: Elsevier, 1987.
35. Yeung, E. S. "Pattern Recognition by Audio Representation of Multivariate Analytic Data." *Anal. Chem.* **52(7)** (1980): 1120–1123.

Stephen A. Brewster, Peter C. Wright, and Alistair D. N. Edwards
HCI Group, Department of Computer Science, University of York, Heslington, York,
Y01 5DD, UK., Tel.: 0904 432765; e-mail: sab@minster.york.ac.uk

A Detailed Investigation into the Effectiveness of Earcons

A detailed experimental evaluation of earcons was carried out to see whether they are an effective means of communicating information in sound. An initial experiment showed that earcons were better than unstructured bursts of sound and that musical timbres were more effective than simple tones. Musicians were shown to be no better than non-musicians when using musical timbres. A second experiment was then carried out which improved upon some of the weaknesses of the pitches and rhythms used in Experiment 1 to give a significant improvement in recognition. From the results some guidelines were drawn up for designers to use when creating earcons. These experiments have formally shown that earcons are an effective method for communicating complex information in sound.

INTRODUCTION

The use of nonspeech audio at the graphical user interface is becoming increasingly popular due to the potential benefits it offers. There are many reasons for this. In everyday life, people communicate using all their senses, with information in one sensory modality being backed up by data from the others. When they come to use computers, the interaction is restricted almost solely to the visual channel and this limitation can cause the interface to intrude into the task that the user is trying to perform. The aim of a multimedia interface is to make the interaction more natural and the interface more transparent by using different forms of input and output. Most current interfaces make little use of sound other than for beeps to indicate errors. It can be used to present information otherwise unavailable on a visual display, for example, mode information,[10,13,14] or information that is hard to discern visually, such as multidimensional numerical data.[5] It is a useful complement to visual output because it can increase the amount of information communicated to the user or reduce the amount the user has to perceive through the visual channel.

To fully understand the design of auditory interfaces, one should have some knowledge of psychoacoustics: the study of the perception of sound. This aims to describe the relationships between the characteristics of a sound which enters the ear and the sensations these produce within the auditory system. In order to create sounds which a listener is able to hear and differentiate, the range of human auditory perception must not be exceeded. Frysinger[5] says: "The characterisation of human hearing is essential to auditory data representation because it defines the limits within which auditory display designs must operate if they are to be effective." Moore[11] gives an overview of the field of psychoacoustics. The work reported here is part of a research project looking at the best ways to integrate auditory and graphical information at the interface. The research aims to find the areas in an interface where the use of sound will be most beneficial and also what types of sounds are the most effective for communication. Sound will be used to present information where the visual interface breaks down and does not tell the user everything that they need to know.

EARCONS AND AUDITORY ICONS

One major question that must be answered when creating an auditory interface is: What sounds should be used? Brewster[3] outlines some of the

A Detailed Investigation into the Effectiveness of Earcons

different systems available. Gaver's auditory icons have been used in several systems, such as the SonicFinder,[6] SharedARK,[7] and ARKola[8] (see Gaver, in this volume). These use environmental sounds that have a semantic link with the object or action they represent. They have been shown to be an effective form of presenting information in sound.

One alternative, and previously untested, method of presenting auditory information are *earcons*.[2,15,16] Earcons are abstract, synthetic tones that can be used in structured combinations to create sound messages to represent parts of an interface. Blattner et al.[2] define earcons as "nonverbal audio messages that are used in the computer/user interface to provide information to the user about some computer object, operation, or interaction." Earcons are composed of motives, which are short, rhythmic sequences of pitches with variable intensity, timbre, and register.

Blattner, Papp, and Glinert define a system of hierarchical earcons in their paper. Each earcon is a node on a tree and inherits all the properties of the earcons above it. In Figure is shown a hierarchy of earcons. There is a maximum of five levels to the tree as there are five parameters that can be varied: rhythm, pitch, timbre, register, and dynamics. In the diagram the top level of the tree is the family rhythm, in this case it is a sound representing error. This sound has rhythm but no pitch; the sounds used are clicks. The rhythmic structure of level one is inherited by level two but this time a second motive is added where pitches are put to the rhythm. At this level the timbre is a sine wave, which produces a "colourless" sound. This is done so that at level three the timbre can be varied. At level three the pitch is also raised by a semitone to make it easier to differentiate from the pitches inherited from level two. Other levels can be created where register and dynamics are varied. To play the final earcon requires three separate motives to be played. In order to speed up the presentation of earcons, this paper suggests that only the last motive need be played. This motive (the one labelled "triangle" in Figure 1) contains all of the information needed. It has the rhythm of the error family, it has the pitch structure of the execution error and the timbre of an underflow error. If only the last motive is used then the length of the whole sound is greatly reduced. The experiments described in this paper test hierarchical earcons used in this manner to discover their effectiveness.

One other powerful feature of earcons is that they can be combined to produce complex audio messages. Earcons for a set of simple operations, such as OPEN, CLOSE, FILE, and PROGRAM, could be created. These could then be combined to produce, for example, earcons for OPEN FILE or CLOSE PROGRAM.

Earcons provide a powerful method of sonification. They can be used for adding sound to both data and interfaces. Related items can be given related sounds, hierarchies of information can be represented. Complex messages made of subunits can be built up. They are a powerful and flexible means of creating auditory messages.

Up to now, no extensive formal experiments have been conducted to see if earcons are an effective means of communicating information using sound.inmxxJonesS. D. Jones and Furner[9] carried out a comparison between earcons, auditory icons, and synthetic speech. Their results showed that subjects preferred earcons but were better able to associate auditory icons to commands. Their results were neither extensive nor detailed enough to give a full idea of whether earcons are useful or not. Barfield,

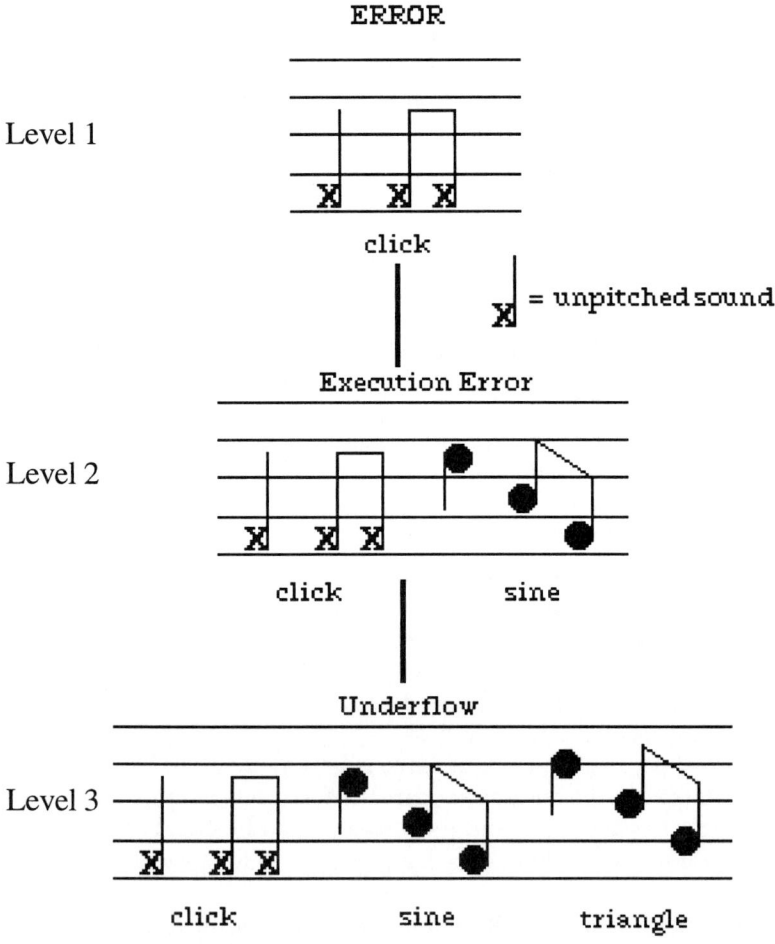

FIGURE 1 Hierarchical earcons (adapted from Blattner et al.[2]).

A Detailed Investigation into theEffectiveness of Earcons

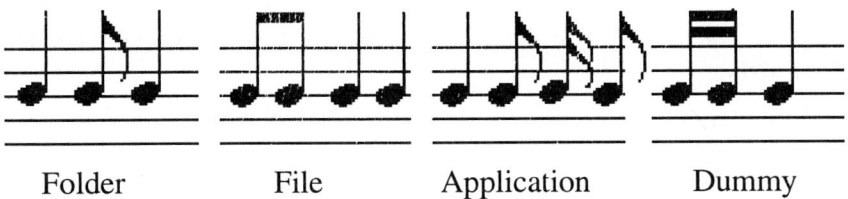

FIGURE 2 Rhythms used in Phase I of Experiment 1.

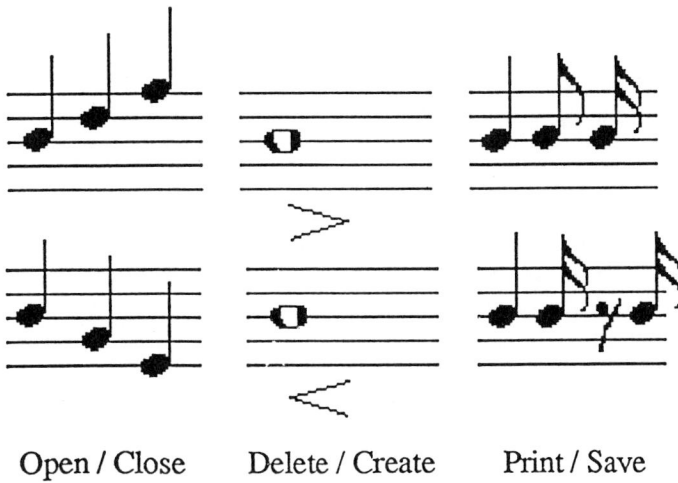

FIGURE 3 Rhythms used in Phase II of Experiment 1.

Rosenberg, and Levasseur[1] also carried out experiments where they used earcons to aid navigation through a menu hierarchy. The earcons they used were very simple, just decreasing in pitch as a subject moved down the hierarchy. These sounds did not fully exploit all the advantages that earcons offer and were also only a secondary part of the experiment. This paper seeks to discover how well complex earcons can be recalled and recognized.

The first experiment described attempts to discover if earcons are better than unstructured bursts of sound and tried to identify the best types of timbres to use to convey information. Blattner et al. suggest the use of simple timbres such as sine or square waves, but psychoacoustics suggests that complex musical instrument timbres may be more effective.[11] The second experiment uses the results of the first to create new earcons to overcome some of the difficulties that came to light. Guidelines are then put forward for use when creating earcons.

THE USE OF INTENSITY AS A PARAMETER FOR MANIPULATING EARCONS

Intensity is one of the parameters put forward by Blattner et al. for differentiating earcons. It is suggested here that care should be taken when using intensity. The intensity of a sound can be thought of as similar to the brightness on a video monitor. On a monitor the user can change the brightness of the display in response to the ambient light level. If the room is light, then the brightness of the display will be increased so that the information on the screen can still be seen. If the room is dark, then the brightness will be turned down so that the screen does not hurt the eyes. The volume control on a monitor acts in a similar way. If the room is noisy then the intensity will be increased to avoid masking. If the room is quiet, then the intensity can be turned down to avoid irritation. If the sounds used vary widely in intensity, then turning up the volume so that the quiet sounds can be heard will cause the loud sounds to become irritating. Conversely, turning down the loud sounds to a pleasant level may cause the quiet ones to fall below the threshold of hearing. One other important psychoacoustic factor is that changing the intensity of a sound can change the perceived pitch of the sound. Brewster[3] reports:

> "Intensity also affects perceived pitch. At less than 2kHz an increase in intensity increases the perceived pitch. At 3kHz and over an increase in intensity decreases the perceived pitch."

These points indicate that intensities used in earcons should be kept within a narrow range and the overall control of intensity given to the user.

EXPERIMENT 1

SUBJECTS

Thirty-six subjects, 3 groups of 12, were used from the University of York. Seventeen of the subjects were musically trained (they could play an instrument and read music). They were randomly allocated to one of three groups so that there was an even mix of musicians and nonmusicians (the simple tone group had only five musicians).

SOUNDS USED

An experiment was designed to find out if structured sounds such as earcons were better than unstructured sounds for communicating information. Simple tones were compared with complex musical timbres. Rhythm and pitch were also tested as ways of differentiating earcons. According to Deutsch,[4] rhythm is one of the most powerful methods for differentiating sound sources. The experiment also attempted to find out how well subjects could identify earcons individually and when played together in sequence. Figures 2 and 3 give the rhythm and pitch structures used in Phase I and II of the experiment. The sounds were based around middle C. In each phase, a dummy rhythm and dummy timbre were inserted into the testing phase. For example, the subject would hear a known rhythm but with a new timbre. This would test to see if subjects could recognize that the earcons had changed.

Three sets of sounds were created:

Musical Sounds: The first set were synthesized musical timbres: piano, brass, marimba, and pan pipes. These were produced by a Roland D110 synthesizer. This set had rhythm information as shown in the figures above.

Simple Sounds: The second set were simple timbres: sine wave, square wave, sawtooth, and a "complex" wave (this was composed of a fundamental plus the first three harmonics; each harmonic had one third of the intensity of the previous one). These sounds were created by SoundEdit on an Apple Macintosh. This set also had rhythm information as shown above.

Control Sounds: The third set had no rhythm information; these were just one-second bursts of sound similar to normal system beeps. This set had timbres made up from the musical group.

EXPERIMENTAL DESIGN

As mentioned, 3 groups of 12 subjects were used. Each of the three groups heard different sound stimuli. The musical group heard the musical sounds described in the previous section. The simple group heard the simple sounds and the control group heard the control sounds. There were four phases to the experiment. In the first phase, subjects heard sounds for icons. In the second they heard sounds for menus. In the third phase, they were tested on the icon sounds from Phase I again. In the last phase, subjects were required to listen to two earcons played in sequence and give information about both sounds heard. Instructions were read from a prepared script.

Write 1
Paint
Paint Folder
Paint 2
Write Folder
Write 2
Draw
HyperCard
Draw 1
Paint 1

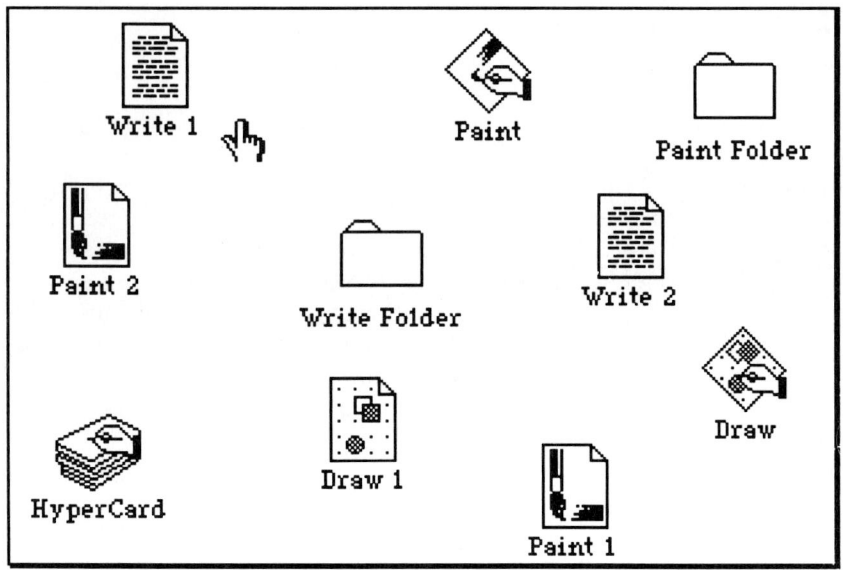

FIGURE 4 The Phase I icon screen.

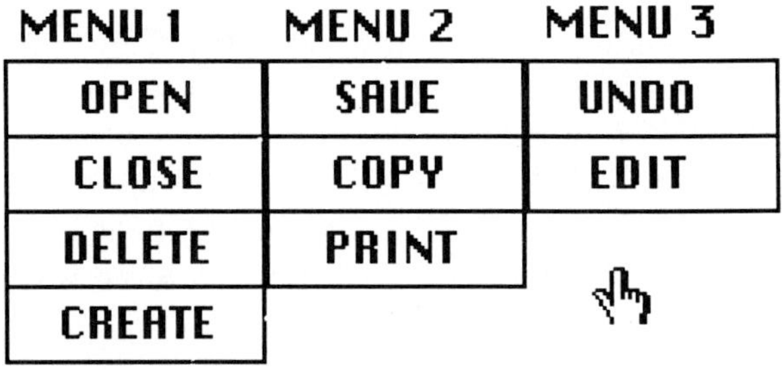

MENU 1
MENU 2
MENU 3
OPEN
SAVE
UNDO
CLOSE
COPY
EDIT
DELETE
PRINT
CREATE

FIGURE 5 The Phase II menu screen.

PHASE I. *Training:* The subjects were presented with the screen shown in Figure 4. Each of the objects on the display had a sound attached to it. The sounds were structured as follows. Each *family* of related items shared the same timbre. For example, the paint application, the paint folder, and paint files all had the same instrument. Items of the same *type* shared the same rhythm. For example, all the applications had the same rhythm. Items in the *same* family and of the *same* type were differentiated by pitch. For example, the first Write file was C below middle C and the second Write file was G below that. In the control group no rhythm was information was given so types were also differentiated by pitch.

The icons were described to each subject. The relationships between types were described. For example, the relationship between applications was indicated by having icons with hands in the graphic. The relationships between families were described and all the members of each family pointed out. The subjects were then asked to learn the names of all the icons. When they thought they had done this they wrote them down. If they were not correct, they were allowed more time to learn them. This meant that, at the end of the training, the subjects knew the names of the icons present. The icons were then played one at a time in random order so the subject could learn them. The whole set of icons was played three times. (Sound example *A* gives the "Paint application," "Write folder," and "Draw file 1" earcons used in the Control, Simple, and Musical groups respectively.)

Testing: When testing the subjects the screen was cleared and a selection of the earcons were played back in random order. Before the sounds were played back, it was indicated to the subject that they might hear some new sounds that they had not heard before. The subject had to supply what information they could about type, family, and, if it was a file, then the number of the file. In this and all of the phases the subject was allowed to hear each sound a maximum of three times. When scoring, a mark was given for each correct piece of information supplied.

Tr 93

Tr 94

PHASE II. In this phase, earcons for menus were tested. Each *menu* had its own timbre and the *items* on each menu were differentiated by rhythm, pitch, or intensity. The screen shown to the users for them to learn the earcons is given in Figure 5. The training was similar to Phase I. The subjects were tested in the same way as before but this time had to supply information about menu and item. (Sound example *B* gives the "Open," "Save," and "Undo" earcons used in the Control, Simple, and Musical groups respectively.) (CD Track 94)

PHASE III. This was a retest of Phase I but no further training time was given and the earcons were presented in a different order. This was to test if the subjects could remember the original set of earcons after having learned another similar set.

PHASE IV. This was a combination of Phases I and II. Again, no chance was given for the subjects to relearn the earcons. The subjects were played two earcons, one followed by another, and asked to give what information they could about each sound they heard. The sounds they heard were from the previous phases and could be played in any order (i.e. it could be menu, then icon; icon, then menu; menu, then menu; or icon, then icon). This would test to see what happened to the recognition of the earcons when played in sequence. A mark was given for any correct piece of information supplied. (Sound example *C* gives the "Open Paint application" and "Save Write folder" earcons used in the Control, Simple, and Musical groups respectively.)

RESULTS AND DISCUSSION OF EXPERIMENT 1

From Figures 6 and 7 it can be seen that the musical earcons came out best overall and in each of the phases. Unfortunately these differences did not reach statistical significance.

PHASE I

The breakdown of scores can be seen in Figure 8. A between-groups ANOVA was carried out on the family scores (family was differentiated by timbre) and showed a significant effect ($F(2, 33) = 9.788$, $p < 0.0005$). A Sheffe F-test showed that the family score in the musical group was significantly better than the simple group ($F(2, 33) = 6.613$, $p < 0.05$). This indicates that the musical instrument timbres were more easily recognized than the simple tones proposed by Blattner et al.

There were no significant differences between the groups in terms of type (differentiated by rhythm). The control group should have performed the worst as they had no rhythm information. However, the results show that the simple and musical groups performed no better. Therefore, the rhythms used did not give any better performance over the straight bursts of sound. This indicates that the chosen rhythms were ineffective. The file scores should have been the same as all groups were differentiated by pitch. A wide variation in results occurred indicating that pitch alone is not an effective means of differentiation.

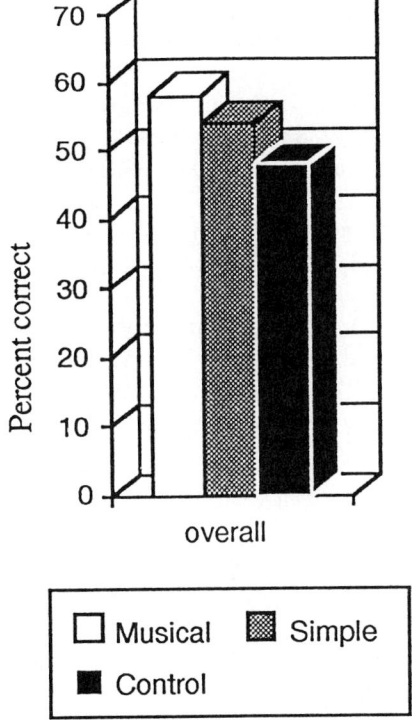

FIGURE 6 Overall scores in Experiment 1.

Within groups the type scores were significantly worse than the family scores for the musical and control groups ($T(11) = 2.96$, $p < 0.05$ and $T(11) = 3.55$, $p < 0.05$, respectively). This again shows that the rhythms chosen were difficult to use. In the simple group the type score was significantly better than the family score. Again this could be because the simple sounds are hard to remember so that the scores were lower.

PHASE II

The overall scores were significantly better than those for Phase I (see Figure 7). An ANOVA on the overall scores showed a significant effect ($F(2, 33) = 5.182$, $p < .011$). This suggests that the rhythms used were more effective, as the timbres were similar to the previous phase. Figure 9 shows the simple and musical groups performed similarly which was to be expected as both used the same rhythms. A Sheffe F-test showed both were significantly better than the control group (musical vs. control $F(2, 33) = 6.278$, $p < 0.05$, simple vs. control $F(2, 33) = 8.089$, $p < 0.05$). Again, this was to be expected as the control group had only pitch to differentiate items. This shows that if rhythms are chosen correctly then they can be very important in aiding recognition. It also shows that pitch alone is very difficult to use.

A Sheffe F-test showed that overall in Phase II the musical group was significantly better than the control group ($F(2, 33) = 4.5$, $p < 0.05$). This would indicate that the musical earcons used in this group were better than unstructured bursts of sound.

An ANOVA on the menu scores between the simple and musical groups showed an effect ($F(1, 22) = 3.684$, $p < 0.68$). A Sheffe F-test showed that the musical instrument timbres just failed to reach significance over the simple tones ($F(1, 22) = 3.684$, $p < 0.10$). A within-groups t-test showed that in the musical group the menu score (differentiated by timbre) was still significantly better than the item score ($T(11) = 2.69$, $p < 0.05$). This seems to indicate, once more, that timbre is a very important factor in the recognition of earcons.

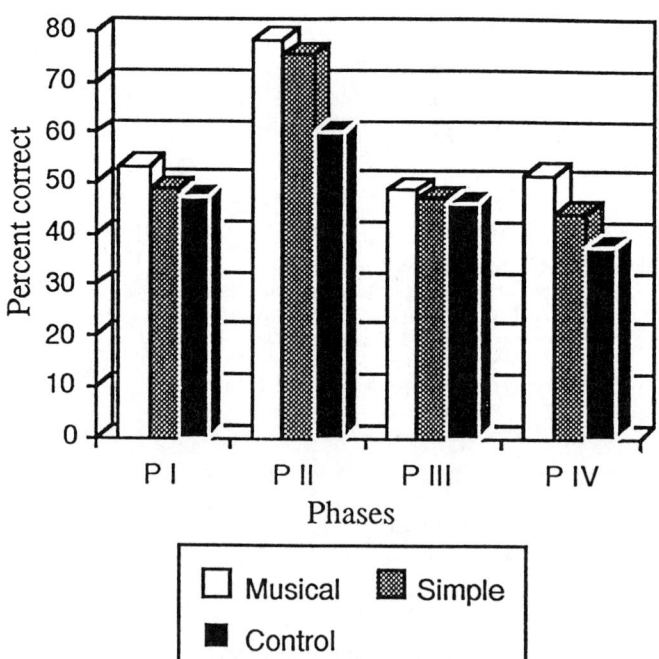

FIGURE 7
Breakdown of overall scores per phase for Experiment 1.

A Detailed Investigation into theEffectiveness of Earcons

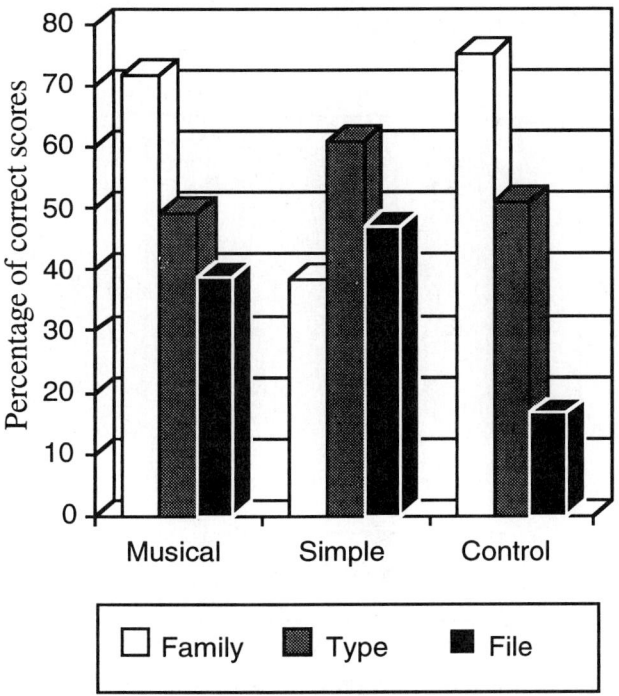

FIGURE 8 Breakdown of scores for Phase I of Experiment 1.

PHASE III

In this phase, the earcons from Phase I were tested again. A period of approximately 15 minutes had passed since the subjects learned the earcons. The scores in Phase III were not significantly different to those in Phase I (see Figure 7). This indicates that subjects managed to recall and remember the earcons from Phase I even after learning the sounds for Phase II, which were very similar. This seems to indicate that a subject's memory for earcons is strong.

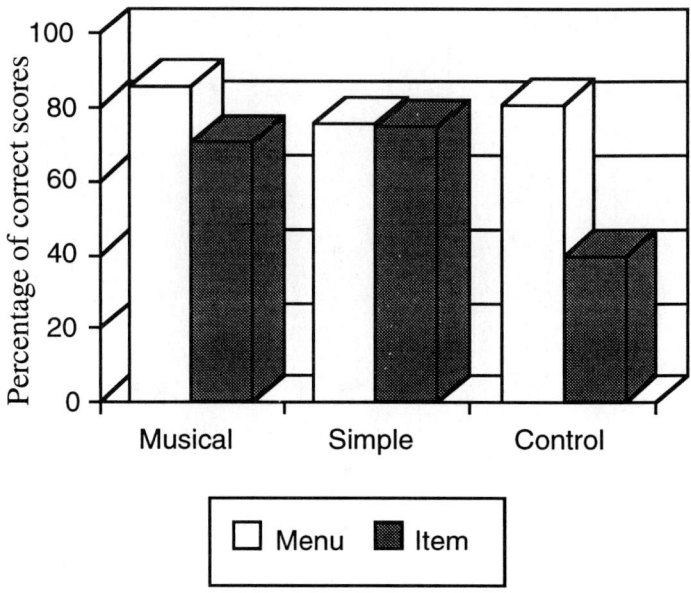

FIGURE 9 Breakdown of scores for Phase II of Experiment 1.

PHASE IV

In Figure 10 is shown the scores in Phase IV where combinations of earcons were tested. A within-groups t-test showed that, in the musical group, the menu/item combination was significantly better than the family/type/file combination ($T(11) = 2.58$, $p < 0.05$). This mimics the results for the musical group from Phases I and II. When comparing Phase IV with the other phases, performance was worse in all groups with the exception of type recognition by the musical group and family recognition by the simple group. This indicates that there is a problem when two earcons are combined together. If the general perception of the icon sounds could be improved then this might raise the scores in Phase IV.

A between-groups ANOVA showed the only significant effect to be on item scores ($F(2, 33) = 4.044$, $p = 0.0269$). A Sheffe F-test showed that the item scores in the musical group was significantly better than the control group ($F(2, 33) = 3.647$, $p < 0.05$). Items were differentiated by combinations of rhythm, pitch, and intensity in the musical group but the control group only used pitch. This indicates again that pitch alone is not a good way of differentiating, but combining it with other variables makes for much better recognition rates.

When comparing the results of previous phases with the corresponding part of Phase IV, it can be seen that in the musical group all scores are significantly worse (except for the Phase I type score which was not significantly different from the type score in Phase IV). In the simple group all the Phase IV scores were worse than the previous phases apart from the family score. In the control group all the scores were worse than the previous phases. This indicates that there is a problem when two earcons are put together.

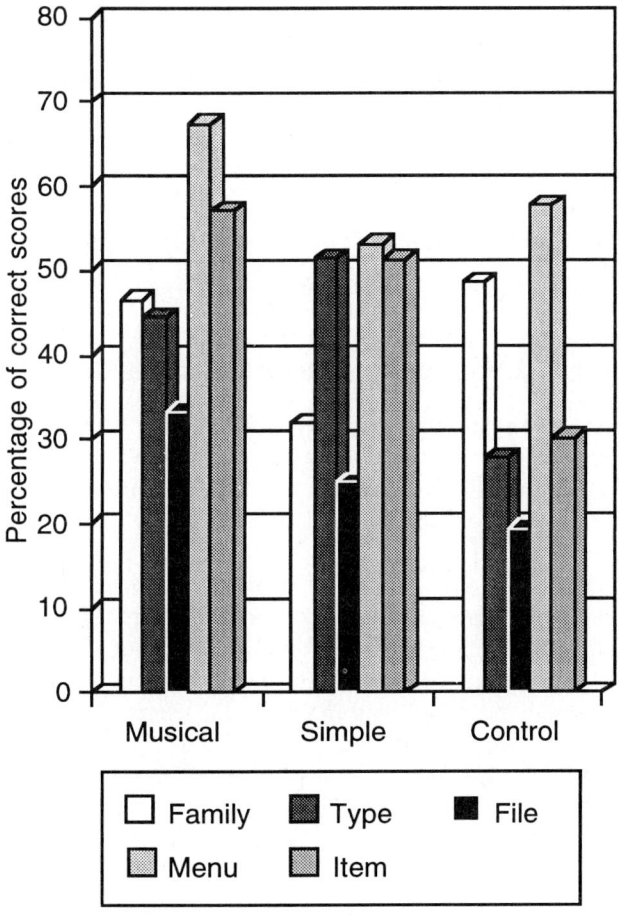

FIGURE 10 Breakdown of scores for Phase IV of Experiment 1.

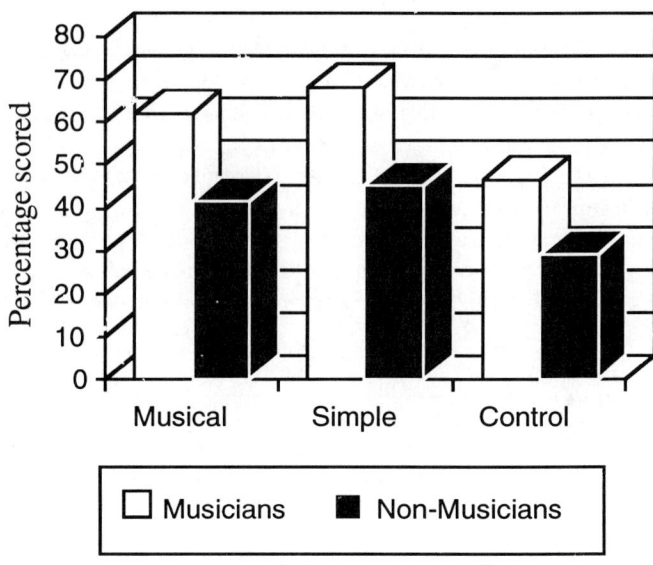

FIGURE 11 Scores of musicians and nonmusicians in Phase IV of Experiment 1.

DISCUSSION OF MUSICIANS VS. NONMUSICIANS

One important factor to consider is that of musical ability. Are earcons only usable by trained musicians or can nonmusicians use them equally as effectively? If the former were true, then it would limit the effectiveness of earcons in the sonification of information.

MUSICAL GROUP. The earcons in the musical group from Experiment 1 were not statistically significantly better recognized by the musicians than the nonmusicians. This means that a nonmusical user of a system involving earcons would have no more difficulties than a musician. The only time a significant difference occurred was in Phase IV where the musicians were better at identifying the file number ($T(10) = 2.83$, $p < 0.05$). This is due to musical training allowing the identification of individual pitches more accurately. The results from Phase IV are shown in Figure 11.

SIMPLE GROUP. When the simple sounds were used the musicians were significantly better than the nonmusicians with the Phase I types and families (types $T(10) = 3.27$, $p < 0.05$, families $T(10) = 2.26$, $p < 0.05$). Again, in Phase IV the musicians were better on type ($T(10) = 3.09$, $p < 0.05$). The problems with the rhythms in Phase I have been discussed above. It seems that the musicians were able to use the difficult rhythm information better as they are more highly trained. In a similar manner the musicians were able to recognize the simple tone timbres which the nonmusicians found hard to differentiate.

CONTROL GROUP. In Phase IV the musicians did significantly better than the non-musicians with the menus ($T(10) = 2.49$, $p < 0.05$), items ($T(10) = 2.48$, $p < 0.05$) and families ($T(10) = 2.85$, $p < 0.05$).

OVERALL. These results seem to indicate that if musical sounds were to be used for the earcons then musically untrained subjects would not be at a disadvantage to musicians. The only case where musicians proved to be better was with things differentiated by pitch alone. If simple tones or bursts of sound are used, then musicians will have an advantage. Therefore, to create sounds that are usable by the general population, musical earcons should be used.

DISCUSSION OF DUMMY EARCONS

In each phase, dummy earcons were introduced in the testing stage to see if the subjects could recognize them. The musical and simple groups had both timbre and rhythm dummies. The control group had only timbre dummies as its earcons contained no rhythm information. An example of the dummy rhythm used in Phase I is given in Figure 2.

MUSICAL GROUP. Recognition of dummy earcons in the musical group is shown in Figure 12. In Phase I it can be seen that dummy timbre recognition was high (83% were recognized) but dummy rhythm recognition was low (only 8% were recognized). This again indicates that the musical timbres were easy to recognize but the rhythms chosen were hard, mirroring the results shown in Figure 8. Phase II recognition rates were high. This shows, as before, that if the rhythms are used carefully, then high rates of recognition can be achieved. In Phase III, the recognition of the dummy rhythms increased to 50%. It is unclear why the subjects could recognize the dummies better after a period of time. The scores in Phase IV mirror the scores shown in Figure 7. The overall rate of recognition was 52%; not significantly different from the overall recognition of the genuine earcons. These results seem to indicate that subjects could recognize dummy earcons with the same level of accuracy as the genuine ones. The implication of this is that subjects can identify earcons not heard before, which makes earcons a more robust means of communication.

SIMPLE GROUP. In Phase I, the simple group had a similar overall score to the musical group but with lower timbre scores. This matches the results from the genuine earcons. These results imply that, with simple earcons, users would find it difficult to identify sounds that they had not heard before. In all of the other phases this group had lower identification rates than the musical group.

CONTROL GROUP. This group had no dummy rhythms. This group had similar timbre scores to the musical group in Phases I, II, and III. In Phase IV the timbre scores were the lowest (4.1%). This could be due to the overall difficulty of the control sounds, as shown in Figure 7.

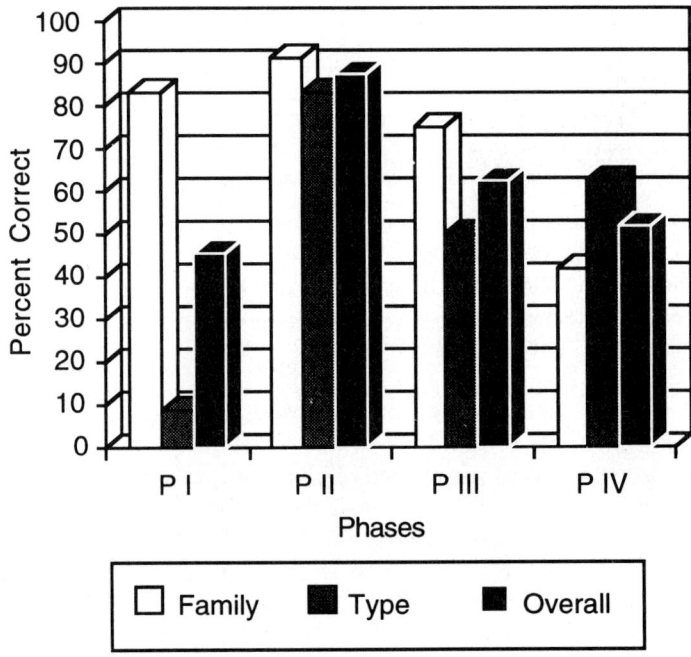

FIGURE 12 Recognition of dummy earcons in Experiment 1.

OVERALL. This result reinforces the power of the musical earcons. Subjects found it easier to recognize the dummies in this group. This again implies that musical earcons are the most effective of the sounds tested.

SUMMARY OF EXPERIMENT 1

Some general conclusions can be drawn from this first experiment. It seemed that earcons were better than unstructured bursts of sound at communicating information under certain circumstances. The issue of how this advantage could be increased needed further examination. Similarly, the musical timbres came out better than the simple tones but often by only small amounts. Further work was needed to make them more effective. The results also indicated that rhythm must be looked at more closely. In Phase I, the rhythms were ineffective but in Phase II they produced significantly better results. The reason for this needed to be ascertained. Finally, the difficulties in recognizing combined earcons had to be reduced so that higher scores could be achieved.

EXPERIMENT 2

From the results of the first experiment it was clear that the recognition of the icon sounds was low when compared to the menu sounds and this could be affecting the score in Phase IV. The icon sounds needed to be improved along the lines of the menu sounds.

SOUNDS USED

The sounds were redesigned so that there were more gross differences between each one. This involved creating new rhythms for files, folders, and applications, each of which had a different number of notes. Each earcon was also given a more complex within-earcon pitch structure. Figure 13 shows the new rhythms and pitch structures for folder, file, and application. No changes were made to the rhythms and pitch structures used for the Phase II menu item sounds.

The use of timbre was also extended so that each family was given two timbres which would play simultaneously. The idea behind multitimbral earcons was to allow greater differences between families; when changing from one family to another, two timbres would change, not just one. This created some problems in the design of the new earcons as great care had

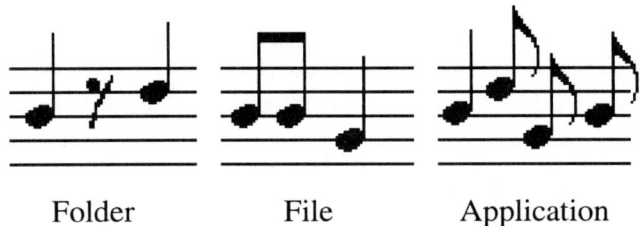

FIGURE 13 New phase I rhythms and pitch structures used in Experiment 2.

to be taken when selecting two timbres to go together so that they did not mask one another.

Findings from research into the perception of sound were included into the experiment. Patterson[12] gives some limits for pitch and intensity ranges. This led to a change in the use of register. In Experiment 1, all the icon sounds were based around middle C (261 Hz). All the sounds were now put into a higher register; for example, the folder sounds were now made two octaves above middle C. In Experiment 1, the "file 1" earcons were an octave below middle C (130 Hz) and the 'file 2's a G below that (98 Hz). These frequencies were below the range suggested by Patterson and were very difficult to tell apart. In Experiment 2, the register of the 'file 1' earcons were three octaves above middle C (1046 Hz) and the 'file 2' at middle C. These were now well within Patterson's ranges.

Tr 96 In response to informal user comments from Experiment 1, a 0.1 second delay was inserted between the two earcons. Subjects had complained that they could not tell where one earcon stopped and the other started. (Sound example *D* gives the new earcons for "Paint application," "Write folder," and "Draw file 1.")

METHOD

The experiment was the same as the previous one in all phases but with the new sounds. A single group of an additional twelve subjects was used. Subjects were chosen from the same population as before so that comparisons could be made with the previous results.

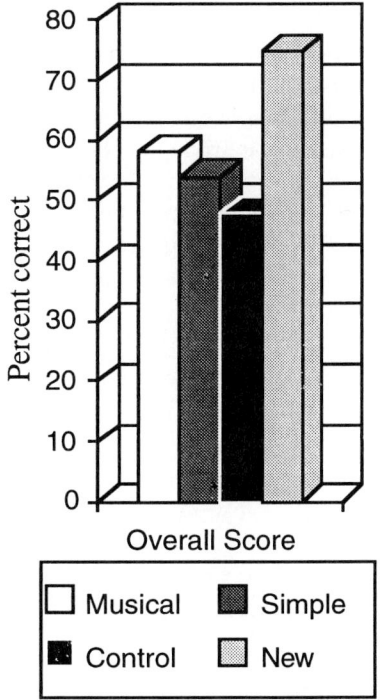

FIGURE 14 Percentage of overall scores with Experiment 2.

RESULTS AND DISCUSSION OF EXPERIMENT 2

As can be seen from Figure 14, the new sounds performed much better than the previous ones. An ANOVA on the overall scores indicated a significant effect ($F(3, 44) = 6.169$, $p < 0.0014$). A Sheffe F-test showed that the new group was significantly better than the control group ($F(3, 44) = 5.426$, $p < 0.05$) and the simple group ($F(3, 44) = 3.613$, $p < 0.05$). This implies that the new earcons were more effective than the ones used in the first experiment. Comparing the musical group (which was the best in all phases of Experiment 1) with the new group, the level of recognition in Phases I and III has been raised to that of Phase II.

PHASE I

The overall recognition rate in Phase I was increased because of a very significantly better type score (differentiated by rhythm) in the new group ($F(1, 22) = 26.677$, $p < 0.05$). The scores increased from 49.1% in the musical group of experiment 1 to 86.6% in the new group (see Figure 15).

This seems to indicate that the new rhythms were effective and very easily recognized.

The wider register range used to differentiate the files made a significant improvement over the previous experiment ($F(1, 22) = 4.829$, $p < 0.05$). This indicates that using the higher pitches and greater differences in register made it easier for subjects to differentiate one from another. The general improvement in recognition in Phase I brought the scores up to the level of the musical group in Phase II of the previous experiment. This indicates that with more careful design of earcons recognition rates can be significantly improved.

PHASES II AND III

The scores in Phase II were unchanged from the previous experiment as was expected. In Phase III, the scores were not significantly different to Phase I, again indicating that the sounds are easily remembered.

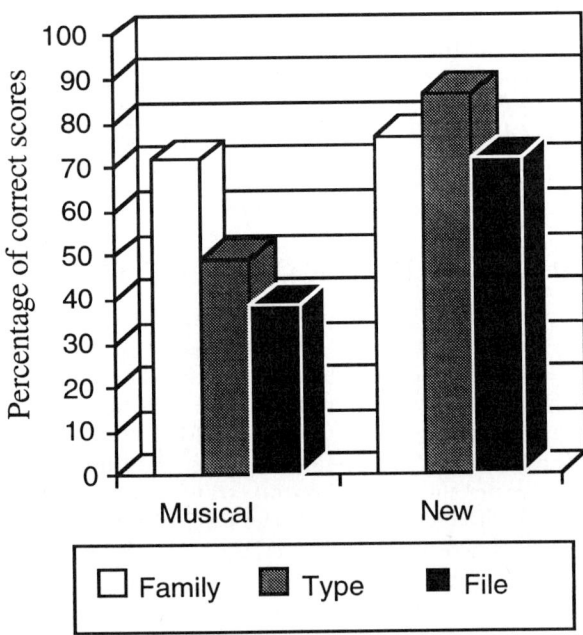

FIGURE 15 Breakdown of scores for Phase I of Experiment 2 compared to the musical group from Experiment 1.

A Detailed Investigation into the Effectiveness of Earcons

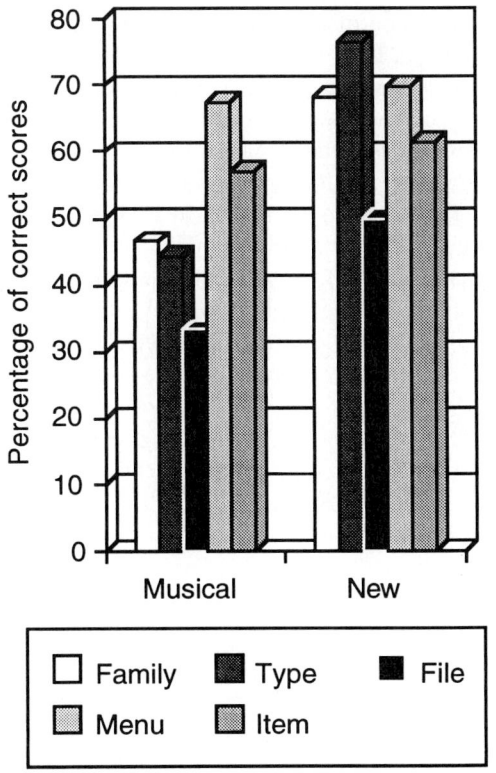

FIGURE 16 Breakdown of scores for Phase IV of Experiment 2.

PHASE IV

In Phase IV (combinations of earcons), the overall score of the new group just failed to reach significance over the musical group ($F(1, 22) = 3.672$, $p < 0.10$). However, the new earcons were significantly better than the musical ones from the previous experiment in terms of type ($F(1, 22) = 9.135$, $p < 0.05$) and family ($F(1, 22) = 4.989$, $p < 0.05$). Figure 16 indicates this. The menu and item scores were not different, as was expected, because the same earcons were used as in Experiment 1. T-tests revealed that recall in Phase IV was still slightly lower than the other phases. The overall Phase I score of the new group was significantly better than the score in Phase IV ($T(11) = 3.02$, $p < 0.05$).

The multitimbral earcons made no difference in Phase I. The family score for the new group was not significantly different to the score in the musical group. There were also no differences in Phases II or III. However, in Phase IV the recognition of icon family was significantly better than in the musical group ($F(1, 22) = 4.989$, $p < 0.05$). A further analysis of the data showed that there was no significant difference between the Phase I and Phase IV scores in the new group. However, the Phase IV score for the musical group was worse than Phase I ($T(11) = 4.983$, $p < 0.05$). This indicates that there was a problem in the musical group that was overcome by the new sounds. It may have been that in Phases I, II, and III only one timbre was heard and so it was clear to which group of earcons it belonged (icons sounds or menu sounds). When two earcons were played together, it was no longer so clear as the timbre could be that of a menu sound or an icon sound. The greater differences between each of the families when using multitimbral earcons may have overcome this.

MUSICIANS AND NONMUSICIANS. The results also show that there was no significant difference in performance between the musicians and nonmusicians with the new sounds in Experiment 2. This seems to indicate that musical earcons are the most effective way of communicating complex information for general users.

GUIDELINES

From the results of the two experiments and studies of literature on psychoacoustics some guidelines have been drawn up for use when creating earcons. These should be used along with the more general guidelines given in Sumikawa[15] and Sumikawa et al.[16] One overall result which came out of the work is that much larger differences than those suggested by Blattner et al. must be used to ensure recognition. If there are only small, subtle changes between earcons, then they are unlikely to be noticed by anyone but skilled musicians.

- *Timbre:* Use musical instrument timbres. Where possible use timbres with multiple harmonics. This helps perception and avoids masking. Timbres should be used that are subjectively easy to tell apart, e.g., use "brass" and "organ" rather than "brass1" and "brass2."
- *Pitch:* Do not use pitch on its own unless there are very big differences between those used (see register below). Complex intra-earcon pitch structures are effective in differentiating earcons if used along with rhythm. Some suggested ranges for pitch are: maximum 5 kHz (four octaves above middle C) and minimum 125 Hz to 150 Hz (an octave below middle C).

- *Register:* If this alone is to be used to differentiate earcons which are otherwise the same, then large differences should be used. Two or three octaves difference give good rates of recognition.
- *Rhythm:* Make them as different as possible. Putting different numbers of notes in each rhythm was very effective. Patterson[12] says that sounds are likely to be confused if the rhythms are similar even if there are large spectral differences. Small note lengths might not be noticed so do not use notes less than eighth notes or quavers. In the experiments described here these lasted 0.125 seconds.
- *Intensity:* Although intensity was not examined in this test, some suggested ranges (from Patterson) are: maximum 20dB above threshold and minimum 10dB above threshold. Care must be taken in the use of intensity. The overall sound level will be under the control of the user of the system. Earcons should all be kept within a close range so that if the user changes the volume of the system, no sound will be lost.
- *Combinations:* When playing earcons one after another, use a gap between them so that users can tell where one finishes and the other starts. A delay of 0.1 seconds is adequate. If the above guidelines are followed for each of the earcons to be combined, then recognition rates should be sufficient.

FUTURE WORK

No work was done in this paper to test the speed of presentation of earcons. The earcons took around 1.5 seconds to play. In a real application of earcons, they would need to be presented so that they could keep up with activity in the interface. A further experiment would be needed to test the maximum speed of display attainable whilst still retaining recognisability. Work is now being undertaken to investigate the presentation of earcons in parallel to speed up the rate of display.

The subjects only heard each of the earcons three times in the training parts of the experiment but reached 80% recognition rates. A more long-term study would show what levels of recognition could be reached when subjects had more time to learn the sounds. Work is also needed to look at the intensity of presentation. In many existing systems the sounds are played much too loud and so become intrusive.

CONCLUSIONS

The results indicate that earcons are an effective means of communication. The work shown has experimentally demonstrated that earcons are better for presenting information than unstructured bursts of sound. This gives a formal basis for their use in future systems. Musical timbres for earcons proved to be more effective than the simple tones proposed by Blattner et al. The subtle transformations suggested by Blattner have been shown to be too small to be recognized by subjects and that gross differences must be used if differentiation is to occur. The results of Experiment 1 indicated that earcons were effective but needed refinements. The results from Experiment 2 show that high levels of recognition can be achieved by careful use of pitch, rhythm and timbre. Multitimbral earcons were put forward and shown to help recognition under some circumstances. A set of guidelines has been suggested, based on the results of the experiments, to help a designer of earcons make sure that they will be easily recognizable by listeners.

This research now means that there is a strong experimental basis to prove that earcons are effective. This work has shown that earcons can be individually recognized rather than recognition being based on hearing a relative change between two sounds. Earcons could therefore be used as landmarks in an auditory space where they give absolute information about events. Developers can create sonifications of data or interfaces that use earcons and can be safe in the knowledge that they are a good means of communication.

ACKNOWLEDGMENTS

We would like to thank all the subjects for participating in the experiment. Thanks also go to Andrew Monk for helping with the statistical analysis of the data. This paper is a more detailed version of one entitled "An Evaluation of Earcons for Use in Auditory Human-Computer Interfaces" which appeared in *INTERCHI 1993*.

REFERENCES

1. Barfield, W., C. Rosenberg, and G. Levasseur. "The Use of Icons, Earcons, and Commands in the Design of an Online Hierarchical Menu." *IEEE Trans. on Prof. Comm.* **34(2)** (1991): 101–108.
2. Blattner, M., D. Sumikawa, and R. Greenberg. "Earcons and Icons: Their Structure and Common Design Principles." *Human Comp. Interaction* **4(1)** (1989): 11–44.
3. Brewster, S. A. "Providing a Model for the Use of Sound in User Interfaces." Technical Report Number YCS 169. University of York, Department of Computer Science, 1992.
4. Deutsch, D. "The Processing of Structured and Unstructured Tonal Sequences." *Percept. & Psychophys.* **28(5)** (1980): 381–389.
5. Frysinger, S. P. "Applied Research in Auditory Data Representation." In *Proceedings of the SPIE,* Vol. 1259, 1990.
6. Gaver, W. "The SonicFinder: An Interface that Uses Auditory Icons." *Human Comp. Interaction* **4(1)** (1989): 67–94.
7. Gaver, W., and R. Smith. "Auditory Icons in Large-Scale Collaborative Environments." In *Interact 1990,* edited by D. Diaper, 735–740. Holland: Elsevier, 1990.
8. Gaver, W., R. Smith, and T. O'Shea. "Effective Sounds in Complex Systems: The ARKola Simulation." In *CHI 1991,* edited by S. Robertson, G. Olson, J. Olson, 85–90. New Orleans: ACM Press, Addison-Wesley, 1991.
9. Jones, S. D., and S. M. Furner. "The Construction of Audio Icons and Information Cues for Human-Computer Dialogues." In *Contemporary Ergonomics.* Proceedings of the Ergonomics Society's 1989 Annual Conference. Reading, MA: Addison-Wesley, 1989.
10. Monk, A. "Mode Errors: A User-Centered Analysis and Some Preventative Measures Using Keying-Contingent Sound." *Intl. J. Man-Mach. Studies* **24** (1986): 313–327.
11. Moore, B. C., ed. *An Introduction to the Psychology of Hearing,* 1–10. London: Academic Press. 1989.
12. Patterson, R. D. "Guidelines for Auditory Warning Systems on Civil Aircraft." CAA Paper Number 82017. Civil Aviation Authority, London, 1982.
13. Sellen, A. J., G. P. Kurtenbach, and W. Buxton. "The Role of Visual and Kinesthetic Feedback in the Prevention of Mode Errors." In *Interact 1990,* edited by D. Diaper, 667–673. Holland: Elsevier, 1990.
14. Sellen, A., G. P. Kurtenbach, and W. Buxton. "The Prevention of Mode Errors Through Sensory Feedback." *Human Comp. Interaction* **7** (1992): 141–164.

15. Sumikawa, D. A. "Guidelines for the Integration of Audio Cues into Computer User Interfaces." Technical Report No. UCRL 53656. Lawrence Livermore National Laboratory, 1985.
16. Sumikawa, D., M. Blattner, K. Joy, and R. Greenberg. "Guidelines for the Syntactic Design of Audio Cues in Computer Interfaces." Technical Report No. UCRL 92925. Lawrence Livermore National Laboratory, 1986.

Jonathan Cohen
ATG Human Interface Group, Apple Computer, Inc., One Infinite Loop, MS 301-3H, Cupertino, CA 95014; (408)974-2884; e-mail: cohenj@applelink.apple.com

Monitoring Background Activities

Dedication:
To the memory of John Cage

Too many things happen in the foreground, and we don't know anything about what happens in the background.
— Heinrich Böll

The personal computer is increasingly becoming a center for delegated and autonomous background activity. How can users be notified about this activity without having their foreground task disrupted? The audio channel offers a number of advantages for notification. ShareMon, a prototype application, employs audio—sound effects or text-to-speech—or graphical messages to notify users about file sharing (a type of background activity). Informally, and in the course of a user study with ShareMon, users found all three modalities informative, but they found that all of the modalities disrupted their foreground activity to some extent. These reactions raise two issues: when is the use of a particular modality appropriate, and how can a sound be designed to be simultaneously informative, pleasant, and/or unobtrusive? Although I discuss a theoretical approach to these issues, I argue that an empirical design-and-test methodology provides a powerful way to resolve them.

1. INTRODUCTION

This paper is motivated by the trend towards computational environments becoming increasingly overloaded. The computer screen is not spacious enough to represent all the ongoing computer-mediated activity that may be relevant. Much of this activity will take place "behind the user's back," some of it initiated by the user, and some targeting the user. What are the best ways to represent this activity? How can this activity be represented without disrupting the user's foreground activity? One approach that has been taken is to supplement the visual channel with the audio channel. There are a number of software environments that notify users about background activities via audio.

ShareMon, the subject of this paper, is a prototype software application that notifies users about one specific kind of background activity. However, it is intended to be a test-bed for general issues about notification of background activity. For this reason, the notifications can be in the form of "metaphorical" or "nonmetaphorical" sound, text-to-speech, or graphics. Preliminary and more long-term reactions from users, as well as those elicited from a user study, raise two issues: when is the use of a particular modality appropriate, and how can a sound be designed to be simultaneously informative, pleasant, and unobtrusive?

This introductory section has three subsections. Section 1.1 presents a scenario of one user's background activity. The purpose of the scenario is to informally motivate the work. Section 1.2 offers some evidence that users will experience an increasing amount of background activity about which they will want to be notified. Section 1.3 discusses the utility of audio—how it can inform users without being disruptive.

1.1 SCENARIO

In the not-too-distant future, a user is sitting at his personal computer, filling out a spreadsheet with the expenses from his last business trip. He prints the spreadsheet, then sends copies of it to his secretary and his project manager. In the meantime, he receives new e-mail messages, and new announcements are being added to his favorite electronic bulletin board (one of which is relevant to his work with the spreadsheet), but he has a few things to do before he can read them. He browses the network to find a supercomputer with available cycles, and uploads a large simulation program. Then he asks his computational agent, Phil, to search the On-line Library of Congress for any articles on Audio Notification published in the last five years.

He remembers some items he needs to add to his spreadsheet, so he returns to that application. While he is occupied with it, the printer runs out of paper, the net goes down, his simulation is suspended because the supercomputer is commandeered by a higher priority task submitted by a government agency, the On-line Library of Congress is shut down due to lack of funds, and Phil, the agent, and the entire International Brotherhood of Virtual Persons go on strike.

Although this scenario is intentionally silly, its point is that the user has set a number of processes in motion and, while he is working on his foreground task, he completely loses track of them. When he interrupts his foreground task to determine the status of his background activities, he will find a series of unpleasant surprises.

1.2 THE PROBLEM: WHAT IS GOING ON BEHIND MY BACK

The personal computer is becoming a hub of activity taking place both locally and over the network. This activity includes both direct manipulation tasks such as working with a spreadsheet, and delegated tasks such as sending something to the printer. Another way to characterize this activity is to distinguish between foreground tasks—that occupy the center of a user's attention—and background tasks—that execute without user intervention.[29] By definition, direct manipulation tasks are foreground tasks.[21,35] Delegated tasks, once initiated, become background tasks (but note that some "autonomous" background tasks may run without having been initiated by the user).

The range of delegated and background tasks potentially available to or affecting users is increasing (and since more CPU power will be available, more users are willing to have background processes running):

- One new source of background activity is distributed computing, where machines "borrow" CPU cycles from other machines to help perform complex calculations.[33]
- Another source is in computers, like the Newton™, which spontaneously network and share data with each other without requiring users to attach cables.[2]
- Work exploring intelligent agents could eventually make delegation interfaces commonplace. For example, the program "Eager" monitors a user's activities and, when it detects an iterative pattern, it writes a program to complete the iteration.[11] "Guides" are anthropomorphic agents that recommend paths through a database of American history.[10] Library servers (agents) could notify users whenever relevant new articles have been published, based on profiles of user interest.[37]
- The increasing availability of speech input as an interface modality is also making delegation interfaces more commonplace.[36]
- Office environments, where workers communicate with each other through workstation-mediated video and/or audio connections, generate both delegated and autonomous background activity around the management of communications. Examples include Media Space connecting Xerox PARC and a Portland site,[31] and RAVE[18] and Khronika[25] at EuroPARC (more about these in Section 2).

It seems clear that an increasing problem will be how to apprise users of the status of their delegated or autonomous background activities without greatly disrupting their foreground activities.

1.3 A SOLUTION: THE AUDIO CHANNEL

For an excellent discussion of the benefits of using the audio channel in human-computer interaction, read Buxton[8] or Gaver.[15] Jenkins writes about "useful dimensions of acoustic information" in a readable and thought-provoking way.[23] These workers argue that the audio channel offers an attractive method for notifying users about background activities because it can inform without being disruptive, since:

- Audio does not take up screen space.

---

[1]This reminds me of the story of the lighthousekeeper, which takes place sometime in the nineteenth century. At the top of the lighthouse there is a cannon that fires once an hour, 24 hours a day. One night, it is about three A.M. and the lighthousekeeper is asleep. The cannon does not fire. The lighthousekeeper jumps up, exclaiming "What was that!"

- Audio fades nicely into the background, but users are alerted when it changes.[8][1]
- People can process audio information while simultaneously engaged in an unrelated task. For example, I'm listening to the radio as I write this passage.
- There is the well-known "Cocktail Party Effect," the ability of a person to selectively attend to one conversation in the midst of others, or switch attention from one conversation to another (see Arons for a recent review[3]). Thus it is possible for users to monitor multiple background processes via the audio channel, as long as the sounds representing each process are distinguishable.[8]
- Since most current direct manipulation tasks are graphical (and mildly haptic), the audio channel is often free—available as a channel for notification.

Furthermore, the technical obstacles to the use of audio are disappearing. As Baecker and Buxton wrote in 1987, introducing a section about audio at the interface:

> ...audio input and output have presented technical and cost effectiveness difficulties which discouraged their use in most applications. Recently, these problems have been greatly reduced, and audio is now a feasible design alternative.[4]

Five years later, the problems continue to be reduced with the popularity of MIDI-controllable devices,[2] built-in audio hardware on most personal computers, and the increasing availability of DSP hardware. It is clear that the audio channel is a useful and feasible channel for notification.

2. PRIOR ART

This section describes some early work by John Cage and collaborators. It also describes four more recent software environments (three by Bill Gaver and collaborators). Section 2.1 describes some remarkably prescient work by John Cage. This work may not be directly relevant to the problem of notification; nevertheless, Cage is a spiritual father of Auditory Display. The other four subsections describe software environments that employ audio to notify users about events or remind them of the state of a system. Users can work with all of these software environments as their main (foreground) task. However, two of the systems, Khronika and RAVE, can also run in

---

[2]Scaletti and Craig do not agree that MIDI makes things easier.[34]

the background while a user is engaged in some other foreground task. The description emphasizes the way the systems use audio when they run in the background.

2.1 CAGE'S *MUSIC OF CHANGES* AND *REUNION*

The field of Auditory Display of Data is a relatively young one,[22] so, in reading through its literature, one can fairly easily point to the pioneers. However, citations of John Cage's work appear to be nonexistent (though Buxton, making a point about the ubiquity of sound, recounts the anecdote about Cage's visit to an anechoic chamber[8]). This is an unfortunate omission because Mr. Cage invented techniques for auditory display of data in the early 1950s. Cage's method of composition of the 1952 work *Music of Changes* seems strikingly familiar—the "data" were generated from the results of coin tosses, and mapped to pitch, duration, amplitude and timbre of sound.[9] Even the higher-order structure of the piece was determined using "chance operations"—coin toss data mapped to changes in tempo, or the number of measures in a section of the score.

In 1968 Cage performed a work called Reunion that took the form of a chess game with Marcel Duchamp.[30] The chessboard was also a photoelectric switching mechanism, which selected audio output from music being created by Lowell Cross, David Behrman, Gordon Mumma, and David Tudor. In a static sense, the configuration of the chessboard was represented in sound; in a dynamic sense, the "event" of moving a piece on the board triggered a change in sound.[3]

Scaletti [speaking at ICAD, October 28, 1992] and Kramer [personal communication, March 15, 1993], both suggest that Cage did not compose Music of Changes or Reunion with the intent to communicate the underlying data structures, but for other esthetic reasons. However, intent is in the ear of the beholder, not just in the mind of the creator: one may listen to a sound for either its musical quality or its informational value. For example, in the work of Lunney or Jackson, demonstrated at ICAD, sound is informative, yet pleasing enough to be heard as music. And one could imagine mentally attempting to position chess pieces while listening to Reunion.

2.2 SONICFINDER

With the SonicFinder, Gaver took a unified and pervasive approach to adding sound to a computer operating system (Macintosh): the data that SonicFinder maps to sounds are "computer events."[15] The mapping of an event to sound is based on Gaver's theory of "everyday listening" that

---

[3] Perhaps Reunion anticipated SonicFinder by twenty years?

suggests that people in everyday situations mostly attend to attributes of the source of a sound (e.g., its size or material), as opposed to "musical listening" during which people mostly attend to attributes of the sounds themselves (e.g., pitch or loudness). For example, the act of selecting an object in the Finder^{TM} causes a tapping sound.

> When an object is selected, its type (i.e., whether it is a file, folder, etc.) is represented by the material of the sound-producing object, so that files make wooden sounds, applications sound like metal, and other object types make other kinds of sounds. The size of the object is also conveyed by the sound, so that large objects make lower pitched sounds than small objects (as they do in the everyday world).[15]

We can define metaphorical versus nonmetaphorical sound by comparing Gaver's approach with that used by Sara Bly. Bly displayed six-dimensional computer-generated random normal data using attributes of sound like pitch, duration, volume, attack, and fundamental and overtone waveshape. She found that audiovisual displays were better comprehended than audio-only or video-only displays, and that comprehension was similar for audio-only and video-only displays.[6] Gaver characterizes the difference between these two approaches using the terms "everyday listening" and "musical listening." However, in this paper the preferred terms are **metaphorical** and **nonmetaphorical** sound (even though metaphorical sound means something more specific to Gaver) because "musical listening" also has to do with structure, emotion, and memory.

For the most part, the auditory feedback in SonicFinder was redundant with the visual feedback. Gaver notes that the one exception is the "Copy" command, which makes a pouring sound that increases in pitch as the copy operation proceeds.[15] If a user were copying a large set of files, he could work on something else off-line while the machine was busy with the copy operation, listening for the cessation of the copying sound to know that his machine was free again. This specific use of the copy command can be seen as an example of notifying a user about a background activity.

2.3 SHAREDARK

In the SharedARK environment, Gaver and Smith "suggest ways that auditory icons can be used to enhance the usability of systems which employ multiprocessing and modes, extended or layered displays, and collaborative workspaces."[16] One such auditory icon is associated with SharedARK's law of gravity, which can be turned on or off by users. When gravity is on, users are reminded by a "quiet, low-pitched humming noise reminiscent of an electrical appliance. This sort of sound tends to fade into the perceptual background after a short time, so that it does not annoy or distract users."[16]

2.4 KHRONIKA

Khronika is a server-based event-browsing and notification system.[25] It can notify users about both scheduled events (like "there will be a meeting with so-and-so in one hour") and unscheduled events (see Section 2.5 for an example). The notifications occur while the user is occupied with a foreground task; the user need not switch to the Khronika application to be notified. Notifications take the form of text-to-speech, sound effects, and graphics (which either stay until dismissed by the user, or go away after a predetermined time). Khronika also offers a browser so that users may get a summary of past, present, and (scheduled) future events.

2.5 RAVE

RAVE, the Ravenscroft Audio Video Environment, is a powerful system that supports cooperative work over audio and video channels.[18] Each RAVE user has a camera and a microphone in his office and this audio/video output is connected to a switcher. From one office, a given user can potentially see and hear any other office. A user can also view certain public spaces.

As part of a well-thought-out effort to balance privacy and functionality, RAVE uses audio notification (via Khronika) to let a user know when someone else is "looking." For example, when person A "glances" at person B, the default sound B hears is that of a door opening and, when the video connection is broken, B hears a door-closing sound. A "sweep" is a series of one-second video connections from one user to a group of other users. As a user is being "swept," he might be notified by the sound of footsteps.

2.6 PRIOR ART SUMMARY

ShareMon borrows a number of the ideas embodied in the work described in this section:

- From SonicFinder, the idea that computational events, including operating-system events, can be represented by sounds, particularly everyday sounds that convey information by varying pitch or timbre.
- From SharedARK, a way to use sound that fades into the background of a user's attention to represent status information.

- From Khronika, the idea of general system that can use sound effects, graphics, or text-to-speech to notify users about events.
- From RAVE, a way to notify users about a certain type of background activity.

3. SHAREMON: A PROTOTYPE MONITORING APPLICATION

This six-part section describes Sharemon, a prototype application that serves as a test-bed for issues around the notification of background activities. The first subsection explains "file sharing," the autonomous background activity that ShareMon monitors. Section 3.2 describes the specific events that ShareMon notifies users about. The last four subsections describe the **notification modalities** that ShareMon employs: metaphorical and nonmetaphorical sounds, text-to-speech, and graphics.

One question needs to be answered, however, before going any further. There are already software environments, such as those described in the previous section, that effectively notify users about background activities. Why is ShareMon a project worth pursuing? Firstly, RAVE and Khronika's user base was limited to a small expert population. ShareMon, because it can run on any Macintosh on a network, is aimed at a larger and more diverse set of users. Secondly, Khronika uses a number of different notification modalities, but, as far as I know, there has not been a study that compared them in terms of effectiveness or user preference, or approached the problem with a design-and-test methodology. Finally, this field is still so young that new examples of the appropriate use of audio are still of interest.

3.1 FILE SHARING DEFINED

File Sharing is a feature of the Macintosh Operating System that gives users the ability to access files from a remote machine on the network. The user treats those files as if they came from secondary storage connected directly to the local machine.[1] The user who wants to make files available for sharing, the **host**, must select a directory on his machine, and specify that directory as "sharable." The host can make the sharable directory available to anyone, or restrict access to specific people by setting up a username and a password for each person.

The person who wants access to the shared files, the **guest**, brings up an application called the Chooser™, which lets him navigate through the network, eventually finding the sharable directory in a list of available "volumes." Once he has logged on, the icon shows up on his desktop, just as if it were another disk drive (see Figure 1). He can open the icon and access the files in the same way that he accesses files on his local mass storage volumes.

From the host's point of view, guests sharing files create autonomous background activity. Hosts have three methods currently available to determine if a guest is logged on to their machine and sharing a file.

The first method requires the host to switch his attention from what he is doing, and pull up an application called the File Sharing Monitor^{TM}. As shown in Figure 2, there are two guests, called "<Guest>" and "Smedley GoodHeart," logged on to the host machine called "Jonathan's IIci." The guests are either reading or writing files on the host machine because the File Sharing Activity indicator at the lower left of the File Sharing Monitor window is about two-thirds full.

(Note the difference between "<Guest>" and "guest." "<Guest>" is the file-sharing equivalent of "John Doe." Anyone on the net can access Sharable Folder without a password or user name—that passwordless person is named "<Guest>." On the other hand, "guest" is the generic term, and may refer to either a registered user or a passwordless person.)

The other two methods are somewhat indeterminate, but the user does not have to switch application environments. Firstly, system response time may degrade, but only if the guest is copying a large file over a slow network. Secondly, hosts can sometimes hear their disk drive whirring and, if they know they are not currently accessing the drive, then a guest must be.

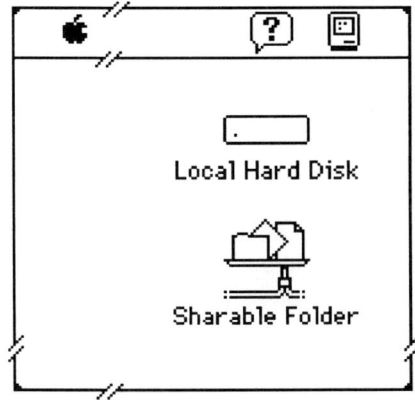

FIGURE 1 A Macintosh desktop (scrunched to fit). "Local Hard Disk" is a local volume. "Sharable Folder" is a directory on a remote machine.

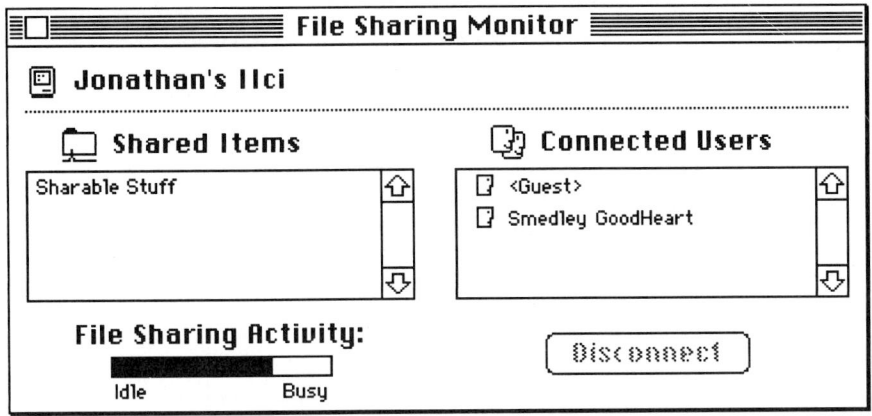

FIGURE 2 The File Sharing Monitor. Two guests are logged on. Some file-sharing activity is indicated at the lower left.

3.2 SHAREMON DEFINED

ShareMon is a prototype application that attempts to give a host a way to find out about file-sharing events without having to interrupt his foreground task. ShareMon is essentially an audio equivalent of the File Sharing Monitor application except that it runs in the background. Once the host sets a couple of parameters to get ShareMon started, it does not require any screen space, and he can go back to working with his foreground application.

ShareMon notifies hosts about three events: a guest logs in, a guest accesses files (expressed as a percentage of CPU time devoted to file-sharing activity—call it **%CPU time**), and a guest logs out. Note that the highest %CPU time occurs when a guest is copying a file to or from the host's machine. At present, if there are multiple guests, there is no way to tell the %CPU time devoted to each, only the overall total.

Shortly after writing ShareMon, I learned about "Nok Nok," a commercially available application for the Macintosh.[7] Nok Nok only notifies users about logins. It employs the Macintosh Notification Manager, which blinks an icon, brings up an alert window, and/or plays a sound.[1] Nok Nok can also notify users by bringing up the File Sharing Monitor application.

While ShareMon is an attempt to solve a particular problem, it is also intended to be a test-bed for issues about notification. For example, how should users be notified about particular events? What notification modalities do users prefer? Which modalities are least intrusive? Which are most intrusive? Which are least informative? Which are most informative?

TABLE 1 ShareMon's default event-to-message mapping

Event	Message
log in	knock-knock-knock
connection reminder	"ahem"
low %CPU time	slow walking
medium low %CPU	medium walking
medium %CPU	fast walking
high %CPU	jogging
very high %CPU	running
log out	door slam

3.3 METAPHORICAL NOTIFICATION

The first version of ShareMon mapped events to messages based on Gaver's idea of everyday listening. Much like RAVE, when a guest logged in, ShareMon played a knock-knock-knock sound; on logout, ShareMon played a door-slam sound (see Table 1). %CPU time was split into six ranges, each range mapped to a sound. For very low %CPU (i.e., a guest was connected but not accessing files), ShareMon occasionally played an "ahem" sound.[4] Low %CPU was mapped to footstep sounds at three different walking speeds. High %CPU was mapped to jogging footsteps and, at the highest level of %CPU time, ShareMon played the sounds of running footsteps.[5]

3.4 NONMETAPHORICAL NOTIFICATION

The %CPU time might be well represented by an aspect of sound, in the same way that the x-coordinate is mapped to pitch in Mansur, Blattner,

---

[4] The throat-clearing sound was the least obnoxious I could think of that a person might make when standing still.

[5] ShareMon also allows a person on one machine to monitor the file-sharing activity of a second (remote) machine somewhere on the net. In other words, the first machine is an **eavesdropper**, and the second machine is the host, which allows access to other guests. The default event-to-message mappings for remote file-sharing events also used the knock/walk/slam sounds, except that each of them had some reverb added, in order to suggest distance or the indirection of the monitoring.

Monitoring Background Activities

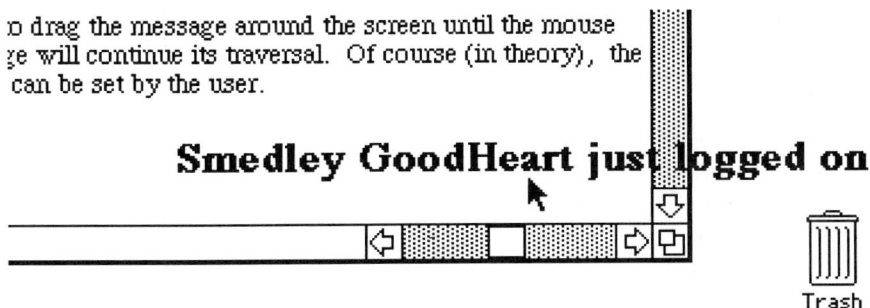

FIGURE 3   A message from ShareMon. The message is centered just above the cursor.

and Joy.[27] On the other hand, one might try something like the way the status of the law of gravity is represented by presence or absence of background hum in SharedARK.[16] The latter example seems especially appropriate to one typical pattern of file sharing: a guest may log on to a host, access a few files, and then remain logged for many hours without doing anything else. Thus no further login or logout events are triggered, and the %CPU time stays at zero. Nevertheless, the host might want to be reminded occasionally that someone is still connected.

Based on these considerations, ShareMon got some new messages that mapped %CPU time to the pitch of either a humming sound or pink noise (filtered white noise). Silence meant that no one was logged in. These sounds were intended to fade easily into the background: if a guest was connected, but not accessing files, the hum or pink noise was to play at a low pitch with an amplitude just above the threshold of audibility. An unfortunate side-effect of the implementation of these sounds, however, was that the loudness also increased along with the pitch.

Another experiment was to add a message that could be thought of as a variation on pink noise. While a guest is logged on, a host hears the sound of ocean waves. As %CPU time increases, the waves break more often. A second message of this type uses a pulse of sound (a short hum lasting about half a second, with a slowly rising, then steady, then slowly falling envelope). The pulse plays more often as %CPU time increases.

3.5 TEXT-TO-SPEECH NOTIFICATION

ShareMon can also notify users with text-to-speech. For logins and logouts they hear messages like "X logged in" or "X logged out," where X is the guest's username. For %CPU time messages, users hear such phrases as "active," "busy," or "very busy." If a guest is connected, but the %CPU time is zero, the message is "Z users connected" where Z is the number of guests currently logged on.

ShareMon users can also mix the modalities but, at present, users cannot have more than one message per event. For example, they cannot have both a sound effect and a text-to-speech message telling them that a guest has logged in, but they can have a text-to-speech login message, and a sound-effect logout message.

3.6 VISUAL NOTIFICATION

Working with audio or video on a personal computer is becoming an increasingly common activity. If a user is in the middle of a difficult sound-editing task, for example, he will not want to be interrupted by background sound effects or text-to-speech messages. How should a user be notified about background activities when his foreground task relies on audio? What if a user is hearing impaired?

Because there are not any other spare sensory channels available,[6] ShareMon allows users to overload the visual channel and be notified graphically. The goal is to use graphics more informatively than blinking an icon, but less intrusively than making a user switch applications and bring up a window. The method is to move a text message and/or picture quickly across the screen until it is centered at the current cursor location, hold it there ("there" is a best guess about where a user's eye might be focused) for a beat of time, and then continue to move the message across the screen until it disappears (see Figure 3). It is a bit like a sprite moving across the screen in a video game. At any time during the message's traversal, a user can hold down the mouse button, causing the message to stick to the cursor. This allows a user to hold on to the message if he needs a longer look, or move it to a less cluttered place on the screen for reading. When the user releases the mouse button, the message will continue its traversal.

If this notification modality does not prove too disturbing for users, it means that the whole graphical arsenal might be available for distinguishing messages of background activity from the foreground: color, font, size, screen location, trajectory, animation, and icons alone or in combination with text.

---

[6] One suggestion is to use the haptic channel, though hardware to support this is not yet generally available. One can imagine a "texture patch," worn on the arm, that could pulse and form patterns (possibly including braille) that users could learn to associate to background events. See Chapter 8 of Baecker and Buxton's book for an introduction to haptics and interface.[4] As far as I know, there are not any olfactory peripherals on the market.

4. REACTIONS TO SHAREMON

In this section, users' reactions to ShareMon are reported. Some of these reactions are preliminary, coming from people who only used ShareMon for a short while. However, some reactions come from a handful of long-term ShareMon users.

This section has three subsections. Section 4.1 describes a surprising difficulty that a user had with some sounds. Section 4.2 discusses some general objections to the use of audio. These objections are echoed by other users in Sections 6.2 and 6.5. Section 4.3 describes users' strong negative reactions to ShareMon's use of humming and pink noise, contrasted with positive reactions to the wave sounds.

4.1 LITERAL OBJECTIONS

One user suggested that the sounds were too literal, that it was too easy to confuse *actual* footsteps and throat-clearings with the ones generated by ShareMon. Originally viewed with some skepticism, this objection is supported by the following e-mail from Tom Erickson (who is the "I" relating the following story; the other names have been changed to protect the innocent):

> By the way, an unexpected side effect of the ShareMon coughing [the "ahem" sound]. Mr. Manners [a well-known author and critic of interface design] was talking with me in my cube, while ShareMon was hacking away, and Mr. Vector came by and said "Ah hah!" (Perhaps Mr. Manners was late for his meeting with Mr. Vector.) Had I not turned around and looked at Mr. Vector, Mr. Manners would have assumed it was ShareMon [making the sound]. Mr. Manners said he was about to ask me what it meant when ShareMon said "Ah hah!"
>
> [Tom Erickson, personal communication, 8/28/92]

4.2 GENERAL OBJECTIONS

The next objection had less to do with ShareMon in particular than it had to do with the use of audio in general. To quote Buxton:

> Let us address a question that inevitably comes up. It usually goes something like, "I work in a crowded office and the last thing I need is more noise to distract me."

Buxton's argument in favor of audio is twofold: audio is already ubiquitous and, if we had more control over audio in our environment, we would find it less objectionable.[8] I certainly agree with him, but the people who objected to the use of audio for notification seemed *angry*, as if I had *dared* to even *think* of the idea of adding more noise to their work environment. They said they would *never* use ShareMon, and perhaps there would be no way to please them using audio. (Perhaps they felt audio was too public in an open-cubicle environment—graphics keeps things private.)

Overall, I have been impressed by the extremely strong reactions people have to audio: the warm fuzzy feeling some people got when they heard their disk drive purring and knew it was okay, the shock and horror when it made unusual sounds, the strong objections to my choice of sounds in ShareMon, and the vehement objections to the idea of audio notification in general. Although supported only by anecdotal evidence, this audio partisanship is also true of the SonicFinder: people who like SonicFinder *really* like it and people who dislike it *really* dislike it. While smell is the most evocative sensation, sound may be nearly as powerful.[7]

4.3 OBJECTIONS TO PINK NOISE—KUDOS TO WAVES

The experiment with humming and pink noise was a disaster, even though users did not have any trouble understanding that pitch correlated with %CPU time. The problem was that most people thought these sounds were obnoxious.

Other workers in this field have mentioned annoyed reactions to sound. Scaletti and Craig algorithmically generated audio accompaniments for data visuaizations.[34] Their observations could be used as rules-of-thumb for decreasing obnoxiousness: "high frequency sounds were more irritating than low frequency sounds" and "complex timbres are easier to locate in space and are less tiring to the listener than are sine waves." Francioni, Albright, and Jackson noted a failed experiment for portraying processor idle time in a parallel processing debugging environment: they played a note for a processor some number of times per beat.

> As an idle burst got longer, the rate of playing that note increased. The idea was that the rapidness of a particular note's playing would draw the listener's attention to it as it got much faster than any of the other notes. In actuality the playback turned out to be quite annoying....[12]

---

[7] This leads me to propose a new unit of measure—the "proust," a measure of a person's emotional response to a stimulus.

In a data visualization system called Exvis, Williams, Smith, and Pecelli use audio textures to sometimes reinforce, and sometimes amplify, visual textural information. They found that some people were severely disturbed by the sound, and they suggest it might have been caused by "an auditory texture produced by a multitude of sounds."[39] Perhaps a similar situation exists with ShareMon: users are displeased by the timbral quality of the humming and pink noise sounds (at one demo, given in a small auditorium, the gentlemen in the control booth nearly had a fit when they started hearing pink noise injected into their sound system).

On the other hand, preliminary reactions to the wave sounds have been positive: one person described it as soothing; during conversations in my office, some people, when asked, said they had not even noticed the sound of waves breaking; and one long-term user chose it to be the notification modality on his office machine.

5. CHOOSING A NOTIFICATION MODALITY

How do we choose between text-to-speech, sound effects, physical attributes of sound, attributes of sources of sound, and numerous graphical possibilities for notification? The last word should be user preference, which may be a matter of taste, physical ability, or mood. However, some rules-of-thumb may help designers choose reasonable defaults for users who do not choose to personalize their event-to-message mappings.

This section has three subsections. Section 5.1 looks at notification modalities with the aim of distinguishing foreground tasks from background notifications. Section 5.2 looks at the characteristics of the events being monitored and attempts to match those to notification modalities. Section 5.3 looks at notification modalities from an information-processing standpoint.

Following Gaver,[13] I think that there are rules of thumb that can be derived from a general notion of **affordance**. Gibson gives this definition of affordance: "The *affordances* of the environment are what it *offers* the animal, what it *provides* or *furnishes*, for good or for ill" [italics in original].[19] Note that the affordance itself is not for good or for ill: these are value judgments that we apply depending on the circumstances. Thus, the same affordance can be viewed in a positive or negative light. Two examples of this are in Section 5.2.

For a discussion of the affordances of audio, and a comparison of cognitive and ecological approaches to perception, see Gaver.[14,17] Also see Gaver for a list of rules-of-thumb about audio at the interface.[13]

5.1 FOREGROUND MODALITIES VS. BACKGROUND MODALITIES

A first rule-of-thumb is to look at the modalities used in the foreground task, and pick distinct modalities for notifying the user about background activity. This affords the user with clear notification from the background that does not interfere with the foreground task. For example, if the foreground task uses only graphics, notify with audio; if the foreground task uses sound in a monochrome graphical environment, notify with color graphics; if the foreground task involves sound effects, notify with text-to-speech (if it is technically feasible to do both at once).

One useful differentiating attribute for text-to-speech is voice quality. At the simplest this could be male versus female, although people are remarkably good at distinguishing many different voice qualities. Pitch and location are good differentiators for sound in general. Location cues can be created from stereo panning, simple reverb, and more complex space reflection models, including recent work on modeling three-dimensional location.[38]

One flaw in this rule-of-thumb is that even if the foreground and background modalities are maximally distinguishable at one instant, the foreground task or its modalities may change. Some operating systems might allow a background application to sense the change in the foreground task, and automatically change the notification modality. For example, on the Macintosh, an application could find out if another application was using sound,[1] and so could switch from notifying with audio to notifying with graphics.

5.2 EVENT CHARACTERISTICS

A notification modality might be chosen by looking at the characteristics of the events to be monitored. Buxton, for example, suggests three categories: alarms and warnings, status and monitoring indicators, and encoded messages (used to present quantitative data).[8] Here a classification of events along three axes is used: frequency, continuity, and priority. For example, in ShareMon, a guest logging off is a relatively infrequent discontinuous event, but %CPU time is continuous and frequent (Buxton might view this as an encoding). In general, an alarm would be triggered by an infrequent, discontinuous, high-priority event, and a warning by a more frequent, discontinuous, lower-priority event. Events that are continuous and infrequently changing would be mapped to status and monitoring indicators.

How is the notion of affordance relevant to these categories? Certain modalities (or characteristics of modalities) stand out more than others— they "afford interruption." The priority axis is an example of the way an affordance can be seen as good or ill: interruption is a virtue for emergencies, but an annoyance for non-emergencies. High-priority events, like alarms, can be mapped to louder, bigger, brighter, and more rapidly changing notifications (and using more modalities) than low-priority events.

The "interruption affordance" varies from good to ill along both the repetition and frequency axes. If the event in question is an infrequent one, a user might appreciate the specific information available in a lengthy text-to-speech message as opposed to a brief sound effect or musical motive. If the event were frequent or repetitious, the user would be annoyed at the repeated interruption. Affordance provides a justification for why continuous sounds, which fade from a user's conscious attention, should be employed to notify users about continuous events (this is not a new idea, but perhaps it is a different way of articulating it).

Another reason for not using text-to-speech to notify users about continuous events is that if the continuous event were changing frequently enough, some status information would be lost. This may be why Lunney and Morrison, in their pioneering work on adding sound to laboratory instruments for use by the blind, provided redundant text-to-speech and tone-varying output for some instruments in a titration experiment. Quick changes in pH (qualitative data) could be monitored by varying tone, and final readings (quantitative data) could be determined by using text-to-speech.[26]

Although text-to-speech versus continuous sounds have been addressed here, this discussion also applies to the visual domain (space) in a way analogous to the audio domain (time). That is to say: text-to-speech is to text as sound effects are to graphical icons. A repeated text message is annoying when compared to the repeated presentation of an icon; text takes more time to read than an icon takes to see, but text might be more appreciated if the event is an infrequent one. (See Julesz and Hirsch for a discussion of perceptual analogies between vision and audition.[24])

5.3 NOTIFICATION MODALITIES AND INFORMATION PROCESSING

A third way to think about affordance is to look at characteristics of the modalities used for notifications in terms of information processing. For example, there is evidence that pictorial icons are more meaningful than text for representing objects and that combined textual and pictorial icons are more effective than either alone.[20] There is also evidence that pictures are more comprehensible than text-to-speech messages (at least in cockpit emergency conditions)[32] but, of course, graphics take up screen real estate.

An advantage of audition over vision is that a sound can be heard anywhere near its output hardware, but an image can only be viewed if the user is looking at the screen at the right time. On the other hand, a user can glean more information from the visual channel: the maximum number of background messages a user can *hear* reaches a limit of discrimination.

Perhaps the maximum number of background messages a user can *see* first reaches a limit of distraction.

Finally, there are issues of directness: text-to-speech is more direct than sound effects, but text is more direct than icons (the analogy between the visual and the auditory continues to hold). Here "direct" is used in the sense that text and text-to-speech directly remind users of what they mean but the user must rely on memory to find the meaning of graphical or auditory icons (see Hutchins, Hollan, and Norman for a discussion of "directness"[21]). There is a limit to the number of event-to-message mappings a user can keep track of, even though this limit might be increased using Gaver's metaphorical approach.[15] At ICAD, Meera Blattner suggested that earcons could be used to increase this limit. Once this limit is reached, the only way for a user to keep track of what a notification means would be for it to use text-to-speech or text to explain itself.

6. USER STUDIES

This section has seven subsections. Section 6.1 describes the first round of a user study of ShareMon, and Sections 6.2 and 6.3 describe the results. Section 6.4 describes a second round of the user study, using a ShareMon that was modified based on results from the first round, and Sections 6.5 and 6.6 describe the new results. Section 6.7 takes another look at the rules-of-thumb discussed in Section 5, to see how well they held up.

6.1 USER STUDY: ROUND ONE

In order to test some of the rules-of-thumb discussed in Section 5, I conducted a user study. The participants were selected from a group of people who responded to a notice, on an Apple internal electronic bulletin board, that requested file-sharing hosts to participate in a user study. Each participant received a t-shirt and a copy of ShareMon. The eight participants were all employees of Apple Computer, Inc., were all highly Macintosh literate (i.e., had been using the Macintosh for more than two years), and all had been file-sharing hosts. Their occupations varied from engineer to technical writer to secretary to product tester to project manager. Five of the participants were male and three of them were female.

There were eight sessions in all, one session per participant. Each session lasted about one hour. Each participant was told that I was interested in how people could be notified about background activity, and that file sharing was one kind of background activity. Each participant sat in front of a computer (call it the **host machine**), with access to a number of issues of an on-line magazine (each issue in the form of a Hypercard^{TM} stack). The participant was assigned the foreground task of characterizing each article that they came across with a few key words or phrases. They typed the keywords into a blank document opened in a word-processing program. Each participant was told that they could examine the issues in any order, and that it did not matter whether they skimmed the articles or read them in detail.

Before any participant began the task, they were introduced to the file-sharing notification by logging on to their machine, copying a file, and logging off. During this introduction, the participant was notified with either text-to-speech, sound effects, or graphics. After the introduction, the session consisted of three parts, each lasting about ten minutes. In each part, the participant was notified about file-sharing events with either text-to-speech, graphics, or sound effects. There were no file-sharing events generated during the first few minutes of each part, so that participants would have time to become absorbed in the foreground task. Then file-sharing events were generated with the participant's machine as the host. At intervals, the participant was asked about the status of file sharing, for example, if the participant knew how many people were currently logged on to his machine.

The file-sharing events were automatically generated by three **guest machines** in my office. These machines were connected to the host machine via Ethernet. At times, one, two, or three of these guest machines could be logged on to the participant's machine and could be copying a file over to the participant's machine. Each of the guest machines logged on as a registered user. The events were automatically generated for three reasons:

- The participant would not be able to correlate my actions during the session with file-sharing events.
- It would give me the freedom to observe the study, and not be busy generating file-sharing events.
- Each participant got the same amount and type of file-sharing events (again, given the anecdotal nature of the study, this point is probably unimportant).

During a typical ten-minute interval there might be five logins (and logouts), and five files copied. A single guest might stay connected to the participant's machine from anywhere between fifteen seconds and five minutes. Note that in these three 10-minute periods, most of the participants were subjected to more file-sharing events than they usually got in a month (most of them got a couple of logins a week).

During each ten-minute period, the participant was notified with a single notification modality. The text-to-speech notification is described in Section 3.5. The graphical notification is described in Section 3.6. The sound effects were the metaphorical sounds listed in Section 3.3.

The sound effects were presented without any explanation of which sounds were mapped to which events. After some time, if the participant was unable to guess the mappings and was frustrated, I explained the event-to-message mapping. Later during the session, I tried to determine whether the participant remembered the mapping.

The order of presentation of the notification modalities was varied between participants. For all the modalities, the connection reminder messages occurred at intervals of 30 seconds.

6.2 ROUND ONE RESULTS: SOUND EFFECTS

In general, the participants found all the notifications distracting: they reported it and I could see them stop working. On the other hand, they did get some work done: they were able to read a number of the articles, and take a few notes. All the users took a copy of ShareMon to play with and a few of them became regular ShareMon users (I think this is a very positive indicator). Although these numbers are not in any way definitive, four of the participants liked the sound effects best (especially if the sounds could be modified), one preferred text-to-speech, two would have liked different or combined modalities for different events, and one preferred graphics.

On an event-by-event basis, how intuitive were the sounds? How easy were they to learn? How informative were they? This level of detail is relevant because it is the level of detail that a software designer must reach to create a usable product. Compare the design of a GUI, which requires paying attention to an equivalent level of detail: each icon is tested for comprehensibility, the fonts are carefully chosen, and each line of text is worded carefully.

The knock-knock-knock sound was not entirely intuitive. It was intended to mean "a guest has logged in." Most of the participants interpreted it as "someone is trying to log in"; whether successfully or not they were not sure. Once it was explained, they seemed to remember it as a login message.

Only one of the participants guessed that the "ahem" sound meant someone was connected but otherwise inactive (the only reason she guessed correctly is because she used software that employed the same sound and had a similar meaning). Some participants suggested it meant that somebody wanted their attention for some reason, but they could not figure out why, particularly in the context of file sharing. One participant heard the sound as if it were coming from outside the room, even though the sound originated from a speaker directly in front of him (in the computer sitting under the desk). All the users thought that the reminders occurred too often. Rather than setting ShareMon to remind them at 30-second intervals, they suggested intervals ranging from a few minutes to a quarter of an hour.

Only one participant spontaneously guessed that the walking sounds meant %CPU time. Some did notice that the intensity increased (i.e., faster walking sounds, higher pitched), but were unable to correlate that to increased %CPU time. Again, they had no trouble remembering what it meant, once it was explained. However, all the participants complained about the sounds, saying they were far too disruptive and happened far too often (during a typical file copying operation, they might hear 15 continuous seconds of walking and running sounds). A number of participants said that they could "feel" the machine slow down when a copy operation was in progress, so that the walking sounds were redundant as well as annoying (however, see Section 6.5 for some counterarguments).

All the participants spontaneously understood that the door slam sound meant "log off." One participant, however, asked whether it meant that only one person had logged off or whether everyone had logged off (he was able to answer his own question later in the session). More about this point in Section 6.5

There was a surprising interaction effect from participants' interpretations of combinations of sounds. For example, here is one participant's interpretation of knocking followed by walking sounds—intended to mean that a guest had logged on and was (immediately) copying a file: "Because right after you're knocking you're walking. I'm outside—I'm knocking the door and you're not responding. So I'm walking back and forth. When [laughs]... when a father is expecting a baby—something like that is the idea I got." This is fine example of the affective power of audio. [query author]

Another participant thought that knocking followed by walking meant that a guest was "coming in" (logging on), but that walking sounds alone meant that a guest was "going out" (logging out). These participants were telling stories based on the sounds they heard and, unfortunately, the stories they were telling were not the ones ShareMon was trying to communicate.

6.3 ROUND ONE RESULTS: TEXT-TO-SPEECH AND GRAPHICS

The main problem with the text-to-speech was intelligibility, even though during the ten minutes that each participant was exposed to the text-to-speech, they were increasingly able to understand the messages. Nevertheless, most of the participants remained uncomfortable with text-to-speech, and were greatly distracted by it. One person, however, preferred the text-to-speech, saying he liked it because he did not have to translate it to make sense of it—he did not care for the sound effects because they required that translation.

A couple of participants said that the problem was that the speech went by too quickly— by the time they turned their attention from the task at hand to the speech being uttered, they had already missed half of it. One participant suggested prefacing each utterance with an "ahem" sound as a warning; another said "tell me twice."

One participant's session began with graphical notification, and in the next part of the session he got text-to-speech notifications. He said that since he had already seen the names in print, it was much easier to understand them when he heard the text-to-speech.

As before, participants complained that the connection reminder message occurred too often, and the %CPU time messages, occurring even more frequently, were even more annoying.

In this round of the study, users found that the graphical messages traveled too quickly across the screen to read. They had to click on a message to read it, which disrupted their work. On the other hand, they liked the small size of the message, and they certainly had the least trouble of all the modalities understanding what a particular message meant.

6.4 USER STUDY: ROUND TWO

Based on the results described in Sections 6.2 and 6.3, I made some changes to ShareMon, and then conducted ShareMon User Study 2.0. This time there were twelve participants, six from ShareMon User Study round one and six new ones from the list of volunteers mentioned above. The task was similar: there were some new issues of the on-line magazine, and participants were to tell me what they liked or did not like about the articles, typing this information into a blank word-processing document.

As before, each session was broken into three parts. This time, however, if the participant ran out of things to say about a particular modality, the ten-minute part was cut short. Also, as before, file-sharing events were automatically generated by three guest machines. However, this time the guest machines could log in to the host machine as either a registered user or as a guest (named "<Guest>").

The main change to ShareMon for round two was the addition of two new events that were mapped to messages: ShareMon could notify the host if a guest opened or closed a file on the host machine. These events were an attempt to replace the %CPU time messages that participants said were redundant, since data transfer (copying a file) was preceded by opening the file, and followed by closing it—most of the %CPU time devoted to file sharing occurred during the data transfer between the file-opening and file-closing events.

Here is a list of changes to the sound effects (see Table 2):

- For logins, the knock-knock-knock sound was followed by the sound of a door creaking open.
- The knock-knock-knock/door creak sounds were only used for <Guest> logins; for registered user logins the notification consisted of the sound of keys jingling followed by the sound of a key turning in a lock followed by the sound of door creaking open.
- The sound of a chair creak was used for the connection reminder message, instead of "ahem." The number of creaks was tied to the number of users logged on. For example, "creak-creak" meant two users and "creak-creak-creak" meant three users. Even if there were more than five users logged on, however, the maximum number of creaks was set at five. In any case, during the user study there were never more than three users logged on to the host machine at any one time, and in real circumstances file-sharing hosts typically have a small number of guests.
- There were no notifications of %CPU time devoted to file sharing (the walking sounds were turned off).

TABLE 2 ShareMon User Study 2.0 event-to-message mappings.

Event	Message
log in (<Guest>)	knock-knock-knock door creak
log in (registered user)	keys jingling key-in-lock door creak
connection reminder	chair creak
%CPU time	—
file open	drawer open
file close	drawer close
log out	door slam

- The sound of a drawer opening notified the host that a guest had opened a file on the host machine; the sound of a drawer closing notified the host that a guest had closed a file on the host machine. The drawer-opening sound had a latch-clicking sound at the beginning, followed by a rolling noise, and the drawer-closing noise had a rolling sound, followed by a latch-clicking sound.

The changes to the text-to-speech notifications were:

- A new higher-quality version of the text-to-speech software was incorporated.
- Users could set the speech rate (words per minute) of the utterances.
- Users could prepend words or phrases such as "um," "it's like," or "oy" to all the utterances.
- There were no %CPU time messages.

The changes to the graphical modality were:

- The message trajectory started in the lower right corner of the screen and moved to the upper right. The message waited a beat when it reached the height of the user's mouse (making a best guess that the user's eye might be at the same height).
- Users could set the speed of the message as it moved up the screen and they could also set the amount of time the message waited at the mouse height.
- Login and logout messages took the form described above. File-opening and file-closing messages were added, also taking the form described above.
- There were no %CPU time messages.

6.5 ROUND TWO RESULTS: SOUND EFFECTS

Generally, people seemed happier this time. As before, they were able to work but still occasionally distracted. Three of the new participants preferred sound effects, one graphics, one text-to-speech, and one wanted to mix modalities. Two of the returning participants preferred the graphical messages, but the other six preferred to mix them. Again, these numbers do not have any great significance. Possibly more significant are the electronically logged choices of five regular ShareMon users: two prefer sound effects, one text-to-speech, and one uses graphics for some messages and sound effects for others. The fifth ShareMon user has three machines: one using sound effects, one text-to-speech, and one using the pulse (described in Section 3.4).

From my observations, I do not think the %CPU time messages were redundant after all. Even though copying a file was bracketed by file-opening and file-closing messages, when participants noticed their machine slowing down, they tended to blame ShareMon rather than file sharing. (Another reason the %CPU time messages are not redundant was given by a long-term ShareMon user. He uses a second machine in his office for file sharing. He runs ShareMon on the second machine, and can hear its messages from the first machine. He cannot "feel" the slowdown of the second machine from the first one, so the auditory cue is the only one he has.)

Several of the people who participated in both rounds of the user study said they were less annoyed by ShareMon in round two than in round one, and I believe that a good part of the cause was the absence of the %CPU time messages. So even though ShareMon was more pleasant without the %CPU time messages, they were useful and should probably be put back in. However, the user should be able to control the amount of time between repetitions of these messages.

As before, participants' reactions to the sound effects will be discussed event-by-event. The knock-knock-knock/door opening combination was a success—all the participants understood that this meant someone had logged in. Unfortunately, no one identified the key turning in the lock sound and, therefore, no one gathered that there were different sounds for registered user and <Guest> logins. Once it was explained, however, participants were able to keep track of the difference, so perhaps the key-in-lock sounds only need to be re-recorded.

The creaking sound fared as poorly as the "ahem" sound. One participant tentatively thought it meant "someone is just sitting there," and not doing anything. None of the other eleven participants intuited its meaning as a connection reminder. Once it was explained, they had no trouble remembering.

One participant "spontaneously" (i.e., with a little prompting) understood that the number of creaks was tied to the number of users, though a second somehow figured it out retrospectively when he was working with text-to-speech messages. A couple of participants thought that the sound had subparts, that is "creak-creak" was a single message. Again, once they knew it was a count, they were able to keep track of the number of users. However, a number of participants expressed dissatisfaction with the need to count, because the attempt to keep track would move their focus away from the foreground activity.

Another problem with the creaking sound was that two of the users heard it as originating from somewhere other than the computer. One person thought it was his chair, and another person thought the sound was coming from outside the room. Since this was the third time his problem was mentioned (see Sections 6.2 and 4.1), I think it is a serious one.

The drawer-opening and drawer-closing sounds were heard variously as thunder, a wave, a toilet flushing, a door sliding open and closed, and as what they were intended to be. The people who heard the sounds as drawers or doors opening and closing guessed that it had to do with files or folders being opened or closed. (However, when the drawer-opening sound immediately followed the door-opening sound, two of these people were not able to identify it as the sound they had previously identified as a drawer opening. Is this is another interesting interaction effect?) Once it was explained, all the participants remembered that these sounds represented file-opening and file-closing messages, but several of them had trouble distinguishing the two sounds. Intriguingly, a couple of participants could tell the difference between the drawer-opening and drawer-closing sounds, but were not able to articulate why. Based on these reactions, the metaphor is apt, but the particular sounds need to be re-recorded.

The storytelling problem discussed in Section 6.2 seemed to be fixed in this round. However, a couple of the participants brought up another problem. One said that, when he heard the door slam sound, it meant "somebody got pissed and then logged off" because the door was slammed and not just closed. Another, hearing the creaking sound, said "something needs oiling; better call facilities and have them fix it." This could be the explanation for the person in the first study who asked whether the door slam meant one person logged off, or everyone did: because the door slam sounded so "final" it was as if file sharing was shut down for the day. The point is that not only can sounds tell the wrong story, they can express the wrong emotional value.

6.6 ROUND TWO RESULTS: TEXT-TO-SPEECH AND GRAPHICS

Overall, people seemed to understand the text-to-speech better in round two than round one. Most of this was due to the new release of the text-to-speech software. I asked six participants to try adjusting the speech rate. All of them found that changing the rate improved intelligibility. Surprisingly however, as many users chose to increase the words-per-minute rate from the default of 175 wpm as decrease it. All six of them tried prepending a phrase onto the text-to-speech utterances. One participant disliked the prepended phrase and canceled it, but the others left it on, two of them noting that it increased intelligibility by warning them that speech was coming.

Participants slowed the graphical messages down so they were readable as they crossed the screen. However, most of the users found the motion of the graphical messages quite distracting, and many said that they would prefer to see the information in an ordinary window, which they could cover or uncover at will and which would stay put. Some of them, however, wanted a window with some additional auditory notification that there had been a change in file-sharing status.

However, one participant, who used ShareMon on a long-term basis, liked the motion of the graphical messages. He used a second computer for file sharing, positioned so its monitor could just be seen out of the corner of his eye. The messages moving along the periphery attracted his attention just enough to be useful without being annoying.

6.7 CHOOSING A NOTIFICATION MODALITY II

How did the rules-of-thumb discussed in Section 5 hold up? The one about contrasting foreground with background modalities (Section 6.1) did not get much play. A couple of participants mentioned in passing that they generally listened to music through headphones while they worked, or that they worked with QuickTime (desktop audio/video software), so that text-to-speech and audio messages would be hard to hear.

The rule-of-thumb that evaluated messages based on the event characteristics of frequency, continuity, and priority (Section 6.2) seemed quite relevant. Most participants were annoyed by the repetition of messages about events they considered low priority: how many users were connected and what was the %CPU time. Several users only wanted to be notified if a particular file was accessed or a particular user logged on; otherwise, they found the messages annoying. Which events a particular user gave priority seemed to be quite task dependent.

The third rule-of-thumb that looked at information-processing characteristics of notification modalities (Section 6.3) was also relevant. For example, one ShareMon user wanted to know the names of the people that were logging in, but did not care about the names of the files they accessed (since he already knew which files were available). This person uses graphical messages for logins and logouts, but the drawer-opening and drawer-closing sounds for file accesses. Another user just wants to get a general sense of activity on his server, so he uses sound effects. Again, which information a particular user wanted was task dependent.

7. CONCLUSION

The fact that ShareMon now has a number of faithful adherents suggests that users might welcome the ability to be notified about background activities. ShareMon, not to mention RAVE and Khronika, also serve as proof that users can monitor background activities over the audio channel.

One surprising aspect of ShareMon is that it was a straightforward application that lent itself well to the exploration of difficult interface issues such as: when to use a particular notification modality, and how can a sound be designed to be simultaneously informative, pleasant, and unobtrusive.

One thing discovered in this process is just how difficult the art of sound design really is. Some of the nastier pitfalls, besides exposing users to obnoxious sounds, were how the interactions of the sound effects led users to hear different stories from the ones intended; how some users thought the sounds had negative connotations; and how sounds that were not characteristic enough or were not recorded properly either mislead or completely mystified users. Whatever progress was made beyond these pitfalls was due to the design-and-test methodology—participants were able to describe problems extremely articulately (even pointedly), and often they suggested the solutions.

My future work in this area will have three main directions. Firstly, users were only notified about one kind of background activity. How would the results be different if users were monitoring a number of different kinds of background activity? For example, how would they keep track of which messages belonged to which events? Secondly, the foreground activity in all these user studies is purely text-based. How will users' preferences for certain kinds of messages change when they are working with images and sounds in the foreground? Third, at ICAD, Bill Buxton mentioned the notion of an "acoustic ecology," a seamless and information-rich, yet unobtrusive, audio environment. ShareMon is a far cry from that subtle and powerful ambiance.

ACKNOWLEDGMENTS

I would like to thank the members of the Human Interface Group in the Advanced Technology Group (HIG/ATG) at Apple for moral support, useful suggestions and discussions, and for being ShareMon guinea "higs." Also, thanks to Gregory Kramer and Beth Mynatt for editorial assistance.

REFERENCES

Newton, Macintosh, Chooser, Finder, SonicFinder, File Sharing Monitor, Sound Manager, Hypercard, and Notification Manager are all trademarks of Apple
Computer, Inc.

1. Apple Computer, Inc. *Inside Macintosh*, Vol. VI. Reading, MA: Addison-Wesley, 1991. In particular see chapter 24, "The Notification Manager"; chapter 6 for a discussion of file sharing; and chapter 22, "The Sound Manager."
2. Apple Computer, Inc. *Newton Technology: An Overview of a New Technology from Apple*. Cupertino, CA: Apple Computer, Inc., 1992.
3. Arons, B. "A Review of the Cocktail Party Effect." *J. Am. Voice I/O Soc.* **12** (July 1992): 35–50.
4. Baecker, R. M., and W. Buxton, eds. *Readings in Human-Computer Interaction*. San Mateo, CA: Morgan Kaufmann, 1987. In particular read the introduction to chapter 9 on the audio channel (393–399). Also see chapter 8 which discusses "haptics" (the use of touch and gesture).
5. Blattner, M. M., D. A. Sumikawa, and R. M. Greenberg. "Earcons and Icons: Their Structure and Common Design Principles." *Hum.-Comp. Inter.* **4** (1989): 11–44.
6. Bly, S. A. "Presenting Information in Sound." *Proceedings of the CHI '82 Conference on Human Factors in Computer Systems* (1982): 371–375. New York: ACM, 1982.
7. Bonner, Pace, and Jeff Amfahr. *Nok Nok* (software). Woburn, MA: Trik Computing, 1992.
8. Buxton, William. "Introduction to This Special Issue on Nonspeech Audio." *Hum.-Comp. Inter.* **4** (1989): 1–9.
9. Cage, John. *Silence*. Middletown, CT: Wesleyan University Press, 1986. See the articled entitled "Composition as Process" (18–34) for how Cage generated the structure of *Music of Changes*: The article entitled "Composition" (57–59) details Cage's method for choosing single sounds.
10. Cypher, Allen. "EAGER: Programming Repetitive Tasks by Example." *Proceedings of the CHI '91 Conference on Human Factors in Computer Systems* (1991): 33–39. New York: ACM, 1982.
11. Don, A., T. Oren, and B. Laurel. "Guides 3.0." *Proceedings of the CHI '91 Conference on Human Factors in Computer Systems* (1991): 447–448. New York: ACM, 1982.
12. Francioni, J. M., L. Albright, and J. A. Jackson. "Debugging Parallel Programs Using Sound." *SIGPlan Notices* **26** (December 1991): 68–75.
13. Gaver, W. "Auditory Icons: Using Sound in Computer Interfaces." *Hum.-Comp. Inter.* **2** (1986): 167–177.
14. Gaver, W. "Everyday Listening and Auditory Icons." Ph.D. dissertation, University of California, San Diego, 1988. Chapter two discusses the relevance of ecological theories of perception to the use of sound at the interface.

15. Gaver, W. "The SonicFinder: An Interface that Uses Auditory Icons." *Hum.-Comp. Inter.* **4** (1989): 67–94.
16. Gaver, W., and R. B. Smith. "Auditory Icons in Large-Scale Collaborative Environments." *Interact '90: Proceedings of the IFIP TC 13 Third International Conference* (1990): 735–40. North Holland: Amsterdam, 1990.
17. Gaver, W. "Technology Affordances." *Proceedings of the CHI '91 Conference on Human Factors in Computer Systems* (1991): 79–84. New York: ACM, 1982.
18. Gaver, W., T. Moran, A. MacLean, L. Lövstrand, P. Dourish, K. Carter, and B. Buxton. "Realizing a Video Environment: EuroPARC's RAVE System." *Proceedings of the CHI '92 Conference on Human Factors in Computer Systems* (1992): 27–37. New York: ACM, 1982.
19. Gibson, J. J. *The Ecological Approach to Visual Perception*, 127. Hillsdale NJ: Lawrence Erlbaum, 1986.
20. Guastello, S. J., M. Traut, and G. Korienek. "Verbal Versus Pictorial Representations of Objects in a Human-Computer Interface." *Intl. J. Man-Machine Stud.* **31** (July 1989): 99–120.
21. Hutchins, E. L., J. D. Hollan, and D. A. Norman. "Direct Manipulation Interfaces." In *User-Centered System Design*, edited by D. A. Norman and S.W. Draper, 87–124. Hillsdale, NJ: Lawrence Erlbaum, 1986.
22. Iversen, Wes. "The Sound of Science." *Comp. Graph. World* **15** (1992): 54–62.
23. Jenkins, J. J. "Acoustic Information for Object, Places, and Events." In *Proceedings of the First International Conference on Event Perception*, edited by W. H. Warren, 115–138. Hillsdale, NJ: Lawrence Erlbaum, 1985.
24. Julesz, B., and I. J. Hirsch. "Visual and Auditory Perception—An Essay of Comparison." In *Human Communication: A Unified View*, edited by E. E. David, 283–340. New York: McGraw-Hill, 1972.
25. Lövstrand, L. "Being Selectively Aware with the Khronika System." In *Proceedings of the Second European Conference on Computer-Supported Cooperative Work*, edited by L. Bannon, M. Robinson, and K. Schmidt, 265–277. Dordrecht, Netherlands: Kluwer, 1991.
26. Lunney, D., and R. C. Morrison. "High-Technology Laboratory Aids for Visually Handicapped Chemistry Students." *J. Chem. Educ.* **58** (1981): 228–231.
27. Mansur, D. L., M. M. Blattner, and K. I. Joy. "Sound-Graphs: A Numerical Data Analysis Method for the Blind." *Proceedings of the 18th Annual Hawaii Conference on System Sciences* (1985): 198–203. Hawaii: Western Periodicals, 1985.

28. Mezrich, J. J., S. Frysinger, and R. Slivjanovski. "Dynamic Representation of Multivariate Time Series Data." *J. Am. Stat. Assoc.* **79** (March 1984): 34–40.
29. Miyata, Y., and D. A. Norman. "Psychological Issues in Support of Multiple Activities." In *User-Centered System Design*, edited by D. A. Norman and S. W. Draper, 265–284. Hillsdale, NJ: Lawrence Erlbaum, 1986.
30. Nyman, Michael. *Experimental Music: Cage and Beyond*, 83–84. New York: Schirmer, 1974.
31. Olson, M. H., and S. A. Bly. "The Portland Experience: A Report on a Distributed Research Group." *Intl. J. Man-Machine Stud.* **34** (1991): 211–228.
32. Robinson, C. P., and R. E. Eberts. "Comparison of Speech and Pictorial Displays in a Cockpit Environment." *Human Factors* **29** (February 1987): 31–44.
33. Sawitzki, Günther. "The NetWork Project: Distributed Computing on the Macintosh." *Develop* **11** (August 1992): 82–105.
34. Scaletti, C., and A. B. Craig. "Using Sound to Extract Meaning from Complex Data." *Proceedings of the SPIE Conference* **1459** (1991): 207–219.
35. Schneiderman, B. "Direct Manipulation: A Step Beyond Programming Languages." *IEEE Computer* **16(8)** (1983): 57–69.
36. Stifleman, L. J., B. Arons, C. Schmandt, and E. Hulteen. "VoiceNotes: A Speech Interface for a Hand-Held Voice Notetaker." Submitted for publication to CHI '93.
37. Story, G. A., L. O'Gorman, D. Fox, L. L. Schaper, and H. V. Jagadish. "The RightPages Image-Based Electronic Library for Alerting and Browsing." *IEEE Computer* **25** (September 1992): 17–26.
38. Wenzel, E. M., F. L. Wightman, and D. J. Kistler. "Localization with Non-Individualized Virtual Acoustic Display Cues." *Proceedings of the CHI '91 Conference on Human Factors in Computer Systems* (1991): 351–359. New York: ACM, 1982.
39. Williams, M. G., S. Smith, and G. Pecelli. "Computer-Human Interface Issues in the Design of an Intelligent Workstation for Scientific Visualization." *SIGCHI Bulletin* **21** (April 1990): 44–49.

Elizabeth D. Mynatt
Graphics, Visualization & Usability Center, College of Computing, Georgia Institute of Technology, Atlanta, GA 30332-0280; beth@cc.gatech.edu

Auditory Presentation of Graphical User Interfaces

The majority of work in auditory user interfaces has focused on adding auditory cues to a visual interface. This paper presents work in designing interactive auditory-only interfaces where the design of the interface is driven by a challenging task—providing access to graphical user interfaces for people who are blind. This task requires that the auditory-only interface must be able to provide the same functionality supported by a graphical user interface. A prototype system called Mercator is described as well as some of the design strategies for Mercator's auditory interfaces. The results from a small user study are also presented with a concluding discussion on future research efforts.

1. INTRODUCTION

The graphical user interface is, at this time, the most common vehicle for presenting a human-computer interface. There are many times, however, when a graphical user interface is inappropriate or unusable. One example is when the task requires that the user's visual attention is somewhere besides the computer screen. Another example is when the computer user is blind or visually impaired.[1,3]

The process of transforming a graphical user interface into a purely auditory interface while maintaining the semantics of the application has not been addressed by research in user interface toolkits or auditory interface design. This paper presents the design and evaluation of auditory interfaces that provide transparent access to X Windows applications for computer users who are blind or severely visually impaired. The process of creating an interactive auditory interface to replace a graphical user interface forces the examination of many difficult problems in auditory interface design.

What does it mean to interact in an audible world? Building on concepts introduced by previous work in auditory interfaces, this research explores conveying symbolic information via auditory cues, navigating in a complex auditory space, and presenting coherent feedback in an interactive auditory interface. Although these issues are explored in the context of providing access to graphical interfaces for computer users who are blind, the results from this work are generally applicable to other auditory-only interfaces as well as visual interfaces that employ auditory cues.

This paper is organized in the following manner. The following section presents background information on access systems used by people with visual impairments and, on X Windows, the graphical user interface common on Unix workstations. The Mercator system which provides the needed information about a running graphical application is also briefly described. The design and evaluation of a prototype auditory interface is discussed in the third section. This discussion includes the design considerations and interface strategies examined in the prototype development as well as the execution and results from a small user study. A description of planned future research efforts concludes the paper.

2. BACKGROUND

2.1 ACCESS TO GRAPHICAL USER INTERFACES

One important breakthrough in HCI was the development of graphical user interfaces. These interfaces provide graphical representations for system objects such as disks and files, interface objects such as buttons and scroll bars, and computing concepts such as multitasking. Unfortunately, these graphical user interfaces (GUIs) are all but completely inaccessible for computer users who are blind or severely visually disabled.[1,3,17] This critical problem has been recognized and addressed in recent legislation

(Title 508 of the Rehabilitation Act of 1986, 1990 Americans with Disabilities Act) which mandates that computer suppliers ensure the accessibility of their systems and that employers must provide accessible equipment.[9]

Work on this project began with a simple question: How can access to X Windows applications be provided to blind computer users. Historically, blind computer users had little trouble accessing standard ASCII terminals. The line-oriented textual output displayed on the screen was stored in the computer's frame buffer. An access program could simply copy the contents of the frame buffer to a speech synthesizer, a braille terminal, or a braille printer. Conversely, the contents of the framebuffer for a graphical interface are simple pixel values. To provide access to GUIs, it is necessary to intercept application output before it reaches the screen. This intercepted application output becomes the basis for an off-screen model of the application interface. The information in the off-screen model is then used to create alternative, accessible interfaces.

The typical scenario to providing access to a graphical interface is as follows: While an unmodified graphical application is running, an outside agent collects information about the application interface by watching objects drawn to the screen and by monitoring the application behavior. This outside agent (or screen reader) then translates the graphical interface into all auditory and/or tactile interface. Not only does the screen reader translate the graphical presentation into an auditory presentation, but the screen reader often provides different user input mechanisms which are more intuitive with the new interface.

The goal of the Mercator[1] Project is to provide *transparent* access to X Windows applications for computer users who are blind or severely visually impaired.[14] In order to achieve this goal, two major problems needed to be addressed. First, in order to provide transparent access to applications, a framework that would support monitoring, modeling, and translating graphical interfaces of X Windows applications without modifying the applications was built. Second, given these application models, a methodology for translating graphical interfaces into nonvisual interfaces was required. This methodology is essentially the implementation of a *hear-and-feel* standard for Mercator interfaces. Like a look-and-feel standard for graphical interfaces, a hear-and-feel standard provides a systematic presentation of nonvisual interfaces across applications.

[1] Named for Gerhardus Mercator, a cartographer who devised a way of projecting the spherical Earth's surface onto a flat surface with straight-line bearings. The Mercator Projection is a mapping between a three-dimensional presentation and a two-dimensional presentation of the same information. The Mercator Environment provides a mapping from a two-dimensional graphical display to a three-dimensional auditory display of the same user interface.

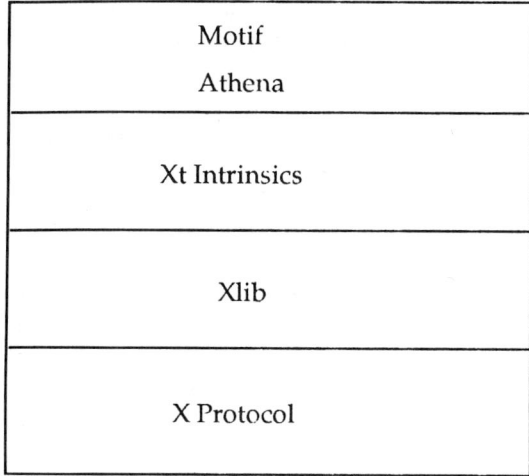

FIGURE 1 X Windows Hierarchy.

2.2 X WINDOWS AND MERCATOR OFF-SCREEN MODELS

X Windows is the de facto standard graphical user interface for Unix environments. The following diagram shows the layers in the X Windows hierarchy. X Windows is based on a client-server architecture, where X applications communicate with a display server over a network protocol. This protocol is the lowest layer of the X hierarchy. Xlib and the Xt Intrinsics provide two programming interfaces to the X protocol. Xlib establishes the concept of events and provides support for drawing graphics and text. The Xt Intrinsics establishes the concept of widgets or programmable interface objects and provides a basic set of widgets. Most people who develop X Windows applications use X toolkits such as Motif or Athena. These toolkits build on top of the Xt Intrinsics and provide many generic interface objects or widgets.

The first goal of the Mercator Project is to build a framework that can monitor, model, and translate application GUI interfaces transparently to the application. The details of this framework can be found in a paper by Mynatt and Edwards.[15] The resulting off-screen models are based on the X widgets which are used to implement the X interface. The application widget hierarchy is stored along with each widget's attributes, resources, and properties.

Essentially this off-screen model provides information about the types of interface objects used in the application interface as well as attributes of these objects. The widget hierarchy provides some information about the relationships between these objects but this information is incomplete. For example, a menu button may be the child of the box containing the menu buttons, but the actual menu associated with the menu button may

Auditory Presentation of Graphical User Interfaces

be located in an unrelated portion of the widget hierarchy. On the whole, the widget specification of the application interface is too low level for a straightforward translation from graphical objects to Mercator objects. Many widgets used in an X application interface are essentially invisible during the visual presentation of the interface and should likewise be invisible in Mercator interfaces as well.

In addition to providing information about the graphical interface, the Mercator architecture provides mechanisms for translating user input and transforming events in the graphical interface into events in the auditory interface. The overall Mercator architecture is shown in the following figure. Mercator communicates with the X application along a protocol that hooks into the Xt Intrinsics and X lib libraries. Information collected about the graphical interface is stored by the Model Manager. The Rules Engine, which is the heart of the Mercator architecture, handles requesting information about the graphical interface, presenting the replacement auditory interface and transforming user input in the interactive environment. For example, the Rules Engine collects information about a graphical menu, triggers the rules that present an auditory menu, and translates user keyboard input to trigger mouse events in the graphical interface.

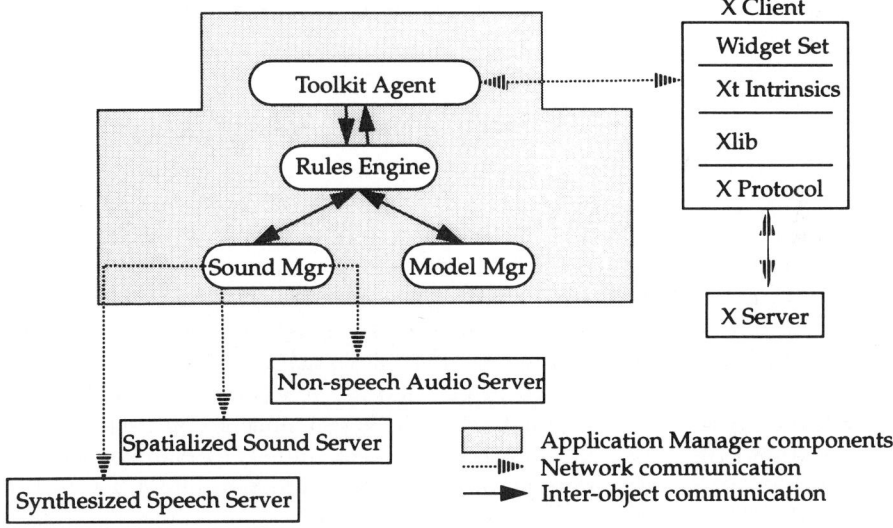

FIGURE 2   Mercator Architecture.

This paper contains the design of Mercator interfaces which are based on the Mercator off-screen model of application GUI interfaces. The strategies for presenting the auditory interface are codified as rules in the Mercator Rules Engine. Likewise the Mercator architecture supports the implementation of different interaction techniques, described in the following sections, such as eliminating the need for mouse input and supporting interface navigation based on a logical structure instead of a spatial representation.

3. AUDIO GUIS
3.1 DESIGN CONSIDERATIONS

This work addresses providing access to graphical user interfaces for computer users who are blind, or more concretely, providing an accessible interface based on a derived model of the graphical interface. In the first phase of this work, there were two main design considerations that had to be addressed. First, what should be the presentation modality for the accessible interface? Second, how much of the visual interface should be modeled in the accessible interface?

WHY USE AUDITORY OUTPUT? The only assumption in this work is that the user will have little to no visual capabilities. The obvious choices for a nonvisual interface modality are auditory and/or tactile. Both modalities are capable of displaying text, either by synthesized speech or by braille output. Some users can even read raised lettering. But, with the exception of text display, the two modalities have different capabilities.

Previous research in auditory interfaces has demonstrated many capabilities. Auditory displays are capable of conveying symbolic information à la Bill Gaver's use of auditory icons in the SonicFinder.[6] Multiple layers of information can be encoded in one auditory cue by manipulating the physical dimensions of the target sound. This technique is common in data sonification tasks (see Bly and others, in this volume). Human beings are capable of monitoring multiple auditory streams. This well-known cocktail party effect[4] supports the implementation of an interactive interface for multiple active objects or processes. The auditory modality is ideal for alerting a user and conveying changes in an interactive environment.

Tactile interfaces offer other advantages. The resolution in a tactile interface is closer to the capabilities of visual interfaces. Theoretically, a tactile pad could present a raised image corresponding to a portion of, or the entire visual display. A tactile display is less intrusive. Like a visual display, the user can stop and restart interaction, quickly regaining the past context.

A tactile display also offers the greatest opportunities for representing pure graphical information that cannot be encoded symbolically into an auditory cue.

Mercator interfaces currently use only auditory output. As a whole, the benefits of the auditory modality are closer to the capabilities of visual interfaces. Multiple streams of symbolic auditory cues and synthesized speech provide the basis for a complex, interactive interface. Two other factors preclude the use of tactile interfaces as a standard part of the interface. First, it is important to note that a significant portion of people who are blind also suffer from diabetes which may cause a reduction in their sensitivity to tactile stimuli.[8] Second, the majority of tactile devices are still prohibitively expensive while audio hardware is becoming ubiquitous in the computing industry.

Given the complementary benefits of auditory and tactile displays, an ideal system would make use of both modalities. At some point in the future, current research in tactile displays for the blind by the Trace Research and Development Center[16] will be integrated with this work to form a multimodal accessible interface. Before that time, Mercator will have one tactile interface. Given that speech synthesizers are notoriously bad for listening to syntactically strict, encoded text such as programming languages, a braille terminal will provide redundant access to textual information.

MODELING THE VISUAL INTERFACE. A second major design question for building access systems for visually impaired users is deciding the degree to which the new system will mimic the existing visual interface. At one extreme the system can model every aspect of the visual interface. For example, in Berkeley System's Outspoken^{TM}, which provides access to the Macintosh, visually impaired users employ a mouse to search the Macintosh screen.[16] When the mouse moves over an interface object, Outspoken reads the label for the object. In these systems, visually impaired users must contend with several characteristics of graphical systems which may be undesirable in an auditory presentation, such as mouse navigation and occluded windows.

At the other extreme, access systems can provide a completely different interface which bears little to no resemblance to the existing visual interface. For example, a menu-based graphical interface could be transformed into an auditory command line interface.

Both approaches have advantages and disadvantages. The goal of the first approach is to ensure compatibility between different interfaces for the same application. This compatibility is necessary to support collaboration between sighted and nonsighted users. Yet if these systems are too visually based they often fail to model the inherent advantages of graphical user interfaces such as the ability to work with multiple sources of information simultaneously. Additionally interface techniques that may be intuitive in the visual interface may only confuse the user of an auditory interface.

The second approach attempts to produce auditory interfaces that are best suited to their medium.[5] This approach allows a new interface design which is based solely on the application's functionality, not on its graphical interface. Therefore the entire basis of the interface can be substantially different. Where graphical interfaces rely on the constant presentation of a vast information space, an auditory interface could rely on a serial presentation of a constrained, highly context-dependent information space.

One result of this analysis is that there are many features of graphical interfaces that *do not need* to be modeled in an auditory interface. Many of these features are artifacts of the relatively small two-dimensional display surfaces typically available to GUIs, and do not add richness to the interaction in an auditory domain. If the GUI screen is considered, in many regards, as a limitation, rather than something to be modeled exactly, then there are more degrees of freedom in creating an accessible auditory presentation.

Mercator interfaces are based on a compromise between the two approaches outlined above. Compatibility between visual and nonvisual interfaces is ensured by translating the interface at the level of the interface components. For example, if the visual interface presents menus, dialog boxes, and push buttons, then the corresponding auditory interface will also present menus, dialog boxes and push buttons. Only the presentation of the interface objects will vary.

By performing the translation at the level of interface objects, rather than at a pixel-by-pixel level (like Outspoken), the auditory interface can be unencumbered by some of the limitations of modeling the graphical interface exactly. In this approach, only the structural (or logical) features of the application interface are modeled, rather than its pixel representation on screen.

3.2 INTERFACE STRATEGIES

Addressing the two previous design considerations formed the basis of the design of Mercator interfaces. But what are the components or building blocks of these auditory interfaces and how are these components presented to the user? Moreover, how is interaction supported in the auditory interface? How does the user navigate the application interface and what feedback is presented to the user as the interface changes due to user or application events? The following sections outline possible answers to these questions. Finally, a prototype implementation of these interface strategies is described.

AUDITORY INTERFACE COMPONENTS. The previous section began to address the issue of how much of the visual interface should be carried over into the auditory presentation. The overall strategy is to maintain the object-based semantics between the two presentations. The hypothesis behind this decision is that the user's mental model of the application interface is also in terms of the objects in that interface and the relationships between those objects. This conceptual model of the application interface is the basis for the graphical and auditory presentations.

The derivation of the conceptual model begins with processing the graphical presentation. The graphical presentation contains the overall model of the application interface, but it also contains artifacts of its visual presentation. These artifacts are the result of presenting the interface in the graphical modality. Strategies such as conserving screen real estate and creating neat, spatially organized layouts can result in graphical objects that are not solely based on the conceptual model of the application interface.

In order to process the graphical presentation, it is necessary to understand how the low-level interface building blocks (or X widgets) are combined to form the higher level interface objects. Some of these building blocks will be visual artifacts of the graphical presentation.

One example is the graphical presentation of a pull-down menu. In X Windows, a number of widgets are required to create this one object. Many of these widgets do nothing but control the overall geometric layout of the menu. Only the conceptual object of the pull-down menu needs to be transferred to the auditory domain. The various building blocks which are used to create the menu do not. The user of a graphical interface is not aware of the nine or so widgets needed to present a menu; therefore, the user of an auditory interface does not need to know about them either. If the auditory interface was created by naively presenting each X widget, then the user would be overwhelmed (and confused) by numerous artifacts of the visual presentation. Even in the graphical interface, these widgets are *hidden* by the user mentally grouping aspects of the constant visual presentation into larger conceptual objects.

Another important question is what are the relationships between the objects in the conceptual model. Knowledge about these relationships will most likely be extremely limited by the information in the off-screen model. Since most interface toolkits do not explicitly model relationships between interface components, it is difficult to glean such information by watching a graphical application, or even by querying the application itself. But some analysis is possible. Hierarchical relationships can be partially determined by the X widget hierarchy. Some dynamic cause and effect relationships can be determined by watching the application behavior. If manipulating one interface object causes changes in another interface object, then these changes need to be conveyed in the auditory presentation as well. For example, if selecting a menu button causes a dialog box to appear, then this relationship must also be conveyed in the auditory interface. Not only must the user be informed that a new object has *appeared*, but the user must be allowed to easily navigate to the new object.

Once the objects in the conceptual model have been identified, the next step is to determine how to convey these objects in an auditory presentation. Like its graphical counterpart, additional information is added to the conceptual model in order to present it effectively in the auditory modality.

One possible example is a container object. Often a number of related objects are grouped together into a container, i.e., a collection of push buttons. In the graphical domain, this relationship can be simply conveyed by drawing a box around the related interface controls. The user generally does not even perceive the container as a separate object; it simply serves to add information about the related push button controls. Due to the serial nature of audio, it may be necessary to make the container more explicit. If the user simply navigated to a push button, the user would be unaware that nearby interface objects are related to the current object because the user would interact with each object one at a time. If the user first entered a container object, then the user would know that any objects within that container were related. For example, the user could be presented with the sound of opening a wooden (slightly creaky) box to convey a container. This sound could be followed by the auditory cue for the objects located in the container.

Another example is the use of scrollable objects. Scroll bars serve two purposes. First, they conserve limited screen real estate by presenting only a portion of a list. Second, they provide a mechanism for the user to quickly search the list. The first use of the scroll bar most likely does not need to be conveyed in the auditory interface since there is no limited display real estate. The second use does need to be transferred to the auditory domain. The question is how should the auditory scroll-bar work. What should be the granularity of the movement? Having the scroll bar move in terms of the information presented in the visual interface does not seem intuitive. It would be difficult to convey to the user how far the scroll bar had moved: down 4 items, down 10 items, or perhaps down 25 items. Another possibility is making the scroll-bar motion consistent across all scrollable items. Controls could allow the user to quickly conduct a binary search on a list. These controls, in addition to controls that allow the user to move to the top and bottom of the list and to step through the items one by one, would support efficient scanning which would be consistent from one scrollable object to another. Just the implementation of a generic auditory scroll bar is a challenging design problem.

3.3 CONVEYING INTERFACE OBJECTS

The next question to be addressed is how to convey the components of the interface model in the auditory interface. One important piece of information that needs to be presented is the identity of the interface component—"What is this object?" Numerous strategies for conveying objects in auditory interfaces have already been suggested by previous work. Possible strategies include using speech, pure tones, earcons (see Blattner et al., this volume), or auditory icons (see Gaver, this volume). For example, an auditory cue to convey a text-entry field could be:

- a synthesized voice saying "text-entry,"
- a pure tone, G-sharp (≈ 415.3047 Hz),
- a musical timbre such as a violin (a simple earcon), or
- the sound of an old-fashioned typewriter.

Each of these approaches has advantages and disadvantages. The speech message is unambiguous and reasonably efficient, but may be confused with other speech messages, i.e., reading the label on the field. A pure tone is easy to produce and takes minimal time to hear, but may be confused with other pure tones. Also the mapping of the note G-sharp to a text-entry field would be difficult to remember. Various musical timbres would also be easy to produce and would be easier to discriminate than pure tones but, again, the mapping from violin to text entry is hardly intuitive.

The author believes that auditory icons offer the most promise for producing discriminable, intuitive mappings. The concept of auditory icons is based on the premise that people describe sounds relative to the objects interacting to make the sounds (see Ballas, in this volume). Computers and other man-made devices make a variety of beeps, buzzes, and other artificial noises that otherwise we would never hear. Conversely, we typically hear things crumbling, things sloshing, things colliding—we hear the results of objects interacting with each other. When asked to describe a sound, we tend to describe it in terms of the objects that generated the sound such as a door slamming, stairs creaking, or glass breaking. In other words, we describe sounds in terms of their sources, not in classical terms such as pitch and duration.

What this realization means to an interface designer is that we can use sounds to remind the user of an object or concept from the user's everyday world. Gaver first introduced the term auditory icon with his interface called SonicFinder.[6] SonicFinder added auditory cues to the graphical Finder environment on the Macintosh. Dragging a file icon with the mouse sounded like dragging something along the floor. Dropping an icon into the trashcan sounded like dropping a large object into a metal trashcan.

In the previous example, the sound of an old-fashioned typewriter maps easily to a text-entry field. The user is reminded of typing or entering text.

In other cases, the mappings will not be as obvious. One implicit question in this work is evaluating whether the concept of auditory icons breaks down with objects that do not have a clear counterpart in the everyday world.

CONVEYING OBJECT ATTRIBUTES. Interface objects typically have many attributes that also need to be conveyed in the auditory interfaces. Text-based attributes can be presented via synthesized speech. For example, the auditory icon for a push button can be presented simultaneously with its text label. Other attributes can be presented by modifying the base auditory icon.

Auditory icons are not limited to simply reflecting categories of events and objects, but can be *parametrized* to reflect their relevant dimensions as well. For example, the auditory icon for a file can be manipulated to convey the size of the file. This approach is useful for conveying icon specific attributes such as the length of a menu. Often these attributes are not explicitly presented in the graphical interface but are simply part of the static graphical presentation. Although Gaver has successfully parametrized auditory icons in a number of interfaces, he notes the difficulty in parametrizing auditory icons due to the difficulty in controlling the sampled sounds along the relevant dimensions (see Gaver, in this volume).

Another interesting question is how to convey attributes which are common across different types of interface objects. For example, the concept of highlighting and greying-out interface objects is common across push buttons, generic icons, and windows. Ludwig, Pincever, and Cohen[11,10] have experimented with adding auditory cues to graphical window systems. Ludwig et al. suggest that various sound effects or *filtears* can be used to create a hierarchical distribution among nearly arbitrary (not pure sine waves) audio source. Filtears are essentially auditory filters that provide a systematic manipulation of an auditory cue without losing the identifiability of the original auditory cue. Filtears provide a means for conveying multiple levels of information via a single sound source. Some of the filtears suggested by Ludwig et al. are listed below. These filtears can be used to systematically manipulate a set a auditory icons in an application interface.

- Self-animation: massive frequency-dependent phase distortion which can make a source sound more lively by accentuating frequency variations in the source signal.
- Distortion: nonlinear wave-shaping (amplitude distortion) which produces a strained perhaps "excited" sound at the cost of intelligibility and listener fatigue.
- Thickening: a chorused "doubling" effect via pitch signals produces a "thicker" sound with possible costs in intelligibility.

- Peaking: linear band-emphasis filtering which is useful in amplifying speech (1 kHz range).
- Distancing: reverberation and echo cues used create a fuller more spacious sound but with reduced intelligibility.
- Muffling: linear, low-pass filtering can create impressions of confinement or distance.
- Thining: linear, high-pass filtering.

NAVIGATION. The navigation paradigm for auditory interfaces must support two main activities. First, it must allow the user to quickly "scan" the interface in the same way as sighted users visually scan a graphical interface. Second, it must allow the user to operate on the interface objects, push buttons, enter text, and so on.

To support both of these activities, the user must be able to quickly move through the interface in a structured manner. Standard mouse navigation is unsuitable since the granularity of the movement is in terms of graphic pixels. Auditory navigation should have a much larger granularity where each movement positions the user at a different auditory interface object. To support navigation from one auditory interface object to another,

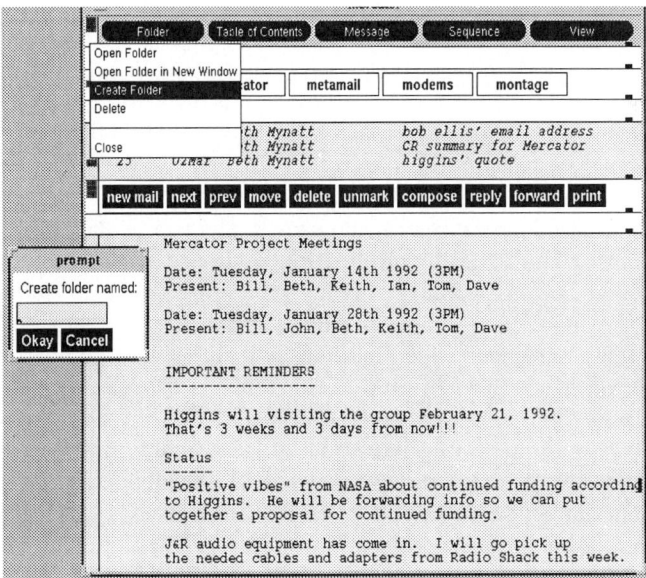

FIGURE 3   A sample X Application (XMH).

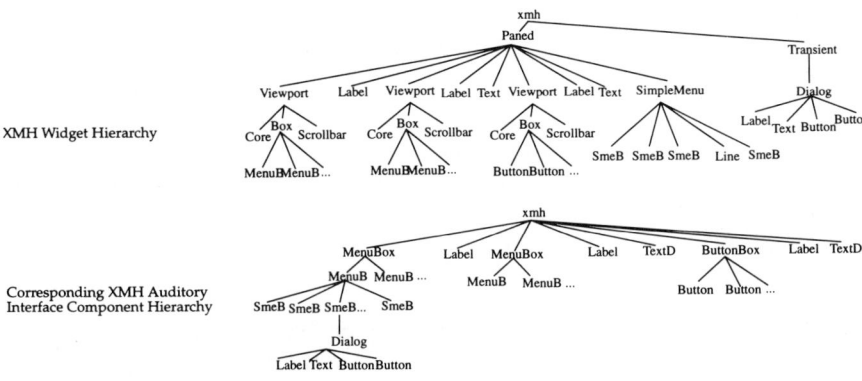

FIGURE 4 (a) XMH Widget Hierarchy. (b) Corresponding XMH Auditory Interface Component Hierarchy.

the user interface is mapped into a tree structure which breaks the user interface down into smaller and smaller auditory objects. This tree structure is derived from the conceptual model of the application interface where that model is partially determined by the X widget hierarchy. The tree structure represents hierarchical relationships as well as dynamic relationships. In Figure 3 is shown a screen shot of the graphical interface for XMH, an X-based mail application. Figures 4(a) and 4(b) show a portion of the XMH widget hierarchy and the corresponding interface tree structure, respectively.

To navigate the user interface, the user simply traverses the interface tree structure. The numeric keypad is used to control navigation. Small jumps in the tree structure are controlled with the arrow keys. Other keys can be mapped to make large jumps in the tree structure. For example, one key on the numeric keypad moves the user to the top of the tree structure. It is worth noting that existing application keyboard shortcuts should work within this structure as well.

USER FEEDBACK. The remaining issue is providing feedback to the user in the auditory interface. This feedback will be closely coupled with the attributes of interface objects (and their identity) as well as with the navigation strategy. The primary purpose of feedback is to acknowledge user input. Since user input manipulates the objects in the interface as well as changing the user's position in the interface, the feedback for these actions will be dependent on object and navigation information.

For example, if the user "selects" an object, such as pushing a button, then that object may become highlighted (graphically and auditorially). The feedback for this action should convey the attribute change in the object. Since the navigation scheme is based on a tree structure, feedback for navigation controls could convey the location or depth within the tree structure.

Some feedback to the user will be based on other aspects of the state of the application interface. If the interface is busy (often graphically conveyed with a clock cursor), then the user may hear a clock sound (or perhaps the Jeopardy song). If the state changes without user action (such as a message appearing on the screen), then an auditory alert might notify the user of the state change.

PROTOTYPE SPECIFICS. A prototype interface builder that demonstrated some of these strategies was constructed. The primary demonstration interface was for the application XMH, a text-based e-mail application. This application is relatively complex with a number of different interface controls and top-level windows.

Auditory icons were used to represent the type of an interface object as well as interface events. As an example, some of the auditory icons included:

- touching a window sounds like tapping on a glass pane,
- searching through a menu creates a series of shutter sounds,
- a variety of push button sounds are used for radio buttons, toggle buttons, and generic push buttons,
- touching a text field sound like an old-fashioned typewriter, and
- going out of bounds in the interface (an error condition) sounds like a basketball bouncing against a wall.

Only one type of filtear was used. A muffling filtear (low-pass) was used to denote that a certain object was "greyed-out" or unavailable. This filtear was applicable for the menu buttons and all push buttons.

TABLE 1 Summary of Navigation Controls

Key	Action
8/up arrow	Move to parent
2/down arrow	Move to first child
6/right arrow	Move to right sibling
4/left arrow	Move to left sibling
0/Ins	Move to top of tree
5	Context-sensitive help
Enter	Activate selection action
=/*−	Can be mapped to other movements

Tree-structured navigation was used. The user entered navigation commands via the numeric keypad. The navigation controls available are summarized in Table 1.

Little additional feedback was presented to the user. As the user navigated to the object, the user would hear the auditory icon for that object (perhaps modified by the muffling filtear) and a synthesized speech message. In most cases the speech message was the label on the object. When no label was available, the type of the object was also voiced. This condition (full speech plus auditory icons) would be appropriate for a new user. As the user became more experienced, the level of speech output should be decreased.

3.4 USER STUDY

A simple user study was conducted to evaluate one possible presentation of Mercator interfaces. Both sighted and nonsighted participants were used in the study. Sighted users were able to see the corresponding visual presentation of the application interface although they only interacted with the interface via Mercator controls The study consisted of the following four parts.

AUDITORY ICON IDENTIFICATION TEST. Participants were asked to play seven sounds and select what they thought the sound represented. The possible choices for each sound were:

- Menu Button: A button that you would select to make a menu appear.
- Command Button: A simple push button.
- Box (container): An object which contains other objects such as a box of push buttons.
- Text Area: An area where you would read and/or enter text.
- Label: A simple label (read-only text).
- Navigation Error: The sound made when a user attempts to navigate out of the bounds of the available area.
- Window Mapped: The sound made when a window appears (pops up) on the screen.

Participants were then asked to play two related sounds that represented the same interface object, and choose what information was conveyed by modifying the first sound to produce the second sound. The second sound was the result of low-pass filtering, the first sound. The possible choices were:

- the object is already selected (highlighted),
- the object is unavailable (greyed out), or
- the object has a sub-object (i.e., a cascading menu).

Auditory Presentation of Graphical User Interfaces

MERCATOR NAVIGATION TASK. Participants were asked to experiment navigating a Mercator interface for the X application XMH, a mail reading tool. Participants heard combinations of speech and nonspeech audio to convey information about different interface objects. The nonspeech audio cues were the same as those presented in the Auditory Icon Identification Test.

WRITTEN FEEDBACK. Participants were asked to record their impressions of the navigation strategy used by Mercator, the auditory icons used by Mercator, and any other comments.

RETAKE AUDITORY ICON IDENTIFICATION TEST. Participants were asked to retake the same test, they had taken earlier.

3.5 TEST RESULTS, FEEDBACK, AND FURTHER SPECULATION

The number of participants precludes the use of formal statistical analysis of the results. The following section details observations from test results, written comments, watching the participants use Mercator, and later discussions with the participants.

CONVEYING INTERFACE OBJECTS AND OBJECT ATTRIBUTES Although participants fared better when retaking the auditory icon identification test, the overall test results indicated that the goal of designing intuitive auditory icons was not satisfied in the prototype interface.

There are many factors that may affect the usability of auditory icons. These factors are:

- Identifiability
  The user must be able to recognize the sound. The ecological frequency of the sound (how common is the sound) versus the relative uniqueness of the sound (how close is this sound to other known sounds) help determine the overall identifiability of the sound.
- Conceptual Mapping
  How well docs the sound map to the aspect of the user interface represented by the auditory icon? Is the mapping intuitive? Can the mapping be easily learned?
- Physical Parameters
  The physical parameters of the sound such as duration and bandwidth may affect the usability of the sound. If the auditory icon is too long, where length is defined as the amount of time that the user must listen to the sound to identify the auditory icon, then the auditory icon may adversely affect the performance of the interface. The bandwidth of the sound may affect how the auditory icon can be manipulated to

convey more information in the interface (see the following discussion on filtears).
- Sound Quality
  The overall sound quality of the auditory icon may affect the identifiability of the sound as well as the user's subjective impression of the interface.
- User Preference
  How the user responds emotionally to the auditory icon is also important. Does the user like the sound? Is the sound harsh or too cute? Does the sound remind the user of unpleasant things?

The most significant problem was the users' inability to identify the auditory cues. This problem resulted from short, synthesized sounds being used. At the time of this experiment, the audio driver for Mercator could not quickly stop a sound once it began playing. To support quick and coherent interaction, the auditory cues had to be extremely short. Most sampled sounds did not meet this criterion. In order to make shorter auditory cues, the base sampled sounds were highly edited which diminished the overall recognizability of the original sound.

Although Mercator can now stop a sound while it is playing, an important design criteria is that an experienced user be able to identify the sound without listening to the entire sound. This assessment must be conducted against other sounds also used in the interface. The auditory icons used in an interface must also be evaluated as a cohesive set. The auditory icons must be relatively unique from each other. The auditory icons should not sound too similar and their conceptual mappings should not counterintuitive. The physical parameters of the auditory icons such as duration, intensity, and sound quality should be roughly equal as the user might attempt to attribute some meaning to any perceivable differences.

If the test subject was able to accurately identify the auditory cue, then the conceptual mappings seemed to be somewhat intuitive or at least easily learned. One issue with evaluating interfaces with auditory icons is whether the test measurements should include initial performance results as well as the time needed to learn the auditory cues.

Because the auditory cues were fairly short, the auditory filtears were difficult to perceive. The individual cues had a large bandwidth which supported the generic use of filtears. In addition to the factors listed above for auditory icons, there are additional factors that affect the usability of filtears. First, does the filtear distort the base sound so that the user can no longer recognize the base auditory icon? Second, does the filtear provide simply a binary mapping or can a range of values be expressed with the filtear?

Most participants judged the 8 kHz sound quality as harsh and the low-frequency auditory icons too "industrial sounding." Although many auditory interface designers appear to believe that high-quality sound (i.e., CD quality) should not be a requirement for auditory interfaces, it was interesting to note that some participants relied on previous experience with other poor sound quality devices, such as video games, to provide clues to identifying the sound.

NAVIGATION. Overall impressions of the navigation strategy were favorable. Participants rarely tried to move outside of the interface structure. Some confusion for sighted participants was caused by contradictions in navigation controls and the visual presentation. For example, moving across (right) the interface tree often corresponded to the pointer moving down the visual presentation. Nonsighted users found the navigation scheme easy to use. Both sets of users requested navigation controls that would provide larger jumps in the interface tree such as moving to the last object in a particular direction as well as the ability to set markers as navigation points. Both of these features were not implemented in the version of Mercator used in this study.

4. FUTURE DIRECTIONS

The first task, which is already in progress, is to create more demonstration interfaces with new sets of sounds and test these interfaces on new users. Since a major problem with the first prototype was the actual identification of an auditory cue, simple experiments which test the identification and possible conceptual mappings of numerous sampled sounds will be conducted independent of any demonstration interface. Like the auditory icon identification test, subjects will be asked to identify a sound and choose a possible mapping from a list of choices. Results from these experiments should indicate whether there are any natural mappings between generic interface objects and auditory icons.

The next step will be to expand on the interface strategies discussed earlier. For example, parametrizing auditory icons could be used as a technique to convey hidden attributes of interface objects. Hidden attributes can be defined as information conveyed indirectly in the visual interface. For example, highlighting is directly conveyed to the user while the size and organization (alphabetical, frequency of use) of a list is indirectly conveyed to the user. While filtears can be used to convey explicit attributes, individual auditory icons can be parametrized to convey these hidden attributes. Likewise, a great deal of design and experimentation is needed to identify general auditory filtears which can be used in various auditory interfaces.

New forms of auditory feedback will be needed to convey different events in the interface. These auditory messages will build on previous work in auditory icons and earcons, as well as research in auditory displays of multivariate data (see Gaver, Blattner, Bly and Kramer, all in this volume). One area of difficulty, common to the evaluation of any complex interface, will be designing experiments that correctly assess the use of these different techniques in a complex auditory interface. Designing and evaluating techniques that can be used across various types of auditory interfaces is the end goal of this research.

4.1 NEW INTERFACE STRATEGIES

In addition to the interface strategies already discussed, there are other possible ways to increase the power of the auditory interfaces provided by Mercator.

SPATIAL ORGANIZATION VIA SPATIALIZED SOUND. One of the cognitive benefits of graphical desktop environments is the ability to organize objects in a two-dimensional space. Related work at Georgia Tech is investigating uses of spatialized sound to create a three-dimensional auditory environment. Some preliminary design work on a system called Audio Rooms—a three-dimensional auditory presentation of the Rooms metaphor introduced by researchers at PARC[7,13]—is aimed at creating an intuitive "desktop" computing environment for nonsighted users. Due to the costs of commercial spatialized sound systems such as the Convolvotron (see Wenzel, in this volume), a low-cost implementation of spatialized sound based on the algorithms implemented in the Convolvotron was built. This system, which runs on a standard DSP 56000, provides two-dimensional auditory cues and is augmented by other auditory filters to create distance cues as well.[2]

PERIPHERY INFORMATION VIA AUDITORY CONTEXTS. Graphical user interfaces allow users to not only see the current object of interest, but permit the users to also see other objects that are related to the current object. This ability allows the users to employ peripheral information to understand the object of focus. In contrast, most auditory interfaces play only one sound at a time. A more powerful scheme would be to present auditory cues for peripheral information as well as auditory cues for the current focus. These secondary cues could be presented simultaneously with the primary cues, or presented spatially around the primary cue. In either case, the user would be presented with an auditory context which would represent the current *area* of focus, not just the current object of focus.

POSSIBLE USES OF TACTILE OUTPUT. The strategy of translating graphical interfaces into auditory interfaces begins to break down as the visual interfaces become more like pictures and use fewer standard representations for interface objects. The extreme case is a picture such as a freehand drawing or a scanned image. Tactile pads which allow users to feel a picture provide some possibilities. Combinations of auditory and tactile output might allow a user to *hear* different colors, or could be used to inform the user of changes in the tactile presentation.

ACKNOWLEDGMENTS

The Mercator project is a joint effort by the Georgia Tech Multimedia Computing Group (a part of the Graphics, Visualization, and Usability Center) and the Center for Rehabilitation Technology. This work has been sponsored by the NASA Marshall Space Flight Center (Research Grant NAG8-194) and Sun Microsystems Laboratories.

A number of people have contributed a great deal of time and energy to the development of the Mercator; without their hard work this system would not have been possible. I would like to thank Gerry Higgins (NASA), Earl Johnson (Sun), Keith Edwards, Dave Burgess, Tom Rodriguez, Ian Smith, and John Goldthwaite (Georgia Tech).

REFERENCES

1. Boyd, L. H., W. L. Boyd, and G. C. Vanderheiden. "The Graphical User Interface: Crisis, Danger and Opportunity." *J. Visual Impair. & Blind.* (December 1990): 496–502.
2. Burgess, David. "Low Cost Sound Spatialization." In *UIST '92: The Fifth Annual Symposium on User Interface Software and Technology Conference Proceedings*, held November 15–18, 1992. Monterey, CA: ACM Press, 1992.
3. Buxton, William. "Human Interface Design and the Handicapped User." In *CHI'86 Conference Proceedings*, edited by M. Mantei and P. Orbeton, 291–297, 1986.
4. Cherry, E.C. "Some Experiments on the Recognition of Speech with One and Two Ears." *J. Acous. Soc. Am.* **22** (1953): 61–62.
5. Edwards, Alistair D. N. "Modeling Blind Users' Interactions with an Auditory Computer Interface." *Intl. J. Man-Mach. Stud.* **30** (1989): 575–589.
6. Gaver, W. W. "The SonicFinder: An Interface that Uses Auditory Icons." *Hum.-Comp. Inter.* **4** (1989): 67–94.
7. Henderson, D. Austin, Jr., and Stuart K. Card. "Rooms: The Use of Multiple Virtual Workspaces to Reduce Space Contention in a Window-Based Graphical User Interface." *ACM Trans. Graph.* (July 1986): 211–243.
8. HumanWare, Artic Technologies, ADHOC, and The Reader Project. "Making Good Decisions on Technology: Access Solutions for Blindness and Low Vision." Paper presented at the Closing the Gap Conference, October 1990 (industry experts panel discussion).
9. Ladner, Richard E. "Public Law 99-506, Section 508, Electronic Equipment Accessibility for Disabled Workers." In *CHI'88 Conference Proceedings*, edited E. Soloway, D. Frye, and S. B. Sheppard, 219–222, 1988.
10. Ludwig, Lester F., and Michael Cohen. "Multi-Dimensional Audio Window Management." *Intl. J. Man-Mach. Stud.* **34(3)** (1991): 319–336.
11. Ludwig, Lester F., Natalio Pincever, and Michael Cohen. "Extending the Notion of a Window System to Audio." *Computer* (August 1990): 66–72.
12. Mountford, S. Joy, and William W. Gaver. "Talking and Listening to Computers." In *The Art of Human-Computer Interface Design*, edited by Brenda Laurel, 319–334. Reading, MA: Addison-Wesley, 1990.
13. Mynatt, Elizabeth, and Keith Edwards. "New Metaphors for Nonvisual Interfaces." In *Extraordinary Human-Computer Interaction*,

editedy by A. D. N. Edwards. Cambridge, MA: Cambridge University Press, 1991.
14. Mynatt, Elizabeth, and W. Keith Edwards. "The Mercator Environment: A Nonvisual Interface to X Windows and Unix Workstations." Tech Report GITGVU-92-05, Graphics, Visualization & Usability Center, Georgia Institute of Technology, February 1992.
15. Mynatt, Elizabeth, and W. Keith Edwards. "Mapping GUIs to Auditory Interfaces." In *UIST '92: The Fifth Annual Symposium on User Interface Software and Technology Conference Proceedings*, held November 15–18, 1992. Monterey, CA: ACM Press, 1992.
16. Vanderheiden, G. C. "Nonvisual Alternative Display Techniques for Output from Graphics-Based Computers." *J. Visual Impair. & Blind.* **83** (1989): 383–390.
17. York, Bryant W., ed. *Final Report of the Boston University Workshop on Computers and Persons with Disabilities*, Boston University, November 1, 1989.

APPENDIX I:
Comments on ICAD'92

SARA BLY

ICAD '92 was an exciting experience for me. Many other people are now tackling problems in sonification, and it was particularly meaningful for me to participate in this community of researchers. At the same time, it was discouraging that we've not made greater advances in the area of sonification. While ten years ago I was able to get away with showing the *potential* for audio representations to convey useful information about multivariate data, today we need to be able to make definitive statements about such mappings. In my mind, there are two gaping holes. One is the lack of any real systematic attack on the space of data-to-sound parameter mapping. A second hole is the lack of application to actual multivariate data problem solving. Although Lunney et al.'s work has always used real data, the focus was on representing it for visually impaired scientists rather than on exploration. Thus, the work by Fitch and Kramer in this volume was most exciting to me as they are placing sonification into a real problem setting with productive results.

ROGER POWELL

For me, ICAD '92 provided a stimulating forum for exchange of ideas in the relatively new field of sonification. Since attendees represented divergent interests from both the academic and industrial worlds, ICAD '92 was productive in showing how we approach the use of sound to convey data from theoretical as well as practical viewpoints.

My job at a graphics-oriented computer company is to assist with the incorporation of audio modality into the user interface. With the wealth of talent and research experience gathered at ICAD, I was able to come back to work with new enthusiasm and ideas. In addition, the effectiveness of the conference in recognizing sonification as an integral part of "perceptual processing" has led to greater support for my project, as I'm sure it has for other attendees.

The conference presented a rich canvas of problems and solutions. I was struck mostly by the similarity of requirements for audio research tools and overlap of design in meeting those needs. We all want fast, powerful yet affordable signal-processing systems for doing this exciting work. We'd all like to be able to share methods and data to accelerate the process. I feel ICAD '92 was a watershed event in identifying the best pathways for us to follow to accomplish great things with auditory display systems.

BILL BUXTON

I found the time worthwhile a lot of the stuff interesting. There was work there that I was not familiar with. The most pleasant surprise was Johnathen Cohen, whose work was a joy to hear.

Generally, I found the quality of work very uneven. Secondly, I have to say that I continue to be disappointed at the lack of science and experimental/human validation of so much of the work. As a community we are going out in an advocate position, promoting this new area of work—an area in which there is a lot of skepticism within the community—and we have not done our homework. Much of what is advocated is very shaky and unsupported, and we have not been very critical of our own work, again, from a methodological and/or scientific perspective. If we want to be taken seriously, this has to change.

Hopefully that which is supported by solid validation and quality design will push out that which is untested, or pseudo-tested. My hopes are that the next time around, presentations are strongly and critically refereed before being accepted, and that there be fewer presentations, and more time for those active in the field to interact.

I started on the positive, and want to end that way: workshops like this are an excellent way to foster a community, and to provide the catalyst for growth and development. My main hopes for the next meeting are that the ingredients be such that the magnitude of the catalytic effect be even stronger.

TOM RETTIG

The arrival of ICAD comes at an interesting time in the personal computer industry. Audio is becoming an increasingly integral component of the computing experience. There is currently an explosion of support both in terms of applications development and market acceptance. Additionally, tremendous resources are being applied to pioneer new digital and interactive media, most of which include sound. ICAD '92 presented an exciting opportunity to interact with people from disparate backgrounds who are wrestling with some of the difficult but promising areas of auditory display.

I was surprised to find that some of the work to date involves individuals or groups that have little background in sound design. People trained at creating sound and music often have a special understanding of the effect on the listener. While much of our sound work at Broderbund simply involves design, audio professionals have had a profound impact on our understanding of the use of sound in the interface. Musicians, sound editors, recording engineers, and producers have been trained to communicate using sound. It is my belief that we can facilitate progress in this field by increasing the involvement of people trained in the craft of making music and other sound. They have a perspective that is important to this field. I look forward to seeing more research which takes advantage of this expertise. I also look forward to continued involvement with this new community, and the opportunity to meet again at the next ICAD.

NAT DURLACH, DANIEL LING, AND GEORGE ZWIG

1. INTRODUCTION

Although each of us has background and interests relevant to the study of auditory displays, most of our work has focused on matters other than "sonification." Thus, the following comments, like those we made at the conference itself, should be viewed as comments by outsiders (kibitzers).

The meaningful use of sound has lagged far behind the development of sophisticated visual imagery in computer user interfaces. The enormous popularity and rich history of SIGGRAPH, and lack of any similar conference for sound, underscores this huge difference.

We believe that this first ICAD had considerable significance in that it was one of the first general meetings dedicated exclusively to the use of sound in human-computer interfaces. Furthermore, the attendees were smart, enthusiastic, creative, and friendly, and the group was sufficiently small to permit extensive one-on-one communication. We felt privileged to be included among the attendees.

Generally speaking, we see three major problem areas in this field: Selection of Display Schemes (Section 2), Real-Time Generation of the Acoustic Waveforms (Section 3), and Search for Useful Applications (Section 4). Accordingly, our comments on ICAD '92, as well as further thoughts about auditory displays (some of which have been stimulated by our attendance at ICAD), will be presented under these headings. As in the conference itself, our attention is focused primarily on the first area.

As will become immediately obvious when reading these comments, we have made no attempt to discuss the individual papers presented at the conference; we have merely attempted to express some general reactions to the conference as a whole. Furthermore, our comments tend to be critical rather than laudatory. Although, as indicated above, our overall reaction to the conference was positive, we feel that we can be most helpful by pointing out areas and issues for which significantly increased attention would, in our opinion, be highly beneficial.

2. SELECTION OF DISPLAY SCHEMES

By "display scheme" we refer to the selection of acoustic waveforms to be used in the display as well as the selection of the mapping from the information to be displayed onto these waveforms. Much of the material presented at ICAD '92 concerned these issues in one form or another. The prevalence of such terms as "sonification," "audification," "auralization," "icons," "earcons," etc. in the various ICAD papers gives testimony to current efforts to classify different kinds of display schemes.

The material covered at the meeting and the profile of background and interests of the participants showed considerable strength in many areas relevant to the selection of display schemes. However, essentially no attention was given to the use of sound in the form of speech. Admittedly, the culture of speech communication experts is rather separate from the cultures of media experts or musicians (or even with most individuals working on auditory perception). This does not imply, however, that such exclusion is beneficial. On the contrary, the interaction and integration of these disparate cultures is likely to prove valuable. In our opinion, if this exclusion (or oversight) were to become a permanent feature of this series of meetings, it would be unfortunate. Not only can speech displays play a major role in human-computer interaction and effectively complement the use of nonspeech sounds, but the physical/computational modeling of speech generation by the human articulatory mechanism is the most advanced example of physical/computational modeling that currently exists in the domain of acoustics (e.g., see Fant,[6] Klatt,[9] and Stevens[18]).

In subsequent discussions with the organizers of the conference, we were informed that this omission was the result of a conscious decision to focus the first meeting on nonspeech audio because of the relative lack of organized attention being given to nonspeech audio. Apparently, the organizers are open to including representatives of the speech community at some of the subsequent meetings.

Another topic that received relatively limited attention concerns the integration of auditory and visual displays. We agree that, at this stage of development, it probably makes sense to explore the relative effectiveness of various types of auditory displays in isolation and without consideration of how this effectiveness might be altered when combined with different types of visual displays. Not only is this important for a full analytic understanding of display effectiveness, but there will always be special applications for which visual displays are meaningless or inappropriate (e.g., when the user is blind). Nevertheless, it will be important in future meetings to devote greater attention to this issue.

In considering combined displays, it might be valuable to exploit the large amount of experience that has already been acquired in the area of movie soundtracks. Even when movies were silent, they were not viewed in silence but accompanied by music. It is well known in the industry that the quality of the soundtrack is crucial and that good sound can even make up for weak visuals. (In this connection, it is interesting to note that essentially all productions shown at the SIGGRAPH Theater include a soundtrack). Studies at the MIT Media Lab and Stanford have also shown that improvement in sound fidelity is often perceived by viewers as improvement in visual quality (e.g., see Newman et al.[10]).

An examination of movie soundtracks provides an interesting basis for reviewing many of the topics discussed at the conference. For example, consider the uses of sound in movie soundtracks and the uses described at ICAD.

Both cases may include the use of everyday environmental sounds to signal events. Similarly, both cases may include the use of relatively abstract, musically based, sound signals to convey event information. In both movies and computer interfaces these abstract-sound signals (e.g., a suspenseful musical phrase or an earcon) must be established by convention and learned by the audience. However, whereas in the ICAD case the emphasis is on conveying complex information by means of abstract sound streams with relatively little affect, in movie soundtracks the emphasis is on conveying relatively simple information by means of musical themes with a great deal of affect (e.g., consider the "shark-is-near" theme in the movie *Jaws*). Also, of course, in most movie soundtracks, speech sounds are a major factor with respect to both (symbolic) information content and affect.

A further area that received relatively little attention concerns interaction between the auditory display and the human user. In essentially all cases, control of the display by the human was envisioned, either explicitly or implicitly, as occurring only through traditional human-computer interfaces (e.g., keyboard, mouse, etc.). Also, the elements and/or characteristics of the display under human control were relatively restricted. No attention was given to the use of interfaces that would involve the user's sensorimotor system in a manner that accelerates the learning of new auditory codes or that facilitates the communication of affect (as when an analog device or virtual instrument is used in the generation of computer synthesized music).

An additional point to be noted concerns the "spatialization" of sound. Although almost all papers on spatialization use the term three-dimensional to describe the work, almost all such work is actually concerned only with two-dimensional, azimuth and elevation. The inclusion of distance, which may prove to be of substantial value, is almost always ignored. (Indeed, most spatialization systems do not even reliably produce sound images that appear to be located outside the head.) The fact that distance tends to be judged rather poorly in the real world (unless source intensity is held fixed) is beside the point. There is no reason to restrict the coding of distance (as well as the coding of elevation and azimuth) to localization cues that are identical to those used in the real world. Even if radically changed codings are ruled out because of the increased learning time required to use them effectively, codes that accentuate or exaggerate normal cues may prove useful. (For a discussion of various auditory spatialization issues, see Wenzel[19] and Durlach[3,4,5]).

Another point to be noted concerns the topic of "auditory scene analysis" and the construction of displays in such a way that individualized "sound streams" are perceived (e.g., see Bregman[1] and Yost[20]). Although auditory displays that separate perceptually into a number of distinct streams may be useful with respect to certain attentional or masking/interference factors, they may be harmful with respect to certain other factors. In particular, it is important to note that the same perceptual mechanisms that cause the auditory scene to resolve itself into separate streams are likely to make comparisons across these streams exceptionally difficult. For example, temporal resolution is likely to be degraded if the temporal comparison involves elements from different streams. Finally, it is important to stress the importance of developing methods to evaluate the effectiveness of various auditory display schemes on a comparative basis (both among themselves and in comparison to vision-only displays). As stated at the meeting, it is important that the field advance beyond the "show-and-tell" stage. If it does not, the contributions provided by this field will be severely limited. An example that we believe to be of particular importance in this area concerns the extent to which the sounds and their coding are "natural" in terms of previous experience and, therefore, require little or no learning to reach asymptotic performance. Clearly, the results of comparing different display schemes (either within or across modalities) that differ in their "naturalness" will depend strongly on the amount of exposure and practice with the displays. In order to advance beyond the "show-and tell" phase of auditory display work, it is necessary to develop procedures for measuring a display's "degree of naturalness" and for separating out effects associated with this variable from those associated with other variables that influence performance with the display. Until such procedures are developed, the only meaningful way to compare different displays will be by means of laboriously obtained task-performance learning curves.

3. REAL-TIME GENERATION OF THE ACOUSTIC WAVEFORMS

The second problem area concerns the generation of the required acoustic waveforms. Two basic methods to effect such generation for real-world acoustic events are based on (1) samples, in which the sounds of real-world events are recorded and then reproduced (possibly after various transformations are applied), and (2) physical modeling, in which the sounds are synthesized by the computer according to physical models (which may be greatly simplified) of the acoustic sources. Both methods are being used extensively for both speech and music. A third method (3) commonly used in the computer music area, involves direct simulation of acoustic waveforms by systematic waveform generation techniques. For environmental everyday sounds, the sampling method has been used the most frequently; however, as illustrated by work presented at the conference, the other two methods are now beginning to be applied to this case, too.

At present, there appear to be no general guidelines to help one determine which of the various methods will be the most cost effective for a given application. Clearly, however, the emphasis on physical modeling (and perhaps the third of the above-mentioned methods as well) is likely to increase as the work on virtual environments becomes more advanced and more attention is given to multimodel immersive environments and user interaction with these environments. Note also that the increased use of physical modeling will accelerate the need for research that is truly interdisciplinary. Not only will increased collaboration with physical acousticians be required (to construct the physically based models), but also increased collaboration with psychoacousticians (to help determine which simplifications of the physical models will lead to sounds that are adequate perceptually).

Finally, and to a certain extent independent of the advantages and disadvantages of a particular sound-generation method when considered in isolation, the selection of an appropriate sound generation method (as well as many other features of the virtual environment system) will be strongly influenced by the need to dynamically generate integrated and synchronized audio/visual images.

4. SEARCH FOR USEFUL APPLICATIONS

The third major problem area concerns the search for applications and determination of effective "application criteria." Although issues relevant to this problem area were considered, at least indirectly, at ICAD (the applications reported upon were presumably chosen for study because the investigators thought they would be successful), there was relatively little explicit discussion of the criteria that should be used for selecting applications. A brief summary of possible criteria is given in the following list.

A. Eyes and hands cannot be used (either because they are busy with other tasks or because the user is physically handicapped).

B. Performance would be significantly enhanced by the inclusion of spoken speech in the display.

C. The function in question involves warning the user, directing the user's attention, monitoring background activities, or judging temporal relationships.

D. There is a need to present information in a realistic manner and the situation portrayed would normally include acoustic sources.

E. There is a need to personalize the computer and/or to introduce affect.

F. It is easy to employ sounds that can be heard and recognized as gestalts and connected with significant events.

G. The population of users towards which the application is directed clearly both need and want help, and there is an opportunity to work closely with these potential users.

With respect to the last item, it might prove useful in some future ICADs to include specific user populations as well as designers of auditory displays. Although many of the displays considered in this first ICAD were directed towards aiding scientists to comprehend and digest complex data structures, the population of scientists who need such help was not well represented at the conference.

ACKNOWLEDGMENTS

Support for this work was provided, in part, by the Air Force Office of Scientific Research (Grant AFOSR-90-0200).
[Note: for references, see the combined list at the end of the appendix.]

DANIEL STEINBERG
1. AUDITORY INTERFACES: DREAM OR NIGHTMARE?

Talking and beeping computers have existed for years in science fiction literature and cinema, while our desktop computers remain, for the most part, silent.[1] Is this because they have nothing meaningful to say to us, or is it because we place such high value on our acoustic space that we will not tolerate sonic intrusions?

Desktop computers currently take up significant space in the office. They occupy a large volume on and above the desk. Their displays form a visual focus (many people spend more time looking at their square foot of CRT glass than at all the other objects in their office combined). Computers require an awkward combination of gross and minute gestures (move the hand from the keyboard to the mouse, position the cursor over the little slider bar). They engage many of our cognitive abilities, from text comprehension to the interpretation of graphic metaphors. Moreover, they are often involved in the performance of nearly every task in the workplace.

The human nervous system is finely tuned to derive enormous amounts of information from subtle acoustic cues. Our ability to recognize aperiodic air pressure variations as spoken language is nothing short of astounding. Considering that we already devote so much of our sensory and psychic attention to the computer, it seems surprising that we are not more anxious to integrate auditory modes into our computing experience.

Researchers and developers of acoustic interfaces seem to be searching for The Compelling Application that will somehow make an unarguable case for acoustic display. I suggest that there is no Compelling Application, as such. Sound in motion pictures was not justified by a particular

---

[1] I say "for the most part" because many desktop systems are quite noisy in their own way, with fans and disk drives humming along. A single computer can add anywhere from 2 to 10 dB of background noise in an office.

movie; it was initially accepted because it was a natural extension to the medium, and it became a basic cinematic requirement because its pervasive deployment raised the public's expectations. In a similar way, we must build sonic interfaces that extend the existing paradigms in natural ways, and we must ensure that, together, they fit into a coherent and consistent model for multisensory computing.

2. WHAT'S SO SPECIAL ABOUT AUDIO?

We tend to be so visually oriented that it is easy to forget the important ways in which hearing differs from seeing. The most obvious difference is that we cannot easily shut out sounds, whereas we can close our eyes or divert our gaze. Since we do not have this self-control within our ears, we tend to exert more control over our aural environment and are less tolerant of auditory "clash" and "noise" than their visual counterparts. We scan visual images in a somewhat nested fashion, zeroing in on detail after forming an overall impression, whereas sound is processed serially. This means that the order of presentation of auditory information is very important, as is the case with movies and video. However, we can recognize, and even attend to, simultaneous audio sources, while we multiplex more slowly in the visual domain.

In some ways our hearing is more sensitive than vision. We tolerate jitters and low resolution in video much more easily than audio delays and breakup. We can easily pick particular sounds out of a noisy background, while it is easy to overlook details in visual displays. On the other hand, our ears tend to fatigue more quickly and our aural attention span is generally shorter.[2] We also have less patience for repetition, especially with respect to spoken messages, which accounts for the failure of a number of commercial attempts to use auditory feedback.[3] Interestingly, we acclimate rather easily to repetitive background sounds, particularly if they are rhythmic or instantly recognizable as familiar sounds in the environment.

It is also important to acknowledge the significant neurophysiological impact on the way we perceive and respond to acoustic events. Despite a great deal of research in psychoacoustics, we do not really understand why our emotional state can be manipulated by music, though composers and musicians know instinctively how to do it. The enormous power of rhythm, though obvious to any rock drummer, has been largely overlooked

---

[2] This may be somewhat attributed to environment and socialization. People used to listen attentively to the radio for long periods, though perhaps not as long as they watch television these days.

[3] For example, consumers detested the talking automobile ("Your door is ajar"). A more interesting use of synthesized speech might have been to say "Your left rear brake light is out" only once each time the engine was started.

in the field of acoustic display. Our brain waves are entrained by rhythmic excitation, principally through auditory stimulus. This suggests that we respond more positively to frequent acoustic events if they are constrained to predictable time slots. We can also detect tempo changes and arrhythmic events in an otherwise constant stream fairly easily, while this is more difficult, and requires greater attention, in visual modes.

3. VALUE AND QUALITY

Many of the characteristics of the human auditory system suggest that we consider our acoustic environment to be more personal, intimate, and, therefore, more sacred than our visual environment. If this is true, then it follows that our auditory displays are measured against somewhat higher standards of quality and aesthetics than have traditionally been applied to visual displays.

In fact, there are at least two distinct criteria that are useful to consider: information value and sound quality. The value of the information provided by a sonic interface may be judged in terms of information content, timeliness of delivery, and entertainment value (that is, the extent to which audio increases the enjoyment of using the computer). The sound quality may be thought of as a combination of signal characteristics (signal-to-noise ratio, frequency response) and aesthetics.

Though it is clearly preferable to have the best possible quality of sound for all applications, our standards of quality will vary widely, depending on the perceived value added by the sonic dimension. When the information value is deemed to be high, the quality of the audio display will not be judged too harshly. However, many of the applications for auditory interfaces utilize audio simply to augment the display of information available visually. In order for these interfaces to be widely accepted, the sounds they generate must be extraordinary; in a sense, they must be so good that the user is normally unaware of their existence.[4]

Our challenge is to bring sonic interfaces above the threshold of acceptance. This may be accomplished by using audio to provide information that is either unavailable or difficult to interpret visually, or by employing sounds that we take for granted. In either case, we must understand more about the particular criteria used to judge the value and quality of audio information.

[4] In motion pictures, for instance, an enormous number of sound effects are layered to produce the sound track. These sounds work entirely on the subconscious level to enhance the perceived realism of the visual action.

4. APPLICATIONS FOR AUDIO INTERFACES

Computer manufacturers have already decided that desktop audio is important, as evidenced by the fact that most systems nowadays are equipped with, or can be upgraded to include, a speaker and microphone.[13] These interfaces are sadly underutilized at present, yet they are starting to be viewed by the industry as a basic requirement of the platform, practically on a par with the mouse. The drive towards multimedia-equipped desktop computers suggests that people are beginning to expect that audio will be integrated in the computing environment, even if they do not yet understand how this will be accomplished.

There are a wide range of application areas for which audio interfaces are being explored. Though it is not necessary to take a full inventory here, it is worth considering some of these areas from the standpoint of information value.

5. INTERPERSONAL AUDIO

It is ironic that one essential element of the office environment is an audio interface whose use significantly predates the personal computer: the telephone. Nobody seems to question the importance of the telephone in the office the way we scrutinize the viability of sonic displays. There are several reasons for this: the phone is only mildly intrusive (it rings from time to time, but it is otherwise silent until we choose to attend to it); we are quite accustomed to it as it is part of our everyday environment both at work and home; phone service is extremely robust and reliable; it provides a valuable service for both business and social functions; and, most importantly, the telephone is viewed primarily as a vehicle for human-to-human communication.

The use of technology to facilitate human collaboration has become a crucial part of the computing environment, initially in the form of electronic mail. Voice annotation for electronic mail and documents extends this message-oriented communication into the audio domain.[17] The deployment of desktop audio and video capabilities promises to enable office-to-office video-conferencing,[12,16] as well as other more informal collaborative modes.[8] This is particularly exciting because it adds an entirely new dimension to human communication.[5]

---

[5] As an interesting aside: users of multimedia mail systems have observed that essentially no "flaming" seems to occur in audio/video mail messages, though the same groups of collaborators do sometimes engage in somewhat heated arguments when using text-only electronic mail. This makes a strong case for the value of the information conveyed by tone of voice, facial expression, etc. It also demonstrates how the medium helps to inhibit anti-social behavior, in that it feels more like interpersonal contact than do purely text-based interfaces.

Since we depend so heavily on interpersonal communication, and because the alternatives (composing mail and faxes, face-to-face meetings) are often quite time-consuming, this application of audio technology can be considered to have very high information value. Consequently, we are willing to tolerate the 4-kHz bandwidth limitations of telephone-quality audio, half-duplex speakerphones, and noisy connections.[6]

6. ACCESS

Before the proliferation of desktop computers, audio interfaces were developed primarily for situations in which mobility and hands-free operation were important. Speech recognition and voice synthesis are often key components of such systems. These technologies are useful in the desktop environment as well, though they tend to be viewed as supplemental interfaces. One notable exception is in the area of access for the physically challenged. Sonic displays are an obvious choice for the visually impaired. Speech recognition also shows promise as an alternative input method for people whose motor functions are limited. In fact, trained speaker-dependent recognition systems are quite effective for people whose speech production is severely impaired. These users place a high value on audio interfaces because they provide access to information and services that otherwise would be inaccessible.

Computer telephone interfaces also provide a form of access. Touchtone and speech recognition are being used to navigate information databases, using a combination of speech synthesis and voice record/playback to deliver data.[14,15,17] The next generation of mobile computing devices will also feature audio as a primary interface mode. As computing moves farther away from the keyboard and screen, auditory interfaces take on an entirely new significance.

7. PERCEPTUALIZATION

The use of sonic display to present data in ways that promote intuitive understanding is a topic of a number of ICAD papers, and need not be discussed here. From the point of view of information value, data sonification is generally considered to be successful if it provides a means of recognizing or comparing patterns and anomalies that are not immediately

---

[6] Time delays, however, are not often tolerated, as evidenced by the fact that many video-conference rooms use a telephone connection for audio transmission, despite the fact that this results in poor audio-video synchronization.

obvious from visual presentations of the information. Sonification often enhances visual representations, as it tends to spread out over time data that is normally compressed into a static image.

Due to the abstract relationship between most numeric data and its sonic representation, a significant effort may be required to learn to extract meaningful information from a data sonification. This effort is not expended lightly, implying that the developer has a responsibility to use sound only when it adds significant value or is intuitively easy to use.

8. AWARENESS

As our computer systems and networks become more complex, we must routinely be aware of many background conditions and events. These include the notification of appointments and incoming messages, printer and network status, and the availability of system resources. When one's attention focus is on a particular task, visual notices can be distracting and often require screen space that is disproportionate to their importance. Audio cues provide this out-of-band information in a manner that increases awareness without visual distraction.

The sounds used for such cues may consist of spoken messages (either pre-recorded or synthesized), or they may be nonspeech sounds that are aural metaphors for particular conditions.[2,11] In the latter case, the metaphor may be obvious (e.g., a camera shutter click when a video snapshot is being taken[8]), but it is more likely to be somewhat ambiguous (for instance, footsteps signaling that a remote user has accessed your system). It would be interesting to study how many aural metaphors an average individual can learn to intuitively recognize, but we should assume that the number is not very large. Therefore, nonspeech audio is not necessarily preferable for every situation. For example, the voice message "printer out of paper" may be more appropriate than an audio cue, considering that the condition does not arise very often, and that there is no obvious sound that a printer makes when it runs out of paper.

9. SONIC FEEDBACK

Nearly every action in the physical world produces a sound of some sort. Sound reinforces our other senses, and it is often the absence of a subconsciously expected sound that triggers our attention, hence the expression "waiting for the other shoe to drop." It would seem natural that physical manipulations in the computer environment (button clicks, menu selections, dragging icons on the screen) should also be reinforced by sound, and certainly there have been several attempts to do this.[7] Because the information is entirely redundant, sound quality becomes extremely important; users will rarely tolerate sound effects that are not aesthetically pleasing.[7]

Note that sonic feedback, where deliberate actions are accompanied by sound, is somewhat different than the use of sound to provide awareness of asynchronous events. Though they both use aural metaphors to convey information, sonic feedback generally has more stringent requirements in terms of synchronization and sound quality.

10. BACKGROUND MUSIC

It is not unusual to encounter people who routinely have music playing while they are working. The most common use for CD-ROM data drives seems to be to play music discs when they are not otherwise engaged. Music broadcasts over wide area networks are beginning to appear, as well. While considering the many ways in which audio is used on the desktop, this phenomenon should not be overlooked.

11. PUTTING IT ALL TOGETHER

Let us suppose, for a moment, that we actually succeeded in developing auditory interfaces for all of the above-mentioned application areas. Could we live with such a noisy computer? Clearly, we would require fine-grained control over the auditory environment. By combining our knowledge of the human sensory system with an understanding of the range of ways in which audio may be utilized, we can derive some basic precepts that should guide our future development:

1. The platform must be capable of delivering reasonably good quality audio signals, both in terms of frequency response and noise levels. Multichannel audio is also desirable, as this allows the use of some spatial cues.[8]

    Moreover, the quality of sounds we present must be high. Sampled sounds should be recorded cleanly, with reasonably high data precision

---

[7] Audible keyclicks in computer keyboards are a case in point. Since they often sound "artificial," users usually disable them after a short while.

[8] Though true three-dimensional spatialization requires significant signal processing and the use of headphones, more modest stereo and multispeaker configurations can still be useful in localizing sounds.

and frequency response.[9] Synthesized sounds and speech should be able to scale to the available bandwidth. All digitized sound data should conform to a uniform volume level, so that the user does not have to adjust the playback volume for every message.

2. Several styles of audio resource management should be available. Some applications will require exclusive use of the audio resources, while others may be willing to share. Mixing and layering of sounds must be supported. Each contributing audio stream should be able to specify volume scaling and stereo (or spatial) position.

3. The user must be able to configure, to some extent, all applications that generate sound. This is critical from the standpoint of privacy, but is also important when sonic interruptions below a certain threshold of importance are undesirable. It would be ideal if our computers could detect our presence, suspending or disabling audio output when we are not there to hear it.[10]

4. In addition to simple on/off controls, some user-tailoring should be available. For instance, users may want to choose between speech messages and nonspeech audio cues, or specify their own sound files, for particular events.

5. Applications that generate asynchronous auditory feedback should support a range of sound initiation policies. Some applications simply play their message whenever a status change occurs. This is appropriate when the timeliness of the message delivery is important. If the user is either absent or inattentive when the message is played, however, the information is lost. Messages whose content is important regardless of delivery time may require different treatment. One approach would be to queue audio messages to a central facility that plays them according to user-configurable policies. Each message might include attributes describing urgency, spatial position, lifetime, repetition rate, etc. The key is to enable the user to select a message playback policy that suits the needs of the moment.

6. Timing and synchronization must be addressed by the platform. When using audio for voice communication, there should be very low latency and no dropouts. Synchronization and real-time scheduling facilities

---

[9] Data recorded with fewer than 12 bits of precision or a sample rate lower than 16 kHz can seem "noisy," particularly if the sounds are meant to fit into the everyday environment.

[10] It is reasonably easy to detect if the user has typed at the keyboard or moved the mouse, but additional sensing is required to know whether or not they are nearby. Even this is not entirely sufficient, as they may be engaged in conversation on the telephone or with a visitor.

must exist for applications, such as sonic feedback, in which the auditory display must be closely coordinated with visual events.

7. It is important to consider the cognitive load that each new sonification places on the user. When audio is used to augment a task that requires foreground attention, one can reasonably expect the user to invest learning time proportional to the value of the information conveyed by the sonification. Audio that is generated asynchronously or used in the background must be easily and intuitively recognized, otherwise it will be just as distracting as a visual notice.

8. The onset and duration of related audio cues should be rhythmically synchronized. Once a rhythm is established, aperiodic sonic events will automatically get the user's attention.

9. In general, we should study the principles of music cognition and apply them to our auditory interfaces. Just as you do not have to be a musician to enjoy listening to music, a sonification does not have to be music to benefit from musical paradigms.

10. Sounds that are heard often should either be extremely natural-sounding or aesthetically appealing. Speech cues in particular should only be used for infrequent events, and should be tailored dynamically to provide event-specific information.

    We may want to consider other approaches to the problem of aural fatigue. Soft drink manufacturers recognize the importance of varying their recipes so that their products do not become "stale." Perhaps we should provide sonic flavors that allow people to vary their environment from time to time.

11. CONCLUSIONS

Audio capabilities are becoming pervasive in the personal computer and workstation platforms. The challenge we now face is to find ways to fully utilize the resources that will soon be present on every desk. Applications that can benefit from the use of audio range from teleconferencing to sonified graphical interface events. If we are to achieve a fully sonified desktop environment, then we must not only ensure that all of the modalities are supported, but that they coexist and add a minimum cognitive load on the human who uses them. Though the responsibility lies with individual developers to provide high-quality auditory interfaces, we must also work together to establish consistency between interfaces and policies that allow them all to integrate into a single multisensory computing environment.

REFERENCES

1. Bregman, A. S. *Auditory Scene Analysis: The Perceptual Organization of Sound.* Cambridge, MA: MIT Press, 1990.
2. Buxton, W., W. W. Gaver, and S. Bly. *Auditory Interfaces: The Use of Nonspeech Audio at the Interface.* Cambridge, MA: Cambridge University Press, in preparation.
3. Durlach, N. I. "Auditory Localization in Teleoperator and Virtual-Environment Systems: Ideas, Issues, and Problems." *Perception* **20** (1991): 543–554.
4. Durlach, N. I., A. Rigopulos, X. D. Pang, W. S. Woods, A. Kulkarni, H. S. Colburn, and E. M. Wenzel. "On the Externalization of Auditory Images." *Presence* **1** (1992): 251–257.
5. Durlach, N. I., B. G. Shinn-Cunningham, and R. M. Held. "Supernormal Auditory Localization. I. General Background." *Presence* **2** (1993): in press.
6. Fant, G. *Acoustic Theory of Speech Production.* The Hague: Mouton, 1960.
7. Gaver, W. W. *Sonic Finder.* Apple Macintosh application, 1989.
8. Gaver, W. W., T. Moran, T., et al. "Realizing a Video Environment: EuroPARC's RAVE System." In *Proceedings of the ACM Conference on Human Factors in Computing Systems.* ACM Press, 1992.
9. Klatt, D. H. "Review of Text-to-Speech Conversion of English." *J. Acous. Soc. Am.* **82** (1987): 737–793.
10. Newman, W. R., A. Krickler, and B. M. Wove. "Television, Sound, and Viewer Perceptions." In *Proceedings Joint IEEE and Audio Engineering Society Meeting*, held February 1991 in Detroit, Michigan.
11. Patterson, R. D. *Guidelines for the Design of Auditory Warning Sounds.* In *Proceedings of the Institute of Acoustics Spring Conference*, held 1989.
12. Pearl, A. "System Support for Integrated Desktop Video Conferencing." Technical Report, Sun Microsystems Laboratories, 1992.
13. Rossum, G. V. "FAQ: Audio File Formats." Electronic posting to comp.dsp newsgroup, 1992.
14. Schmandt, C. "Speech Synthesis Gives Voiced Access to an Electronic Mail System." *Speech Technol.* **Aug/Sept** 1984.
15. Schmandt, C. "Caltalk: A Multimedia Calendar." In *Proceedings of the AVIOS Conference* held in Bethesda, 1990.
16. Schooler, E. M., and S. Casner. "A Packet-Switched Multimedia Conference System." *ACM SIGIOS Bull.* **10(1)** (1989).
17. Steinberg, D., and M. Rua. "Desktop Audio at Sun Microsystems." In *Proceedings of the AVIOS conference*, held in Minneapolis, 1992.

18. Stevens, K. N. "Theoretical Aspects of Speech Production." *Volta Rev.* **94** (1992): 5–32.
19. Wenzel, E. M. "Localization in Virtual Acoustic Displays." *Presence* **1** (1992): 80–107.
20. Yost, W. A. "Auditory Image Perception and Analysis: The Basis for Hearing." *Hearing Res.* **56** (1991): 8–18.

APPENDIX II:
Resources

JOURNALS

Journal of the Acoustical Society of America
500 Sunnyside Boulevard
Woodbury, NY 11797
> Publishes original studies in several areas of acoustics including psychological and physiological acoustics, speech production, perception and processing, music and musical instruments, noise effects, and control. It also publishes information on acoustics news, standards, and patents.

Journal of the Audio Engineering Society
Audio Engineering Society
60 East 42nd Street, Rm. 2520
New York, NY 10165-2520
> Original papers covering all aspects of audio engineering, including such topics as digital recording, speaker and microphone design, sound recording techniques, sound synchronization and tranmission in various media such as film and television, sound reinforcement, and audio measurement and analysis. In addition to the journal, the proceedings from the AES conferences are another information resource.

Perception and Psychophysics
Psychonomic Society Publications
1710 Fortview Road
Austin, TX 78704
> Publishes original studies of sensory processes, perception, and psychophysics. Primarily empirical studies of fundamental perceptual phenomena.

Human Factors and Ergonomics Society
Publications Division
Box 1369
Santa Monica, CA 90406-1369
> Publishes original studies about people in relation to machines and environments. Auditory display studies are published only occasionally but usually address the effectiveness of the auditory display within the context of a human-machine system.

Computer Music Journal
MIT Press
Cambridge, MA 02142
> CMJ is the most important source for information on new sound synthesis alogrithms, computer music composition techniques, computer-assisted music analysis programs, and a host of other issues.

Computer Human Interface (CHI)
Proceedings of ACM-SIGCHI
Addison-Wesley Publishing Company, Order Department
One Jacob Way
Reading, MA 01867, USA
> The CHI conference proceedings are a primary source for the best, most recent papers describing work on human-computer interaction (including the design of nonspeech audio interfaces).

ELECTRONIC FORUMS

icad@santafe.edu (icad-request@santafe.edu)
> Following ICAD '92, this forum was established to discuss a variety of auditory display issues, including sonification, audification, sound in immersive interfaces (virtual reality), teleoperation and generalized computer interfaces, perceptual issues in auditory display, technologies supporting AD creation, data handling for AD systems, and applications of AD.

sound@acm.org (sound-request@acm.org)
> This forum was established for presentation and discussion of research in sound computation. This is an arena for discussing technical issues in sound software and hardware. While ICAD has a strong emphasis on the perception and application issues, "sound" was established to reflect the ACM's emphasis on computer science issues.

auditory@vm1.mcgill.ca (owner-auditory@vm1.mcgill.ca)
>   This forum is peopled by auditory perception researchers, and the issues that arise include the whole spectrum of their research interests, including auditory stream segregation, pitch perception, spatial hearing, and so on.

BOOKS AND OTHER PUBLICATIONS
SOUND COMPUTATION AND CONTROL AND AUDIO ENGINEERING

Curtis Roads and John Strawn, eds.
Foundations of Computer Music
ISBN 0-262-68051-3
MIT Press, 1985
>   A survey of sound computation and other computer music issues.

F. Richard Moore
Elements of Computer Music
ISBN 0-13-252552-6
Prentice Hall, 1990
>   Covers how to analyze process and synthesize musical sound. A lot of digital signal processing, composition techniques (random numbers, Markov Processes, etc.), and the use of music.

Curtis Roads, ed.
The Music Machine
ISBN 0-262-18131-2
MIT Press, 1989
>   A collection of papers from *Computer Music Journal.*

Ken C. Pohlmann, ed.
Advanced Digital Audio
SBN 0-672-22768-1
SAMS, 1991
>   An excellent book covering a variety of advanced topics in digital signal processing for audio. Covers several formats (optical disk technology, digital audio systems for video and film, data compression, signal processing for audio, and DSP architecture).

Internation Computer Music Conference Proceedings
Published by
International Computer Music Association
P.O. Box 1634
San Francisco, CA 94101 USA
>   Published annually (since 1974), collection of papers from researchers working in computer music. Any topic where a computer and music might intersect (even remotely) is within the scope of these conferences. Papers on synthesis, real-time systems, analysis, composition, alternate controllers, acoustics, and others.

MIDI 1.0 Detailed Specification, Version 4.2 © 1993
Published and distributed by
International MIDI Association
5316 West 57th Street
Los Angeles, CA 90056
(310) 649-6434
> The official specification for the industry-standard Musical Instrument Digital Interface.

Rothstein, Joseph
MIDI: A Comprehensive Introduction
Computer Music and Digital Audio Series, Vol. 7
1 (Series Editor John Strawn)
Madison, WI: A-R Editions, 1992
> A thorough discussion of the basic principles of MIDI. Describes categories of MIDI instruments, accessories, and computer software, and tells how to get it all to work together.

SOUND PERCEPTION AND COGNITION

D. J. Getty and J. H. Howard, Jr.
Auditory and Visual Pattern Recognition
Hillsdale, NJ: Erlbaum, 1981
> This edited volume includes five chapters on perception of complex auditory patterns, two chapters on theoretical approaches to pattern recognition, and three chapters on multidimensional approaches to pattern perception.

Handel, S.
Listening: An Introduction to the Perception of Auditory Events
Cambridge, MA: MIT Press, 1989.
> Includes, in one source, broad and detailed coverage of auditory topics including sound production (especially by musical instruments and by voice), propagation, modeling, and the physiology of the auditory system. Covers parallels between speech and music throughout.

National Research Council
Classification of Complex Nonspeech Sounds
Washington, DC: National Academy Press, 1989
> This report was prepared by the Committee on Hearing, Bioacoustics, and Biomechanics to review and evaluate the literature on the classification of complex nonspeech, nonmusic, transient sounds. Literature on perception, signal processing, auditory object perception, limits of auditory processing, acoustic transients, sonar, and multidimensional analysis is also reviewed.

K. R. Boff, L. Kaufman, and J. P. Thomas
Handbook of Perception and Human Performance
Sensory Processes and Perception, vol. 1
New York: John Wiley & Sons, 1986
> Various parameters of sound are delineated and discussed including their interpretation by individuals having auditory pathologies. An excellent first source for the definition of sound parameters and inquiry into the complexities of sonic phenomena.

K. R. Boff and J. E. Lincoln, eds.
Engineering Data Compendium: Human Perception and Performance
Armstrong Aerospace Medical Research Laboratory
Wright-Patterson Air Force Base, Ohio, 1988
> This three-volume compendium distills information from the research literature about human perception and performance that is of potential value to systems designers. Plans include putting the compendium on CD. A separate user's guide is also available.

W. J. Dowling and D. L. Harwood
Music Cognition
San Diego: Academic Press, 1986
> A general text providing an abundance of information concerning the physical characteristics of musical sound and the processes involved in its perception. Topics covered include basic acoustics, physiology of hearing, music perception (e.g. timbre, consonance/dissonance, etc.), melodic organization, temporal organization, emotion, and meaning, and cultural context of musical experience; abundant references to research in each of these areas is provided for further reading.

B. Truax
Acoustic Communication
Norwood, NJ: Ablex, 1984
> Broad coverages of sound including speech, music, natural sound, and sounds of modern life. Presents a communication model which places sound in a mediating role between listeners and and the environment. Presents results of the World Soundscape Project.

Manfred Clynes, ed.
Music, Mind, and Brain: The Neuropsychology of Music
ISBN 0-306-40908-9
New York: Plenum Press, 1982
> Collection of papers based on the conference on Physical and Neuropsychological Foundation of Music which was in Ossiach (wherever that is!) in 1980. Covers topics such as the nature of the language of music, how the brain organises musical experience, perception of sound and rhythm, and how computers can help contribute to a better understanding of musical processes. [British spelling intentional?]

Diana Deutsch, ed.
The Psychology of Music
ISBN 0-12-213562-8
New York: Academic Press, 1982
> A well-known book that covers perception, analysis of timbre, rhythm and tempo, timing, melodic processes, and others.

MULTIMEDIA

Meera M. Blattner and R. B. Dannenberg, eds.
Multimedia Interface Design.
Reading, MA: ACM Press/Addison-Wesley, 1992
> Eight of the 21 chapters of this book are focused on sound in the multimedia interface. Many of the other chapters consider the role of sound as one of the elements of components of the multimedia interface.

John K. Buford, ed.
Multimedia Systems.
Reading, MA: ACM/Addisoon-Wesley, 1994.
> Provides a technical overview of multimedia systems, including information on sound and video recording, signal processing, system architectures, and user interfaces. Covers fundamental principles, applications and current research in multimedia, as well as operating systems, database management systems, and network communication.

A. D. N. Edeards and S. Holland, eds.
Multimedia Interface Design in Education.
NATO ASI Series F: Computer and Systems Sciences, Vol. 76.
Springer-Verlag, 1992.
> This is a collection of papers from a workshop. As suggested by the title, the emphasis is on use of multimedia in education, though naturally many of the conclusions have a broader significance. The authors came from a wide variety of backgrounds—both theoretical and practical and there is an emphasis on the use of multiple media within the human-computer interface, as well as the use of computers to control multimedia displays.

M. Mayberry, ed.
Intelligent Multimedia Interfaces.
Menlo Park, CA AAAI Press, 1993.

APPENDIX III:
Annotated Bibliography

Contributors to this bibliography included Jim Ballas, Stephen Brewster, Steve Frysinger, Bill Gaver, David Jameson, Gregory Kramer, Stuart Smith, Rory Stuart, Chris Weber, and Sheila Williams. Additionally, many authors provided annotations of their own works.

Albright, L., A. J. Jackson, and J. Francioni. "Auralization of Parallel Programs." *SICHI Bull.* **23(4)** (1991): 86–87.
> Reasons why the auditory systems excels at various tasks are outlined in this article describing parallel program debugging with sound.

Allen, J. B., and D. A. Berkley. "Image Model for Efficiently Modeling Small-Room Acoustics." *J. Acoust. Soc. Am.* **65** (1979): 943–950.
> One of the core papers discussing the image model for the simulation of reverberant rooms which has been applied to the interactive synthesis of virtual acoustic sources.

Arons, B. "Interactively Skimming Recorded Speech." In *Proceedings of the User Interfaces Software and Technology (UIST) Conference.* Reading, MA: ACM Press/Addison-Wesley, 1993 (in press).
> A nonvisual user interface for interactively skimming speech recordings is described. SpeechSkimmer uses simple speech-processing techniques to allow a user to hear recorded sounds quickly, and at several levels of detail. User interaction through a manual input device provides continuous real-time control of speed and detail level of the audio presentation.

Arons, B. "Techniques, Perception, and Applications of Time-Compressed Speech." In *Proceedings of 1992 Conference, American Voice I/O Society*, held Sep. 1992, 169–177.

    A review of time-compressed speech including the limits of perception, practical time-domain compression techniques, and an extensive bibliography.

Arons, B. "A Review of the Cocktail Party Effect." *J. Am. Voice I/O Soc.* **12** (Jul. 1992): 35–50.

    A review of research in the area of multichannel and spatial listening with an emphasis on techniques that could be used in speech-based systems.

Arons, B. "Hyperspeech: Navigating in Speech-Only Hypermedia." In *Hypertext '91*, 133-146. Reading, MA: ACM Press/Addison-Wesley, 1991.

    Hyperspeech is a speech-only (nonvisual) hypermedia application that explores issues of speech user interfaces, navigation, and system architecture in a purely audio environment without a visual display. The system uses speech recognition input and synthetic speech feedback to aid in navigating through a database of digitally recorded speech segments.

Arons, B. *Hyperspeech* (videotape). *ACM SIGGRAPH Video Rev.* **88** (1993). InterCHI '93 Technical Video Program.

    A four-minute video showing the Hyperspeech system in use.

Arons, B. "The Design of Audio Servers and Toolkits for Supporting Speech in the User Interface." *J. Am. Voice I/O Soc.* **9** (Mar. 1991): 27–41.

    An overview of audio servers and design thoughts for toolkits built on top of an audio server to provide a higher-level programming interface.

    Arons describes tools for rapidly prototyping and debugging multimedia servers and applications. He includes details of a SparcStation-based audio server, speech recognition server, and several interactive applications.

Asano, F., Y. Suzuki, and T. Sone. "Role of Spectral Cues in Medial Plane Localization." *J. Acous. Soc. Am.* **88** (1990): 159–168.

    A study of localization cues using simulated transfer functions simplified via autoregressive moving-average models in order to study what cues are critical for median plane localization. The conclusion was that macroscopic patterns above 5 kHz are used to judge elevation, and macroscopic patterns in the high frequencies as well as microscopic patterns below 2 kHz are used for front-rear judgment.

Astheimer, P. "Sonification Tools to Supplement Dataflow Visualization." In *Third Eurographics Workshop on Visualization in Scientific Computing*, held April 1992, in Viareggio, Italy. (Also in *Scientific Visualization—Advanced Software Techniques*, edited by Patrizia Palamidese, 15–36. London: Ellis Horwood, 1993.)

    Astheimer presents a detailed concept for the integration of sonification tools in dataflow visualization systems. The approach is evaluated with an implementation of tools within the apE-system of the Ohio Supercomputer Center and some examples.

Astheimer, P. "Realtime Sonification to Enhance the Human-Computer Interaction in Virtual Worlds." In *Proceedings Fourth Eurographics Workshop on Visualization in Scientific Computing, Abingdon*, held April 1993, in England.

An overview of IGD's virtual reality system "Virtual Design." Several acoustic rendering algorithms are explained concerning sound events, direct sound propagation, a statistical approach, and the image source algorithm.

Astheimer, P. "Sounds of Silence—How to Animate Virtual Worlds with Sound." In *Proceedings ICAT/VET*, held May 1993, in Houston, Texas, USA.

The author presents a concept for an audiovisual virtual reality environment. The facilities of IGD's virtual reality demonstration center and the architecture of the proprietary system "Virtual Design" are introduced. The general processing and data interpretation schema is explained.

Astheimer, P. "What You See is What You Hear—Acoustics Applied to Virtual Worlds." IEEE Symposium on Virtual Reality, held October 1993, in San Jose, California, USA. Los Alamitos, CA: IEEE Computer Society Press, 1993.

This paper concentrates on the realization and problems of the calculation of sound propagation in arbitrary environments in real time. A brief overview over IGD's virtual reality system "Virtual Design"; the basic framework is given. The differences between graphic and acoustic models and rendering algorithms are discussed. Possible solutions for the rendering and subsequent auralization phase are explained. Several examples demonstrate the application of acoustic renderers.

Ballas, J. A. "Delivery of Information Through Sound." In *Auditory Display: Sonification, Audification, and Auditory Interfaces*, edited by G. Kramer. Santa Fe Institute Studies in the Sciences of Complexity, Proc. Vol. XVIII. Reading, MA: Addison Wesley, 1994.

Ballas presents an overview of how different forms of information can be effectively delivered through nonspeech sound. The coverage is organized by linguistic devices. In addition, some details are presented on the importance of listener expectancy, and how it may be measured.

Ballas, J. A. "Common Factors in the Identification of an Assortment of Brief Everyday Sounds." *J. Exp. Psych.: Hum. Percep. & Perf.* **19** (1993): 250–267.

Ballas presents five experiments conducted to investigate factors that are involved in the identification of brief everyday sounds. In contrast to other studies, the sounds were quite varied in type, and the factors studied included acoustic, ecological, perceptual, and cognitive. Results support a hybrid approach to understanding sound identification.

Ballas, J. A., and T. Mullins. "Effects of Context on the Identification of Everyday Sounds." *Hum. Perf.* **4(3)** (1991): 199–219.

The authors present the results of four experiments conducted to investigate the effects of context on the identification of brief everyday sounds. The sounds were nearly homonymous (i.e., similar sounds produced by different causes). Results showed that context had a significnt effect, especially in biasing listeners against a sound cause that was inconsistent with the context.

Ballas, J. A., and J. H. Howard, Jr. "Interpreting the Language of Environmental Sounds." *Envir. & Beh.* **19** (1987): 91–114.

The authors present some comparisons between the perceptual identification of environmental sounds and well-studied speech perception processes. Comparisons are made at the macro level, as well as in the details.

Begault, D. R., and E. M. Wenzel. "Headphone Localization of Speech." *Hum. Factors* **35(2)** (1993): 361–376.
> An empirical study of subjects judging the position of speech presented over headphones using nonindividualized HRTFs. Subjects expressed their judgments by saying their estimate of distance and direction after each speech segment was played. Patterns of errors are described, and it is concluded that useful azimuth judgements for speech are possible for most subjects using nonindividualized HRTFs.

Bidlack, R. "Chaotic Systems as Simple (but Complex) Compositional Algorithms." *Comp. Music J.* **16(3)** (1992): 33–47.
> Bidlack describes his portrayal of nonlinear mathematical events within his music much the same as earlier composers utilized such phenomena as prime numbers and the Fibonaci series.

Blattner, Meera M., Ephraim P. Glinert, and Albert L. Papp, III. "Sonic Enhancements for Two-Dimensional Graphic Displays." In *Auditory Display: Sonification, Audification, and Auditory Interfaces*, edited by G. Kramer. Santa Fe Institute Studies in the Sciences of Complexity, Proc. Vol. XVIII. Reading, MA: Addison Wesley, 1994.
> By studying the specific example of a visually cluttered map, the authors suggest general principles that lead to a taxonomy of characteristics for the successful utilization of nonspeech audio to enhance the human-computer interface. Their approach is to divide information into families, each of which is then separately represented in the audio subspace by a set of related earcons.

Blattner, Meera M. "Sound in the Multimedia Interface." In *The Proceedings of Ed-Media '93*, held June 23–26, 1993, in Orlando, Florida, 76–81. Association for the Advancement of Computing in Education, 1993.
> The focus of this article is on recent developments in audio; however, the motivation for the use of sound is to provide a richer learning experience. This article begins with a description of the flow state, that state of mind in which we are deeply involved with what we are doing, and proposes some techniques for achieving the flow state through our use of audio.

Blattner, M. M., and R. M. Greenberg. "Communicating and Learning Through Non-speech Audio." In *Multimedia Interface Design in Education*, edited by A. Edwards and S. Holland. NATO ASI Series F, 133–143. Berlin: Springer-Verlag, 1992.
> This article begins with an examination of the way structured sounds have been used by human beings in a variety of contexts and goes on to discuss the how the lessons from the past may help us in the design and use of sound in the computer-user interface. Nonspeech sound messages, called earcons, are described with an application to the study of language.

Blattner, M. M., R. M. Greenberg, and M. Kamegai. "Listening to Turbulence: An Example of Scientific Audiolization." In *Multimedia Interface Design*, edited by M. Blattner and R. Dannenberg, 87–102. Reading, MA: ACM Press/Addison-Wesley, 1992.
> The authors discuss some of the sonic elements that could be used to represent fluid flow.

Blattner, Meera M., and R. B. Dannenberg, eds. *Multimedia Interface Design.* Reading, MA: ACM Press/Addison-Wesley, 1992. To be published in Chinese by Shanghai Popular Press, 1994.
> Eight of the 21 chapters of this book are focused on sound in the multimedia interface. Many of the other chapters consider the role of sound as one of the elements of components of the multimedia interface.

Blattner, M. M., D. A. Sumikawa, and R. M. Greenberg. "Earcons and Icons: Their Structure and Common Design Principles." *Hum.-Comp. Inter.* **4(1)** (1989): 11–44.
> This article describes earcons, auditory messages used in the computer-user interface to provide information and feedback. The focus of the article is on the structure of earcons and the design principles they share with icons.

Blauert, J. *Spatial Hearing: The Psychophysics of Human Sound Localization.* Cambridge, MA: MIT Press, 1983.
> The author provides a thorough overview of the psychophysical research on spatial hearing in Europe (particularly Germany) and the United States prior to 1983. Classic text on sound localization.

Blauert, J. "Sound Localization in the Median Plane." *Acustica* **22** (1969): 205–213.
> The author describes a series of experiments that demonstrated the role of linear distortions caused by pinna filtering in localizing sound in the median plane. He demonstrates that the "duplex theory," which postulated interaural time and intensity differences as the cues for localization, was not sufficient to explain all localization phenomena.

Bly, S. "Sound and Computer Information Presentation." Unpublished doctoral dissertation, University of California, Davis, 1982.
> Bly evaluates auditory displays for three classes of data: multivariate, logarithmic, and time-varying. A series of formal experiments on multivariate data displays were conducted, demonstrating that in many cases auditory displays elicited human performance equal to or greater than that elicited by conventional visual displays.

Bly, S., S. P. Frysinger, D. Lunney, D. L. Mansur, J. J. Mezrich, and R. C. Morrison. "Communication with Sound." In *Readings in Human-Computer Interaction: A Multidisciplinary Approach,* edited by R. Baecker and W. A. S. Buxton, 420–424. Los Altos: Morgan Kaufmann, 1987.
> Contributors discussed their approaches to communicating data via sound at CHI '85 and this chapter is a result of that presentation. This is the first time that there was a national conference session dedicated exclusively to the general use of nonspeech audio for data representation.

Boff, K. R., L. Kaufman, and J. P. Thomas. *Handbook of Perception and Human Performance.* Sensory Processes and Perception, Vol. 1. New York: John Wiley & Sons, 1986.
> Various sound parameters are delineated and discussed, including their interpretation by individuals having auditory pathologies. An excellent first source for the definition of sound parameters and inquiry into the complexities of sonic phenomena.

Boff, K. R., and J. E. Lincoln, eds. *Engineering Data Compendium: Human Perception and Performance.* Ohio: Armstrong Aerospace Medical Research Laboratory, Wright-Patterson Air Force Base, 1988.
> This three-volume compendium distills information from the research literature about human perception and performance that is of potential value to systems designers. Plans include putting the compendium on CD. A separate user's guide is also available.

Borin, G., G. De Poli, and A. Sarti. "Algorithms and Structures for Synthesis Using Physical Models." *Comp. Music J.* **16(4)** (1993).
> This is the introductory article to two special issues on physical modeling for sound synthesis in this excellent journal; this article reviews techniques.

Bregman, A. S. "Auditory Scene Analysis." In *Proceedings of the 7th International Conference on Pattern Recognition,* held in Montreal, 168–175, 1984.
> Classic paper in which the concept of a positive assignment of components of a complex acoustic signal into multiple perceptual streams was first introduced

Bregman, A. S., and Y. Tougas. "Propagation of Constraints in Auditory Organization." *Percep. & Psycho.* **46(4)** (1989): 395–396.
> The authors present psychoacoustic evidence that grouping occurs on the basis of all evidence in the acoustic signal. This is not consistent with grouping as a consequence of the output from particular filters.

Bregman, A. S. *Auditory Scene Analysis.* Cambridge, MA: MIT Press, 1990.
> Bregman provides a comprehensive theoretical discussion of the principal factors involved in the perceptual organization of auditory stimuli, especially Gestalt principles of organization in auditory stream segregation.

Brewster, S. A., P. C. Wright, and A. D. N. Edwards. "A Detailed Investigation into the Effectiveness of Earcons." In *Auditory Display: Sonification, Audification, and Auditory Interfaces,* edited by G. Kramer. Santa Fe Institute Studies in the Sciences of Complexity, Proc. Vol. XVIII. Reading, MA: Addison Wesley, 1994.
> The authors carried out experiments on structured audio messages, earcons, to see if they were an effective means of communicating in sound. An initial experiment showed earcons to be better than unstructured sounds and that musical timbres were more effective than simple tones. A second experiement was carried out to develop ideas from the first. A set of guidelines are presented.

Bronkhorst, A. W., and R. Plomp. "The Effect of Head-Induced Interaural Time and Level Differences on Speech Intelligibilty in Noise." *J. Acous. Soc. Am.* **83** (1988): 1508–1516.
> Spoken sentences in Dutch with noise were recorded in an anechoic room with a KEMAR manikin, and the role of interaural time delay and headshadowing on intelligibility was studied.

Brown, M. H. "An Introduction to Zeus: Audiovisualization of Some Elementary Sequential and Parallel Sorting Algorithms." In *CHI '92 Proceedings,* 663–664. Reading, MA: ACM Press/Addison-Wesley, 1992.
> Visualization and sonification of parallel programs demonstrate that sound can reinforce, supplant, and expand the visual channel.

Brown, M. L., S. L. Newsome, and E. P. Glinert. "An Experiment into the Use of Auditory Cues to Reduce Visual Workload." In *CHI '89 Proceedings*, 339–346. Reading, MA: ACM Press/Addison-Wesley, 1989.
> Sound is presented as a means to reduce visual overload. However, subject testing revealed a doubled reaction time for sound cueing vs. visual cueing. The authors recommend aural training for effective implementation.

Buford, J. K. "Multimedia Systems." Reading, MA: ACM/Addison-Wesley, 1993.
> Provides a technical overview of multimedia systems, including information on sound and video recording, signal processing, system architectures, and user interfaces. Covers fundamental principles, applications and current research in multimedia, as well as operating systems, database management systems, and network communication.

Burdic, W. S. *Underwater Acoustic System Analysis*. Englewood Cliffs, NJ: Prentice-Hall, 1984.
> This book provides a good general background in the fundamentals of sonar systems, including a historical background, basic acoustics, transducers, ocean acoustics, sonar signal processing, decision theory, beamforming, and active and passive systems. Although the presentation is sometimes mathematically technical, no specific background is assumed.

Burgess, D. "Techniques for Low Cost Spatial Audio." In *UIST '92: User Interface Software and Technology*. Reading, MA: ACM Press/Addison-Wesley, in press.
> Burgess describes a technique for synthetic spatialization of audio and the computational requirements and performance of the technique.

Burns, E. M., and D. Ward. "Intervals Scales and Tuning." In *The Psychology of Music*, edited by Diana Deutsch. New York: Academic Press, 1982.
> A discussion of sensory consonance and dissonance reviews the work of earlier researchers and the conclusions regarding pure and complex tone interpretation. A good base reference for further exploration into consonance and dissonance.

Butler, R. A., and R. A. Humanski. "Localization of Sound in the Vertical Plane With and Without High-Frequency Spectral Cues." *Percep. & Psycho.* **51**(2) (1992): 182–186.
> Noise bursts were played over seven loudspeakers spaced 15 degrees apart in the vertical plane, and subjects judged the position of the sources. The authors conclude from the results that, without pinna cues, subjects can still localize low-pass noise in the lateral vertical plane using binaural time and level differences, but that pinna cues are critical for accurate localization in the median vertical plane.

Buttenfield, B. P., and C. R. Weber. "Visualization and Hypermedia in GIS." In *Human Factors in GIS*, edited by H. Hearnshaw and D. Medyckyj-Scott. London: Belhaven Press, in press.
> An overview of sonification types is presented for their implementation into cartographic displays and Geographic Information Systems.

Buxton, W., W. Gaver, and S. Bly. "The Use of Non-speech Audio at the Interface." Tutorial no. 10, given at CHI '89.
>   A good overview of the use of nonspeech audio, the psychology of everyday listening, alarms and warning systems, and pertinent issues from psychoacoustics and music perception. A number of classic papers are reproduced.

Buxton, B. "Using Our Ears: An Introduction to the Use of Nonspeech Audio Cues." In *Extracting Meaning from Complex Data: Processing, Display, Interaction*, edited by E. J. Farrel, Vol. 1259, 124–127. SPIE, 1990.
>   An overview of the classes of audio cue and their utility.

Buxton, W., and T. Moran. "EuroPARC's Integrated Interactive Intermedia Facility (iiif): Early Experiences." In *Proceedings of the IFIP WG8.4 Conference on Multi-User Interfaces and Applications*, held September, 1990, in Herakleion, Crete.
>   The authors review the design, technology, and early applications of EuroPARC's media space; a computer-controlled network of audio and video gear designed to support collaboration.

Calhoun, G. L., G. Valencia, and T. A. Furness. "Three-Dimensional Auditory Cue Simulation for Crew Station Design/Evaluation." In *Proceedings of the Human Factors Society 31st Annual Meeting*, 1398–1402. Santa Monica, CA: Human Factors Society, 1987.
>   Researchers from Armstrong Aerospace Medical Research Laboratory at Wright-Patterson Air Force Base compared two methods of generating cues to simulate three-dimensional auditory display for cockpit simulation. The described use of mechanical means to simulate localization cues preceeded DSP-based simulation.

Calhoun, G. L., W. P. Janson, and G. Valencia. "Effectiveness of Three-Dimensional Auditory Directional Cues." In *Proceedings of the Human Factors Society 32nd Annual Meeting*, 68–72. Santa Monica, CA: Human Factors Society, 1988.
>   The authors compare auditory cues for directing visual attention to peripheral targets. The cues tested were visual symbol, coded aural cue, speech cue, three-dimensional nonspeech audio (spatially cued white noise), and three-dimensional speech. Ordering these cues by mean reaction times, from fastest to slowest, they found: visual bar, three-dimensional tone, three-dimensional speech, speech, and coded tone. The superiority of spatial audio display to coded tone for this task is noted.

Cazden, N. "Sensory Theories of Musical Consonance." *J. Aesthetics & Art Crit.* **20** (1962): 301–319.
>   In these two articles, Cazden addresses the cultural and historical preconceptions which underly the consonance interpretation of sound by listeners. Cazden notes that the approach to the problem of consonance given by sensory theories entails fundamental error in its isolation of sound from cultural and historical context.

Chambers, J. M., M. V. Mathews, and F. R. Moore. "Auditory Data Inspection." Technical Memorandum 74-1214-20, AT&T Bell Laboratories, 1974.
>   The authors investigate the use of sound to represent quantitative data using multiple parameters of sound to encode those dimensions of multidimensional data which were not displayed on a conventional scatter plot. Their auditory display was based on three parameters: frequency, spectral content, and amplitude modulation.

Without formal experimentation, they found that their auditorily enhanced scatter plot display system promoted the classification of multivariate data.

Cherry, E. C. "Some Experiments on the Recognition of Speech with One and with Two Ears." *J. Acous. Soc. Am.* **25** (1953): 975–979.
   A classic paper on the cocktail-party effect, demonstrating the role of attention in the ability to track one voice from a crowd.

Chowning, J. "The Synthesis of Complex Audio Spectra by Means of Frequency Modulation." *J. Audio Engr. Soc.* **21** 526–534.
   The article that introduced frequency modulation as an efficient and powerful way to synthesize complex spectra. Provides a detailed explanation of the mathematics of frequency modulation.

Clynes, Manfred, ed. *Music, Mind, and Brain: The Neuropsychology of Music.* ISBN 0-306-40908-9. New York: Plenum, 1982.
   A collection of papers based on the conference on Physical and Neuropsychological Foundation of Music which was in Ossiach (wherever that is!) in 1980. It covers topics such as the nature of the language of music, how the brain organises musical experience, perception of sound and rhythm, and how computers can help contribute to a better understanding of musical processes.

Cohen, J. "Monitoring Background Activities." In *Auditory Display: Sonification, Audification, and Auditory Interfaces*, edited by G. Kramer. Santa Fe Institute Studies in the Sciences of Complexity, Proc. Vol. XVIII. Reading, MA: Addison Wesley, 1994.
   This paper describes a design-and-test approach to the question of what kinds of notifications are appropriate for users to monitor background activities on a computer. Users' reactions to sound effects, text-to-speech, and graphical notifications of background "file sharing" events suggested ways to improve the notifications. A second study confirmed the utility of some of these changes, but pointed out other areas needing improvement.

Computer-Human Interface: The Proceedings of ACM-SIGCHI (series). Reading, MA: Addison-Wesley.
   The CHI conference proceedings are a primary source for the best, most recent papers describing work on human-computer interaction (including the design of nonspeech audio interfaces).

Computer Music Association. *International Computer Music Conference Proceedings* (series). San Francisco, CA: ICMA.
   Published annually (since 1974), this collection of papers from researchers working in computer music address any topic, within the scope of these conferences, in which a computer and music might intersect (even remotely). Papers on synthesis, real-time systems, analysis, composition, alternate controllers, acoustics, and others.

Computer Music Journal. Cambridge, MA: MIT Press.
   CMJ is the most important source for information on new sound synthesis algorithms, computer music composition techniques, computer-assisted music analysis programs, and a host of other issues.

Crawford, C. "Lessons from Computer Game Design." In *The Art of Human-Computer Interface Design*, edited by B. Laurel. Reading, MA: Addison-Wesley, 1990.
> Viewer expectation of sound has primarily been a result of the popularity of television for which the sound track is an essential part of the viewing experience, and through which most of the information is carried. This can be illustrated by watching a program with the sound off, and then listening to a similar program with the picture brightness turned all the way down. The program is usually more comprehensible without picture than without sound.

Crispien, K., and H. Petrie. "Providing Access to GUI's for Blind People Using a Multimedia System—Based on Spatial Audio Presentation." Audio Engineering Society 95th Convention Preprints, October 1993. New York: AES Press, in press.
> This paper deals with the acoustic aspects addressed in the GUIB project (Textual and Graphical User Interfaces for Blind people). General aspects and strategies for the use of audio for the development of a nonvisual, acoustic display are introduced and discussed. Technical acoustic developments for a spatial representation of the auditory display, namely a multiple loudspeaker device and initial experiments with a HRTF-based headphone system are described. The results of an initial acoustic evaluation study on 18 blind and sighted subjects are presented. Finally an outlook for future research activities is given.

Dannenbring, G. L., and A. S. Bregman. "Streaming vs. Fusion of Sinusoidal Components of Complex Tones." *Percep. & Psycho.* **24(4)** (1978): 369–376.
> The authors describe perceptual consequences of the interaction between relative intensity and onset/offset asynchrony of partials of complex tones.

Das, S., and R. Bargar. "Sound for Virtual Immersive Environments." Notes for Course 23, Applied Virtual Reality, Chapter 4, SIGGRAPH '93, 1993.
> The authors discuss issues involved in implementing and using sound in virtual reality, as well as providing a description of the CAVE audio system, from both hardware and software standpoints.

Davies, J. B. *The Psychology of Music*. Stanford, CA: Stanford University Press, 1978.
> Davies provides a clear and decipherable account of the state of research up until the publication date, covering all of the fundamental areas: physics of sound, early psychophysical studies, melody perception, musical aptitude, as well as the basic musical parameters (pitch, loudness, timbre, duration) and their physical correlates. Particularly interesting are Davies' human perspective, e.g., "music exists in the ear of the listener, and nowhere else," and his final chapter on specific musical instrument families and character traits of the individuals who play them.

Davis, D. *Computer Applications in Music: A Bibliography*. ISBN 0-89579-225-7. A-R Editions, 1988.
> A collection of references to other papers in the computer/music domain, over 500 pages of them. Covers aesthetics, composition, music in education, digital audio and signal processing, MIDI, programming languges, synthesis, and many others.

Deutsch, D. "Music Recognition." *Psych. Rev.* **76(3)** (1969): 300–307.
> Knowing what we perceive about harmonic intervals is dependent upon how we perceive them. This causality is important if they are to be utilized as words of a natural language for data displays. Harmonic intervals are basic operatives of

musical abstraction, and the question arises as to whether or not their recognition is innate or learned.

Deutsch, D. "Organizational Processes in Music." In *Music, Mind and Brain: The Neuropsychology of Music*, edited by M. Clynes. New York: Plenum, 1982.
    The elements of music may be isolated through decomposition but, in practice, are dependent upon each other. They are multicolinear in their perception by the listener.

Diana Deutsch, ed. *The Psychology of Music*. ISBN 0-12-213562-8. New York: Academic Press, 1982.
    A well-known book, covers perception, analysis of timbre, rhythm and tempo, timing, melodic processes, and others.

Deutsch, D. "The Tritone Paradox: An Influence of Language on Music Perception." *Music Percep.* **8** (1991): 335–347.
    Of particular interest, as the author presents evidence that individuals not only perceive the same musical intervals between complex tones differently but also that the perception of each individual is related to his or her own customary speech patterns.

DiGiano, Christopher J., and Ronald M. Baecker. "Program Auralization: Sound Enhancements to the Programming Environment." In *Proceedings of the Graphics Interface '92*, 44–52, 1992.
    The authors identify classes of program information suitable for mapping to sound and suggest how to add auralization capabilities to programming environments. they describe LogoMedia, a sound-enhanced programming system which illustrates these concepts.

DiGiano, Christopher J. "Visualizing Program Behavior Using Non-speech Audio." M.Sc. Thesis, University of Toronto, 1992.
    DiGiano addresses the use of sound for software visualization and considers it in concert with the other modalities. The potential of sound to illucidate a program's behavior is investigated. A programming environment is presented which supports the ability to trace control and data flow during program execution using audio.

Doll, T. J., and D. J. Folds. "Auditory Signals in Military Aircraft: Ergonomic Principles Versus Practice." *Appl. Ergo.* **17** (1986): 257–264.
    The authors studied and compared the auditory signals used in a variety of aircraft and found no standardization. They found also that a relatively large number of signals were used to make it difficult for the crew to recall the meaning of the messages.

Doll, T. J., and T. E. Hanna. "Enhanced Detection with Bimodal Sonar Displays." *Human Factors* **31** (1989): 539–550.
    This paper is an examination of the visually and aurally enhanced sonar displays. Signal uncertainty was found to cause significantly greater decrement in performance for detectability in visual displays than in auditory displays.

Doll, T. J., T. E. Hanna, and J. S. Russotti. "Masking in Three-Dimensional Auditory Displays." *Human Factors* **34(3)** (1992): 255–265.
    The authors study masking in a three-dimensional display for a simulated sonar task. Found detectability of a tonal signal is greater when background noise is uncorrelated. Head coupling of the three-dimensional display had no significant effect

given that the task was simple signal detection rather than localization, classification, or tracking.

Doughty, J., and W. Garner. "Pitch Characteristics of Short Tones II: Pitch as a Function of Duration." *J. Exp. Psych.* **38** (1948): 478–494.
  One of the earliest issues in the psychology of hearing was how long a tone must be in order to have an identifiable pitch. The authors show that when a tone is long enough to have a perceptible pitch, the actual pitch has little or no dependence on duration.

Dowling, W.J., and D. L. Harwood. *Music Cognition.* San Diego: Academic Press, 1986.
  A general text providing an abundance of information concerning the physical characteristics of musical sound and the processes involved in its perception. Topics covered include basic acoustics, physiology of hearing, music perception (e.g., timbre, consonance/dissonance, etc.), melodic organization, temporal organization, emotion and meaning, and cultural context of musical experience; abundant references to research in each of these areas are provided for further reading.

Draper, S., K. Waite, and P. Gray. "Alternative Bases for Comprehensibility and Competition for Expression in an Icon Generation Tool." In *Proceedings of Interact '90*, held August 27–31, 1990, in Cambridge, UK. Amsterdam: North Holland, 1991.
  The authors describe a system for systematically generating families of icons. Notable for suggesting the possibilities of parameterizing visual icons.

Durlach N. I., and L. D. Braida. "Intensity Perception I: Preliminary Theory of Intensity Resolution." *J. Acous. Soc. Am.* **46(2)** (1969): 372–383.
  Durlach, Braida, and their colleagues in a series of papers have proposed a general model of acoustic intensity resolution which incorporates the noise of sensory and memory processes. The model addresses factors that affect memory noise, such as the stimulus range, timing of experimental events, and the type of task.

Durlach, N. I., and X. D. Pang. "Interaural Magnification." *J. Acous. Soc. Am.* **80** (1986): 1849–1850.
  A brief examination of the issues involved in super-localization display, i.e., enhancing the normal cues used in localization. Problems with the use of an "enlarged head" (with greater distance between the ears) are addressed, and a signal-processing scheme for interaural magnification is described.

Durlach, N. I., A. Rigopulos, X. D. Pang, W. S. Woods, A. Kulkarni, H. S. Colburn, and E. M. Wenzel. "On the Externalization of Auditory Images." *Presence* **1(2)** (1992): 251–257.
  The authors discuss some of the important factors involved in synthesizing virtual acoustic sources beyond the simulation of pinna cues.

Durlach, N. I. "Auditory Localization in Teleoperator and Virtual Environment Systems: Ideas, Issues, and Problems." *Perception* **20** (1991): 543–554.
  The author discusses the use of auditory localization cues for virtual environments and teleoperations, with special attention to the potential for superlocalization (i.e., providing enhanced cues). Schemes for encoding position are described and their difficulties are discussed. This paper is a review of the literature and a position statement, rather than a presentation of empirical results.

Edwards, A. D. N. "Adapting User Interfaces for Visually Disabled Users." Ph.D. Thesis, The Open University, July 1987. (Available on microfiche from the British Library, Shelf number DX 80409.)
> Edwards describes how a graphical user interface can be adapted to be accessible to blind people through the use of speech and nonspeech sounds.

Edwards, A. D. N., and S. Holland, eds. *Multimedia Interface Design in Education.* NATO ASI Series F: Computer and Systems Sciences, Vol. 76. Berlin: Springer-Verlag, 1992.
> This book includes several chapters on the use of nonspeech sounds, including earcons and music.

Edwards, A. D. N. "Modeling Blind Users' Interactions with an Auditory Computer Interface." *Intl. J. Man-Mach. Stud.* **30(5)** (1989): 575–589.
> Edwards describes a model of the interaction between blind users and a mouse-based interface using nonspeech sounds.

Edwards, A. D. N. "Soundtrack: An Auditory Interface for Blind Users." *Hum.-Comp. Inter.* **4(1)** (1989): 45–66.
> Edwards describes how a graphical user interface can be adapted to be accessible to blind people through the use of speech and nonspeech sounds.

Edwards, A. D. N. "Graphical User Interfaces and Blind People." In *Proceedings 3rd International Conference on Computers for Handicapped Persons*, held July 1992 in Vienna, 114–119.
> A summary of developments in making GUIs accessible to blind people by the addition of an auditory channel.

Edwards, A. D. N. "Evaluation of Outspoken Software for Blind Users." Technical Report YCS150, Department of Computer Science, University of York, 1991.
> The evaluation of a commercial product that makes the Macintosh accessible to blind users through the addition of speech and nonspeech sounds.

Edwards, A. D. N., and S. Holland (eds.) *Multimedia Interface Design in Education.* NATO ASI Series F, Computer and Systems Sciences, Vol. 76. Berlin: Springer-Verlag, 1992.
> This is a collection of papers from a workshop. As suggested by the title, the emphasis is on use of multimedia in education, though naturally many of the conclusions have a broader significance. The authors come from a wide variety of backgrounds—both theoretical and practical and there is an emphasis on the use of multiple media within the human-computer interface, as well as the use of computers to control multimedia displays.

Edworthy, J., S. Loxley, and I. Dennis. "Improving Auditory Warning Design: Relationship Between Warning Sound Parameters and Perceived Urgency." *Human Factors* **33** (1991): 205–232.
> The authors examine the role of both spectral and temporal parameters in conveying urgency. They identify nine parameters that contribute to perceived urgency and show how selected combinations of these parameters could convey varied levels of urgency. The parameters include spectral and envelope properties of sound bursts as well as temporal and melodic patterns across several bursts which are joined to formed an urgency alarm.

Edworthy, J., and R. D. Patterson. "Ergonomic Factors in Auditory Systems." In *Proceedings of Ergonomics International '85*, edited by I. D. Brown. Taylor and Frances, 1985.
> An important paper on the design of speech and nonspeech sounds for use in aircraft cockpits.

Evans, B. "Correlating Sonic and Graphic Materials in Scientific Visualization." In *Extracting Meaning from Complex Data: Processing, Display, Interaction*, edited by E. J. Farrel, Vol. 1259, 154–162. SPIE, 1990.
> Evans generated variable-pitch domain sonifications from a mathematical abstraction similar to that which generates fractal Julia Sets. He notes, in an analogy to cartographic color selection, that selection of a pitch domain may affect the aesthetic and informative quality of a "sonic map." He sonifies the mathematical model with quarter-tone (24 pitches per octave), chromatic (12 pitches per octave), diatonic (7 pitches per octave), and hexatonic (6 pitches per octave) scales. Informal results reveal that the quarter-tone scale better reflects the actual event, though the diatonic and hexatonic scales, which were more pleasing to listeners, sonified the process with less detail.

Fisher, P. "Hearing the Error in Classified Remotely Sensed Images." Unpublished manuscript in review, University of Leicester, 1993.
> Fisher reports using auditory data representations for error detection in classified, remotely sensed images. One of the few applications of sonification to cartography and GIS.

Fisher, S. S., E. J. Wenzel, C. Coler, and M. W. McGreevy. "Virtual Interface Environment Workstations." In *Proceedings of the 32nd Annual Meeting of the Human Factors Society*, held in Anaheim, CA, 91–95, 1988.
> This paper on NASA-Ames Research Center's Virtual Interface Environment Workstation includes an early description of the center's work using binaural auditory display, synthesis of three-dimensional sound cues, speech synthesis and recognition, and associating "sound signatures" with objects or types of information display in a virtual environment.

Fitch, T., and G. Kramer. "Sonifying the Body Electric: Superiority of an Auditory over a Visual Display in a Complex, Multivariate System." In *Auditory Display: Sonification, Audification, and Auditory Interfaces*, edited by G. Kramer. Santa Fe Institute Studies in the Sciences of Complexity, Proc. Vol. XVIII. Reading, MA: Addison Wesley, 1994.
> The authors present an eight-variable auditory interface for anesthesiologists which uses self-labeling streams with data variables "piggy-backed" upon that stream by manipulation of selected acoustic variables. Subjects using the display demonstrated faster and more accurate response using the auditory display than with the visual and the combined auditory/visual displays.

Fletcher, H., and W. A. Munson. "Loudness: Its Definition, Measurement and Calculation." *J. Acous. Soc. Am.* **5** (1933): 82–88.
> The authors refer to "dynamic" as the perceived loudness of a passage of music. This perception of amplitude is discussed in detail.

Forbes, T. W. "Auditory Signals for Instrument Flying." *J. Aeronautical Soc.* May (1946): 255–258.
    After finding that combinations of tones created a confusing display that was difficult to use, the author turned to one signal in which multiple data variables were represented by multiple auditory variables. He found that pilots were able to use the display as well as a visual display after only an hour of training. Four key design points were suggested: (1) Pilots have certain habitual methods of thinking about the airplane, and the signals must be designed to fit these habits of thought. (2) Because most fliers are accustomed to using visual indicators, the auditory indicators must be as simple and self-explanatory as possible. (3) When multiple signals were used, there was a tendency for one signal to "capture" the attention of the pilot, to the exclusion of the other signals. This phenomenon should be avoided. (4) The display should be designed to fit the capabilities of the average pilot and should be subjected to unbiased psychological testing.

Francioni, J. F., L. Albright, and J. A. Jackson. "Debugging Parallel Programs Using Sound." In *Proceedings of the ACM/ONR Workshop on Parallel and Distributed Debugging*, 68–73. Reading, MA: ACM Press/Addison-Wesley, 1991.
    These two articles describe the same research: the mapping of parallel processor activity to sound parameters. By building structures such as jazz-like chords whose notes' pitch, attack, and crescendo describe the activity of various processors, the authors are able to analyze processor loads, flow of processor control, and processor communication.

Francioni, J. F., J. A. Jackson, and L. Albright. "The Sounds of Parallel Programs." In *Proceedings of the Sixth Distributed Memory Computing Conference*, held in Portland, OR, 570–577, 1991a.
    This paper introduces auralization techniques as a means for studying the run-time behavior of parallel programs. Examples are described of simple sound mappings that directly map run-time events of parallel programs to MIDI sound events. Although the sound playbacks discussed in this paper are not synchronized with any graphical representations, the basic feasibility of the auralization idea is demonstrated.

Frantii, G. E., and L. A. Leverault "Auditory Discrimination of Seismic Signals from Earthquakes and Explosions." *Bull. Seis. Soc. Am.* **55(1)** (1965): 1–26.
    Twenty-one observers classified 200 time-compressed, audibly displayed seismic events as either earthquakes or explosions correctly 2/3 of the time (where 1/2 corresponds to chance performance). Experiments were done to determine the receiver operating characteristics of listeners, the effect of training on performance, the effect of epicentral distance, and the effect of dual (horizontal and vertical) component playback. Among the significant conclusions was that observers reached plateau performance with the 1500 decisions and that the performance could be improved by using multiple component (stereo) playbacks.

Freed, D. J., and W. L. Martens. "Deriving Psychophysical Relations for Timbre." In *Proceedings of the International Computer Music Conference*, held October 20–24, 1986, in The Hague, The Netherlands, 1986.
    The authors present acoustic analyses and experiments on the auditory perception of mallet hardness; one of the few examples of studies of everyday listening.

Frysinger, S. P. "Pattern Recognition in Auditory Data Representation." Unpublished Thesis, Stevens Institute of Technology, Hoboken, 1988.

Frysinger, S. P. "Applied Research in Auditory Data Representation." In *Extracting Meaning From Complex Data—Proceedings of the SPIE/SPSE Symposium on Electronic Imaging*, held February 1990, edited by E. J. Farrell. Springfield, VA: SPIE, 1990.

> These two papers include an investigation of auditory/visual representations of multivariate time-series data. Two forced-choice experiments were conducted in which subjects determined which of two data sets was correlated. Subjects' data interpretation performance was found to depend upon detection task. For correlation detection, time-series dimensionality was a significant variable in display performance, and the combined auditory/visual display proved superior to the auditory-only display, while for trained pattern detection, dimensionality was not a factor, and the performance of the auditory/visual display was essentially the same as the auditory-only display.

Gaver, W. W. "Everyday Listening and Auditory Icons." Ph.D. Dissertation, University of California, San Diego, 1988.

> A two-part dissertation: Part One explores the basic psychology of everyday listening (a.k.a. auditory event perception) via physical analyses, protocol studies, and a study of hearing the length and material of struck bars. Part Two consists of a collection of papers about auditory icons.

Gaver, W. W. "The SonicFinder: An Interface that Uses Auditory Icons." *Hum.-Comp. Inter.* **4(1)** (1989).

> The authors describe the SonicFinder, a modification of the most commonly used Macintosh program which incorporates auditory icons. The paper also contains a discussion of mapping sounds to underlying computer events.

Gaver, W. W. "Sound Support for Collaboration." In *Proceedings of the Second European Conference on Computer-Supported Collaborative Work*, held September 24–27, 1991, in Amsterdam. Dordrecht: Kluwer, 1991.

> Gaver suggests that sound offers a new dimension for awareness in collaborative systems. He uses as examples the ARKola study (which is also described in Gaver et al., 1991) and the EAR system.

Gaver, W. W. "Using and Creating Auditory Icons." In *Auditory Display: Sonification, Audification, and Auditory Interfaces*, edited by G. Kramer. Santa Fe Institute Studies in the Sciences of Complexity, Proc. Vol. XVIII. Reading, MA: Addison Wesley, 1994.

> This paper reviews the author's work on auditory icons from 1988 to his present research in parameterizing the icons by a variety of synthesis techniques. A number of systems are described which illustrate the functions that auditory icons can perform.

Gaver, W. W. "Synthesizing Auditory Icons." In *Proceedings of INTERCHI '93*, held April 24–29, 1993, in Amsterdam. Reading, MA: ACM Press/Addison-Wesley, 1993.

> Gaver escribes a series of algorithms for synthesizing everyday sounds specified in terms of their causal event; he suggests that these algorithms might be useful in creating parameterized auditory icons.

Gaver, W. W. "What in the World Do We Hear? An Ecological Approach to Auditory Source Perception." *Ecol. Psych.* **(5)1** (1993).
> Gaver suggests everyday listening as a field of study and develops a framework for describing everyday sounds via physical analyses and protocol studies.

Gaver, W. W., and R. B. Smith. "Auditory Icons in Large-Scale Collaborative Environments." In *Proceedings of Human-Computer Interaction: Interact '90*, held August 27–31, 1990, in Cambridge, UK, 735–740. Amsterdam: North Holland, 1990.
> The authors describe SoundShark, an auditory interface to a large-scale collaborative system designed for distance education. They give examples of sounds used to confirm user actions, to convey information about ongoing processes and modes, to aid navigation, and to support collaboration.

Gaver, W. W., T. Moran, A. MacLean, L. Lvstrand, P. Dourish, K. Carter, and W. Buxton. "Realizing a Video Environment: EuroPARC's RAVE System." In *Proceedings of CHI '92*, held May 3–7, 1992, in Monterey, CA. Reading, MA: ACM Press/Addison-Wesley, 1992.
> A review of EuroPARC's RAVE system, a computer-controlled audio-video network that supports remote collaboration (see also Buxton and Moran, 1990). The authors discuss the nature of collaboration, the emergent functionality of the system, and issues concerning privacy; and describe related systems for collaboration.

Gaver, W. W., R. B. Smith, and T. O'Shea. "Effective Sounds in Complex Systems: The ARKola Simulation." In *Proceedings of CHI '91*, held April 28–May 2, 1991, in New Orleans. Reading, MA: ACM Press/Addison-Wesley, 1991.
> The ARKola bottling factory was a software simulation designed for testing auditory icons in a complex, cooperative task. Observations of participants running the plant with and without sound suggested that auditory icons affected their perception of the plant and their collaboration (see also Gaver & Smith, 1990).

Getty, D. J., and J. H. Howard, Jr. *Auditory and Visual Pattern Recognition*. Hillsdale, NJ: Erlbaum, 1981.
> This edited volume includes five chapters on the perception of complex auditory patterns, two chapters on theoretical approaches to pattern recognition, and three chapters on multidimensional approaches to pattern perception.

Gibson, J. J. *The Senses Considered as Perceptual Systems*. Boston: Houghton Mifflin, 1966.
> Gibson is the founder of the ecological approach to perception. In this book he argues that the senses should be considered systems, extending beyond the primary sensory mechanisms, for picking up information in the environment. This is Gibson's only major work that discusses audition.

Gibson, J. J. *The Ecological Approach to Visual Perception*. Boston: Houghton Mifflin, 1979.
> Gibson's last book includes a classic description of the information available for visual perception, the concept of affordances, and experiments showing how people pick up and use information in the world. As with all of Gibson's work, this book is extremely well written and a pleasure to read.

Glavin, S. "Creating Sound Symbols from Digital Terrain Models: An Exploration of Cartographc Communication Forms." Unpublished Master's Thesis, Carleton University, 1987.
> In the cartographic application realm, Glavin has sonified three-dimensional landscapes by creating sound symbols from Digital Terrain Models. This is the earliest citation of sonification in cartography.

Grantham, D. W. "Detection and Discrimination of Simulated Motion of Auditory Targets in the Horizontal Plane." *J. Acous. Soc. Am.* **79** (1986): 1939–1949.
> Using a technique of simulating auditory motion in the horizontal plane with two fixed loudspeakers which is described in some depth in the appendix, Grantham tested subjects' ability to detect and discriminate motion of 500-Hz tones. It is concluded that, for the range of simulated velocities simulated, subjects used spatial change rather than velocity per se in these detection and discrimination tasks.

Green, D. M. "Audition: Psychophysics and Perception." In *Stevens' Handbook of Experimental Psychology*, edited by R. C. Atkinson, R. J. Herrnstein, G. Lindzey, and R. D. Luce, 377–408. New York: Wiley, 1988.
> Covers psychophysical performance in detection and discrimination of intensity and frequency, sound localization, and perception of loudness and pitch.

Grinstein, G., and S. Smith. "The Perceptualization of Scientific Data." In *Proceedings of the SPIE/SPSE Conference on Electronic Imaging*, Vol. 1259, 190–199. Santa Clara, CA: SPIE, 1990.
> This paper was the first to fully describe "Exvis," the integrated visualization and sonification system developed at University of Massachusetts' Lowell group. The accompanying video shows a style of interaction pioneered in Exvis and provides several examples of typical sounds generated by Exvis.

Haddad, Richard A., and Thomas W. Parsons. *Digital Signal Processing: Theory, Applications, and Hardware.* ISBN 0-7167-8206-5. Computer Science Press, 1991.
> An excellent (but sometimes difficult) book covering a wide variety of topics including numerical operations (convolution, Fourier Transform), digital representation of speech, filters (FIR and IIR), FFTs, DSP algorithms, applications (speech, synthesis, recognition, image processing), and some descriptions of DSP chips (mainly from Texas Instruments).

Handel, S. *Listening: An Introduction to the Perception of Auditory Events.* Cambridge, MA: MIT press, 1989.
> This book includes in one source broad and detailed coverage of auditory topics including sound production (especially by musical instruments and by voice), propagation, modelling, and the physiology of the auditory system. It covers parallels between speech and music throughout.

Hawkins, H. L., and J. C. Presson. "Auditory Information Processing." In *Handbook of Perception and Human Performance*, edited by K. R. Boff, L. Kaufman, and J. P. Thomas, Chap. 26. New York: Wiley, 1986.
> The authors focus on topics related to the capacity to process auditory information including attention and memory, and factors that mediate processing capacity such as noise and aging.

Hayward, Chris. "Listening to the Earth Sing." In *Auditory Display: Sonification, Audification, and Auditory Interfaces*, edited by G. Kramer. Santa Fe Institute Studies in the Sciences of Complexity, Proc. Vol. XVIII. Reading, MA: Addison Wesley, 1994.
> Techniques for auditory monitoring and analysis of seismic data are described. Previous published work is limited to two papers now nearly 30 years old. This paper broadens the applications from the discrimination problem of earthquakes and nuclear explosions to training, quality control, free oscillation display, data discovery, large data set display, even recognition, education, model matching, signal detection,a nd onset timing. Simple processing techniques including interpolation, time compression, automatic gain control, frequency doubleing, audio annotation and markers, looping, and stereo are used to create seven audio data sets.

Heine, W-D., and R. Guski. "Listening, the Perception of Auditory Events? An Essay Review of *Listening: An Introduction to the Perception of Auditory Events* by Stephen Handel." *Ecol. Psych.* **3(3)** (1991): 263–275.
> This short essay criticises Handel's book from an ecological perspective and offers suggestions of what an ecological approach to audition might consider.

Helmholtz, H. von *Selected Writings of Hermann von Helmholtz*, edited by Russell Kahl. Middletown, CT: Wesleyan University Press, 1971.
> A classic text on the work of the nineteenth-century scientist into the realm of audiology and sonic phenomena. A necessity for historical continuity of research.

Hirsh, I. J. "Auditory Perception and Speech." In *Steven's Handbook of Experimental Psychology*, edited by R. C. Atkinson, R. J. Herrnstein, G. Lindzey, and R. D. Luce, 377–408. New York: Wiley, 1988.
> Hirsh organizes his coverage of audition into single sounds, sound sequences, and speech, covering the important perceptual attributes of each type of sound.

Howell, P., R. West, and I. Cross, eds. *Representing Musical Structure*. London: Academic Press, 1991.
> The authors present a range of studies of musical structure from a perceptual approach.

Howard, J. H., Jr., and J. A. Ballas. "Syntactic and Semantic Factors in the Classification of Nonspeech Transient Patterns." *Percep. & Psycho.* **28** (1980): 431–439.
> The authors present the results of three experiments conducted to assess the role of syntactic (i.e., temporal) and semantic (i.e., knowledge) factors in the classification of sequences of brief sounds. Their results indicate that both factors and their interaction are important. Previous research has shown these to be important in speech and language perception. These studies demonstrate their importance in nonspeech sound perception.

Human Factors Journal. Human Factors and Ergonomics Society Publications Division, Box 1369, Santa Monica, CA 90406-1369.
> This journal publishes original studies about people in relation to machines and environments. Auditory display studies are published only occasionally but usually address the effectivness of the auditory display within the context of a human-machine system.

Hutchins, E. L., J. D. Hollan, and D. A. Norman. "Direct Manipulation Interfaces." In *User-Centered System Design: New Perspectives on Human-Computer Interaction*, edited by D. A. Norman and S. W. Draper, 87–124. Hillsdale, NJ: Lawrence Erlbaum, 1986.
> This paper argues that direct manipulation systems (now often referred to as "GUI") are valuable because they minimize semantic and articulatory distances between humans and computers.

International MIDI Association. *MIDI 1.0 Detailed Specification*, Version 4.2. Los Angeles, CA: MIDI, 1993.
> The official specification for the industry-standard Musical Instrument Digital Interface.

Jaffe, D., and J. Smith. "Extensions of the Karplus-Strong Plucked String Algorithm." *Comp. Music J.* **7(2)** (1983).
> The authors explore and extend the Karplus-Strong algorithm, an extremely efficient approximate model of the physics of plucked strings.

Jameson, D. "Sonnet: Audio-Enhanced Monitoring and Debugging." In *Auditory Display: Sonification, Audification, and Auditory Interfaces*, edited by G. Kramer. Santa Fe Institute Studies in the Sciences of Complexity, Proc. Vol. XVIII. Reading, MA: Addison Wesley, 1994.
> Jameson describes a visual programming language for attaching run-time actions to running programs. The run-time actions allow highly controlable sounds to be attached both to programs and to data. Examples include differentiating among different sorting algorithms by their auditory characteristics as well as tracking trends in variables over time.

Jenkins, J. J. "Acoustic Information for Objects, Places, and Events." In *Persistence and Change: Proceedings of the First International Conference on Event Perception* edited by W. H. Warren and R. E. Shaw. Hillsdale, NJ: Lawrence Erlbaum, 1985.
> An exploration of audition from an ecological point of view. Jenkins summarizes the benefits of acoustic information over visual information, particularly in natural settings. The advantages include unobtrusive monitoring, no requirement for an external energy source if natural events are producing the sound, provision of information about the cause of the sound and its source in space, and interrupt capability because sound does not require oriented receptors for effective delivery of the information.

Jones, S. D., and S. M. Furner. "The Construction of Audio Icons and Information Cues for Human-Computer Dialogues." In *Contemporary Ergonomics: Proceedings of the Ergonomics Society's 1989 Annual Conference*, edited by T. Megaw. Reading, MA: Addison-Wesley, 1989.
> Some early experiments into the effectiveness of earcons and auditory icons.

Kanizsa, G. *Organization in Vision: Essays on Gestalt Perception*. New York: Praeger, 1979.
> In this classic gestalt text, Kanizsa gives examples of the effects of gestalt processes.

Karsenty, S., J. A. Landay, and C. Weikart. "Inferring Graphical Constraints with Rockit." In *Human-Computer Interaction: Proceedings of CHI '92*, held in 1993 at the University of York, UK. Also as Research Report 17, Digital Equipment Corporation, Paris Research Laboratory.
> This paper describes a graphical tool that helps in creating constrained graphical objects. Constraints are inferred by the system and shown to the user both graphically and sonically.

Kistler, D. K., and F. L. Wightman. "A Model of Head-Related Transfer Functions Based on Principal Components Analysis and Minimum-Phase Reconstruction." *J. Acoust. Soc. Am.* **91** (1992): 1637–1647.
> One of the first papers to thoroughly describe a systematic approach to simplifying the head-related transfer function.

Koffka, K. *Principles of Gestalt Psychology.* London: Kegan Paul, 1936.
> In this classic gestalt text, Koffka identifies applicable major grouping principles, particularly in vision.

Kramer, G., ed. *Auditory Display: Sonification, Audification, and Auditory Interfaces.* Santa Fe Institute Studies in the Sciences of Complexity, Proc. Vol. XVIII. Reading, MA: Addison-Wesley, 1994.
> A collection of 21 papers, defining the state of auditory display research at the time of its publication. It includes an extensive introductory chapter, foreword by A. Bregman, annotated bibliography, audio CD of sound examples, resources appendix, and informal comments by several ICAD participants.

Kramer, G. "An Introduction to Auditory Display." In *Auditory Display: Sonification, Audification, and Auditory Interfaces*, edited by G. Kramer. Santa Fe Institute Studies in the Sciences of Complexity, Proc. Vol. XVIII. Reading, MA: Addison Wesley, 1994.
> Kramer provides an overview of the field, including history, other uses of nonspeech audio, advantages and difficulties with auditory display, its relationship to music, and possible applications. Introduces the symbolic/analogic continuum as a means of comparing and analyzing display techniques.

Kramer, G. "Some Organizing Principles for Representing Data with Sound." In *Auditory Display: Sonification, Audification, and Auditory Interfaces*, edited by G. Kramer. Santa Fe Institute Studies in the Sciences of Complexity, Proc. Vol. XVIII. Reading, MA: Addison Wesley, 1994.
> Kramer provides a broad set of techniques for sonification display design. He describes parameter nesting and introduces beacons and dynamic beacons, affective and metaphorical association, and data type/data family association.

Kramer, G. "Sound and Communication in Virtual Reality." In *Communication in the Age of Virtual Reality*, edited by F. Biocca and M. Levy. Hillsdale, NJ: Lawrence Earlbaum, 1994.
> Kramer discusses the use of sound in virtual environments from the standpoint of Biocca's Communications Design Matrix. He provides an overview of auditory implementations in virtual reality and suggests a number of extensions of these techniques. He introduces the concept of audible objects as a factor in VR displays and describes work on sonification displays for enriching the formation of mental models.

Kramer, G. "Sonification of Financial Data: An Overview of Spreadsheet and Database Sonification." In *The Proceedings of Virtual Reality Systems '93, SIG Advanced Applications*, held 1993 in New York, NY. New York: SIG, 1993.
> A brief description of sonification, along with a case study of the sonification of five- and ten-dimensional financial data.

Kramer, G. "Sonification and Virtual Reality I: An Introduction." In *VR Becomes a Business, the Proceedings of Virtual Reality '92*. Westport, CT: Meckler, 1992
> An introductory paper on sonification, relating auditory data representation to immersive interfaces.

Kramer, G. "Audification: Using Sound to Understand Complex Systems and Navigate Large Data Sets." Proceedings of the Santa Fe Institute Science Board, Santa Fe Institute, 1990.
> Kramer describes his auditory display concepts and research from 1989–1990 and relates them to comprehending complexity and navigating large data sets.

Kramer, G. "Audification of the ACOT Predator/Prey Model." Unpublished research report prepared for Apple Computer's Advanced Technology Group, Apple Classrooms of Tomorrow, 1990.
> Kramer describes his work with Apple Computer to bring sonification techniques to a predator-prey model; the integrated hardware/software system for producing the sonifications is described. Realistic and abstract sonifications for the same data set are presented, and their possible impact on the formation of mental models and on students with different learning styles are discussed.

Kramer, G., and S. Ellison. "Audification: The Use of Sound to Display Multivariate Data." In *The Proceedings of the International Computer Music Conference*, 214–221. San Francisco, CA: ICMA, 1991.
> The authors introduce parameter nesting, a technique for developing high-dimensional displays, and introduce the Clarity Sonification Toolkit, an object-oriented research and development tool for developing and testing sonification techniques. They provide an in-depth description of how to use these tools to sonify a nine-dimensional Lorenz equation.

Krumhansl, C. L. *Cognitive Foundations of Musical Pitch*. Oxford Psychology Series, Vol. 17. Oxford: Oxford University Press, 1990.
> Study of musical pitch from a perceptual perspective. Krumhansl presents a range of models of pitch phenomena and considers how these might be encoded and remembered.

Lakoff, G., and M. Johnson. *Metaphors We Live By*. Chicago: University of Chicago Press, 1980.
> The authors provide a powerful description of how we use metaphors in everyday language without even knowing that we are doing so. They investigate how these metaphors are not only a reflection of our thinking processes, but how they shape those very processes.

Lamb, H. *The Dynamical Theory of Sound*, 2nd ed. New York: Dover, 1960.
> A classic book on the physics of sound and sound-producing events.

Laurel, B. "Interface as Mimesis." In *User-Centered System Design: New Perspectives on Human-Computer Interface*, edited by D. A. Norman and S. W. Draper. Hillsdale, NJ: Lawrence Erlbaum, 1986.
> The author introduces the analogy between using a computer and attending a drama; she suggests that users' engagement with a system is an important dimension to be considered in design.

Loomis, J. M., C. Hebert, and J. G. Cicinelli. "Active Localization of Virtual Sounds." *J. Acous. Soc. Am.* **88** (1990): 1757–1764.
> As part of a larger project to produce the user interface for a personal navigation system, the authors developed a low-cost, computer-controlled, analog, virtual sound display system that does not use direction-dependent pinna cues. Using only interaural time difference and interaural intensity difference coupled with head orientation, they found that subjects could home to virtual sound sources quite well, and found some indications that the virtual sounds were perceived as being externalized. While the use of HRTFs may produce more realistic displays, this study shows the potential of simple virtual auditory displays, e.g., for navigation tasks.

Lovstrand, L. "Being Selectively Aware with the Khronika System." In *Proceedings of ECSCW '91*, held September 25–27, 1991, in Amsterdam, The Netherlands.
> The author describes an event database that uses nonspeech audio cues amongst its techniques for notifying users about events.

Lunney, D., and R. C. Morrison. "High-Technology Laboratory Aids for Visually Handicapped Chemistry Students." *J. Chem. Ed.* **58(3)** (1981): 228–231.
> The authors present analytical chemistry data from infrared spectral to visually impaired students. The pitch of a tone was made proportional to the frequency location of the infrared peak it represents. In informal tests, subjects were able to accurately identify a range of learned compounds.

Madhyastha, T. M., and D. A. Reed. "A Framework for Sonification Design." In *Auditory Display: Sonification, Audification, and Auditory Interfaces*, edited by G. Kramer. Santa Fe Institute Studies in the Sciences of Complexity, Proc. Vol. XVIII. Reading, MA: Addison Wesley, 1994.
> The authors describe Porsonify, a toolkit that provides a uniform network interface to sound devices through table-driven sound servers. All device-specific functions are encapsulated in control files, so that user interfaces to configure sound devices and sonifications can be generated independently of the underlying hardware. Creation of some example sonifications using this toolkit is discussed.

Mansur, D. L. "Graphs in Sound: A Numerical Data Analysis Method for the Blind." Unpublished Thesis, University of California, Davis, 1984.
> The author tested the ability of subjects to make certain judgements about x-y "plots" using continuously varying pitch to represent the dependent variable (y), and time the independent variable (x). He was primarily concerned with the development of displays to make exploratory data analysis possible for visually impaired analysts. He found that with limited training, subjects were able to recognize key features of the data, such as linearity, monotonicity, and symmetry, for between 79 and 95 percent of the trials.

Mansur, D. L., M. M. Blattner, and K. I. Joy. "Sound-Graphs: A Numerical Data Analysis Method for the Blind." In *Proceedings of the 18th Annual Hawaiian International Conference on System Science*, held January 1985 in Honolulu, Hawaii. Los Alamitos, CA: IEEE Computer Society Press, 1984. Also in *J. Med. Sys.* **9** (1985): 163–174.

> Sound-Graphs are composed of three-second periods of continuously varying pitch. They were developed and used to provide the blind with a rapid and intuitive understanding of numerical data (x–y graphs). This work is primarily from the M.S. thesis (with the same name) by Douglass L. Mansur, University of California, Davis, 1984; also published as Lawrence Livermore Technical Report UCRL-53548.

Mansur, D. L., M. M. Blattner, and K. I. Joy. "The Representation of Line Graphs Through Audio Images." Technical Report UCRL-91586, Lawrence Livermore National Laboratory, Livermore, CA, September 1984.

> Holistic sound and graphical images bear certain resemblances to the way we manipulate them. This article examines tools that manipulate both graphical and line graphs and their sonic equivalents.

Mansur, D. L., M. M. Blattner, and K. I. Joy. "Sound-Graphs: A Numerical Data Analysis Method for the Blind." *J. Med. Sys.* **9** (1985): 163–174.

> The authors describe how simple line graphs can be translated into nonspeech sounds for presentation to blind people.

Matlin, M. W. *Sensation and Perception*, 2nd ed. Massachusetts: Allyn and Bacon, 1988.

> A good introductory text to perception that distinguishes between the physical responses to stimuli and the perceptual effects.

Mayer-Kress, G., R. Bargar, and I. Choi. "Musical Structures in Data From Chaotic Attractors." In *Auditory Display: Sonification, Audification, and Auditory Interfaces*, edited by G. Kramer. Santa Fe Institute Studies in the Sciences of Complexity, Proc. Vol. XVIII. Reading, MA: Addison Wesley, 1994.

> The authors exhibit parallels between structures of chaotic dynamical systems and music and indicate the possibility of using this connection to enhance the perception of recurrent features in complex signals. They describe three auditory representations of chaotic systems.

McAdams, S. "Spectral Fusion and the Creation of Auditory Images." In *Music, Mind and Brain*, Chap. XV. New York: Plenum, 1982.

> McAdams discusses aspects of musical perception beyond the boundaries of acoustic analysis; and, in evoking an auditory image from an acoustic signal, the roles of familiarity, of learning and context, of synthetic and analytic listening, and of interacting with the primitive grouping processes of harmonicity and coordinated modulation.

McCabe, R. K., and A. A. Rangwalla. "Auditory Display of Computational Fluid Dynamics Data." In *Auditory Display: Sonification, Audification, and Auditory Interfaces*, edited by G. Kramer. Santa Fe Institute Studies in the Sciences of Complexity, Proc. Vol. XVIII. Reading, MA: Addison Wesley, 1994.

> Direct simulation and parameter mapping techniques are discussed in the context of how they can be used to enhance the understanding of data from computational fluid dynamics simulations. Two case studies are presented. The first case describes how parameter mapping techniques were used to help analyze the results

from a simulation of the Penn State artificial heart. The second case shows how direct simulation was used to better understand the tonal acoustics of rotor stator interactions inside a jet turbine.

McIntyre, M. E., R. T. Schumacher, and J. Woodhouse. "On the Oscillations of Instruments." *JASA* **74** (1983): S52.
>An account of temporally based physical modeling techniques with examples; an excellent work.

Meyer, L. B. *Emotion and Meaning in Music*. Chicago: University of Chicago Press, 1956.
>Forming the basis of Meyer's past 35 years of research and theoretical writing, this text is doubtlessly a classic in the field of music psychology. Proceeding from John Dewey's (1894) "conflict theory of emotion," the author provides results of experimental investigations and musical examples to support his premise that "emotion or affect is aroused when a tendency to respond is arrested or inhibited."

Mezrich, J. J., S. P. Frysinger, and R. Slivjanovski. "Dynamic Representation of Multivariate Time-Series Data." *J. Am. Stat. Assoc.* **79** (1984): 34–40.
>Dynamic data representation employing both auditory and visual components for multivariate time-series displays, such as for economic indicators. In their scheme, the analyst is confronted at any moment with one multivariate sample from the time series, rather than the whole data set, producing samples which are displayed in succession rather like frames in a movie. The results of their experiment indicate that the dynamic auditory/visual display outperforms the static visual displays in most cases for the correlation detection task.

Monk, A. "Mode Errors: A User-Centered Analysis and Some Preventative Measures Using Keying-Contingent Sound." *IJMMS* **24** (1986): 313–327.
>Monk uses sound to reduce the number of mode errors in an interface.

Moore, F. Richard. *Elements of Computer Music*. ISBN 0-13-252552-6. Prentice Hall, 1990.
>Moore covers how to analyze process and synthesize musical sound. A lot of digital signal processing, composition techniques (random numbers, Markov Processes, etc.), and the uses of music.

Mulligan, B. E., D. K. McBride, and L. S. Goodman. "A Design Guide for Non-speech Auditory Displays." Pensacola, FL: Naval Aerospace Medical Research Laboratory, 1987.
>The authors provide algorithms that assist the designer in designing auditory signals, especially in ways to enhance detectability of signals in noise and to increase loudness without increasing signal length.

Mynatt, E. "Auditory Presentation of Graphical User Interfaces." In *Auditory Display: Sonification, Audification, and Auditory Interfaces*, edited by G. Kramer. Santa Fe Institute Studies in the Sciences of Complexity, Proc. Vol. XVIII. Reading, MA: Addison Wesley, 1994.
>Mynatt presents work in designing interactive, auditory interfaces that provide access to graphical user interfaces for people who are blind. She discusses a prototype system called Mercator which explores conveying symbolic information and supporting navigation in the auditory interface.

National Research Council. *Classification of Complex Nonspeech Sounds.* Washington, DC: National Academy Press, 1989.
> This report was prepared by the Committee on Hearing, Bioacoustics, and Biomechanics to review and evaluate the literature on the classification of complex nonspeech, nonmusic, transient sounds. Literature on perception, signal processing, auditory object perception, limits of auditory processing, acoustic transients, sonar, and multidimensional analysis is also reviewed.

Oldfield, S. R., and S. P. A. Parker. "Acuity of Sound Localization: A Topography of Auditory Space. I. Normal Hearing Conditions." *Perception* **13** (1984a): 581–600.

Oldfield, S. R., and S. P. A. Parker. (1984b). "Acuity of Sound Localization: A Topography of Auditory Space. II. Pinna Cues Absent." *Perception* **13** (1984b): 601–617.

Oldfield, S. R., and S. P. A. Parker. (1986). "Acuity of Sound Localization: A Topography of Auditory Space. III. Monaural Hearing Conditions." *Perception* **15** (1986): 67–81.
> These three studies examine the ability of subjects under different sets of conditions to localize white noise played through a speaker that varied in position. The blindfolded subjects pointed a special gun at the perceived source of sound played in an anechoic chamber over a boom-mounted speaker. In the first study subjects listened normally; in the second study pinna cues were removed by inserting individually cast pinnae molds with access holes to the auditory canal into subjects' ears; and in the third study monaural conditions were created by inserting "ear defenders" into subjects' right ears and covering their right ears with fitted earmuffs. These studies demonstrate the importance of pinna cues for determining elevation and reducing front/back reversals, and show that elevation discrimination was good under monaural conditions, but that azimuth discrimination was reduced.

O'Leary, A., and G. Rhodes. "Cross-Modal Effects on Visual and Auditory Object Perception." *Percep. & Psycho.* **35** (1984): 565–569.
> Using a display that combined a stimulus for auditory stream segregation with its visually apparent movement analog, these Stanford University researchers demonstrated cross-modal influences between vision and audition on perceptual organization. Subjects hearing the same auditory sequence perceived it as two tones if a concurrent visual sequence was presented that was perceived as two moving dots, and one tone if a concurrent visual sequence perceived as a single object was presented.

Oppenheim, Alan V., ed. *Applications of Digital Signal Processing.* ISBN 0-13-039115-8. Prentice Hall, 1978.
> Collection of papers on DSP applications, A couple of chapters are dedicated to processing audio signals and speech. The rest is exotic (RADAR, SONAR, Geophysics) and there is one chapter on digital image processing.

Parncutt, R. *Harmony: A Psychoacoustical Approach.* Berlin: Springer-Verlag, 1989.
> Parncutt develops a model of Western tonal music on the basis of psychoacoustics and psychomusicology and applies the model to the identification of the specific effects of musical conditioning and to the analysis of musical compositions. Included are interesting studies of analytic vs. synthetic, etc. listening preferences for

both musicians and nonmusicians (categorizing listeners according to their listening styles).

Patterson, R. D. "Guidelines for Auditory Warning Systems on Civil Aircraft." Paper No. 82017, Civil Aviation Authority, London, 1982.
    One of the first and best papers to discuss the proper structure of auditory warning signals based on psychoacoustical principles.

Peacock, K. "Synesthetic Perception: Alexander Scriabin's Color Hearing." *Music Percep.* **2(4)** (1985): 483–506.
    A curious phenomena which has surfaced repeatedly since the late Baroque era has come to be known as synaesthesia. It was used by the Romanticists of the nineteenth century as an effective means to enrich their accounts of sensuous impressions. Other names for the phenomena include chromesthesia, photothesia, synopsia, color hearing, and color audition. People who have this characteristic, experience a crossover between one or more sensory modes. Thus, they might be blessed with the ability to hear colors or odors, or see sounds. People who habitually perceive stimuli in this manner are often surprised when told that not everyone shares this faculty. Color hearing, though only one form of synaesthesia, is probably the commonest

Perrott, D. R., K. Saberi, K. Brown, and T. Z. Strybel. "Auditory Psychomotor Coordination and Visual Search Performance." *Percep. & Psycho.* **48** (1990): 214–226.
    The authors postulate that the primary function of the auditory spatial system is to direct the eyes. They studied the visual search time for targets presented with and without associated spatial audio cues and found that presenting a 10-Hz click train from the same location as the visual target substantially reduced the time for visual search.

Perrott, D. R., T. Sadralodabai, K. Saberi, and T. Z. Strybel. "Aurally Aided Visual Search in the Central Visual Field: Effects of Visual Load and Visual Enhancement of the Target." *Human Factors* **33** (1991): 389–400.
    Visual search performance in displays with distracters was studied with and without spatially correlated audio cues. The value of audio in directing visual search was found to be particularly great when there were a large number (63) of distracter images. Potential in applications such as enhanced cockpit displays is noted.

Pitt, I. J., and A. D. N. Edwards. "Navigating the Interface by Sound for Blind Users." In *People and Computers VI: Proceedings of the CHI '91 Conference*, edited by D. Diaper and N. Hammond, 373–383. Cambridge: Cambridge University Press, 1991.
    A description of experiments on using sounds to guide navigation in graphical user interfaces.

Plomp, R., and W. J. M. Levelt. "Tonal Consonance and Critical Bandwidth." *J. Acous. Soc. Am.* **38** (1965): 548–560.
    Musicians' rank harmonic intervals in terms of their consonance and dissonance differently than naive subjects. For nonmusicians, consonance and "pleasantness" are one and the same. This is not true for trained musicians who often consider dissonances more pleasant for reasons of harmony or aesthetics.

Plomp, R. "Acoustical Aspects of Cocktail Parties." *Acustica* **38** (1977): 186–191.
> Plomp considers cocktail parties in terms of signal-to-noise ratio and speech intelligibility. He discovers the importance of room acoustics—the height of the hall and the sound absorption characteristics of the ceiling and walls.

Pollack, I., and L. Ficks. "Information of Elementary Multidimensional Auditory Displays." *J. Acous. Soc. Am.* **26** (1954): 155–158.
> The authors consider two different mappings of multidimensional data onto the parameters of sound. Using these two display types, they measure the information transmitted to subjects as the sum of the number of bits in each correctly identified dimensional level. Their results indicate that multidimensional displays, in general, outperformed unidimensional displays measured elsewhere, and that subdivision of display dimensions into finer levels does not improve information transmission as much as increasing the number of display dimensions does.

Pohlmann, Ken C., ed. *Advanced Digital Audio.* ISBN 0-672-22768-1. SAMS, 1991.
> An excellent book covering a variety of advanced topics in digital signal processing for audio. It covers several formats (optical disk technology, digital audio systems for video and film, data compression, signal processing for audio, and DSP architecture).

Pratt, C. C. "Music as the Language of Emotion." Lecture delivered in the Whittall Pavilion of the Library of Congress, December 21, 1950. Washington, DC: U.S. Government Printing Office, 1952.
> In a study involving 227 college students, the author finds a strong consensus as to the mood conveyed in four distinct pieces of music that cannot be accounted for by the status of music as a language of emotions.

Rabenhorst, D. A., E. J. Farrell, D. H. Jameson, T. D. Linton, and J. A. Mandelman. "Complementary Visualization and Sonification of Multidimensional Data." In *Extracting Meaning from Complex Data: Processing, Display, Interaction*, edited by E. J. Farrel, Vol. 1259, 147–153. SPIE, 1990.
> Data enhancement is only the first level of auditory data representation. Utilization of the auditory channel to present unseen data is represented in this paper. The goal of these researchers was to use sound in a manner that "is intuitive enough and readily learnable enough to be effective as an additional sensory input to a mental model."

Rakowski, A. "Intonation Variants of Musical Intervals in Isolation and in Musical Contexts." *Psych. Music* **18** (1990): 60–72.
> Rakowski presents experimental evidence to demonstrate that different musical intervals have different perceptual salience for musicians, some having more distant category boundaries and some more accurate recognition (strong and weak intervals). Further evidence is presented that musicians tune intervals to match perceptual categorizations. Three factors—acoustic (tuning to beats arising from its relationship with preceding note), psychological (perceptual) and aesthetic (accentuation of intervals)—all participate is this phenomenon.

Rasch, R. A., and R. Plomp. "The Perception of Musical Tones." In *The Psychology of Music*, edited by Diana Deutsch. New York: Academic Press, 1982.
> Musicians' rank dyads in terms of their consonance and dissonance differently than naive subjects.

Richards, W. *Natural Computation*. Cambridge, MA: MIT Press, 1988.
> This book includes seven chapters on sound interpretation including: representing acoustic information, models of binaural localization and separation, schematizing spectrograms, acoustics of the signing, recovering material properties from sound, perception of breaking and bouncing events, and perception of melodies.

Risset, J. C., and D. L. Wessel. "Exploration of Timbre by Analysis and Synthesis." In *The Psychology of Music*, edited by D. Deutsch. New York: Academic Press, 1982.
> The authors describe analysis and synthesis as a technique for data reduction in computer music. Seminal work by founders of the field.

Roads, Curtis, and John Strawn, eds. *Foundations of Computer Music*. ISBN 0-262-68051-3. Cambridge, MA: MIT Press, 1985.
> A survey of sound computation and other computer music issues.

Roads, Curtis, ed. *The Music Machine*. ISBN 0-262-18131-2. Cambridge, MA: MIT Press, 1989.
> A collection of papers from *Computer Music Journal*.

Roederer, J. G. *Introduction to the Physics and Psychophysics of Music*. New York: Springer-Verlag, 1973.
> A good introduction for the "intelligent layman."

Rothstein, Joseph. *MIDI: A Comprehensive Introduction*. Computer Music and Digital Audio Series, Vol. 7. John Strawn, Series Editor. Madison, WI: A-R Editions, 1992.
> A thorough discussion of the basic principles of MIDI. Rothstein describes categories of MIDI instruments, accessories, and computer software, and tells how to get it all to work together.

Scaletti, C. "Sound Synthesis Algorithms for Auditory Data Representation." In *Auditory Display: Sonification, Audification, and Auditory Interfaces*, edited by G. Kramer. Santa Fe Institute Studies in the Sciences of Complexity, Proc. Vol. XVIII. Reading, MA: Addison Wesley, 1994.
> Scaletti provides a working definition of sonification. Models of data as continuous streams and as discrete events are illustrated using the author's sound specification language Kyma. Several sound synthesis algorithms are outlined, each with an example application, and there is a summary of which synthesis algorithms are best applied to which kinds of data. The paper concludes with an enumeration of some of the open questions and future research directions in the field of auditory display.

Scaletti, C., and A. B. Craig. "Using Sound to Extract Meaning from Complex Data." In *Extracting Meaning from Complex Data: Processing, Display, Interaction*, edited by E. J. Farrel, Vol. 1259, 147–153. SPIE, 1990.
> Because sound is an inherently time-variable phenomena, Scaletti and Craig concentrated their ADR work on the representation of time-varying data mapped to animated graphics and sound. Examples discussed include sonifications of forest fire data, Los Angeles pollution levels, and swinging pendula.

Schafer, R. M. *The Tuning of the World.* New York: Knopf, 1977.
> Schafer describes studies of the acoustic environment undertaken in the World Soundscape Project including historical changes in the acoustic environment, cross-cultural studies of listening preferences and sound interpretation, and studies of references to sound in literature.

Scharf, B., and S. Buus. "Audition I: Stimulus, Physiology, Thresholds." In *Handbook of Perception and Human Performance*, edited by K. R. Boff, L. Kaufman, and J. P. Thomas, Vol. 1, 14.1–14.71. New York: Wiley, 1986.
> Standard reference to have on your bookshelf.

Scharf, B., and A. J. M. Houtsma. "Audition II: Pitch, Localization, Aural Distortion, Pathology." In *Handbook of Perception and Human Performance*, edited by K. R. Boff, L. Kaufman, and J. P. Thomas, Chap. 26. New York: Wiley, 1986.
> This book covers psychophysical performance in detection and discrimination of intensity and frequency, sound localization, and perception of loudness and pitch.

Scherer, Klaus R., and James S. Oshinsky. "Cue Utilization in Emotion Attribution from Auditory Stimuli." *Motivation & Emotion* **1(4)** (1977).
> The authors describe a study using a MOOG synthesizer in which seven 2-level factors, amplitude and pitch level, pitch contour, pitch variability, tempo, envelope and filtration, and other more complicated stimuli were systematically manipulated and then rated on emotional impact. Inter-judge (naive students) agreement was generally good, some emotions having more reliable cues than others, as might be expected.

Schmandt, C., B. Arons, and C. Simmons. "Voice Interaction in an Integrated Office and Telecommunications Environment." In *Proceedings of 1985 Conference*. American Voice I/O Society, 1985.
> The Conversational Desktop is a conversational office assistant that manages personal communications (phone calls, voice mail messages, scheduling, reminders, etc.). The system engages the user in a conversation to resolve ambiguous speech recognition input.

Schmandt, C., and B. Arons. *Conversational Desktop* (videotape). *ACM SIGGRAPH Video Rev.* **27** (1987).
> A four-minute videotape demonstrating many features of the Conversational Desktop.

Schmandt, C., and B. Arons. "Getting the Word." *UNIX Rev.* **7** (Oct. 1989): 54–62.
> An overview of "Desktop Audio" including the systems and interface requirements for the use of speech and audio in the personal workstation. It includes a summary of the VOX Audio Server, a system for managing and controlling the audio resources in a networked personal workstation.

Schroeder, M. R. "Digital Simulation of Sound Transmission in Reverberant Spaces." *J. Acous. Soc. Am.* **47** (1970): 424–431.
> One of the core papers discussing techniques for the simulation of reverberant acoustic environments.

Sloboda, J. A. "Music Structure and Emotional Response: Some Empirical Findings." *Psych. Music* **19** (1991): 110–120.
> The author presents analysis of experimental results of emotive response to music extracts related to the musical structure of the compositions.

Smith, R. B. "A Prototype Futuristic Technology for Distance Education." In *Proceedings of the NATO Advanced Workshop on New Directions in Educational Technology*, held November 10–13, 1988, in Cranfield, UK.
> Smith describes SharedARK, a collaborative system that was the basis of SoundShark (Gaver & Smith, 1990) and ARKola (Gaver et al., 1991).

Smith, S. "An Auditory Display for Exploratory Visualization of Multidimensional Data." In *Workstations for Experiment*, edited by G. Grinstein and J. Encarnacao. Berlin: Springer-Verlag, 1991.
> Although it was published in 1991, this is actually the earliest paper about the University of Massachusetts' Lowell work in sonification. It shows what their thinking was as they embarked on their investigations in 1988, which is now mostly of historical interest.

Smith, S., R. D. Bergeron, and G. Grinstein. "Stereophonic and Surface Sound Generation for Exploratory Data Analysis." In *Multimedia and Multimodal Interface Design*, edited by M. Blattner and R. Dannenberg. Reading, MA: ACM Press/Addison-Wesley, 1992.

Smith, S., R. D. Bergeron, and G. Grinstein. "Stereophonic and Surface Sound Generation for Exploratory Data Analysis." In *Proceedings of CHI '90,* held 1990, in Seattle, WA. ACM Press, 1990.
> This paper, published in two places, describes the authors' attempt to introduce spatial aspects of sound into sonification. This direction was not pursued further.

Smith, S., G. Grinstein, and R. M. Pickett. "Global Geometric, Sound, and Color Controls for the Visualization of Scientific Data." In *Proceedings of the SPIE/SPSE Conference on Electronic Imaging*, Vol. 1459, 192–206. San Jose, CA: SPIE, 1991.
> The authors argue that users should be able to fine-tune visual and auditory data displays to achieve the optimal presentation of their data. They gives examples of how this can be done with the "iconographic" display techniques they developed. The accompanying video gives one brief sound example.

Smith, S., R. M. Pickett, and M. G. Williams. "Environments for Exploring Auditory Representations of Multidimensional Data." In *Auditory Display: Sonification, Audification, and Auditory Interfaces*, edited by G. Kramer. Santa Fe Institute Studies in the Sciences of Complexity, Proc. Vol. XVIII. Reading, MA: Addison Wesley, 1994.
> The authors outline a starting approach to sonification and argue for psychometric testing as part of the sonification design process.

Sorkin, R. D. "Design of Auditory and Tactile Displays." In *Handbook of Human Factors*, edited by G. Salvendy, 549–576. New York: Wiley & Sons, 1987.
> In this chapter Sorkin addresses factors that must be considered in establishing the level, pitch, duration, shape, and temporal pattern of a sound. In addition, he covers the design of binaural sounds and complex coding for sounds.

Sorkin, R. D., F. L. Wightman, D. S. Kistler, and G. C. Elvers. "An Exploratory Study on the Use of Movement-Correlated Cues in an Auditory Head-Up Display." *Human Factors* **31** (1989): 161–166.
> A sequence of three signals incorporating HRTF cues for auditory localization, was played to subjects over headphones, and subjects had to indicate the location of the source via keypress on a computer. This study focused on the importance of head movement in localization, and there were three conditions presented: (1) source fixed in physical space, head movement allowed; (2) no head movement allowed; and (3) source fixed in position relative to the subject's head. Azimuthal localization was found to be considerably better in the first case (source fixed in physical space/head movement allowed), demonstrating the importance to auditory localization of correlating cues to self-initiated movements of the listener's head.

Sorkin, R. D., D. E. Robinson, and B. G. Berg. "A Detection Theory Method for Evaluating Visual and Auditory Displays." In *Proceedings of the Human Factors Society*, Vol. 2, 1184–1188, 1987.
> This paper describes a signal detection method for evaluating different display codes and formats. The method can be used to assess the relative importance of different elements of the display. The paper briefly summarizes data from different types of auditory and visual displays.

Sorkin, R. D., and D. D. Woods. "Systems with Human Monitors: A Signal Detection Analysis." *Hum.-Comp. Inter.* **1** (1985): 49–75.
> This paper analyses the general system composed of a human operator plus an automated alarm subsystem. The combined human machine system is modeled as a two-stage detection system in which the operator and alarm subsystem monitor partially correlated noisy channels. System performance is shown to be highly sensitive to the decision bias (response criterion) of the alarm. The customary practice of using a "liberal" bias setting for the alarm (yielding a moderately high false alarm rate) is shown to produce poor overall system performance.

Sorkin, R. D., B. H. Kantowitz, and S. C. Kantowitz. "Likelihood Alarm Displays." *Human Factors* **30** (1988): 445–459.
> This study describes a type of multilevel or graded alarm display in which the likelihood of the alarmed condition is encoded within the display. For example, the levels of an auditory alarm could vary by repetition rate or voice quality; and the levels of a visual display could vary by color. Several dual-task (tracking and alarm monitoring) experiments demonstrate the feasibility of Likelihood Alarm Displays.

Sorkin, R. D. "Why are People Turning Off Our Alarms?" *J. Acous. Soc. Am.* **84** (1988): 1107–1108. Reprinted in *Human Factors Soc. Bull.* **32** (1989): 3–4.
> In this short paper Sorkin describes several tragic accidents in which auditory alarms had been disabled or ignored. The author argues that two culprits are high false-alarm rates and excessive sound levels.

Sorkin, R. D. "Perception of Temporal Patterns Defined by Tonal Sequences." *J. Acous. Soc. Am.* **87** (1990): 1695–1701.
> In this study Sorkin describes a general model (the temporal correlation model) for predicting a listener's ability to discriminate between two auditory tone sequences that differ in their temporal pattern. According to the model, the listener abstracts the relative times of occurrence of the tones in each pattern and then computes the correlation between the two lists of relative times.

Speeth, S. D. "Seismometer Sounds." *J. Acous. Soc. Am.* **33** (1961): 909–916.
:   The author audified seismic data (sped up the playback of data recorded by seismometers to place the resultant frequencies in the audible range), and then set human subjects to the task of determining whether the stimulus was a bomb blast or an earthquake (after an appropriate training program). In this experiment, subjects were able to correctly classify seismic records as either bomb blasts or earthquakes for over 90% of the trials. Furthermore, because of the time compression required to bring the seismic signals into the audible range, an analyst could review 24-hours worth of data in about 5 minutes.

Stifelman, L. J., B. Arons, C. Schmandt, and E. A. Hulteen. "VoiceNotes: A Speech Interface for a Hand-Held Voice Notetaker." In *Proceedings of INTERCHI '93*, 179–186. Reading, MA: ACM Press/Addison-Wesley, 1993.
:   VoiceNotes is an application for a voice-controlled hand-held computer that allows the creation, management, and retrieval of user-authored "voice notes"—small segments of digitized speech containing thoughts, ideas, reminders, or things to do. VoiceNotes explores the problem of capturing and retrieving spontaneous ideas, the use of speech as data, and the use of speech input and output in the user interface for a hand-held computer without a visual display.

Stratton, V. N., and A. H. Zalanowski. "The Effects of Music and Cognition on Mood." *Psych. Music* **19** (1991): 121–127.
:   The author presents evidence that although the expected responses to pieces of music, selected for their affect inducing effects, apparently influenced the mood state of subjects performing a concurrent cognitive task (storytelling about a picture), the effect disappeared when specific mood instructions were given with the storytelling instructions. There was evidence that familiarity with the music and subjects individual preferences for the music selected also affected the extent to which their mood was influenced by the music.

Strothotte, T., K. Fellbaum, K. Crispien, M. Krause, and M. Kurze. "Multimedia Interfaces for Blind Computer Users." In *Rehabilitation Technology—Proceedings of the 1st TIDE Congress*, held April 6–7, 1993, in Brussels. ISSN: 0926-9630. IOS Press, 1993.
:   This paper deals with selected aspects of blind peoples' access to GUI computer systems which are addressed by the GUIB project (Textual and Graphical User Interfaces for Blind people). A new loudspeaker-based device for two-dimensional sound output to enable users to locate the position of screen objects is described. In a prototypical application, blind people are given access to a class of computer-generated graphics using the new device in an interactive process of exploration.

Strybel, T. Z., A. M. Witty, and D. R. Perrott. "Auditory Apparent Motion in the Free Field: The Effects of Stimulus Duration and Intensity." *Percep. & Psycho.* **52(2)** (1992): 139–143.
:   The authors find that a minimum duration of 10–50 msec is required for the perception of auditory apparent motion, with the exact time varying from listener to listener.

Stuart, R. "Virtual Auditory Worlds: An Overview." In *VR Becomes a Business: Proceedings of Virtual Reality '92*, held September 1992, in San Jose, CA, 144–166. Westport, CT: Meckler, 1992.
> An overview of issues concerning virtual auditory environments and applications that have been proposed or on which work is proceeding. It includes an extensive bibliography.

Sumikawa, D. A., M. M. Blattner, K. I. Joy, and R. M. Greenberg. "Guidelines for the Syntactic Design of Audio Cues in Computer Interfaces." In *Nineteenth Annual Hawaii International Conference on System Sciences* Los Alamitos, CA: IEEE Computer Society Press, 1986.
> The material for this article is drawn from an M.S. thesis with the same name by Denise A. Sumikawa, University of California, Davis, also published as Lawrence Livermore National Laboratory Technical Report, UCRL-53656, June 1985. The material was later extended and became: "Earcons and Icons: Their Structure and Common Design Principles."

Tenney, J., and L. Polansky. "Temporal Gestalt Perception in Music." *J. Music Theory* **24(2)** (1979): 205–241.
> The authors describe a simple computational system for identifying hierarchical structure in melodies based on frequency proximity and temporal distance, and assuming categorical perception. The model is based to some extent on Tenney's 1961 book, *Meta Hodos*.

Terenzi, F. "Design and Realization of an Integrated System for the Composition of Musical Scores and for the Numerical Synthesis of Sound (Special Application for Translation of Radiation from Galaxies into Sound Using Computer Music Procedures)." Unpublished manuscript, Physics Department, University of Milan, 1988.
> Terenzi describes audification of radio astronomy data and musical use of the results.

Terhardt, E. "Pitch, Consonance, and Harmony." *J. Acous. Soc. Am.* **55(5)** (1974): 1061–1069.
> Stumpf used the listener's judgment of the harmonic interval's "fusion" as the ranking criterion for his perception study; Malmberg used "smoothness" for his. The discrepancies between their studies are a result of the uncontrolled overtone series used in tone production, and the bias due to the adjective used to describe the harmonic intervals. Psychoacoustic studies relating to harmonic intervals continued in the first half of the twentieth century. Reviews of their contributions can be found in this article.

Terhardt, E. "Toward Understanding Pitch Perception: Problems, Concepts and Solutions." In *Psychophysical, Physiological and Behavioural Studies in Hearing*, edited by G. van den Brink and F. A. Bilsen. Delft University Press, 1980.
> Terhardt presents a broad overview of pitch perception concepts and terminology.

Terhardt, E. "Pitch of Pure of Tones: Its Relation to Intensity." In *Facts and Models in Hearing*, edited by E. Zwicker and E. Terhardt. New York: Springer-Verlag, 1974.
> The author finds that for some individuals the change in pitch with intensity can be as large as that reported by Stevens; however, when these changes are averaged over many subjects, the changes are insignificant.

Terhardt, E. "Gestalt Principles and Music Perception." In *Auditory Processing of Complex Sounds*, edited by W. A. Yost and C. S. Watson. Hillsdale, NJ: Lawrence Erlbaum, 1987.
> Terhardt identifies contours and categorized representations of sensory objects as the focus of gestalt perception. The theory presented is known as Hierarchical Processing of Categories (HPC) in which perception is organized on hierarchical layers, each of which are concerned only with categorizing or recategorizing. The most peripheral level corresponds to analytic listening and spectral pitch, analogous to primary visual contours, whereas synthetic forms are secondary, including virtual pitch. The multiple levels of perception associated with Western Tonal music are considered within the context of HPC.

Terwoght, M. M., and F. van Grinsven. "Musical Expression of Moodstates." *Psych. Music* **19** (1991): 99–109.
> The authors present an experimental study of emotional response to short extracts from eight pieces of classical music.

Treisman, A. "Properties, Parts, and Objects." In *Handbook of Perception and Human Performance*, edited by K. R. Boff, L. Kaufman, and J. P. Thomas, Chap. 35. New York: Wiley, 1986.
> Treisman explores the information-processing mechanisms that identify the objects and events of subjective experience from physical stimuli. Includes coverage of similarity judgment, perceptual analysis of dimensions/features/parts, and integration of parts and properties. Although most of the coverage and examples are based on visual studies, auditory studies are covered where appropriate.

Truax, B. *Acoustic Communication*. Norwood, NJ: Ablex, 1984.
> Broad coverage of sound including speech, music, natural sound and sounds of modern life. Truax presents a communication model that places sound in a mediating role between listeners and the environment. He presents results of the World Soundscape Project.

Tzelgov, J., R. Srebro, A. Henik, and A. Kushelevsky. "Radiation Detection by Ear and by Eye." *Human Factors* **29(1)** (1987): 87–98.
> Tzelgov et al. found that in a seach task the auditory signal was better than the visual display or the dual mode system. In a detection task, there were no differences between the single modes and no differences between single modes and dual mode. Their interpretation of the results considers a visual bias effect in which the operator's attention is directed away from other aspects of the monitoring task.

Urdang, E. G., and R. Stuart. "Orientation Enhancement Through Integrated Virtual Reality and Geographic Information Systems." In *Proceedings of CSUN's Seventh Annual International Conference on Technology and Persons with Disabilities*, held March 1993 in Northridge, CA, 55–61.
> The authors propose the use of virtual acoustic display in combination with geographic information systems to assist a visually impaired user to navigate in a city.

Vanderveer, N. J. *Ecological Acoustics: Human Perception of Environmental Sounds*. Dissertation Abstracts International. 40/09B, 4543. University Microfilms No. 8004002.
> Probably the earliest example of an ecological approach to audition. Vanderveer describes the result of a protocol study in which people identified events from the sounds they made.

Vicario, G. B. "Some Observations in the Auditory Field." In *Organization and Representation in Perception*, edited by J. Beck. Hillsdale, NJ: Lawrence Erlbaum, 1982.
> Vicario identifies dependency relations, effects of context (embedded figures), the concept of an auditory field, and completion phenomena in the auditory mode of perception.

Walker. R. "The Effects of Culture, Environment, Age, and Musical Training on Choices of Visual Metaphors for Sound." *Percep. & Psycho.* **42** (1987): 491–502.
> The author reports on studies of choices of visual metaphors for sound parameters. In the sound domain he looked at frequency, waveform, amplitude, and duration.

Ward, W. D. "Subjective Musical Pitch." *J. Acous. Soc. Am.* **26(3)** 369–380.
> Ward presents empirical evidence that the pitch of pure tones is a subjective judgment of the listener, not very consistent across subjects, and whose rate of change is independent of the frequency level of the acoustic source.

Wallach, H., E. B. Newman, and M. R. Rosenzweig. "The Precedence Effect in Sound Localization." *Am. J. Psych.* **57** (1949): 315–336.
> In a reverberant room, two similar sounds reach a subject's ears from different directions, with one sound following the other after a short delay; yet the subject fuses them into a single sound and localizes this sound based on the source of the first sound to reach the ears. The authors study this perceptual phenomena, which they term the "precedence effect."

Warren, D. H., R. B. Welch, and T. J. McCarthy. "The Role of Visual-Auditory 'Compellingness' in the Ventriloquism Effect: Implications for Transitivity Among the Spatial Senses." *Percep. & Psycho.* **30** (1981): 557–564.
> The authors study intersensory interactions and find, with sufficiently compelling cues, visual cues can [text missing?].

Warren, W. H., R. R. Verbrugge. "Auditory Perception of Breaking and Bouncing Events: A Case Study in Ecological Acoustics." *J. Exp. Psych.* **10** (1984): 704–712.
> A seminal study of everyday listening which used analysis and synthesis of events to link acoustical information to event perception.

Weber, C. R. "Sonic Enhancement of Map Information: Experiments Using Harmonic Intervals." Unpublished dissertation, Department of Geography, State University of New York at Buffalo, 1993.
> The author has found an relationship among aural variables that is analogous to the hierarchy of visual variables presented by Bertin (1983). Pitch supersedes both texture (consonance) and color (scale position).

Weber, C. R., and M. A. Yuan. "Statistical Analysis of Various Adjectives Predicting Consonance/Dissonance and Intertonal Distance in Harmonic Intervals." Technical Papers, ACSM/ASPRS Annual Convention, New Orleans, Vol. 1, 391–400.
> The authors report successful delineation of relative consonance and intertonal distance selection by subjects associating dyads with various continua of cartographic adjectives. These results seem to hold only when the dyads are presented in isolation.

Welch, R. B. *Perceptual Modification: Adapting to Altered Sensory Environments.* New York: Academic Press, 1978.
> Excellent classic source on adaptation and response to presentation of altered sensory cues. Welch also considers intersensory interactions (more on this can be found in a paper by Welch and Warren (1986)).

Welch, R. B., and D. H. Warren. "Intersensory Interactions." In *Handbook of Perception and Human Performance,* edited by K. R. Boff, L. Kaufman, and J. P. Thomas, chap. 25. New York: Wiley, 1986.
> The authors review and evaluate research on intersensory bias, particularly interactions between vision and audition on detection, spatial localization, and perception of temporal events. They take the view that sensory modalities vary in their appropriateness for the perception of various events.

Wenzel, E. M., F. L. Wightman, and S. H. Foster. "Development of a Three-Dimensional Auditory Display System." *SIGCHI Bull.* **20** (1988): 52–57.
> An early description of the three-dimensional auditory display system created at NASA-Ames that would become the Convolvotron. The authors describe measurement and testing of HRTFs.

Wenzel, E. M., and S. H. Foste. "Real-Time Digital Synthesis of Virtual Acoustic Environments." *Comp. Graphics* **24(2)** (1990): 139–140.

Wenzel, E. M., F. L. Wightman, and D. J. Kistler. "Localization with Nonindividualized Virtual Acoustic Display Cues." in *CHI '91 Proceedings*, 351–359. Reading, MA: ACM Press/Addison-Wesley, 1991.
> Virtual interface research is represented in the work of Wenzel et al. (1988, 1990, 1991) who have developed three-dimensional auditory cues transmitted over user-worn headphones. The authors have found that even simple auditory cues—such as a sound signaling a direction, distance, and, finally, contact with a virtual object—can aid the user in manipulating the virtual world.

Wenzel, E. M. "Localization in Virtual Acoustic Displays." *Presence: Teleop. & Virtual Environ.* **1** (1992): 80–107.
> Wenzel provides an overview of the acoustical, psychoacoustical, and technological bases for the synthesis of spatial sound in virtual displays, with an emphasis on the work conducted at NASA-Ames Research Center.

Wenzel, E. M. "Spatial Sound and Sonification." In *Auditory Display: Sonification, Audification, and Auditory Interfaces*, edited by G. Kramer. Santa Fe Institute Studies in the Sciences of Complexity, Proc. Vol. XVIII. Reading, MA: Addison Wesley, 1994.
> Wenzel provides a brief description of three-dimensional sound synthesis and describes the performance advantages that can be expected when these techniques are applied to sound streams in sonification displays. Specific examples, and the lessons learned from each, are discussed for applications in telerobotic control, aeronautical displays, and shuttle launch communications.

"What's That Noise." *Home Mechanix* (May 1986): 81–107.
> This article includes descriptions of the sounds that can be useful in diagnosing problems with automobiles.

Wiener, F. M., and D. A. Ross. "The Pressure Distribution in the Auditory Canal in a Progressive Sound Field." *J. Acous. Soc. Am.* **18** (1946): 401.
> The authors took sound measurements using probe microphones. According to Blauert (1969), they were the first to measure the linear distortions caused by pinna, head, and ear canal.

Wightman, F. L., and D. J. Kistler. "Headphone Simulation of Free-Field Listening I: Stimulus Synthesis." *J. Acous. Soc. Am.* **85** (1989): 858–867.
> The authors describe a technique for measuring head-related transfer functions and synthesizing static virtual sound sources, which forms the basis of current approaches to spatial sound displays.

Wildes, R., and W. Richards. "Recovering Material Properties from Sound." In *Natural Computation*, edited by W. Richards. Cambridge, MA: MIT Press, 1988.
> Using analytical physics, the authors suggest that auditory identification of material involves judging damping and partial bandwidth, which together specify the internal friction-characterizing materials.

Williams, M. G., S. Smith, and G. Pecelli. "Experimentally Driven Visual Language Design: Texture Perception Experiments for Iconographic Displays." In *Proceedings of the IEEE 1989 Visual Languages Workshop*, held in Rome, Italy, 62–67. Rome: IEEE, 1989.
> In this report the authors describe the only formal experiment conducted to date by the University of Massachusetts' Lowell group to evaluate their "iconographic" approach to both visualization and sonification. Like many similar experiments conducted during the 1980s, it showed that subjects' performance on a data analysis task improved modestly when the subjects used a combined visual-auditory data display rather than just a visual data display.

Williams, M. G., S. Smith, and G. Pecelli. "Computer-Human Interface Issues in the Design of an Intelligent Workstation for Scientific Visualization." *SIGCHI Bull.* **21 (4)** (1990): 44–49.
> The Exploratory Visualization project presents sonification of anatomic map data through an iconic technique. See annotations to S. Smith and G. Grinstein.

Williams, S. M. "STREAMER: A Prototype Tool for Computational Modelling of Auditory Grouping Effects." Research Report No: CS-89-31, Department of Computer Science, University of Sheffield, 1989.
>   Williams presents a simple gestalt-based computational model of auditory streaming together with a proposition for a framework of auditory gestalt.

Williams, S. M. "Perceptual Principles in Sound Grouping." In *Auditory Display: Sonification, Audification, and Auditory Interfaces*, edited by G. Kramer. Santa Fe Institute Studies in the Sciences of Complexity, Proc. Vol. XVIII. Reading, MA: Addison Wesley, 1994.
>   Overview of auditory perception from a gestalt viewpoint, presenting examples of phenomena which may influence the interpretation of Auditory Displays.

Witten, M. "Increasing Our Understanding of Biological Models Through Visual and Sonic Representations: A Cortical Case Study." *Intl. J. Supercomp. Appl.* **6(3)** (Fall 1992): 257–280.
>   Witten describes the use of integrated sonification and visualization in representation of digitized image data.

Yeung, E. S. "Pattern Recognition by Audio Representation of Multivariate Analytical Data." *Anal. Chem.* **52** (1980): 1120–1123.
>   Audible display for experimental data from analytical chemistry, designed to incorporate auditory parameters exhibiting continuity in scaling and relative independence from each other. His display consisted of data vectors, each dimension of which corresponded to the detected levels of various metals in a given sample, with one vector per sample. The analysis task involved classifying a given vector as belonging to one of four sets, after having been trained with vectors from those four sets. Although Yeung did not compare the performance of his subjects using the auditory display with that of any other display, he noted that all of his subjects achieved the 98% correct classification rate after (at most) two training sessions.

Yost, W. A., and C. S. Watson, eds. *Auditory Processing of Complex Sounds.* Proceedings of a workshop in April, 1986. Hillsdale, NJ: Lawrence Erlbaum, 1987.
>   This book includes a wide range of psychoacoustic and physiological studies on the processing of complex sounds.

Zwislocki, J. "Temporal Summation of Loudness." *J. Acous. Soc. Am.* **46** (1969): 413–441.
>   In this study, among several others, Zwislocki shows that the ear averages sound energy over periods up to 200 milliseconds, so that loudness increases by 10 dB with a tenfold increase in duration.

APPENDIX IV:
Annotations to Audio Examples

1. INTRODUCTION

This appendix provides a quick reference to the sound examples contained on the audio compact disk included with this volume. The CD, labeled ICAD '92, contains 96 tracks of audio examples that were submitted by the authors to illustrate points and demonstrate techniques.

Each chapter for which there are sound examples (11 out of the 21 chapters) has the program number of the appropriate sound example marked in the margins (see example, right margin this page). We suggest that the reader/listener refer directly to the relevent chapter for a full explantation of the principles to which the sound examples apply. Again, this written appendix is primarily intended as an orienting reference to these examples.

Tr 0

The use of headphones is strongly advised. For some tracks, most notably Wenzel's spatialization demonstrations, headphones are essential to perceiving the intended phenomena. For others, such as Madyastha et al. and Kramer, the use of headphones will create a more distinct and compelling demonstration of the use of stereo location as a sonification variable.

A note about voice annotations on the CD: To orient the reader and assure a user-friendly result, the editor required that all sound examples be accompanied by some voice annotation. Some of the examples are extensively annotated, e.g., Wenzel, Kramer, and Madyastha. Others are minmally annotated, e.g., Scaletti, Fitch, Blattner and Brewster. The annotations were all specified by the authors. Additionally, orientation and ease of use took precedence over any difficulties that may be encountered in cueing up examples for classroom use.

TRACKS 1–23
Sheila Williams
"Perceptual Principles in Sound Grouping"

All the sounds in the examples provided to illustrate the paper on "Perceptual Principles in Sound Grouping" are created using the MITSYN[1] synthesiser and are described in the text of the paper, together with the percept they are most expected to induce.

Technical details follow:

1. Similarity

    a. Track 1. Pure tones all of same frequency, duration, and amplitude.

    b. Track 2. Pure tones of same duration and frequency as 1(a) alternating 2 high, 1 low amplitude.

2. Proximity

    a. Track 3. Tones of equal amplitude and duration, four different frequencies equally spaced on log scale.

    b. Track 4. Tones of equal amplitude and duration, four different frequencies but lower two separated from upper two by larger distance than relationship between each pair on log scale.

3. Good Continuation

    a. Track 5. Alternating pure tones at different frequencies (equal amplitude and duration) connected by tone glides.

    b. Track 6. Alternating pure tones at different frequencies (equal amplitude and duration) separated by silent intervals.

---

[1] Henke, W. L., *MITSYN: A Synergistic Family of High-Level Languages for Signal Processing* (Version 8.1), 133 Bright Road, Belmont, Massachusets 02178, 1990.

Annotations to Audio Examples

4. Habit or Familiarity

    a. **Track 7**. Music. Pure tones created by single oscillator. Equivalent amplitude envelopes, frequencies, and durations based on the accompaniment to Mozart's "Das Kinderspiel." (Virtual polyphony may be heard during part of this piece.)

    b. **Track 8**. Speechlike sounds created by adjusting the MITSYN "score" input to the demonstration speech synthesiser supplied with MITSYN (i.e., varying the numbers which form the input to the speech synthesis demo).

5. Belongingness

    a. **Track 9**. Three-tone complexes alternating with silence. The pure tone partials of each complex are of equal duration and related by octaves.

    b. **Track 10**. Three-tone complexes alternating with single pure tones of the same frequency as the lowest tone in each complex.

6. Common Fate

    a. **Track 11**. Three-glide complexes alternating with silence. The partials are related by octaves throughout the glide.

    b. **Track 12**. Synchronous components made up of two glides plus one (middle) pure tone of the same centre frequency as the centre glides in 6(a). The outer glide partials same as in 6(a).

7. Closure

    a. **Track 13**. Repeating single pure tone separated by silence, no amplitude shaping at start and stop of each tone.

    b. **Track 14**. Repeating single pure tone separated by relatively loud noise burst.

    c. **Track 15**. Repeating single tone separated by relatively loud noise burst, in which both noise and tone amplitude envelopes are shaped at onset and offset.

8. Stability

    a. **Track 16**. Alternating pure tones with silent intervals which progressively decrease.

    b. **Track 17**. Alternating pure tones with silent intervals which progressively increase.

9. van Noorden's alternating two-tone patterns

   a. Track 18. Relatively fast alternating pure tones close in frequency

   b. Track 19. Relatively fast alternating pure tones at greater frequency disparity

   c. Track 20. Slower alternating pure tones at the same frequency disparity as in 9(b).

10. Three-tone streaming

    a. Track 21. Repeated pattern of three sequential pure tones with middle tone close in frequency to upper tone

    b. Track 22. Repeated three tone pattern with middle tone close in frequency to lower tone

    c. Track 23. Repeated three-tone pattern with middle tone around average streaming midpoint found for our subjects

TRACKS 24–25
Elizabeth M. Wenzel
"Spatial Sound and Sonification"

Track 24. [As described in the section on "Performance Advantages of Virtual Acoustic Displays."]

This example illustrates the advantages of using spatialized sound in the context of speech communications for commercial aviation. Four radio communications channels were originally recorded at Phoenix International Airport. Subsequently, the signals were either mixed simultaneously as in the normal method of presentation in the cockpit, or spatialized by the Convolvotron into four separate locations. Notice that in the spatialized segments, the individual speech messages are much more intelligible and it is easier to switch attention between them than with the nonspatialized signals.

Track 25. [As described in the section on Aeronautical Displays within "Applications Case Studies."]

This example illustrates a spatialized auditory cue for a TCAS, or Traffic Collision Avoidance System. It is presented as it might be used in the cockpit to warn pilots of aircraft in the vicinity that are on a potential collision course. Currently, the word "traffic" is used as an alert in combination with a two-dimensional, radarlike visual display, and the pilot is required to "acquire" the aircraft by visually locating it out the window. In the example, and in the study by Begault (in press), the word "traffic" has simply been spatialized to correspond to one of seven segments of the pilot's out-the-window view. Under these conditions, targets were acquired about 2.2 sec faster than with nonspatialized alerts.

Acknowledgement: These demonstrations were produced by Durand Begault.

TRACKS 26–29
Robin Bargar
"Pattern and Reference in Auditory Display"

Musical technqiues can be used to design auditory representations with differentiated signals. It will be helpful to separate these devices from the musical styles where they commonly appear. The following examples demonstrate the distinction between musical devices and musical style. It can be understood that musical devices do not depend upon the styles where they are commonly heard. Rather, the styles result from applying musical devices in a particular way.

Track 26. Inflection originates in speech. It can be generated and controlled by numerical data. In this example, inflection depends upon amplitude and brightness changes; silences help emphasize inflection by breaking the sound stream into groups that are similar to speech groups. Silences allow the ear some rest from the sound stream, enabling finer judgements upon relative emphasis of various accents. Inflection is used in most musical styles. To help distinguish it from other musical components, this example uses inflection as a more important organizing tool than pitch or meter. Sounds in this example do not stress standard western musical structures of "melody" or "beat," in the traditional sense of emphasizing repeated patterns of predominantly stepwise pitch change (half steps), or regular metric emphasis of divisions and subdivisions of time units (typically 3/4 or 4/4 time signatures).

Data with many irregularites may be represented using inflection techniques, rather than attempting to create a more regular musical framework. Note that melodic and metric characteristics are still present, though irregular, in this example. Pitch changes are mostly by disjunct motion; phrases are articulated by accents of loudness and brightness, separated by pauses of varying durations that are not regular (in most music, meter is constrained to placing new notes on time units that are integer multiples of 2 or 3). This suggests that pitch and meter may provide a secondary platform for representing highly irregular data, while accent, brightness, and silence provide the primary platform.

Track 27. This example shows how constraints can be applied to create more identifiable audio channels, and to couple several channels together. Constraints applied include regular meter (a "beat") and a pitch scale applied to three distinct frequency bandwidths. The material in each band is different but adheres to the same constriants, helping it to combine with the other bands into a single audio information channel. By limiting the new material to three frequency bandwidths, and giving each bandwidth a unique role with a common meter and tonal center, more predictable material emerges against the less traditional inflection materials. The common tonal center is essentially a structure based on octave equivalence and equal divisions of the octave into a traditional musical scale. A sieve applied to incoming data, such as a histogram with pitch-related bins, will force data to adhere to this type of pitch constraint. The ostinato, a repeated pattern of regular pulses in the middle frequency band, provides a time reference for listeners to measure the rate of change of the three parts together. Sieves and histograms can also be used to create this type of metric constraint. Pitch and metric sieves allow simpler patterns that are easier to follow in groups. The inaccuracy of the resulting quantization of data must be anticipated as a trade-off for the reduced complexity of the representation.

Track 28. The grouping techniques of Track 27 are extended in this example. The initial ostinato signals the presence of a regular time unit, and the introduction of percussion corroborates that timing. Pitch and rhythmic changes are aligned with this timing so that a sense of regular meter is reinforced by all parts. The parts are located in the same frequency bands as example 2.

The most relevant concepts here for data representation are those of redundancy and reliability of an auditory channel. Many audio parts may be combined to represent a single data stream, in order to reinforce the identity of that stream with multiple pitch and rhythmic conventions. The use of familiar musical patterns confirms the regularity and reliability of the auditory space (reliability being a function of the listener's ability to predict the next musical event, not a characteristic of the sonic event itself).

Reliability, that is, predictability, can be an important perceptual impression to create, if the representation intends to maximize the likelihood that changes to the data will be audible. In this example, the presence of a "reliable" pattern creates a strong sense of interruption when the pattern ceases

Annotations to Audio Examples

The familiarity of the patterns in this example, and their momentary reliability, suggest that beyond an alternative auditory space, this example establishes a musical syntax. Note that in both examples 2 and 3 the more abstract material of example 1 is present, but that it fades in attention in the presence of a familiar syntax. Data streams rendered with greater degrees of regularity can also be expected to emerge to the fore of a listener's attention.

Track 29. The limit of listeners' capbility to follow multiple channels often presents problems for representing multiple simultaneous data streams. Two devices that can improve listening performance are multiple syntaxes and a judicious use of silence. Creating silence requires establishing a threshold for measuring changes in the data, beneath which no sound is produced. Syntax assists instrumentation in identifying different data variables. In this example, segments of musical syntaxes are given a distribution similar to that of the material in Track 26. Although the material is irregular, it is marked by recognizable features that help the listener to distinguish between several recurring identities. In this case the term "polyphony" might be used to suggest the simultaneous tracking of these separate syntaxes (an extension of its traditional meaning in music which applies to simultaneous auditory channels of greater similarity[2]).

As the example continues the syntaxes are allowed longer durations and greater overlap. In this case their stereo position and the apparent distance of the sound source from the listener, helps to keep the auditory identities separate. The most striking examples of this involve the appearance of a drum part far in the background that traverses the stereo space. Notice that the simplicity of the drum syntax allows it to be reduced to a narrow frequency band and low amplitude, and still maintain a perceived independent identity.

---

[2] It is not unlikely that some musicians would object to this use of the term "polyphony," as it is commonly understood to account for the unification of musical materials rather than the simultaneous presence of distinct auditory spaces. The decision to use this term here reflects upon the historical likelihood that, as modern observers, we tend to stylistically unify those older musics that are referred to as "polyphonic," whereas historically, contemporary listeners may have been able to hear a greater independence of those (then new) juxtapositions of sound.

TRACKS 30–35
Gregory Kramer
"Some Organizing Principles for Representing Data with Sound"

The sonification examples used here represent 4.5 years of weekly financial data rendered in a multidimensional display. In all there are 265 five-dimensional points. The five dimensions are: stocks, bonds, the U.S. Dollar, interest rates, and commodities.[3]

The data covers a period from September, 1987 through March, 1992. See Figure 1 for a graphical representation of the data.

Track 30: Introduction This track introduces the data variable to sound parameter mappings. The basic mappings for this sonification are the following:

- Stocks are mapped to pitch.
- Bonds are mapped to pulsing speed.
- The dollar is mapped to brightness.
- Interest rates are mapped to detuning.
- Commodities are mapped to attack time.

Track 31: Scaling the range of the auditory variable. [Described in Section 7.1 of the paper.]

It is possible to adjust the relative forcefulness of different auditory variables in a variety of ways. First we focus on scaling down the range of the auditory variable by setting the upper and lower limits of that variable in the target sound generator.

We will listen to 28 weeks of data around the 1987 stock market crash. Since stock data controls pitch, a substantial drop in pitch at the crash will be evident. We then narrow the pitch range so that the same data causes a less dramatic pitch change.

To place this in context, we then sonify the same segment with all five data variables represented in the sound. The narrower pitch range should allow us to hear the other dimensions with less distraction than the wider ranging pitch variable we hear first.

---

[3] More specifically, the data include: Stocks—The Dow Jones Industrial Average; Bonds—The Lehman Brothers T-Bond Index; Interest Rates—Federal Funds Rate, NY Fed; U.S. Dollar—J.P. Morgan Index vs. 15 Currencies; and Commodities—CRB Futures Index (1967=100).

Annotations to Audio Examples

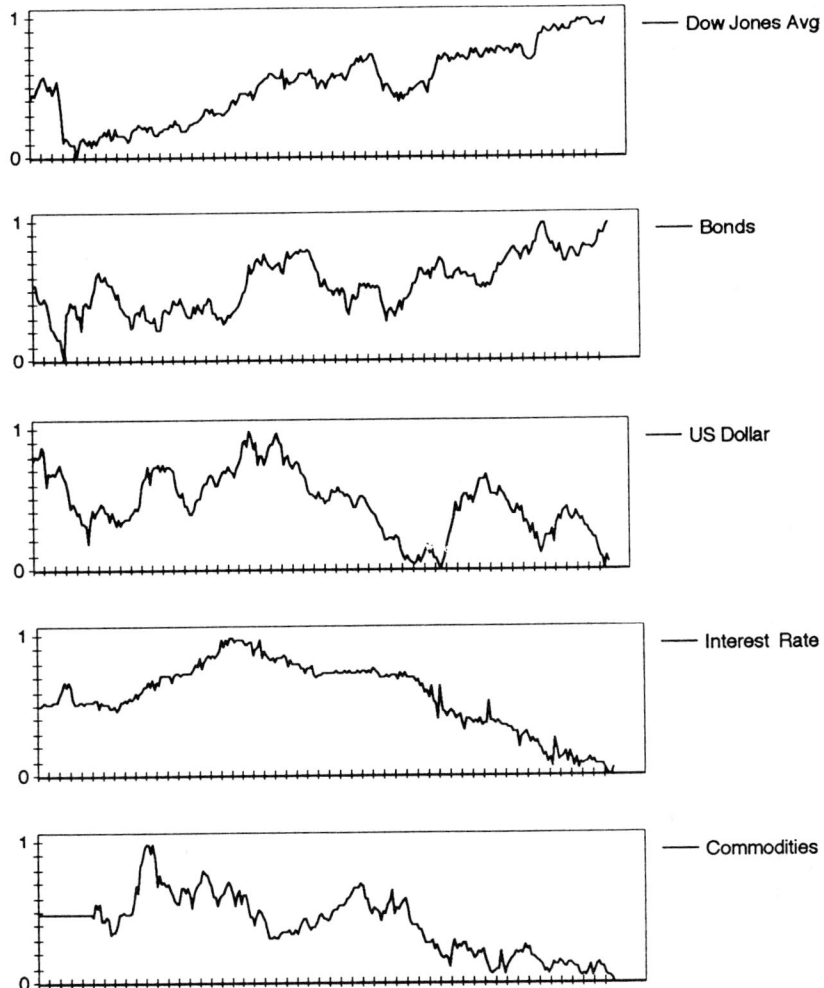

FIGURE 1 Graphical representation of five dimensions: stocks, bonds, interest rates, U.S. dollar, and commodities.

Track 32: Mapping the data to more than one variable. [As described in Section 7.2.]

We now map one of the variables to more than one sound parameter. We will loop a 20-week period in early 1991 when the dollar rose steeply. As we map the dollar to additional parameters, you will hear repeated iterations of the same data. For the first three iterations the dollar controls only attack time. For the next three, we add detune. For the next three, we add brightness. And, for the last three, we add stereo location.

As we add parameters, the contour in the data should become more evident.

Track 33: Weighting by a variety of mappings. [As described in Section 7.3.]
We now take this same 20-week region of data and play the original mapping three times through. The mappings are the following:

- Stocks to pitch.
- Bonds to speed.
- Dollar to brightness, detune, and stereo location.

We then switch the stocks and the dollar mappings, producing a new display:

- Dollar to pitch.
- Stocks to brightness, detune, and stereo location.
- Bonds to speed (as before).

The dollar's rise should be evident in the pitch change. We then switch the dollar and bonds, to produce the mapping:

- Dollar to speed.
- Bonds to pitch.
- Stocks to brightness, detune, and stereo location (as before).

The contour previously controlling pitch is now present an increase in speed of the pulses.

Track 34: Beacons. [As described in Section 8.]
This track demonstrates the use of beacons. The mappings used are as described in the introduction, with the additional mapping of commodities to stereo location.

- Stocks are mapped to pitch.
- Bonds are mapped to pulsing speed.
- The dollar is mapped to brightness.
- Interest rates are mapped to detuning.
- Commodities are mapped to attack time and stereo location.

First, we will play the 1987 market crash and extract a beacon from the bottom of the crash. A tone is played to indicate when the beacon was sampled. Then we listen to data from 4.5 years later, when the market was heating up and we will extract a beacon from that period. Now, we will play these representations of the two system states back to back. Recall that each beacon represents a distinctive multidimensional data point in the 5-dimensional data base as well as in time. Attempts to mentally maintain specific data to sound mappings tend to be quite difficult. At the same time, hearing overall sound qualities, or gestalts, happens spontaneously. Changes in pitch, speed, brightness, location, and so on give each beacon a distinctive quality that can be associated or identified with distinctive system states without mental deconstruction of the component parameters.

Now we use the second beacon, from early 1992, as a reference point in a broader movement of the financial markets. We play through the two year period in 1991 and 92 where the bull market heated up, periodically injecting the selected beacon, which serves as a reference to where the market is headed.[4]

Track 35: Dynamic Beacons. [As described in Section 8.1.]

Dynamic beacons represent a short segment of time-sequenced data. We have extracted these beacons from the financial data set and will use the same mappings as for the static beacons. We will play each beacon three times to make it easier to hear the contour in the data.

Hearing overall sound qualities, or gestalts, is fundamental to the beacons concept. It may, nevertheless, be interesting to focus on specific sound variables to determine if changes other than pitch can be heard during these short beacons. In particular, speed changes should be prominent in the second set of beacons and brightness changes in the third set of beacons. This is because the bonds and dollar data increased substantially in the segments selected.

Once you can easily do this, try attending to combinations of variables, such as speed (bonds) and pitch (stocks). You can try this with the first two dynamic beacons, which focus on a region of the data where there is a rapid change in stocks (pitch), and a concurrent change in bonds. Recall, however, that observing overall system states as a result of auditory gestalt formation, is bound to be a more effective way to use beacons than attending to the internal structure of the beacons.

Six dynamic beacons: First, we will display a large drop in stocks, then a smaller stock drop two years later. The next dynamic beacon represents a segment in which bonds increased. Bonds, you will recall, are mapped to speed. Following that is a beacon representing another segment [one year later] in which the bond market jumped. Finally, we will listen to the representation of an increase in the dollar [mapped to brightness] followed by another segment, 2.5 years later, in which the dollar increased.

---

[4] In some ways this may be compared to the auditory equivalent of seeing the landscape behind a picket fence as one drives by. We can hear the trend, even though it is interrupted by an intermittent obstruction. In this example, we can hear the more or less monotonic movement towards a higher pitch (higher stock values) and towards the right speaker (a reduction in commodities prices), as we approach the point from which the beacon data was extracted.

TRACKS 36–41
Carla Scaletti
"Sound Synthesis Algorithms for Auditory Data Representations"

[See the section "Sound Synthesis Techniques" for a discussion of the methods used in these examples.]

Sound can be used to provide feedback in a matching or calibration task or to assist in detecting when one data stream is the same as or different from another. The following five examples exploit what is known about the results of adding sinusoids and multiplying sinusoids to extract information about the underlying data streams.

Track 36: Matching or detecting a reference value. In these examples, the reference point is constant and mapped to a frequency of 440 hz. A time-varying data stream is mapped to frequencies from 0 to 440 hz.

a. Reference value mapped to 440-hz sinusoid.

b. Sine with data mapped to frequency: Data is one half-period of a sine.

c. Reference * (0 to 440 hz): You can hear both the sum and the difference frequencies. As the changing frequency approaches 440, the difference frequency approaches zero. Unfortunately, it goes below the range of audible frequencies as it drops below 20–30 hertz, so it does not provide a very accurate indication of when the data stream is equal to the reference.

d. Reference + (0 to 440 hz): In this example, both the reference and the data stream are audible. As the values in the data stream approach the reference, you can hear beats. The beat frequency goes to zero when the values are equal to the reference. This could be useful for "tuning in" to the reference; it provides information about how close you are to the reference at all times.

Track 37: Comparing two nearly identical streams.

a. The displacement of a damped pendulum mapped to the frequency of a sinusoid.

b. The same mapping but with noise added to the displacement data. The noise is 0.2 times the full amplitude of the data.

c. The sum of the original mapping and the noisy mapping. You can hear where the two deviate, but it is difficult to hear how far apart they are.

d. The product of the original and the noisy mappings. You can hear the deviations as sum and difference tones. The higher the frequency of the

Annotations to Audio Examples

difference tone, the greater the deviation between the original and the noisy mappings.

e. Amplitude modulation, or the product of the original and noisy mappings with the original added back in (the carrier, the sum, and the difference). This makes it easier to hear where the errors occur with respect to the original mapping.

Track 38: Accentuating small differences by adding two signals.

a. The position of a damped pendulum with noise of one part in one thousand added to it. You can hear small deviations if you listen carefully.

b. The noisy mapping added to the original mapping. You can hear the deviations as beats between the frequencies.

c. The noise here is one part in 10,000, and it is difficult to notice.

d. When this noisy mapping is added to the original mapping, you can hear the differences between the two as phase cancellation, a gradually fading in and out of the amplitude.

Track 39: Difference between two nearly identical streams. By taking the difference between the original mapping and the noisy mapping, you can detect even smaller noise values. When two signals are identical, their difference is 0 or silence. Once the two signals deviate from each other, even if it was just for one sample, you can hear the difference signal for the rest of the duration. Thus, this technique would be good for detecting when the first deviation occurs, but not good for detecting multiple, transient deviations.

a. This is the difference between the original mapping and noise that is one part in 100,000. The difference signal is large enough that it starts clipping.

b. This is the same thing except that the noise is one part in ten million. The difference signal is quiet but still audible.

Track 40: Transient noise. In the following examples, a burst of noise is added to the data about one eighth of the way into the mapping.

a. This is the noisy data mapped to frequency.

b. Here the noisy mapping is the carrier and the original mapping is the modulator in an amplitude modulation. One eighth of the way through, you can hear the difference tones caused by the noise. After that, you hear only the sum (because the signals are identical and the difference frequency is zero). So this technique could be used for detecting the timing and the magnitude of transient differences between data streams.

c. This is the difference between the original and the noisy data mappings. You hear nothing until one eighth of the way through when the noise burst occurs. From then on, you hear a signal, indicating that, once the two mappings get off from one another, they never line up exactly the same again. Thus, this could be used for detecting the timing of the first small deviation, but not for detecting subsequent ones.

Track 41: Resonator as a Grid. This example uses a filter to act as a grid in frequency space. As a time-varying frequency passes through a harmonic of the filter's fundamental delay period, it sounds louder.

TRACKS 42–46
David H. Jameson
"Sonnet: Audio-Enhanced Monitoring and Debugging"

The following used Sonnet to provide auditory representations of software processes. The pictures are bitmap images taken from the screen of an IBM RS/6000 running Sonnet under AIX. The digits in circles are used in the following descriptions to reference particular areas of interest.

Track 42.
Newton-Raphson. The first four examples are based on a scenario for investigating how the Newton-Raphson method for finding roots of equations converges (see Table 1). In the original program, with a guess of –1 and a maximum of 10 iterations, the algorithm did not find a root. The Sonnet scenario was setup to allow me to listen to dx as it (supposedly) approached 0 (see Figure 2). Extra components were present so that the guess and maximum iteration values could be changed on the fly (2, 3). The main sound control components are at the bottom of the picture (7). A reference tone (unchanging) was produced on MIDI channel 1 and a tone whose frequency is controlled (using MIDI Pitch Bend messages) by dx scaled by a factor 600 was produced on channel 0. If the scaled value is within the valid range of the pitch bend message, the Constrain component (10) passes the value through the upper output directly into the Pitch Bend component. Otherwise, the value is sent via the lower output port where, depending on whether it is positive or negative, the maximum or minimum possible pitch bend message is sent respectively.

Annotations to Audio Examples

TABLE 1

Sound #	Guess	Max Iterations
1	-1	10
2	-1	20
3	-1	30
4	0	10

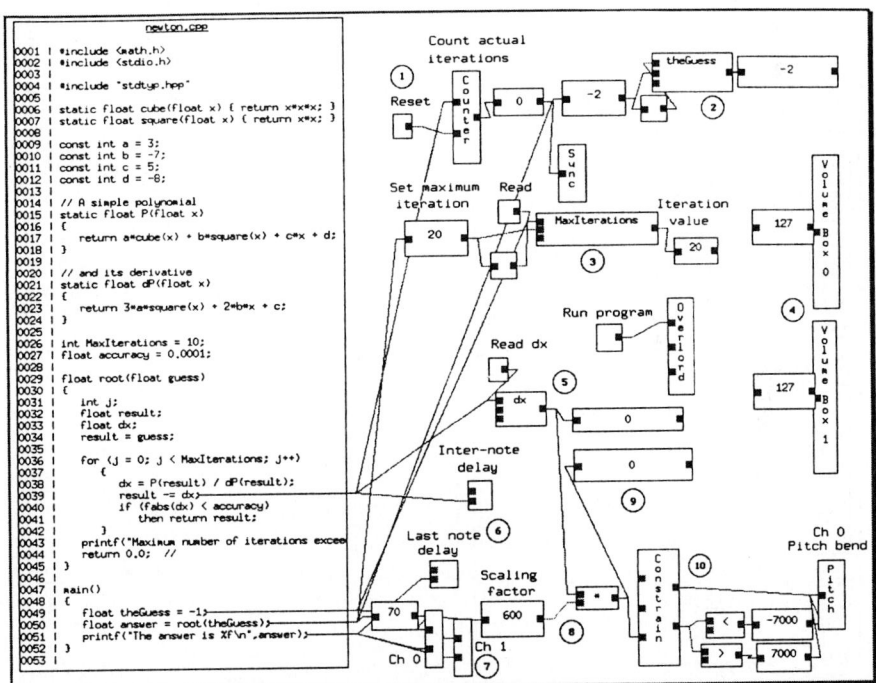

FIGURE 2   Sonnet scenario for Newton-Raphson.

Sorting. The next four examples are the sounds of different sorting algorithms, bubble, insertion, and selection. The scenario was setup to allow the algorithm to be specified from Sonnet by controlling the value of the variable **Algorithm** (see Figure 3). The array variable **Buffer** is played each time a swap occurs. The values of **Buffer** are accessed by means of a **For** component (2) that generates a sequence of integers from 0 to Index-1, where Index is the size of the array (6). The output values of Buffer are used as note numbers to the **Note On** component.

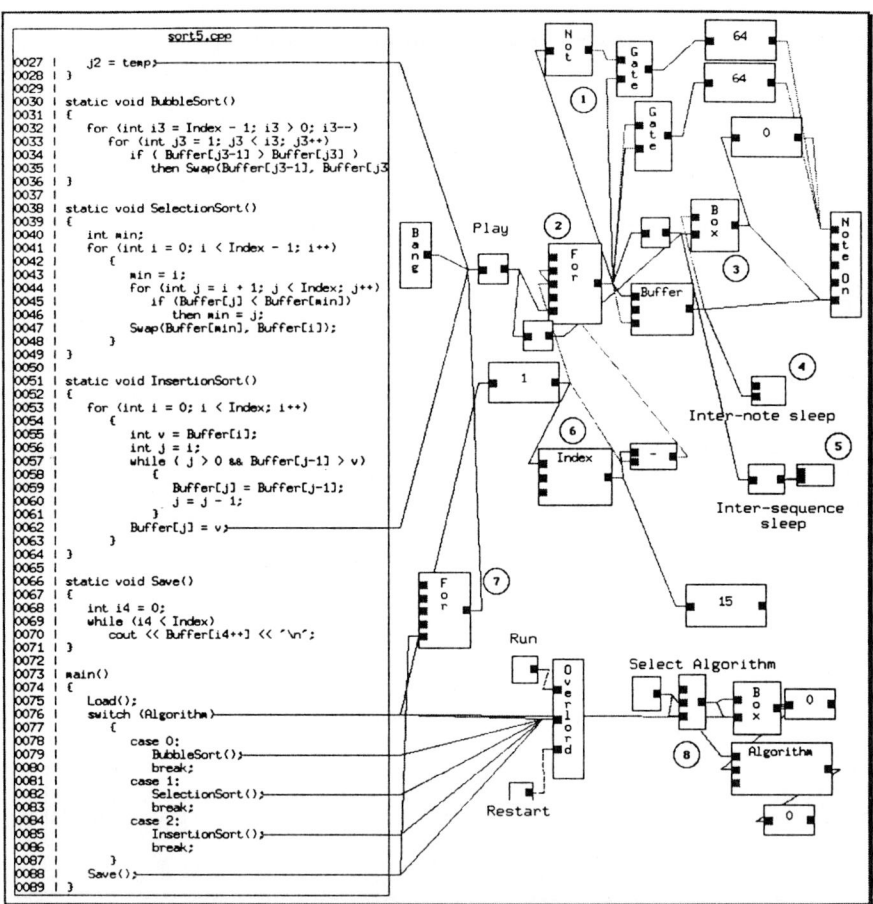

FIGURE 3 The scenario was setup to allow the sorting algorithm to be specified from Sonnet by controlling the value of the variable **Algorithm**.

Annotations to Audio Examples

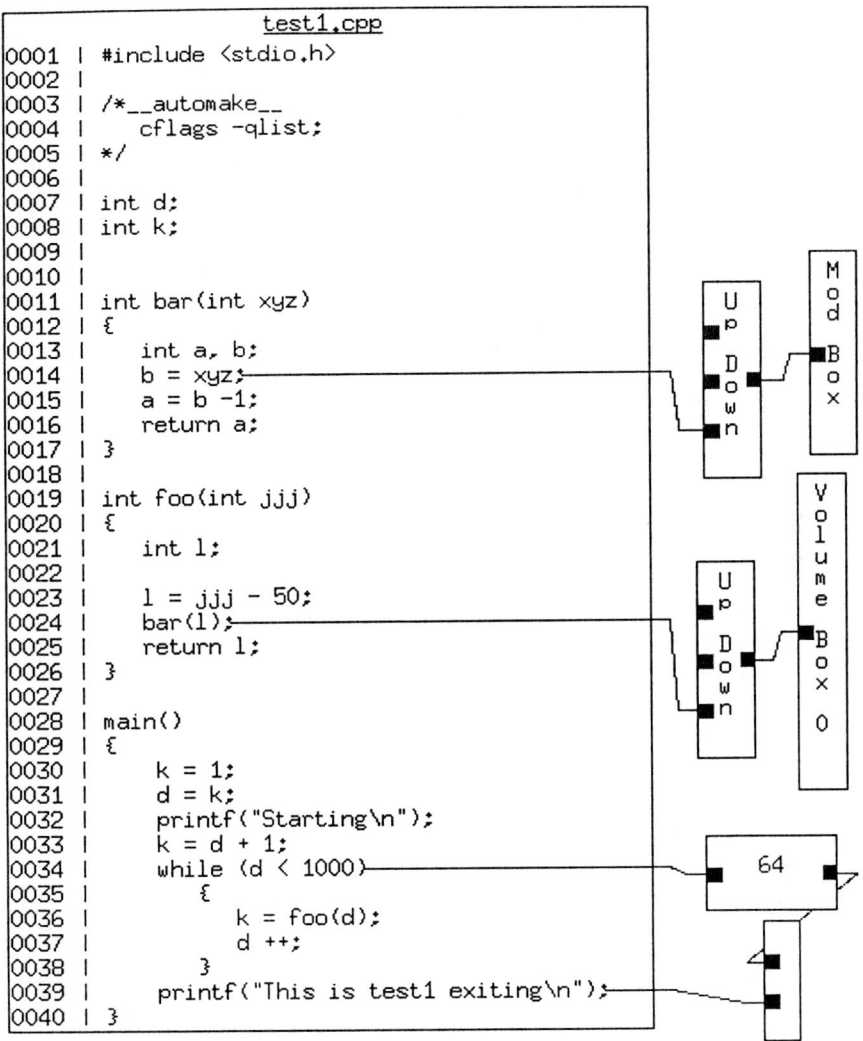

FIGURE 4 A note is triggered when a loop starts and terminated when the loop ends.

Track 43: Bubble Sort.

a. First example: Sequence was played with a 75-ms delay between notes.

b. Second example: This is the final (badly) sorted sequence played a couple of times with a 300-ms delay between notes. The very last note is clearly lower than the rest of the sequence. This hints at an "off-by-one" error in the indexing scheme and this was in fact the problem.

Track 44: Insertion sort.

Track 45: Selection sort. A 1/8 note tuning was used, again with 75-ms delay between notes. The sound is clearly partitioned into two parts, a monotonic ascending part, and a random part. As the sort progresses, the ascending part gets longer and the random part shorter.

Track 46: Overloading Sounds. In the final example, a note is triggered when a loop starts and terminated when the loop ends (see Figure 4). The Up-Down component is a blackbox that generates the next number in a range starting at a specified minimum value, increasing to a maximum value each time an impulse is received. When the maximum value is reached, the value will decrease until it reaches the minimum value again. This produces the sawtooth wave form. Each time the function foo() is called, the volume is changed slightly. Over time, a slow tremolo effect is heard. Each time bar() is called, the cutoff frequency of a lowpass filter is changed slightly.

A blackbox is used to encapsulate some set of components and their interconnections. The concept is similar to a macro or subroutine in a traditional programming language.

TRACKS 47–50
Tara M. Madhyastha and Daniel A. Reed
"A Framework for Sonification Design"

Track 47: Places Rated Data. [As described in Section 3.2.1 "Direct Mapping."]

Each track is a sonification of the rated cities in that state. Each city is represented by a single note whose characteristics reflect changes in rating data. The recreation index is mapped to note pitch (1 of 36 over four octaves in a C major scale). The city population is scaled to generate a duration of up to five seconds for each note. The climate index determines the timbre; three instruments were chosen to represent the range of weather. Bad weather is a raspy digital sound, average was a flute, and above average weather was a pleasant mixture of bells and strings. The volume was controlled by the housing cost index. Longitude is mapped to stereo balance.

Listen for which states sound more "pleasant" to live in than others.

- Illinois
- California
- Florida
- New York
- New Jersey

SONIC SCATTER PLOTS

The following two examples are the sound equivalent of graphical scatter plots. The Places Rated data set is sorted by the first variable, which is used as the time axis for the other variables, which are mapped to note characteristics.

Track 48: Population and the Arts. [As described in Section 3.2.2 "Sonic Scatter Plots."]

Here, population is the sort key and the time axis, and the arts index is mapped to pitch. Notice the high correlation between these variables.

Track 49: Climate and Recreation. [As described in Section 3.2.2 "Sonic Scatter Plots."]

The climate index is used as the time axis, and the recreation index is mapped to pitch.

Track 50: Performance Data. [As described in Section 3.3 "Performance Data."]

This is a sonification of a simplex code running on eight processors. A note begins at a message send and is terminated at the corresponding receive. The pitch of a note corresponds to the processor identifier.

TRACKS 51–59

Jay Alan Jackson and Joan M. Francioni
"Synchronization of Visual and Aural Parallel Program Performance Data"

Each playback was generated based on trace data captured during the runtime of actual parallel programs. A full description of each of the playbacks, as they relate to specific program behavior, is given in Section 5 of the paper.

Track 51 (corresponding to Figure 1 in the paper). Sounds in the left stereo field represent sending message events; sounds in the right stereo field represent receive events. Higher pitch corresponds to higher processor number.

Track 52 (corresponding to Figure 1 in the paper). Same sound mapping as playback 1; slower tempo.

Track 53 (corresponding to Figure 2 in the paper). All sounds represent sending message events. Processors were all assigned a Geiger counter tone: one of indefinite pitch and short duration.

Track 54 (corresponding to Figure 2 in the paper). All sounds represent sending message events. Processors were assigned pitch relative to processor number; voice is piano.

Track 55 (corresponding to Figure 2 in the paper). All sounds represent sending message events. Processors were assigned pitch relative to processor number; however, notes do not correspond to a musical scale. Voice is percussive.

Track 56 (corresponding to Figure 3 in the paper). All sounds represent sending message events. Processors were assigned pitch relative to processor number; voice is flute; tempo is relatively fast.

Track 57 (corresponding to Figure 3 in the paper). Same sound mapping as in the preceding playback, but tempo is slowed down considerably. Only the two subregions of the figure are included.

Track 58 (corresponding to Figure 4 in the paper). Sound corresponds to idle bursts of each processor, which is depicted as white space in the figure. Higher pitch corresponds to higher processor number. Duration of sound is proportional to length of idle burst. Bell tone signifies beginning of each idle burst.

Track 59 (corresponding to Figure 5 in the paper). Same sound mapping as in the preceding playback. Playback generated for a different program's trace file.

Annotations to Audio Examples

TRACKS 60–62
W. Tecumseh Fitch and Gregory Kramer
"Sonifying the Body Electric: Superiority of an Auditory over a Visual Display in a Complex, Multivariate System"

OVERVIEW

This auditory display is a hybrid: it combines realistic "base streams" (a heart sound and breathing noise) with abstract "piggy-back" variables which modify the pitch, timbre, and rhythm of the base streams. The first track introduces the base streams, while the second gives some examples of changes in the "piggy-back" variables alone. The final track has examples of changes in several variables at once.

The sounds were generated under the control of a program written in Opcode's Max. The heart sound was produced by a custom patch on a Yamaha DX100 synthesizer. The breathing sound was synthesized in real time on a Digidesign Audiomedia board (with a Motorola 56001 DSP chip) under the control of John Bate's Unison software. See the text for details.

On each track, a baseline period of normal functioning is followed by a change in one or more variables and then a return to normal. To orient the listener in these examples, each change is preceded by the sound of the pupillary reflex (a higher-pitched heart tone). During the experiment testing this display, the pupillary reflex sound occurred only in response to the subject pressing the "test reflex" button.

Track 60: Realistic variables. This track introduces the base streams, which are "self-labeling" since they sound like what they represent. First you hear an increase in heart rate, then a decrease in breathing rate.

Track 61: Piggy-back variables. These variables modify the base streams in ways which are arbitrary; thus, they have only an abstract relationship to their sonic counterparts. First, an increase in the pitch of the heart sound indicates a rise in blood pressure. Then, a change in the timbre of the heart sound indicates a drop in carbon dioxide level. Finally, the increase in the "pitch" of the breathing sound (the mean center frequency of the band-pass filter) indicates a drop in temperature (it is supposed to sound like a cold Arctic wind).

Track 62: Multivariate changes. These three medical complications are separated slightly to make the transitions more obvious.

The first complication is a drop in oxygen level (there is no direct readout for oxygen in this experiment, so oxygen level must be inferred from other variables). A decrease in blood oxygen results in an increase in heart rate, a rise in blood pressure, and the appearance of atrio-ventricular dissociation. These are reflected in an increase in the pitch and rate of the heart sound, along with a rhythmic dissociation of the two components of the heart sound. This rhythmic dissociation differs from that in the next example in that a steady pulse is maintained, but the delay of the second sound becomes irregular.

The second example on this track is blood loss. This leads to a drop in blood pressure, a decrease in body temperature, and fibrillation of the heart. The sonic result is a drop in pitch of the heart sound, and increase in the pitch of the breathing sound, and a complete randomization of the heart sound (no steady pulse is maintained).

The final example is an overdose of anesthetic, which leads to a decrease in blood pressure and respiratory rate, and a loss of the pupillary reflex. The sonic effects are a slowing of the breathing sound and a drop in heart sound pitch. In the real experiment, pushing the "reflex test" button (which normally triggers the pupillary reflex sound) would give no result in this condition.

TRACKS 63–64
Gottfried Mayer-Kress, Robin Bargar, and Insook Choi
"Musical Structures in Data from Chaotic Attractors"

Two voltages are recorded from the Chua circuit onto each of the audio channels at 44.1-kHz sampling rate. The parameters of the circuit that will determine the attractor and, therefore, the observed signal are described by Mayer-Kress et al.[5]

Track 63 (Example C1). Bassoon sequence: Asymmetric configuration of the Chua circuit produces bassoonlike sounds upon variation of the resistor R. By breaking the symmetry of the system (unequal break points, see the above reference for details) and then changing the same resistor as in C2, one obtains a period-adding sequence of periodic orbits of distinct periods. In this sequence the structure of the orbit can be characterized by the number of small oscillations (loops around one of the unstable foci) that occur, during one period. During each of these periods, the system loops

---

[5] G. Mayer-Kress, I. Choi, N. Weber, R. Bargar, and A. Hvbler, "Musical Signals from Chua's Circuit," Technical Report CCSR-93-4, to appear in *IEEE Transactions on Circuits and Systems*, special issue on "Chaos in Nonlinear Electric Circuits," in press, and the references therein.

around the second unstable focus exactly once per period. This dynamical process produces a waveform similar to that of a bassoon.

Track 64 (Example C2). Double Scroll: The Double Scroll attractor is the standard chaotic attractor of the Chua circuit and it is symmetric with respect to the voltage $v2$. The symmetry of the above case severely limits the class of signals that can be obtained from the circuit.

Track 65 (Example D1). "Divergence of pitch due to a 0.1% difference in initial conditions."

Two properties of chaotic systems are illustrated in this example:

1. the structured irregularity of a chaotic system, and
2. the sensitive dependence to initial conditions.

Left and right audio channel display two different instruments, the pitch of which corresponds to a sequence of numbers generated by the logistic map. The bifurcation parameter is given by $a = 3.9$ for both cases. The initial condition for the first case is given by $x45$, the 45th iterate of $x0 = 0.7435897435897437$, which is close to an unstable fixed point. The second sequence, simultaneously played on the second audio channel, has an initial point of $x45 + 0.001$. One can hear the rapid, exponential separation of the sound streams as well as the recurrence of similar patterns.

In the sequence of examples on Tracks 66 to 73, we illustrate the sensitivity of chaotic systems against small changes in the bifurcation parameter r. In each case we play on one channel the pitch sequence of a periodic attractor and in the second channel the sequence that we obtain by changing the bifurcation parameter by 0.001, using identical initial conditions. The resulting sequence will have recurrences of long patterns that are close to the corresponding periodic orbit. Intermittent bursts of excursions from the periodic attractor become more frequent while the duration of the periodic structure is reduced, when the parameter is further removed from the bifurcation point.

The specific parameters for each case is given below (see Figure 8 of the paper):

Track 66 (Example E1). In all cases, except for period 3, the intermittency parameter rpi for a periodic parameter rp was chosen as $rpi = rp - 0.001$. "Period 3 plus Intermittency" $r3 = 3.83187$; $r3i = 3.8283$ (see Figure 7(a), loc. cit.)

Track 67 (Example E2). "Period 4 plus Intermittency" $r4 = 3.960270$.

Track 68 (Example E3). "Period 5 plus Intermittency, parameter close to 3.7" $r51 = 3.73891$.

Track 69 (Example E4). "Period 5 plus Intermittency, parameter close to 3.9" $r52 = 3.905706$.

This parameter range was used in the composition "Shadowing Lemma: $rC[3.9, 3.905706]$ where period 5 cycle occurs, $x0 = 0.7435897435897437$," by I. Choi, first performed at Composers Forum, Urbana, IL on 4/19/93.

Track 70 (Example E5). "Period 6 plus Intermittency" $r6 = 3.9375364$.

Track 71 (Example E6). "Period 7 plus Intermittency, parameter close to 3.7" $r71 = 3.701769$.

Track 72 (Example E7). "Period 7 plus Intermittency, parameter close to 3.8" $r72 = 3.77421$.

Track 73 (Example E8). "Period 8 plus Intermittency" $r8 = 3.80077$.

Track 74 (Example G). By changing a the value of one resistor as a bifurcation parameter, one obtains a sequence of periodic and aperiodic or chaotic signals as the resistance is continuously decreased over a time of 6 sec.

The beginning of the sequence is marked by a (local) Hopf bifurcation from a fixed point (focus) to a limit cycle. The sequence terminates with a global bifurcation to a large limit cycle.

Note: Because of the global nature of the latter bifurcation, one generally obtains a different sequence by reversing the procedure and increasing the resistance. This bistability or hysteresis effect often has an undesirable effect in applications, especially it makes it more difficult to obtain reproducible results.

Track 75 (Example H1). Histogram accumulation and an associated decrease in new events.

Track 76 (Example H4). Histogram with limited window for updating statistics.

Track 77 (Example I). Timbre shift based upon the changes occurring in the histogram.

Annotations to Audio Examples

TRACKS 78–84
Chris Hayward
"Listening to the Earth Sing"

All audio examples other than Track 78, the synthetic model seismograms, are actual seismic field recordings made to determine the depth and nature of the bedrock or are recordings from well-known seismic observatories.

Raw data was previewed on a Sun SPARCstation 2 and interesting examples were audified. Sample audio segments were reviewed using the built-in Sun 8000-Hz single-channel μlaw output device. Forty-five minutes of sound was selected for scrutiny and critical listening. These interesting segments were transferred to an SGI Indigo Extreme where they were converted from the original seismic data to 44.1-KHz 16-bit stereo AIFF files. The individual examples were edited to meet the time available on the CD and then annotated to produce the DAT used for mastering.

Track 78: Model Traces. A computer program simulates the recordings from a shallow buried explosion in a multi-layered earth. Thirty-nine separate seismometers record the seismic activity at increasing distances from the explosion. When created each seismogram is about six second long is sampled at 166 Hz and contains frequencies as low as 5 Hz.

Track 78: 0:01.6–0:07.0. Thirty-nine simulated seismograms are played 48 times faster than recorded (time-compressed 48 times) resulting in 8 audio seismograms played each second. Each fifth and tenth seismogram begins with a short wood-block tone.

The energy fades away as the seismometers get further from the source. This decay in amplitude so dominates the recording that it is difficult to compare other aspects of the changing nature of the seismograms. Although the power in each seismogram varies by over 600,000, almost all seismograms are audible without doing special variable scaling as would be necessary for plotted data.

Track 78: 0:08.1–0:13.8. The model is repeated with the seismograms individually scaled to remove the effects of energy decay with distance. Each six-second seismogram has been divided by the maximum value in the seismogram.

The gradual shift in dominant in tone from high to low as seismograms move farther from the explosion illustrate the loss in high frequencies with increasing distance from the source. Later seismograms have an echo. This echo occurs at the more distant seismograms where energy traveling at two different velocities becomes well separated in time. On seismogram 30 it is also possible to distinguish the delay between the annotation tone and the initial arrival of the seismic impulse.

Track 78: 0:14.9–0:18.6. The previous normalized seismograms are further modified with a fast automatic gain control (AGC) that adjusts the amplitude each 1/40 second to keep the energy constant even within the individual seismogram. Then the seismograms are frequency-doubled. Frequency doubling preserves the harmonic relationships between fundamentals and harmonics but shifts more of the energy into higher frequencies where small differences are more easily distinguished.

Only the first 20 seismograms are played as an example. The seismograms have a distinctly springy sound. It is sometimes difficult to distinguish the start of individual seismograms.

Track 78: 0:19.4–0:25.8. Frequency-doubled seismograms when played at half speed have the same frequency spectrum as the original version. Annotation tones have been added to the start of each seismogram in addition to the tones at every fifth and tenth seismogram. The tones are overlaid with a slightly different stereo balance than the seismic data to help separate them.

Track 78: 0:28.0–0:34.5. Frequency doubling followed by half-speed playback is repeated to yield audio seismograms with 12-times compression.

Each seismogram is a sliding tone which moves quickly from high to low frequency. The first few seismograms have a constant superimposed sharp metallic tone. This may be an artifact of the modeling program. Later seismograms have a more complex extended shift in frequency. This shift from high to low is a consequence of the rapid attenuation of high frequencies in earth materials. Energy returning from deeper reflections is of lower frequency than that which only travels a short distance through the shallow layers.

Track 78: 0:36.1–0:41.8. One channel is the previous frequency-doubled model traces played at 48-times compression. The other channel is the result of a slightly different model. It takes careful comparison of the plots of the two models to distinguish differences between them.

In the audio display, the differences are not apparent on the first few traces, but become progressively more distinct at later times, finally becoming a bouncing from channel to channel. While it is possible to tell that the two models are different, it is difficult to describe the differences.

Track 78: 0:45.0–1:05.8. The model frequency-doubled twice and played at 12-times compression. There are annotation tones at the start of each seismogram as well as every fifth and tenth seismogram. In addition a tone is added just before the low-frequency surface wave. The annotation tones are separated from each other with slightly different frequencies and balance.

In this case not only do the traces have every fifth and tenth trace marked, but there is also a mark on each trace at the expected time of the surface wave. With the annotation, the surface wave is clearly distinguishable as a short low-frequency tone that becomes progressively delayed from the start of the seismogram. Although the multiple annotation tones are at first listening complex, on repeated listening they are separated sufficiently that surface wave annotation may be distinguished from tones at the start of the seismogram.

Track 79: Shallow Seismic Exploration. These examples use field seismic records acquired with geometry's similar to the model. Each seismogram is at an increasing distance from the explosion or hammer blow. Often in the field, equipment limitations require that a recording of one shot into many geophones or seismometers be simulated with multiple shots into the available geophones. In these examples 20 or 12 geophones are recorded at a time. If all the shots or geophones are not identical, artifacts in the recorded data may obscure information used to interpret geology. In some cases, these artifacts may be misinterpreted as geologic structures.

Recordings with many geophones, such as these, are made during the initial investigation of an area to determine basic seismic and noise characteristics region prior to starting a more detailed seismic investigation.

Track 79: 0:03.5–0:11.5. This seismic section was acquired with 20 RefTek 16-bit A/D recorders sampling at 250 Hz. All gains were set identically to avoid clipping even the strong near signal. The source was a Betsy Seisgun, an industrial shotgun that points down into the ground. Geophones are separated by 1 meter. Sixty records or traces in three sets have been recorded in increasing distances.

Traces are played back at 16-times compression or 8 traces per second. Every fifth, tenth, and twentieth trace is prefixed with an annotation tone. A single scaling factor was used for all traces.

Listen for the overall change in the tone after each twentieth seismogram and the general change in frequency and amplitude with distance. The abrupt changes in tone every twentieth trace are a result of slight differences in the characteristics of the shotgun source from shot to shot. Traces 21 through 40 have a higher background noise (it sounds like traffic noise) relative to the other two shots. This is probably wind noise, and if recognized in the field may have been avoided.

While every fifth, tenth, and twentieth traces are marked at their start, it is difficult to match the marker with the exact trace. It is also difficult to recognize the progressive delay between the annotation marker and the dominant energy although it is evident that the markers and arrivals do not exactly coincide.

Track 79: 0:12.3–0:22.1. This section of 72 traces was recorded with six different sledge hammer blows into twelve geophones. The data was recorded with an EG&G twelve-channel 10-bit recorder sampling at 2000 Hz. The limited dynamic range of the recorder made it necessary to individually adjust the gain of each channel. Gains were adjusted to best record the seismic reflections and, therefore, the strong arrivals are clipped.

The section is played back at a compression of four, or about eight traces per second. Each geophone is played with the six consecutive hammer blows, then the next geophone is played. Every fifth, tenth, and twentieth trace are marked. No other processing has been applied.

The dependence on geophone coupling is evident in the sharp changes from each group of six, while the shot dependence appears as a repeated six-beat pattern. The first three shots have a pronounced ring evident on the first three beats of the first three measures of six (traces 1–18). At more distant offsets from the source the ringiness is no longer distinguishable because most of the high frequencies have been attenuated. A difference in background noise that sounds like something walking through tall grass or heavey-breathing is associated with some shots and is similar to that heard in the previous Betsy Seisgun section. Recognizing the changes from sensor to sensor is more difficult. It is most obvious at the far offsets where there is a distinct change in tone every six beats.

Track 79: 0:22.7–0:37.7. Rapidly repeating each individual trace three or more times emphasizes the differences between the individual traces as well as allowing the traces to be studied more closely. Each trace is annotated once at the start as well as every fifth and tenth trace. The ringiness of the first four shots is clearly audible in the first 18 traces although at longer distances these high frequencies are nearly gone. However, at the farther offset a difference in background noise (sounds a little like walking through tall grass or breathing) associated with some shots similar to that heard in the previous Betsy Shotgun section. Listening for the changes from sensor to sensor is more difficult. It is most obvious at the far offsets where there is a distinct change in tone every six beats.

Track 79: 0:38.8–0:48.6. The same section is taken, but normalized with automatic gain control and played through the left channel. The right channel has the corresponding shear wave section (horizontal hammer blow). In this case annotation has been deleted to avoid complicating the selection.

The breathing sound on the right channel corresponds to different noise conditions at the time of each shot. Theoretically, the direct arrival of the shear waves (right channel) should occur after the compression wave, but the compression Rayleigh (left channel) surface wave should occur after the shear Love surface wave (right channel).

Annotations to Audio Examples 651

Track 79: 0:51.3–1:03.7. The same compression and shear sections are slowed down by 4 and frequency shifted to maintain the section in the audible range. Strong clipping of the compression section (left channel) is obvious (it was clipped in the field). The initial arrival occurs at the same time on each trace, while the later surface waves generally arrive first on the right channel.

Track 80: Shallow Seismic Reflections. Once the initial investigations are complete, the next step is often to profile an area to build up a geologic cross section. This involves recording a number of records as the line is scooted along the profile. In this case each set of 12 traces represents one shot. Data is recorded from 14-Hz horizontal geophones sampled at 2000 Hz with a 10-bit EG&G recorder.

Individual traces are placed in the stereo field to represent their position along the ground.

Track 80: 0:02.7–0:29.9. In this unusual high-frequency seismic section, four records (1, 4, 5, 12) selected from a larger set are audible without frequency shifting or time compression. Each record begins with a marker tone.

All records should sound identical under ideal conditions. Instead the third record is noticeably weaker, and the fourth is stronger but has obvious 60-Hz noise pickup on several traces. The second and third records also contain a small chirp caused by grass rubbing against the geophone. These significant problems were not recognized in the field during recording but are immediately identifiable here.

Track 80: 0:32.0–0:56.8. The set of all 12 consecutive records is played back at 8-times speed. Each record begins with marker tone and traces are panned across the channels to represent their ground placement. At this increased speed, the previous two chirps will be sharp clicks. There are also at least 8 noise spikes, 4 traces with high 60 Hz noise, several records with differences in the shot, and several interesting background tonal noises (on records 3 and 7).

Track 80: 0:58.4–1:08.9. Frequency-doubling the first 5 records of the preceding brings out the background noise and suppresses the dominant thump of the initial arrival. In this manner the traces may be examined to verify that all background white noise is similar.

The dominant frequency is rather high initially and then falls off towards trace 12 on each record. In Record 1 a difference is shown suggesting that there may have been a problem with the first couple of channels. The differences from shot to shot are most obvious when the first few traces of each shot are compared. The second record has generally lower frequency.

Track 81: Seismic Stations. Examples of seismic stations used for earthquake recording have been extracted from the US UN GSE seismic network. A number of high-quality 24-bit ADC sampling at 40 and 10 Hz have been used to collect the data. The wide dynamic range here ensures these stations do not clip, even during large events. For the purposes of audification, most records must be normalized to make the noise audible. Data is played back at 8000/40-times compression, allowing 24 hours of data to be scanned in 7.2 minutes, faster than many stations can scroll the data on a CRT display.

Track 81: 0:01.8–0:18.0. Samples of background noise for four seismic observatories—Blacksburg, Virginia; North Pole, Alaska (near Fairbanks); Piñon Flats, California; and Lajitas, Texas—are played back at 200-times compression. These were all observatories used in a three-month test of a world-wide system for monitoring seismic signals to detect nuclear tests.

All segments have been normalized to the same scale. Only 5 seconds of audio is sufficient to identify each of the stations.

Blacksburg has a strong low-frequency component audible as a low rumbling of 100 Hz and less. Superimposed on this noise are small surface wave (Lg) arrivals (sounding like failed attempts to whistle) from distant mining explosions. A strong tone at 15 Hz is shifted to 3000 Hz and forms an annoying high frequency whistle. It is helpful to vary the settings on a graphic equalizer while looping this sequence.

North Pole has an overall quieter background. A small frequency peak at 12 Hz on the spectra is shifted to 2400 Hz and in the audio display is a set of two repeated whistlelike tones. These are probably a natural harmonic of a manmade structure, such as power poles vibrated into resonance by wind. From the audio section it is obvious that these tones are not present during the whole time but change. They are quite distinctive and would be immediately recognizable. The quiet staticlike sound is a part of the natural noise background and may be distant earth tremors related to volcanic activity.

Piñon Flats has a number of tonals that sound like continuous high-pitched whistles on the recording. These are likely a result of one or more manmade structures. From the recording it is evident that this is more of a problem at Piñon Flats than Alaska. There is also a distant surface wave arrival similar to that recorded at Blacksburg.

Lajitas is almost silent in contrast to Piñon Flats. At the very end of the Lajitas selection the careful listener will hear a small surface wave, again like that of Blacksburg. Lajitas shows none of the tonals present on other stations.

Track 81: 0:19.4–0:43.0. An extended record of 100 minutes of recorded data from North Pole (labeled as Fairbanks on the recording) includes a near local earthquake and two more distant earthquakes. The near earthquake reverberates, like a gunshot in an empty gymnasium. Three smaller distant earthquakes may be recognized by their similarity to a heavy door latching as it closes. These are recognizable underneath the large number of crackling and staticlike sounds that may be a microseismic storm.

Track 81: 0:45.7–1:05.4. The distinctly different record from Piñon Flats is difficult to listen to because of the continuous high-pitched tonals but listening to the selection through an equalizer with the highs above 3000-Hz cut helps. There are three regional earthquakes (each less than 1500 km away), each sounding like the double thump of a heavy door latching and closing. Times between the two thumps are related to the distance so these three events can be ranked by distance during three or four listening.

Along with the impulses are two surface (Lg) waves. These begin with low frequencies and generally end with high frequencies. They are reminiscent of what might be used to simulate wind blowing across the desert or a roadrunner dashing off across the desert. One local event is heard as a sharp pop or click, almost like a production problem.

Track 82: Selected Earthquakes. Recorded earthquake signals vary depending on the size, location, distance, depth, and mechanism of the quake. The geology between the quake and observatory and the recording instrument response also influences the appearance of the record.

Five earthquakes recorded from 27 degrees[6] away at five different stations are played back at 800-times compression. The traces were originally recorded at 10 Hz with 24-bit digitizers.

The recordings are:

Track 82: 0:03.7–0:6.1. 5.1 mb on March 9, 1990 in Southern Alaska recorded in Pinedale, Wyoming (31 degrees).

Track 82: 0:06.1–0:09.9. 5.2 mb on March 13, 1990 in Guerrero, Mexico recorded near Fairbanks, Alaska (27 degrees).

Track 82: 0:09.9–0:13.2. 5.3 mb on July 10, 1990 off Coast of Mexico recorded at Piñon Flats, California (27 degrees).

Track 82: 0:13.2–0:16.8. 5.5 mb on February 28, 1990 in Southern California recorded at Blacksburg, Virginia (30 degrees).

Track 82: 0:16.8–0:20.5. 5.5 mb on February 3, 1990 in Queen Charlotte Islands recorded at Lajitas, Texas (29 degrees).

[6] A degree is approximately 111 km.

Listen for the separation of the initial compression and shear arrival (it sounds a little like a heavy door closing) and the separation between the initial arrival and the surface wave (a low-frequency boing). A careful listener will recognize the stations at Blacksburg, Piñon Flats, and Fairbanks from the characteristic noise. While these events are all similar distances, there is still a wide variety depending on the exact station and event combination.

Track 82: 0:23.3–0:52.8. Eight earthquakes from 12 degrees to 101 degrees away are played back at 800-times compression. All events are preceded by an annotation tone. All events were recorded at Latjias, Texas. See Table 2.

The additional compression is required to make the low-frequency surface wave at 1/20 Hz audible. At 101 degrees the low-frequency surface wave is nearly inaudible. Three distinct arrivals are audible for each record, although at the near distances the first two occur close together. Some stations may be identified by the noise underneath the event.

Track 82: 0:54.0–1:05.7. Three of the preceding earthquakes, the first one at 12 degrees and the last two near 100 degrees, are compressed by 1600 to emphasize the low frequencies. The long duration and dispersion are indicators of extreme distance to the earthquake. These surface waves were not associated with the original even during the first stage of visual analysis. In fact, the surface waves were not immediately apparent on the plots. With audio displays as with video, the wrong form of display can result in missing important information.

TABLE 2 Eight earthquakes, recorded at Latjias, Texas, at 800-times compression; see Track 82: 0:23.3–0:52.8.

Num	Time	Date	Distance (degrees)	Magnitude (mb)
1	23:45:46.6	May 11, 1990	12.31	5.2
2	00:56:34.0	Jul. 10, 1990	20.72	5.3
3	10:01:02.0	Feb. 3, 1990	29.35	5.1
4	15:23:14.0	Aug. 14, 1990	56.65	5.8
5	18:10:17.0	Jul. 4, 1990	61.32	5.1
6	07:34:20.0	May 5, 1990	90.59	5.3
7	03:31:36.0	Jul. 10, 1990	99.45	6.0
8	02:41:48.0	Feb. 17, 1990	101.7	6.1

Annotations to Audio Examples

Track 83: Nuclear Explosions. Seismic data recorded from nuclear explosions have much in common with earthquakes but have a smaller surface wave and shear wave in relation to an equivalent sized earthquake. In this dataset, two U.S. underground nuclear tests were recorded at Lajitas.

Track 83: 0:01.4–0:13.5. Two nuclear explosions (June 13, 1990 and March 10, 1990) of slightly different magnitudes recorded in Lajitas at 40 Hz and played back at 200-times compression.

The reverberations suggest that the travel path from the explosion to Lajitas includes a large number of nearly perfect reflectors, much like a light pipe. The absence of a surface wave is one of the diagnostic features of explosions.

Track 83: 0:15.0–0:21.7. Stereo recordings were created by playing the recorded vertical channel into both channels, the recorded north channel into the left channel and the recorded east channel into the right.

The roominess of the resultant signal is startling. Careful listening will reveal that the sounds making up the seismic signal are placed differently in the stereo space. On repeated listening one can distinguish other features besides the primary event. The background noise is different on the two channels. Near the end of the main signal is a faint signal from an unrelated earthquake.

Track 83: 0:21.7–0:25.7. Time compressing by 400 times was done to listen for longer period surface waves (none are immediately audible). The second event almost masked in the previous recording is clearer. The shift to lower frequencies as the reverberation dies away is obvious and the characteristic two-thump signature is clear.

Track 84: Array Data. The GERESS (GErman Regional Experimental Seismic Station) seismic array includes 25 elements distributed over a 2-Km aperture. Three elements on the outer ring were formed into a stereo image. The first element is played on both channels, the east element into the left channel and the west element into the right channel. The data is played back at 200-times recorded speed. It was expected that the stereo placement of each event should give an indication as to the distance and the direction to the event.

Track 84: 0:01.5–0:03.9. Earthquake 1. Oct. 28, 1991 at 00:31, Yugoslavia 4.0 mb, 773 km distance, 127° azimuth.

Track 84: 0:03.9–0:06.3. Earthquake 2. Oct. 31, 1991 at 09:30, Northern Italy 4.7 mb, 507 km distance, 214° azimuth.

Track 84: 0:06.3–0:08.3. Earthquake 3. Nov. 20, 1991 at 01:53, Switzerland 4.8 mb, 338 km distance, 235° azimuth.

Track 84: 0:08.3–0:10.8. Earthquake 4. Jan. 8, 1992 at 08:48, Austria 1.6 mb, 179 km distance, 149° azimuth.

The same data is frequency-doubled and played with a 100-times compression.

Track 84: 0:13.2–0:17.5. Earthquake 1.

Track 84: 0:17.5–0:21.0. Earthquake 2.

Track 84: 0:20.3–0:25.4. Earthquake 3.

Track 84: 0:25.4–0:30.1. Earthquake 4.

Frequency doubling and expanding are used to increase detail. The quakes are not in order from closest to most distant. Relative distances may be determined by comparing these quakes with those on track 82. The final quake does not have the reverberant nature of the preceding three. The third quake appears to be complex having a double arrival which is duplicated in the later S wave arrival. Background noise on the three channels is different. Portions of each event are stronger on one channel or another. The last quake with its extended high-frequency range is particularly distinctive, appearing to shift from one channel to the other. This should not be the case, as the direction of arrival for both phases should be the same although the velocity will to be different.

Track 84: 0:33.3–1:09.9. Elements of a 100-Km array are recorded at 20 Hz and played back at 400-times compression. The selection represents about 200 minutes of recording. The Lajitas station is played back on both channels, Marathon, Texas (to the east) on the left channel and Fort Shafter, Texas (to the west) on the right channel.

The noise is distinctive, with several clicks and pops that denote data glitches appearing in only the left or right channel. The right channel has the largest problem with glitches. There are a number of events of various sizes, all close to the same distance. None localize very well. There is a large event near the end of the section and during the coda, a number of smaller events. The background noise at the Marathon and Fort Shafter is down just prior to the main event.

Acknowledgments. The earthquake data was collected by the Southern Methodist Geophysics staff as part of Defense Advanced Research Projects Agency contract numers MDA 972-89-C-0054 and MDA 972-88-K0001. Exploration seismic data were collected by Tom Goforth at Baylor University.

TRACKS 85–92
Meera M. Blattner, Albert L. Papp III, and Ephraim P. Glinert
"Sonic Enhancement of Two-Dimensional Graphic Displays"

Track 85. An earcon that is a combination of two earcons. [As described in Section 2.2.]

A motive may be an earcon, or it may be part of a compounded earcon. Let A and B be earcons that represent different messages. A and B can be combined by juxtaposing A and B to form a third earcon AB. For example, if A is an earcon for "file" and B the earcon for "deleted," then AB is the earcon for "file deleted."

Track 86. Two earcons, where the second earcon is a transformation of the first. [As described in Section 2.2.]

Earcon A may be transformed into earcon B by a modification in the construction of A (for example, by changing the pitch in one of its notes). If A were an earcon for "the computer is up," the transformed earcon might instead inform the listener that "the computer is down."

Track 87. Four earcons formed hierarchically. [As described in Section 2.2.]

A family of earcons may have an inherited structure. A family motive A is an unpitched rhythm of not more than five notes. To use A to define a family of messages, the family motive is first elaborated with pitch $A+p \longrightarrow B$. The result B is then preceded by the family motive A, to form a new earcon AB. Hence, the earcon has two distinct components, an unpitched motive followed by a pitched motive with the same rhythm. A third earcon, ABC, can be constructed by combining AB with a third motive C whose pitch and rhythm are identical to those of AB but which possesses an easily recognizable timbre $A + p + t = B + t \longrightarrow C$. If, for example, A were to indicate a printer problem, then AB could indicate the printer is out of paper.

Track 88. Two earcons played sequentially first, then overlaid. [As described in Section 2.2.]

To display two or more earcons simultaneously, we have to identify their temporal locations with respect to each other. Two primary methods are considered below: overlaying one earcon on top of another, and the sequencing of earcons.

Track 89. Sampled sound example of a ball bouncing, followed by the sound of a tom-tom drum—they sound similar. [As described in Section 2.4.]

Often, physical events that are quite different give rise to similar sounds. One might say these are naturally occurring auditory homonyms. For example, a hammer hitting a table may have the same sound as a car door slamming. Hollywood takes advantage of this observation on a regular basis.

Track 90. Earcons for four access levels. [As described in Section 5.]

There is a simple earcon of a tom-tom drum pounding, which represents a building restriction. The pounding is transformed in pitch and frequency to indicate different access restriction levels of buildings. Higher restrictions are represented by faster, higher-pitched knocking.

Track 91. The "computer" earcon and three tranformed computer earcons. [As described in Section 5.]

The computer earcon is a four-note earcon using a flute timbre which is transformed over the x-axis, y-axis, and both axes.

Track 92. A hierarchically structured set of administrative earcons. [As described in Section 5.]

A simple three-note earcon in one pitch using a saxophone timbre indicates an administrative building. The second-level motives inherit all the properties of the first earcon except that they have different pitch changes. It is important to note that each family of earcons shares the same timbre. Since timbre is one of the most easily recognized attributes of sound, one can immediately identify the family of an earcon just by recognizing what instrument is used in playing that earcon.

TRACKS 93–96
Stephane A. Brewster, Peter C. Wright, and Alistair D. N. Edwards
"A Detailed Investigation into the Effectiveness of Earcons"

Track 93. "Paint application," "Write folder," and "Draw file 1" earcons used in the Control, Simple, and Musical groups respectively [as described in section "Experiment 1, Experimental design, phase I"].

Track 94. "Open," "Save," and "Undo" earcons used in the Control, Simple and Musical groups respectively [as described in section "Experiment 1, Experimental design, phase II"].

Track 95. "Open Paint application" and "Save Write folder" earcons used in the Control, Simple, and Musical groups respectively [as described in section "Experiment 1, Experimental design, phase IV"].

Track 96. New earcons for "Paint application," "Write folder," and "Draw file 1" [as described in section "Experiment 2, Sounds used"].

Index

A

absence of persistence, 13-14
absolute values, 13
acoustic compilers, 227
acoustic ecology, 54-55, 153
active listening, 151
aerospace, 38, 44, 144
    shuttle launch communications, 145
    see also spacesuits
affective association, 212, 214, 216
affective response, 8
afford interruption, 516
affordance, 515, 517
AGC,
    see automatic gain control
aircraft warning systems, 44
alarms, 22, 38, 43-45, 83, 137, 159, 517
Albright, L., 514
Alternate Reality Kit, 423
ambiguity, 159-160
amplitude, 236
analogic representation, 21
analogic vs. symbolic, 21
analogic/symbolic continuum, 24, 27
analysis and synthesis of events, 429
analytic perception, 98
analytical chemistry, 38
animation, 512
    with sound, 455
annotation, 387, 402
anthropomorphic agents, 502
ARKola, 42, 424, 440
array data, 400
articulation, 101
artificial heart pump, 327
asymptotically static, 343
attention, 98-99, 101-102, 517
attracting set, 343
attractors, 341-342, 344, 346-347, 350
audiation, 11, 185, 188
Audible Image Software, 156
audification, 185-186, 370, 376, 383
    of seismic data, 378, 382
    vs. sonification, 189
    see also zeroth-order mapping,

audification (cont'd.)
    direct simulation, and data-as-signal
audified seismograms, 378
audio annotation, 330
audio channel, 500, 502
audio-enhanced debugger, 253
audio indicators, 331-332
audio rooms, 552
audio thermometer, 330
audition, 69, 308, 517
auditory characteristics, 460
    taxonomy of, 460-463, 465-466
auditory cues, 449, 541
auditory data representation, 174
auditory design, 50
auditory display, 208, 327, 332, 503
    and music composition, 357, 342
    and related fields, 3
    and two-dimensional maps, 458
    applications of, 17-18, 378
    auditory-only, 533
    benefits of, 6-11
    CFD data for, 327
    collaborative potential in, 151
    developing theory of, 95
    difficulties with, 11-15
    history of, 36
    importance of annotation, 402
    interdisciplinary boundaries of, 68
    introduction to, 1
    measuring effectiveness of, 167
    of large data set, 380
    pitch-encoded, 47
    taxonomy for, 460
    up to eight variables, 308
    virtual three-dimensional, 160
    vs. music, 51-53
    vs. visual display, 40, 307, 309, 317, 322
auditory event perception, 417
auditory feedback, 142, 505
auditory grouping, 95, 97, 103-104, 115
    modeling of, 113-114
auditory homonym, 455
auditory icons, 22, 417, 427, 449-450, 454-455, 505, 543-544, 548, 550-551

Index

auditory icons (cont'd)
    families of, 420
    parametrized, 417, 420, 428
    with animation, 455
auditory information, 448
auditory interface navigation, 545
auditory landmarks, 423
auditory maps, 458
auditory messages, 449
    syntax for, 451
auditory parameters, 341
auditory perception, 63, 95, 97
auditory periphery, 113
auditory pointers, 449
auditory process monitor, 422
auditory representations, 341-342, 357
    of multidimensional data, 171
    of n-cylinders, 362
auditory scene, 153-154
    analysis of, 457
auditory streams, 96-98, 102, 104, 106-109, 112-113, 116, 174
auditory symbology, 141-142
auditory texture, 425, 515
auditory views, 204
aural cursor, 387
aural displays, 298
auralization, 128, 133, 255, 292, 294, 298, 394
    prototype tool for, 296
autocorrelation function, 344
automatic gain control (AGC), 384, 391
automobile parameters, 41
autonomous background activity, 508
autonomous background tasks, 501
axial turbine, 334

B

background activities, 499-500, 507
background tasks, 501
backgrounding, 7
balanced display, 199
Ballas, J. A., 79, 89, 91, 454
Barfield, W., 474
Bargar, R., 151
basin of attraction, 343
beacons, 28, 185, 202-203, 207
    and gestalt, 206
    dynamic, 203
    sequencing, 205
beam-formed, 400
beating phenomena, 343
Beauvois, M. W., 118
Beauvois and Meddis model, 114
Begault, D. R., 26, 144
Behrman, D., 504
belongingness, 106
bifurcation, 350-351, 353
blade passing, 335
Blattner, M. M., 22, 32, 447, 449-450, 452, 458, 473, 475
blind users, 39, 46, 48, 517, 533
    access to computer programs for, 49
    see also vision-impaired users
Bly, S., 39, 176, 231, 279-280, 322, 450, 456, 505
boundedness, 343
Bregman, A. S., 97, 107, 117, 174, 457
Brewster, S. A., 453, 458, 471-472
brightness, 195-196, 199, 216
Broadbent, D. E., 101
Brooks, J. E., 89
Brown, M., 449
Burrows, A. A., 85
Buxton, W., 55, 456, 502

C

Cage, J., 503-504
cartographic data, 459
causal uncertainty, 88-90
Chambers, J. M., 38
chance operations, 504
chaotic attractors, 341
Chernoff, H., 456
Chernoff's faces, 279, 456
CHI '85, 41
Chowning, J., 346
Chua, 347, 354, 356
    oscillator, 346
civil engineering, 373
closure, 107
cocktail party effect, 101, 138, 503
cognition, 110
cognitive expectancy, 86

Cohen, J., 23, 454, 499
Cohen, M., 451, 544
collaborative system, 423
color, 512
common depth point recording (CDP), 392, 394
communicative medium, 2
comodulation masking release, 106
complex sounds, 95
complex timbres, 514
complexity, 341, 356
composition, 153, 341
compressor, 334
computational fluid dynamics, 327, 329
computational models, 113
computer events, 504
computer hardware, 58-61, 269, 311
    Encore Multimax, 286
    Macintosh, 421, 505
    new technology for, 167
    Silicon Graphics, 332
computer music, 344
computer programs, 57
    AD-supported, 56
    loops in programming, 259, 387
    parallel, 291-293
computer software, 269, 421
    and data format, 67
    ARKola, 424
    Audio Image Software, 156
    Choices, 285
    EAR, 425
    FAST, 328
    File Sharing, 507
    Khronika, 506
    Mercator, 533, 535
    network protocol, 271
    Outspoken, 539
    ParaGraph, 294
    parallel, 292-293
    Plot 3D, 328
    Porsonify, 268
    RAVE, 506
    SharedARK, 505
    ShareMon, 500, 507
    SonicFinder, 421, 504
    Sonification Toolkit, 190

computer software (cont'd.)
    Sonnet, 253
    SoundShark, 422
    special-purpose, 169
    SPLASH, 285
    X Windows, 535-536, 541
    see also IRCAM
    see also Kyma
computerized maps, 459
conjunction with other displays, 8
connection reminder, 510
context, 91
continuity, 457, 516
continuous events, 517
Convolvotron, 59, 133-134, 141
correlation, 156
Craig, A. B., 43, 514
cross-modality matching, 85
Cross, L., 504
crystal trace capture library, 284

D

D'Albe, F., 46
data discovery, 380
data-as-signal, 213
data sonification, 342
Data Sonification Project, 225
data exploration, 15, 20, 405
data family, 208-211
data family/stream associations, 209
data formats, 67
data hierarchies, 208
data manipulation, 205
data parameters, piggy-back, 321
data representation, 456
    see also Chernoff's faces
data set display, 380
data streams, self-labeling, 320
data structures, 175-176
data-to-acoustic mappings, 406
data type, 208, 211
data type/parameter associations, 209
data visualization, 171
debugging, 262, 292
delegation interfaces, 501-502
design-and-test methodology, 500
detection, 176

Detweiler, M. C., 454
Deutsch, D., 114, 457
diegesis, 154
directness, 186
direct engagement, 421
direct manipulation, 501
direct mapping, 280
direct perception, 419
direct simulation, 329, 334
discretization, 335, 344, 346
discriminant analysis problems, 406
display hierarchies, 208
distance education, 423
distributed computing, 502
domain of attractivity, 343
Dreyfus, H. L., 448
DSP hardware, 503
Duchamp, M., 504
duplex theory, 128

E

ear training, 342
earcons, 22, 26, 57, 197, 447, 450-453, 455-456, 458, 466, 473, 486
    combining earcons, 452
    compounded, 452
    concurrent, 458
    definition of, 473
    dummy, 487
    effectiveness of, 471
    manipulation via intensity, 476
    role of musical ability, 486, 494
    structure, 452
    vs. sound bursts, 475
earthquakes, 373, 383, 398
ecological approach, 419
education, 381
Edwards, A. D. N., 49, 453, 458, 471
Edworthy, J., 81
electrocardiogram, 242
Ellison, S., 260, 450
empirical evaluation, 97, 112-113, 116-117
encoded messages, 516
engagement, 10
enhanced realism, 10
Environmental Audio Reminders (EAR), 425-426

environmental interfaces, 426
equipment problems, 381
ERB mechanism, 118
eyes-free display, 379
Evans, B., 42
event attributes, 429
event recognition, 380
event-to-sound mappings, 296-297
exploration seismology, 373, 376, 379, 382
Exvis, 42, 245, 515
eyes-free display, 7

F

Familant, M. E., 454
familiarity, 104
FAST, 328, 333
Fechner, G., 111
Ficks, L., 37
fidelity, 159
field data, 392
field quality control, 392
file sharing monitor, 507-508
filtears, 544, 547, 551
filtered white noise, 511
filtering, 240
finite impulse response (FIR) filters, 131, 240
Fitch, W. T., 26, 307
flow solver, 329, 334
FM signal, 385
FM synthesis, 441
Forbes, T. W., 34
foreground activity, 500
foreground tasks, 501
Fourier transforms, 336
fractal structure, 343
frame of reference, 155-157
framing, 479
Francioni, J. M., 23, 103-104, 255, 291, 293, 514
Frantti, G. E., 37, 377
free oscillation data, 379
Freed, D. J., 431
frequency, 516
Frequency Modulation (FM) synthesis, 441
frequency doubling, 385, 391

frequency modulation, 238
Frysinger, S. P., 40-41, 168
Furner, S. M., 453, 474

G

Gantt chart, 295, 298
Gaver, W. W., 22, 86, 215, 417, 422, 449, 453, 456, 502, 504, 543-544
Geiger counter, 21, 330
geographic map, 26
geologic hazard estimates, 373
geologic maps, 373
geophone, 373
gestalts, 27, 95, 97, 99-102, 185, 206, 211
  grouping principles, 457
  formation of, 8
Gibson, J. J., 419, 515
Glinert, E. P., 447, 449
global variable, 211
globally attracting, 343
good continuation, 104
GMIDI, 57
granular synthesis, 191, 238
  algorithm, 409
graphic displays, two-dimensional, 447
graphical user interfaces, 19, 46, 534, 538
Greenberg, R. M., 450, 452, 458
Greiner, H., 38
grid coarseness, 335, 337
grid stretching, 335
Grinstein, G., 450
grouping, 4, 158
  principles of, 95-98, 101-102
guest, 508
  machines, 519

H

Haken, L., 243
harmonics, 335, 337
harmony, 359, 361
Hayward, C., 27
head-related transfer functions (HRTFs), 131
hidden information, 448
high dimensionality, 10
high frequency sounds, 514
histogram, 351

host machine, 507, 518
Hotchkiss, R. S., 255
human auditory system, 323
human-computer interface, 447
human visual system, 323
hydrophone, 30
hypercube, 284

I

icon, 512
  sounds, 489
infinite impulse response (IIR) filter, 240
inflection, 161
InfoProbe, 256
information generators, 2
information processing, 515
information production rate, 343
information receivers, 2
information representation, 21
InfoSound system, 256, 293
intelligent agents, 502
intelligibility, 138, 521
intensity, 476
intention, 53
inter-stream linking, 211
interaction effect, 521
interactive, 381, 387
interface modality, 502
intermittency, 341, 350-352, 357
intermodal correlations, 10
intermodal invariants, 156
interruption affordance, 517
IRCAM Music Workstation, 169
irregularity, 341, 347
iteration, 158

J

Jackson, J. A., 23, 103-104, 291, 514
Jameson, D. H., 253
Jay, K. I., 449
Jenkins, J. J., 502
Jones, S. D., 453
just noticeable differences (JNDs), 110, 200

K

Kaiser, W., 38

Index

Kamegai, M., 458
key turning in a lock, 523
Khronika, 426, 502, 506-507
Kneale, W. C., 80
Kobus, D. A., 322
Korg M1, 332
Kramer, G., 23, 26-27, 42, 86, 185, 260, 307, 450, 459, 504
Krumhansl, C. L., 114
Kurzweil Reading Machine, 47
Kyma, 169, 223, 228, 230, 232-234, 409-410

L

la langue, 152
language, 22, 448, 450
    difference between levels of, 450
    lexical level of, 450
    perception, 91
large data set display, 380
Lauer, H., 46
Levasseur, G., 474
levels of directness, 186
Leverault, L. A., 377
Lewandowski, L. J., 322
library servers, 502
limen, 110
limit cycles, 343
limitations of sampling, 428
limitations of synthesis, 428
linear theory, 336
listening
    active, 151
    analytic vs. synthetic, 98
    everyday listening, 418-419, 504
    musical, 419, 504
    parallel, 8
listening experience, 381
literal, 513
localization, 160, 163
logistic map, 347, 349-350
    coupled, 363-364
looping, 387
loudness, 192, 511
low frequency sounds, 514
Ludwig, L. F., 451, 544
Lunney, D., 48, 451

Lyapunov exponents, 343

M

Madavan, N. K., 334
Madhyastha, T. M., 23, 267
Magellan's view of Venus, 459
Mansur, D. L., 48, 449
mapping, 103, 105, 111, 115, 119, 157, 257, 405
    data-to-acoustic, 406
    comparison via addition, 243
    sequencing of, 201
markers, 387
masking, 424
Marteus, W. L., 431
Mathews, M., 227
Matlin, M. W., 109
MAX, 227
McAdams, S., 105, 114
McCabe, K, 327
Meddis, R., 118
Media Space, 502
medium walking, 510
Mercator, 533, 535-537, 539-540, 548-549
metaphor, 84-85
metaphorical association, 212-213, 216, 500, 505
metawidgets, 456
meter, 161-162
method of loci, 454
Mezrich, J. J., 40
microphone, 329
MIDI, 57, 136, 168-169, 227, 269-271, 296, 331-333, 409, 423, 503
    "general", 332
    "note on", 332-333
    "pitch bend", 332
    "program change", 332
    protocol, 57
    sequencing software, 331
Milroy, R., 83
mimesis, 421
Misenheimer, R., 243
mix modalities, 524
modalities, 507
    auditory vs. tactile, 538
    foreground vs. background, 516

models, 381, 430
　Beauvois and Meddis, 114
　computational, 114
　STREAMER, 114
　Terhardt's, 114
modes, 336, 423
monitoring applications, 18
monitoring noise, 379
Morrison, R. C., 48, 451
Morse Code, 32
motive, 451-452
Mowinski, L., 46
Mullin, T., 91
multidimensional information, 420
multivariate data, 245, 279, 307, 405-406
Mumma, G., 504
music, 49, 51, 357
　chaos in, 357
　composition, 50, 151, 153-155
　conveying quantitative information, 66
　experimental, 341
　N languages, 227
　of changes, 504
　order in, 357
　performance, 50
　vs. audio design, 50
musical messages, 428
musical scale, 194
musical structure, 114, 341
　in data from chaotic attractors, 341
musical texture, 458
Mynatt, E. D., 23, 49, 533

N

nth-order mapping, 330
　see also parameter mapping
Navier-Stokes, 334-335
navigation, 423, 545
neurons, 34
Newsome, S., 449
Newton, I., 502
noise survey, 395
nonlinguistic structures, 342
nonlinear dynamical systems, 341
nonmetaphorical, 500, 505

nonspeech audio, 29, 136-137
normalized pressure, 332
notification modalities, 507
nuclear explosions, 376-377, 399
numerical dissipation, 335, 337
numerical models, 381
Nyquist, 228

O

O'Shea, T., 424
oil well log data, 40
onomatopoeia, 86
onset time, 382
operating systems, 516
Optaudicon, 46
Optophone, 46
ordered systems, 343
ordinary differential equation, 346
organizational principles of music, 342
orthogonality, 13, 198
　lack of, 14
"outspoken", 47, 539

P

P-waves, 372
Pablo Performance Analysis Environment, 267, 277, 283
Papp, A. L., 447
ParaGraph, 294, 298
parallel listening, 8
parallel processing, 448
parallel programs, 291-293
parameter mapping, 234, 327, 330-332
　data-to-sound, 408
parameter nesting, 192, 260, 450
parameter overlap, 198-199
Parncutt, R., 114
password, 507
pattern recognition, 27, 407
Patterson, R. D., 44-45, 65, 83, 490
Pecelli, G., 515
Peirce, C. S., 225
Penn State artificial heart pump, 332
per stream variable, 210
perception, 95, 109
　ecological perspective on, 419
　vs. sensation, 109

Index

perceptual effects, 114
perceptual units, 67
performance data, 283, 291, 293
performance tuning, 292
periodicity, 349-351
physical modeling, 430
physiological data, 310, 312
Pickett, R. M., 167, 450
PICL, 283, 295-296
Pincever, N., 451, 544
pink noise, 511
Pinker, S., 117
pitch, 511
    differentiation of, 361
    nesting, 193-194, 196
    perception of, 345
    shift, 331
planetary seismology, 373, 379
plasma wave data, 35
Plot 3D, 328
point attractors, 343
Pollack, I., 37
polyphony, 161, 163, 212, 452
polyphonic auditory constructions, 341
Porsonify, 267-270, 274, 276, 278-280, 409
    sonic widgets, 275-276
positive feedback, 343
pragnanz, 100
preattentive distortions, 17
primitive segregation, 457
probability distribution, 341, 343, 351
processes, 423
    monitoring of, 206
processing techniques, 382
propagating waves, 335
proximity, 103
psychoacoustics, 419, 472, 475
psychomusicology, 66
psychophysical experiments, 177-178
psychophysics, 111
    creation of experiments in, 177
Ptolemy, 50
Pythagoras, 49

Q

quality control, 379

quasi-periodic tori, 343

R

Rabenhorst, D. A., 451
Rai, M. M., 334-335
random systems, 343
Rangwalla, A. A., 335
RAVE, 426, 502, 506-507
reading machines, 46
recall, 454
recognition, 176
recurrence, 349, 356
Reed, D. A., 23, 267, 293
register, 161
relative pitch, 162
remembered sounds, 90
response time data, 316
retinal dimensions, 451
reunion, 504
reverb, 196, 516
rhythms, 161-162
Richards, W., 434
Risset, J.-C., 346
Rogowitz, B., 16
Rosenberg, C., 474
rotor-stator, 334-335, 337
    interaction, 327
    tonal acoustics, 334-336
rule-of-thumb, 516
run-time behavior, 292

S

S-waves, 372
sampled sounds, 241, 428
samplers, 136
sampling rate, 329
Scaletti, C., 43, 223, 504, 514
scaling, 176, 200
Scarf, F., 35, 38
schemas, 105
scientific visualization, 328, 332
scientific workstation, 304
scraping, 437
    algorithm, 438
    attributes, 437
    sounds, 437
screen location, 512

screen reader, 535
screen space, 502
seismic analysts, 374
seismic data, 37, 39, 48, 369-370, 373, 382-383
seismic storms, 380
self-similarity, 341, 357-358
self-sustained oscillations, 343
semantics, 462
    interpretations of, 462
    of sound, 453
sensation, 109
sensitivity to initial conditions, 343
sensory psychology, 110
sequence, 153, 158, 161
SharedARK, 423, 505
ShareMon, 454-455, 499-500, 506-507, 509-512, 516, 522, 527
    defining, 509
    user reactions, 513
    user studies, 518
Sheffe F-test, 480-481
signal detection, 381
Silicon Graphics, 332
similarity, 102
similes, 84
sine waves, 514
singularities, 343, 352
situational awareness, 138
Slivjanovski, R., 40
slow walking, 510
Smith, R. B., 422
Smith, S., 27-28, 42, 68, 167, 245, 450, 515
Smoliar, S., 33
Software
    see computer software and
    and computer programs
sonar, 28-31, 373
sonic scatter plot, 281-282
SonicFinder, 41, 158, 197, 421-422, 504, 543
SonicHistogram, 244
sonification, 23, 27, 39, 41-42, 49, 51, 64, 79, 127, 135, 187, 224, 246, 267, 292, 342, 370, 383, 406, 410-411
    acoustic variables for, 64

sonification (cont'd.)
    and interactivity, 66
    and music, 50-53
    and sound synthesis algorithms, 62
    areas of research in, 225
    data-as-signal, 231
    data format, 67
    in virtual reality, 245
    integrated with visualization, 277
    of turbulence, 458
    structure for, 274
    Toolkit, 42, 190-191
    using linguistic analogies, 81
    with earcons, 474
    working definition of, 223
    see also Data Sonification Project and Exvis
Sonnenwald, D. H., 293
Sonnet, 253, 255-257, 261
sound, 26, 241, 428, 425, 514
    algorithms, 432
    attributes, 432, 450
    byte, 159
    design, 151-152, 154-155
    device servers, 271
    effects, 520, 524, 544
    enhanced realism, 10
    grouping, 95-96
    identification, 86-87
    in virtual reality, 224
    localization, 128
    metaphorical vs. nonmetaphorical, 505
    perception of, 490
    production of, 454
    process, 298
    properties of, 254
    realistic vs. abstract, 196
    scales, 112
    semantics of, 453
    separation of complex faces, 456
    sequential combinations of, 462
    spatialization, 26
    structuring as a language, 448
    synthesized, 136, 342
    text-to-speech, 500
    with animation, 455

sound synthesis, 169, 223, 234, 345
    algorithms, 62
    recent history of, 227
    three-dimensional, 127
sound waves, 372
sound-producing event, 420
SoundShark, 422-423
Space-Time display, 295
spacesuits, 35
spatial sound, 128, 552
spatial synthesis, 135
spectral dominance theory, 118
spectrograms, 379
speech rate, 524
speech signals, 113
Speeth, S. D., 37, 377
spinning modes, 334
stability, 100, 108
Stanford Parallel Applications for Shared Memory (SPLASH), 285
status and monitoring indicators, 516
stereo, 387-388, 395, 400, 403
    panning, 516
stereotoner, 46
stereotyped sounds, 426
Stevens, S. S., 111
stochastic systems, 343
storytelling, 526
stream segregation, 138
stream, simultaneous, 211
STREAMER, 95, 114-117
structural composition, 461
structure of earcons, 452-453
structured grid, 328
Sumikawa, D. A., 450, 452, 494
symbolic, 362
    dynamics, 362
    representation, 21
    vs. analogic, 22, 27
symbols repertoire, 152
synesthesia, 11, 35
synthesis algorithms, 417
synthesis parameters, 433
synthesized sound, 136, 342
synthetic perception, 98

synthetic seismograms, 388

T

talking drums, 32
tangibility, 425
task dependence, 15
taxonomy
    of auditory characteristics, 460-463, 465-466
TCAS, 27, 144
telerobotic control, 141-143
temperature measurements, 242
template matching, 15
temporal correlations, 343
temporal resolution, 10
Terhardt, E., 99, 118
Terhardt's theoretical model, 114
text-to-speech, 511, 521, 526
three-dimensional sound, 128
timbral changes, 391
timbre, 234, 337, 451
time-compression, 384
time-dependency, 334
time-varying multidimensional data, 407
tonal acoustics, 334-337
tonal center, 156, 161
tone salience, 118
trace, 295
training, 65, 314, 316, 377, 379
    see also ear training
training paradigms, 17
trajectory, 343, 512
transfer functions, 239, 275
Truax, B., 54
Tudor, D., 504
turbomachines, 335
Tzelgov, J., 322

U

universal aspects, 341
user feedback, 546

V

Veitengruber, J. E., 83
Verbrugge, R. R., 174, 435
VIEW, 141, 143
virtual engine, 187

virtual acoustic displays, 135, 137
virtual environments, 19
virtual reality, 267, 330, 439
    sound in, 224
    see also Alternate Reality Kit
vision-impaired users, 18, 36, 39, 46, 48-49, 533
vision MIDI sequencing software, 333
visotoner, 46
visual mapping, 314
visual programming language, 258
visualization, 332
voice, 331, 516
volcanoes, 373
vortex shedding, 337
vorticity, 332

W

Walker, R., 85
Wampler, C., 255
warning systems
    see alarms

Warren, W. H., 435, 174
wave equation, 372
wave table, 330
waveshaping, 240
Weber, E. H., 111
Weber's Law, 111
Weierstrass function, 344-345
Wenzel, E. M., 26, 127, 141, 160, 451
Wertheimer-Benary effect, 102
Wildes, R., 434
Williams, M. G., 167, 515
Williams, S. M., 95, 457
Wright, P. C., 453, 458, 471

X

x–y plots, 336-337, 339
X Windows, 534-536, 541

Y

Yeung, E. S., 38, 450

Z

zeroth-order mapping, 329